Springer Series in Geomechanics and Geoengineering

Series Editor

Wei Wu, Universität für Bodenkultur, Vienna, Austria

Geomechanics deals with the application of the principle of mechanics to geomaterials including experimental, analytical and numerical investigations into the mechanical, physical, hydraulic and thermal properties of geomaterials as multiphase media. Geoengineering covers a wide range of engineering disciplines related to geomaterials from traditional to emerging areas.

The objective of the book series is to publish monographs, handbooks, workshop proceedings and textbooks. The book series is intended to cover both the state-of-the-art and the recent developments in geomechanics and geoengineering. Besides researchers, the series provides valuable references for engineering practitioners and graduate students.

**** Now indexed by SCOPUS, EI and Springerlink****

More information about this series at http://www.springer.com/series/8069

Erkan Topal
Editor

Proceedings of the 28th International Symposium on Mine Planning and Equipment Selection - MPES 2019

Springer

Editor
Erkan Topal
Mining Engineering and Metallurgical Engineering
Curtin University
Perth, WA, Australia

ISSN 1866-8755 ISSN 1866-8763 (electronic)
Springer Series in Geomechanics and Geoengineering
ISBN 978-3-030-33953-1 ISBN 978-3-030-33954-8 (eBook)
https://doi.org/10.1007/978-3-030-33954-8

This Springer imprint is published by the registered company Springer Nature Switzerland AG
The registered company address is: Gewerbestrasse 11, 6330 Cham, Switzerland

Foreword

It is a privilege to release the proceedings of the 28th International Symposium on Mine Planning and Equipment Selection - MPES 2019 Conference. MPES symposium has been recognised as the leading forums for promoting international technology transfer, with the main focus on novel development and application of all aspects of mine planning as well as mining equipment.

All the papers in the proceedings have been independently peer reviewed and edited by experts in the field, to ensure the highest relevance and quality. The coverage of this conference proceedings includes but it is not limited to the following areas:

- Advances in Surface and Underground Mine Planning, Design and Computer Application,
- Mine Equipment Selection, Utilisation and Maintenance,
- Artificial intelligence and Mine Automation,
- Resource Modelling and Geometallurgical Planning,
- Mine Health, Safety and Environment,
- Sustainable Development of Mining Resources,
- Drilling, Blasting and Excavation Engineering,
- Rock Mechanics and Geotechnical Applications,
- Mining Economics.

I would like to express my sincere appreciations to all the authors and presenters, keynote speakers, editorial/review panel for their valuable contributions, the Organising Committee for their hard work and sponsors and exhibitors for making this conference as a successful and enjoyable event.

On behalf of Organising Committee, I welcome all of you—the participants of MPES 2019—to Western Australia and wish you a successful conference. I also hope that the proceedings will serve as an important source of knowledge and references to all participants and other interested readers.

Erkan Topal
Conference Chair

Organisation

MPES 2019 Committees

Erkan Topal (Chair)	Curtin University

Local Organizing and Technical Committee

Waqar Asad	Curtin University
Mehmet Cigla	Curtin University
Eric Lilford	Curtin University
Hyong Doo Jang	Curtin University
Micah Nehring	The University of Queensland
Mahinda Kuruppu	Curtin University

Reviewers' Committee

Apurna Gosh
Ataç Başçetin
Behzad Ghodrati
Bunda Besa
Brian White
Cuthbert Musingwini
Eric Lilford
Erkan Topal
Ernest Baafi
Fidelis Suorineni
Guang Xu
Hakan Basarir
Hooman Askari Nasab
Mahinda Kurruppu
Mehmet Cigla
Micah Nehring
Mohan Yellishetty
Morteza Osanloo
Mostafa Sharifzadeh
Mustafa Kumral
Nasser Madani
Nelson Morales
Nuray Demirel
Radoslaw Zimroz
Simon Dominy
Steven Rupprecht
Tom Jang
Waqar Asad
Youhei Kawamura

Contents

Mine Equipment Selection, Utilisation and Maintenance

Mining Economics

Advances in Surface Mine Planning Design and Computer Application

Keynote Paper: Some Perspectives on Implications of Top Globally Ranked Factors Affecting Mining on the Future of Mine Planning

Cuthbert Musingwini(✉)

School of Mining Engineering, University of Witwatersrand, Johannesburg, South Africa
Cuthbert.Musingwini@wits.ac.za

Abstract. As of 2019, the 'Big Four' accounting and auditing global firms have been reporting annually for about a decade, on perceived top 10 factors affecting the mining industry. The reports refer to the factors as issues, trends or risks faced by the mining industry. The factors and their ranking order are both dynamic as both the list and ranking order change annually and across the firms undertaking the surveys. The factors have a direct or indirect bearing on the mine planning process since mine planning is central in informing high-level or strategic decision making for mining companies. Mining companies must continually adapt since the business environment changes due to the impacts of these factors. This paper considers how the most recently top-ranked factors can affect mine planning into the future and why it is important for the mine planning fraternity to keep track of these factors. This ensures that the mine planning process continues to deliver outputs that enable mining companies to continue to make robust business decisions.

Keywords: Mine planning · Risk · Strategic decision making · 'Big Four' accounting and auditing global firms

1 Introduction

As of 2019, the 'Big Four' accounting and auditing global firms namely, Ernest & Young Global Limited (EY), KPMG, Deloitte & Touché (Deloitte) and PricewaterhouseCoopers (PwC) have been providing guidance annually for about a decade on strategic or high-level key factors faced by the mining industry. For example, the 2019 Deloitte and EY reports are the 11th editions since inception of the report series by the two firms. Each of the firms generally reports these factors as being among the top 10 issues, trends or risks affecting the mining industry. The list of factors and their relative ranking change both annually and across the firms undertaking the surveys in any given year. The factors identified in these reports, inform strategy and decision making at mining company board, Chief Executive Officer (CEO) and executive management levels. Some of the factors either directly or indirectly affect how mining companies undertake mine planning. This is due to the central role that mine planning plays in

E. Topal (Ed.): MPES 2019, SSGG, pp. 3–9, 2020.
https://doi.org/10.1007/978-3-030-33954-8_1

informing decision making in mining companies as CEOs of publically listed mining companies make public announcements predominantly based on the results of mine planning [1].

The factors straddle across the broad categories of technical, economic, social, legal, environmental and political factors. These categories reflect the categories of Modifying Factors used in the mine planning process to convert Mineral Resources to Mineral Reserves. It is therefore important for the mine planning fraternity to keep track of the global top-ranked factors affecting the mining sector as these have a direct or indirect bearing on the mine planning process. This will enable the mine planning process to adapt to future needs, and retain and relevance in informing robust decision-making by mining companies. The next section presents the rankings in 2019 of the different factors identified by the 'Big Four' as mostly affecting the mining industry.

2 Top-Ranked High-Level Factors in Mining in 2019

The accounting and auditing firms mentioned in the previous section, compile the list of factors from surveys that they undertake annually on a global scale. To derive balanced views and well-informed rankings that fairly reflect global perceptions and understanding, the firms conduct the surveys on a wide range of stakeholders. The list of factors and the ranking order change annually. This is due to several reasons such as perspectives of respondents changing with new knowledge and awareness of developments occurring within and outside of the mining sector, differences in the wording of the factors by each of the firms undertaking the surveys, and the different stakeholder databases polled in each survey by each firm. For example, the continual technological developments associated with the fourth industrial revolution (4IR) occurring outside of the mining sector affect how mining processes should be designed to remain aligned with changes in both upstream and downstream industries that have linkages with the mining sector as they affect how the mining industry should conduct its business. Any misalignment with upstream and downstream industries can have undesirable consequences to the mining sector. Table 1 illustrates the ranking order in 2019 of the factors by each of the four firms.

Table 1. Ranking of the perceived top high-level factors in 2019 [2–5].

Factor ranking	Deloitte	EY	KPMG	PwC*
1	Rethinking mining strategy	License to operate	Macro financial risks	Maximizing market returns for investors
2	The frontier of analytics and artificial intelligence	Digital effectiveness	Permitting risk	Safety and the environment
3	Managing risk in the digital era	Maximizing portfolio returns	Community relations and social license to operate	Technology adoption for automation and digitization

(continued)

Table 1. *(continued)*

Factor ranking	Deloitte	EY	KPMG	PwC*
4	Digitizing the supply chain	Cyber	Access to capital, including liquidity	Creating sustainable value for all stakeholders to fix 'brand mining'
5	Driving sustainable shared social outcomes	Rising costs	Economic downturn/uncertainty	Climate change
6	Exploring the water-energy nexus	Energy mix	Ability to access and replace reserves	Energy mix shift away from combustion engines to electricity & renewable energy
7	Decoding capital projects	Future of workforce	Political instability	Optimizing asset portfolios by disposing non-core assets to drive efficiencies and improve productivity
8	Reimagining work, workers, and the workplace	Disruption	Regulatory and compliance changes/burden	Regulatory and political uncertainties
9	Operationalizing diversity and inclusion programs	Fraud	Controlling operating costs	Social licence to operate
10	Demanding provenance	New World commodities	Environmental risks, including new regulations and access to key talent	Changing commodity mix due to changing consumer consumption patterns and increased use of technological devices

Note: *PwC did not provide a ranking order and it was reasonable to assume the order of presentation of factors in the report as a proxy for the ranking order.

It is apparent from Table 1 that each of the four firms can describe the nearly same factor in slightly different ways. In addition, one firm can identify a single factor, while another firm can split the same factor into more factors depending on the level of detail attached to the factor. The lists are dynamic, and so is the associated lexicon. For example, in the 3-year period spanning 2017 to 2019 the list of top factors compiled by Deloitte has changed as indicated in Table 2. It is apparent from Table 2 that the factor "Supporting strategic priorities" in 2017 could be nearly the same factor as "Rethinking mining strategy" in 2019, indicating the change in terminology to describe nearly the same factor.

Table 2. Ranking of the perceived top factors by Deloitte for the past three years (2017–2019) [2, 6, 7].

Factor ranking	2017	2018	2019
1	Understanding the drivers of shareholder value	Bringing digital to life	Rethinking mining strategy
2	Unlocking productivity improvement	Overcoming innovation barriers	The frontier of analytics and artificial intelligence
3	Operating in an ecosystem	The future of work	Managing risk in the digital era
4	The digital revolution	The image of mining	Digitizing the supply chain
5	Mapping the threat landscape	Transforming stakeholder relationships	Driving sustainable shared social outcomes
6	Creating a shared vision for the sector	Water management	Exploring the water-energy nexus
7	Re-earning the social license to operate	Changing shareholder expectations	Decoding capital projects
8	Supporting strategic priorities	Reserve replacement woes	Reimagining work, workers, and the workplace
9	Creating healthy and inclusive workforces	Realigning mining boards	Operationalizing diversity and inclusion programs
10	Adopting an integrated approach to reporting	Commodities of the future	Demanding provenance

It is also evident from Table 1 that mine planning will need some re-thinking to acknowledge the key high-level factors, which, if not addressed properly, pose serious risks to mining companies. Firstly, there is a growing impact of 4IR on the design and operation of mines, including adoption of associated technology. Secondly, there is increasing global emphasis on shared value perceptions by mining stakeholders. Shared value is achievable if mine plans are robust enough to maximize financial returns. If companies are cash-positive, they have the financial flexibility and capacity to address other value expectations from different stakeholders. Thirdly, there is a growing global demand for cleaner energy mixes, which can cause some commodities to experience a decline in demand if they have no role to play in the new or future energy mixes, unless the mining companies invest in downstream technologies to stimulate demand for their commodities. Lastly, mine planning must proactively account for the impact of the broad cluster of factors pertaining to regulatory, environmental, geopolitical and/or governmental factors. Table 1 does not indicate the time trend of some of the factors since it is a snapshot of the key high-level factors at the 2019 date stamp. However, some of the factors have tended to be perennial as can be inferred from Table 2, indicating that the mine planning fraternity cannot afford to ignore these factors when developing mine plans to support robust decisions by mining companies.

Based on the foregoing discussions, each of the factors can fall under broad groups of factors. For example, 4IR related factors include factors listed as “the frontier of analytics and artificial intelligence", "managing risk in the digital era", "digitizing the supply chain", and cyber-related risks indicated as "cyber". The next section outlines some views on this broad categorization of the factors.

3 Implications for Mine Planning

A scan through the factors listed in Tables 1 and 2 shows that mine planning has to evolve in four broad categories. These are the following broad clusters of factors:

- Financial and economic factors to account for such issues as operating cost reduction, market dynamics, capital, funding, maximizing portfolio returns, mergers and acquisitions (M&A), and organic growth.
- 4IR factors that are focused on operational improvement through innovation, adoption of 4IR technologies including automation and digitization, productivity improvement, new energy mix and associated minerals emerging as new world commodities.
- Sustainable shared value factors that can ensure a clear value proposition for 'brand mining' through a commonly understood shared value metric (SVM), which can be measured easily and incorporates well-articulated stakeholder expectations, social license to operate, and address the growing demands for provenance.
- Governmental and geopolitical factors, which can cause disruption, introduce uncertainty arising from political instability, environmental risks and trade wars and/or tariffs.

The implications of these factors to mine planning into the future are:

- It is no longer sufficient to measure the economic or financial benefit accruing from a mining project in terms of the traditional valuation metrics which are net present value (NPV), internal rate of return (IRR), economic valued add (EVA) as these do not represent shared value to a wide range of stakeholders. The different stakeholder views can be contradictory or competing against each other, requiring some trade-offs to be made for a balanced shared value. It is important, as part of the mine planning process, to develop an alternative value measurement metric that addresses diverse stakeholder needs. Practically, such a metric could be the priority score derived from applying multiple criteria decision-making (MCDM) techniques in analysing outputs from mine planning concurrently with different stakeholder value expectations. The metric could also be a combination of a traditional value metric such as using a positive NPV in conjunction with expected utility (EU) from discrete choice analysis of different stakeholders as presented by Awuah-Offei *et al.* [8].
- Mine designs, should as far as possible, incorporate the use of 4IR-inspired equipment that allows for automation and present mine planning systems in ways that enable digitization. This includes incorporation of 4IR-related techniques such as data analytics and artificial intelligence.

- Government and geopolitical factors invariably introduce uncertainty especially when they are often changed requiring stochastic mine planning to be the norm rather than the exception in the future of mine planning.

4 Concluding Remarks

It is possible to draw some inferences from a review of the factors on how mine planning should adapt into the future. Some of these inferences are:

- It is necessary to develop an alternative shared value metric to measure value beyond the traditional value metrics such as NPV, IRR or EVA so that the mine planning process can adequately incorporate shared value among all stakeholders to ensure that 'brand mining' can resonate with all stakeholders.
- Mine planning must incorporate uncertainty since some of the factors indicate that uncertainty is inherent in decision making for the mining sector. Accordingly, stochastic mine planning should become the norm in future mine planning.
- The mine planning process should design energy efficient mining systems that are cognisant of the shifting energy mix towards electricity and renewable energy supplies.
- The mine planning process should cater for designing mining systems that incorporate 4IR technologies and associated techniques for improved automation and digitization. In addition, continual review of cyber risk mitigation ensures secure application of 4IR technologies.

The above inferences are not exhaustive, but indicate that the mine planning process can evolve into the future by tracking and incorporating the findings from the annual surveys undertaken by the 'Big Four' accounting and auditing firms. In this way, the mine planning fraternity can ensure that the mine planning process continues to provide outputs that enable mining companies to make robust business decisions into the future.

References

1. Musingwini, C.: Keynote address: the importance of teaching mineral reporting and valuation codes in mine planning and valuation discourses within a mining engineering curriculum. In: Proceedings of the Society of Mining Professors 6th Regional Conference 2018, Johannesburg, South Africa, 12–14 March 2018, pp. 237–243. The Southern African Institute of Mining and Metallurgy (2018)
2. Deloitte: Tracking the trends 2019 – the top 10 issues transforming the future of mining (2019). https://www2.deloitte.com/content/dam/Deloitte/au/Documents/energy-resources/deloitte-au-er-tracking-the-trends-2019–210119.pdf. Accessed 26 Aug 2019
3. Ernst & Young Global Limited (EY): Top 10 business risks facing mining and metals in 2019–20 (2019). https://assets.ey.com/content/dam/ey-sites/ey-com/en_gl/topics/mining-metals/mining-metals-pdfs/ey-top-10-business-risks-facing-mining-and-metals-in-2019-20_v2.pdf. Accessed 26 Aug 2019

4. KPMG International (KPMG): Risks and opportunities for mining – Outlook 2019 (2019). https://assets.kpmg/content/dam/kpmg/xx/pdf/2019/02/global-mining-risk-survey-2019.pdf. Accessed 26 Aug 2019
5. PricewaterhouseCoopers (PwC): Mine 2019 - resourcing the future (2019). https://www.pwc.com/gx/en/energy-utilities-mining/publications/pdf/mine-report-2019.pdf. Accessed 26 Aug 2019
6. Deloitte: Tracking the trends 2017 – the top 10 trends mining companies will face in the coming year (2017). https://www2.deloitte.com/content/dam/Deloitte/global/Documents/Energy-and-Resources/gx-er-tracking-the-trends-2017.pdf. Accessed 26 Aug 2019
7. Deloitte: Tracking the trends 2018 – the top 10 issues shaping mining in the year ahead (2018). https://www2.deloitte.com/content/dam/Deloitte/mx/Documents/energy-resources/2018/Tracking-the-Trends-2018.pdf. Accessed 26 Aug 2019
8. Awuah-Offei, K., Que, S., Boateng, M.K., Weidner, N., Samaranayake, V.A.: Evaluating mine development alternatives for social risks using discrete choice models. Presentation at DEPLAMIN Workshop, Santa Cruz, Colchagua, Chile, 16–19 November 2018 (2018)

A New Search Algorithm for Finding Candidate Crusher Locations Inside Open Pit Mines

Morteza Paricheh(✉) and Morteza Osanloo

Department of Mining and Metallurgical Engineering,
Amirkabir University of Technology, Tehran, Iran
mpariche@aut.ac.ir, morteza.osanloo@gmail.com

Abstract. Nowadays, In-Pit Crushing and Conveying (IPCC) planning has been addressed as one of the hottest area of research in the mining engineering. Semi-mobile IPCC systems have had the most compatibility and coordination with the open-pit mining strategy. The existing open pit mine production planning problem gets more complex and intractable by concurrently optimizing the open pit mine production and IPCC planning. This is mainly because of the new successor blocks' constraints due to the in-pit crushers. The more candidate locations are defined truly, the less complex the problem would be. This requires deploying efficient search procedures to find the true candidate locations. This paper presents a new search algorithm for defining the most practical and less involving opportunity cost candidates. As well, a new clustering technique was developed to aggregate the blocks within the same azimuth domains, pushbacks and benches. The algorithm searches among the aggregated blocks inside the ultimate pit limit and tries to select the best candidates considering a series of generic and specific rules. The aggregation technique and search algorithm were applied in an open pit copper mine. The results indicated that the ideas provide hopeful tools for defining the candidate points of semi-mobile PCC systems.

Keywords: Open pit mining · Semi-mobile in-pit crushing and conveying system · Crusher location · Search algorithm · Aggregation procedure

1 Introduction

Nowadays, high grade ore deposits at the near-surface have been extracted and mining operations must be continued deeper for extracting lower grade ore deposits. If the mining operation is decided to be extracted using the surface mining methods, open pit mining operation is the only option which gives an opportunity to do large scale mining in deeper grounds. Such open pits should mainly cope with waste management and transportation difficulties. Open pit mining in deeper grounds necessitates higher stripping ratios and more transportation requirements. As mining operation deepens and gets larger, the haulage distances increase. The higher striping ratios highlight the importance of transportation again. As a much sensible way, these mega open pits should be planned in a manner to take advantage of technological advances as much as possible. In-Pit Crushing and Conveying (IPCC) technology-although is not a new

E. Topal (Ed.): MPES 2019, SSGG, pp. 10–25, 2020.
https://doi.org/10.1007/978-3-030-33954-8_2

technology but has received paramount attentions newly-is an idea that provides an opportunity to resolve these dilemmas (both waste management and transportation difficulties). In each unique operation, to be as cost-effectively as possible, the IPCC systems must be carefully planned and designed. Transition time from the shovel-truck system to semi-mobile IPCC system, conveyor design, crusher location and its relocation plan are the main questions that are to be addressed. Additionally, there have been some research works that address optimization of the system design elements in line with production planning and haul fleet sizing [1–5]. There are a series of strong reasons for these combinations. For a deeper and critical overview on IPCC systems and its contributing design elements, the readers are encouraged to have a look on [1] and the references within. The IPCC planning is classified as a long term planning problem, since it deals with a few hundreds of millions of dollars for a period of at least ten years. Therefore, we do not make a short term plan that follows our long term decisions. Instead, we are going to make a long term plan in which the budget planning regarding the IPCC application and its locations during the mine life is done.

Combined form of these problems (i.e., IPCC planning, production planning and truck fleet sizing) is computationally intractable due to large number of variables and constraints. The IPCC planning problem is a kind of dynamic facility location problems which requires some predefined candidate locations [6]. Each candidate location can be a mining block or a set of mining blocks, depending on the crusher size, block size and the operational space required. As well, the size and practicality of the IPCC planning problem can be largely controlled by truly definition of the candidate locations. Suppose that variable $u_b^t \in [0, 1]$ is a binary in an Open Pit Mine Production Scheduling (OPMPS), which represents extraction of a given block b ($b \in B$) at time period t, where B is the set of blocks inside the ultimate pit limit. The variable $v_j^t \in [0, 1]$ also gets 1 if the in-pit crusher is located at location j at time period t and 0 otherwise. The mathematical programming form of the combined optimization problem includes a constraint for each candidate location j as in Eq. 1.

$$\sum_{b \in F_j} \sum_{\tau=1}^{t} u_b^\tau \le |F_j|(1 - v_j^t) \quad \forall j \in J, t \in T \tag{1}$$

Where $t \in T$ is the set of time period, $j \in J$ is the set of candidate locations, $b \in Fj$ is the set of locked blocks beneath the crusher j. This constraint prohibits, considering the wall slope, the extraction of a series of successor blocks below the candidate location (i.e., subsets of F) before and till the existence of crusher at location j. This constraint acts as the slope/precedence constraint in a common production scheduling. Figure 1 indicates these successor blocks when the rule is based on freezing/locking nine (Fig. 1a) and five (Fig. 1b) blocks below a given candidate location j. Besides, the number of variables in the combined production planning problem would be multiplied by |J|. Therefore, whatever the size of set J is reduced, the more tractable the combined optimization model would be.

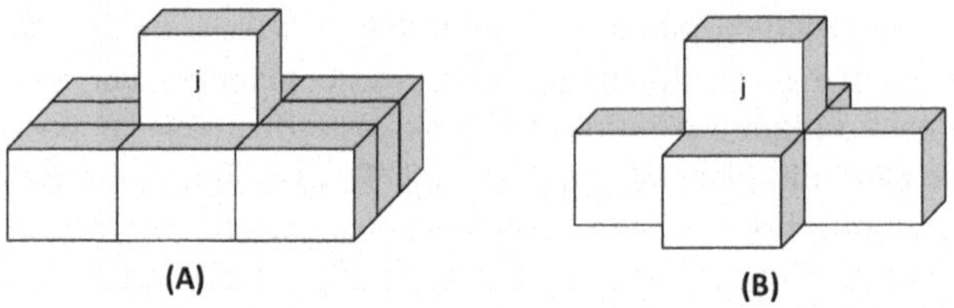

Fig. 1. The successor blocks of candidate point j, (a) nine block pattern (|F| = 9), (b) five block pattern (|F| = 5).

Suppose one block is big enough to lend the area for locating the crusher. Therefore, in the first glance, we have |B| candidate locations inside the pit. Choosing all blocks inside the pit as candidate locations is irrational. In addition, for real cases when the number of blocks exceeds a few hundreds of thousands, |B| candidate points make the problem really unsolvable.

In addition to the above said computational complexities that candidate locations add to the IPCC planning problem, definition of the candidate points is of high interests from practical and economic perspectives since excessive movements of the crusher are not recommended. In addition, some candidate points do not involve sufficient positional, geotechnical, and operational requirements. Therefore, this paper tries to provide a new search algorithm to guide the procedure of candidate point selection. Indeed, in order to pave the way for an efficient solution procedure for the combined optimization problem, we are going to intelligently define a finite set of practical candidate locations. The topic of finding candidate locations for installing the semi-mobile in-pit crusher has never been researched yet. In order to avoid mentioning duplicate content and to shorten the paper, we refer the readers to our overview on the IPCC systems and the references within [1].

This paper is organized as follow: in Sect. 2, the main assumptions are provided and a detailed description of the problem is presented. Section 3 contains all the details concerning the mathematical formulation of the algorithm. Thereafter, implementation details and numerical results are given in Sect. 4. Finally, after a deep discussion in Sect. 5, we will end the paper with the main concluding remarks and perspectives.

2 Problem Description and Assumptions

There is a set of regular rectangular-shaped blocks inside the open pit limits and we are going to identify those blocks or more specifically those discrete sets of blocks holding a series of special characteristics. Practically, the search algorithm should be confined by a series of azimuth domains. These azimuth domains are basically defined based on:

(1) Topography,
(2) Downstream facility locations,
(3) The conveyor exit strategy and
(4) Primary mining schedule or mining direction

Suppose that a primary production schedule (without considering the IPCC plan) has shown that the mining direction must be started from the east and continued to the west. In this situation, in order to expose the candidate locations, the production must be firstly started from the west if the candidate locations have been already defined in the west walls. In high grade ore deposits compared to lower grade deposits, the primary production schedule may create more incomes than changing that due to exposing the candidate locations for the crusher. Figure 2a indicates an ore-body which has the same or near the same inclination as the stable wall slope angle. The ore body has outcrop and continued to the deep. The dashed pits inside the ultimate pit limit shows the way of progress or direction of the mining operation (mining phases/pushbacks). In this case, there is always a wall which has reached its final limit. Now, suppose that the downstream facilities (i.e., secondary crusher or waste dump) are located in the east part of the deposit (Fig. 3a). In this case, the east wall is the best part for locating the candidate crusher locations. Inversely, if the downstream processes have been located in the west part of the pit, and similarly the east wall is selected for locating the candidate locations, there should be existed a lengthier conveyor line for carrying the material from the east to the west of the pit. Otherwise, if the west wall had been selected, the candidate locations will not be available to install an in-pit crusher only if the direct blocks in contact with that candidate location have been extracted before. Hence, the sequence will change if the in-pit crushing could add more value to the project than the primary extraction sequence. Otherwise, the concurrent optimization of the problems will not result in further value. The same situation exists for the second example in Fig. 2b where the grade decreases from the east to the west.

There are some ore deposits where the ore grade decreases toward the sides or the ore body is almost vertical. In these cases, the mining sequences look like onion shells. The extraction is started from the inner sell and continued to the outer shells. In such situations, there is no wall that reaches its final limit until the end of mine life (Figs. 2c and 3b). As a result, the primary production schedule plays a major role for defining the candidate locations. There must be a compatibility with the expected production schedule (integrated OPMPS and IPCC) and the candidate locations defined. Otherwise, the value of in-pit crushing in a combined optimization problem might not exceed the value created by a separate primary production schedule. Anyway, in order to gain the maximum possible value from an open pit project, the candidate locations must truly follow the primary sequences of extractions. Therefore, the methodology must support all types of the deposits, no matter how the shape of the ore body and mining direction/mining sequences are.

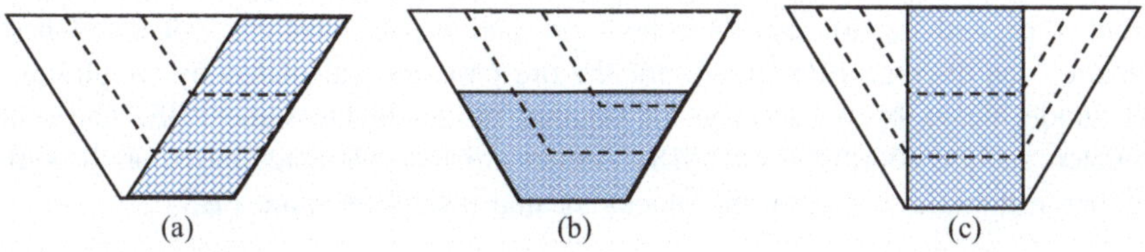

Fig. 2. Cross-sectional views of different ore body shapes and the corresponding mining directions, the numbers indicate the extraction order or mining directions

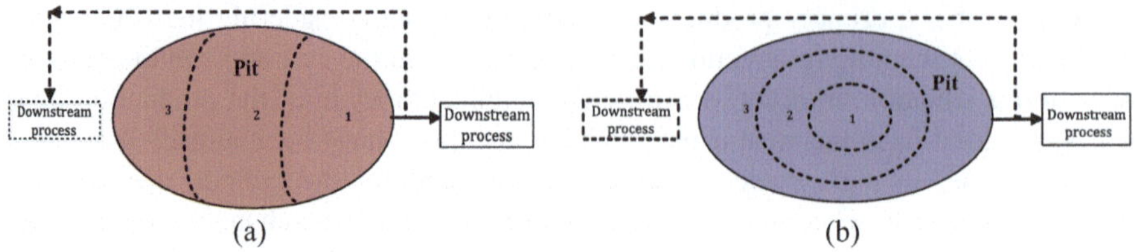

Fig. 3. Plan views of pits in (a) Figs. 2a and b, and (b) Fig. 2c - Arrows represent conveyor lines; numbers represent the extraction orders

In addition, the topography prohibits selecting any type of conveyor exit strategy and also exiting the conveyor from some special parts of the pit. The conveyor exit strategy also usually constrained by truck ramping system and the production schedule. It cannot be installed in some parts since it is in interface with the existing ramping system. Currently, as a general rule, all of these constraints are performed using engineering judgments just by finding a pit wall which reaches its final limit compatible with mining direction and production schedule, location of downstream facilities, topography and operational requirements for installing the conveyor. The constraints which have been mentioned in this section are called general rules. We have some specific rules which are to be addressed in the next section.

3 Search Algorithm

3.1 Aggregation Procedure

As shown in the previous section, all of the general rules except those for the mining direction can be expressed as a series of azimuth domains. The rules for the mining direction or the sequence of extractions can be also considered using the pushback concept. To do this, we developed a new clustering procedure for aggregating the blocks located in the same azimuth domain and pushback. Indeed, we modified the concept of phase-bench/pushback-bench practiced by [7, 8] to Phase-Bench-Slice (PBS). Figure 4a shows a simple definition of phase-bench. In this way, those blocks located at the same bench and phase are grouped and considered as the same unit for further scheduling purposes. We added another limiting factor to identify/classify those blocks located in the same azimuth domain or slice (Fig. 4b). Figure 4c indicates a schematic view of the idea. Here, those blocks inside the pit limit located in the same PBS are clustered. It is worth mentioning that the materials inside a PBS are classified by rock type and the tonnages for both ore and waste parts are calculated firstly. Thereafter, the average PBS grades for the ore parts are calculated by weighting the block grade means by the tonnage of original blocks in the same PBS. The spatial parameters such as coordinates of PBSs' centers are also calculated based on the simple weighted arithmetic means on the blocks located inside the same PBS.

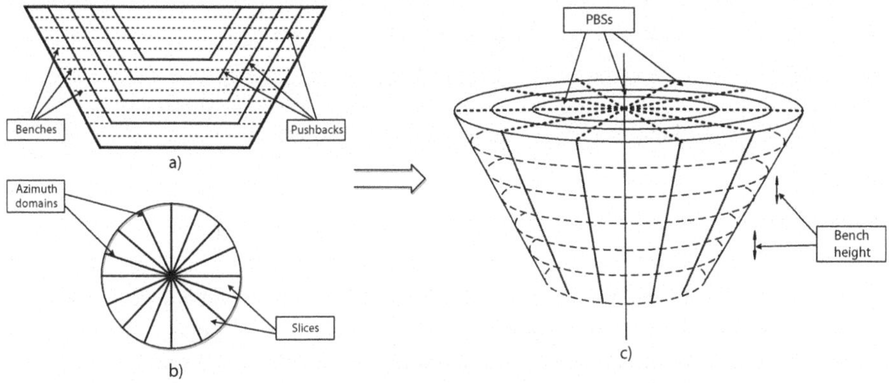

Fig. 4. Schematic view of the proposed aggregation procedure

3.2 Slice Definition

To find those blocks inside the pit and in a predefined azimuth range, one can identify the azimuth of any given block and compare it with the predefined range of azimuth. Note that azimuth is defined as the angle in clockwise direction from the geographical north. Suppose three blocks signed by their x and y coordinates in Fig. 5 (i.e., (x1, y1), (x2, y2), (x3, y3)) are to be classified in term of either they are inside the azimuth domain A1–A2 or not. In Fig. 5, if the azimuth domain is defined to be between A1 and A2, the block 1 would be inside the domain and the two other blocks 2 and 3 are outside the azimuth domain. Similarly, if the azimuth domain is between A2 and A1, the first block is outside and the two others are inside the domain.

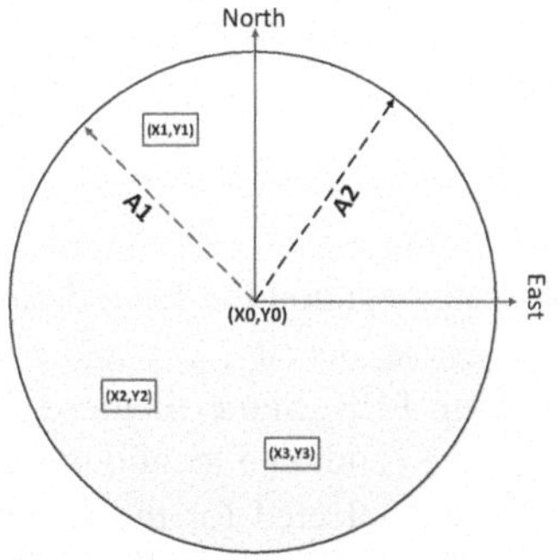

Fig. 5. The azimuth domain (dashed arrows), origin point (x0, y0), north and east directions (arrows) and given blocks (xi, yi)

In order to find those blocks in a special domain of azimuths, the well-known function atan2 has been used (Eq. 2). The value of atan2 is returned in the interval $(-\pi, \pi]$. Indeed, atan2 is signed with counterclockwise angles being positive, and clockwise being negative. Specifically, atan2 (x, y) is in the interval $[0, \pi]$ when $y \geq 0$, and in $(-\pi, 0)$ when $y < 0$.

$$\mathrm{atan2}(x, y) = \begin{cases} \arctan\left(\frac{y}{x}\right) & x > 0 \\ \arctan\left(\frac{y}{x}\right) + \pi & x < 0, y \geq 0 \\ \arctan\left(\frac{y}{x}\right) - \pi & x < 0, y < 0 \\ \frac{\pi}{2} & x = 0, y > 0 \\ -\frac{\pi}{2} & x = 0, y < 0 \\ \mathrm{NA} & x = 0, y = 0 \end{cases} \tag{2}$$

In order to calculating the azimuth of a given block, a benchmark point of origin (x0, y0) must be defined first. Selection of the origin is impressively important, however, we are trying to explain the methodology first. We will discuss the selection of this origin in discussion. After defining the origin, a new relative coordinate of the block (x′, y′) would be defined as in Eqs. 3 and 4.

$$x' = x - x0 \tag{3}$$

$$y' = y - y0 \tag{4}$$

This relative coordinates of the given block are the inputs of function atan2. Then, in order to convert the result of function atan2 (x, y) to the azimuth in degree, the following function is proposed:

$$A(x, y) = \begin{cases} 90 - \mathrm{rad2deg}(\mathrm{atan2}(\frac{y'}{x'})) & \text{if } \mathrm{atan2}(\frac{y'}{x'}) \leq 0 \mid \mathrm{atan2}(\frac{y'}{x'}) \leq 90 \\ 450 - \mathrm{rad2deg}(\mathrm{atan2}(\frac{y'}{x'})) & \text{otherwise} \end{cases} \tag{5}$$

Where, A(x, y) represents the azimuth of block in coordinate (x, y), rad2deg is a function for converting the value of function atan2 from radian to degree.

3.3 Specific Rules

Up to now, a new form of block aggregation has been defined. All the properties of the blocks inside each PBS such as grade, density, x, y and z coordinates can be averaged and assigned to the corresponding PBS. Some of these PBSs located in special and predefined azimuth domains can be candidate locations for the crusher. Suppose three azimuth domains of 24° in size are selected for the search procedure as the primary solution space. Depending on the number of benches and pushbacks, there may be more than a few hundreds of candidate locations. But, this is not short enough for the subsequent IPCC planning. There are some specific rules listed below by which the number of candidate locations can be reduced. The specific rules are explained in detail in the following subsections.

- Depth
- Pushback
- Required space

- Radius of influence
- Frozen economic values

Depth Restriction

We firstly confined the search domain by a predefined depth from which downward the candidate locations can be logically defined. This minimum depth is a threshold that brings about a maximum haulage distance that the trucking system might not economically operate. Indeed, those benches near the pit rim are prohibited from contributing to further search. The depth can be easily known using the distance each bench adds to the truck haulage and comparing the cumulative distances with a maximum threshold in mind (Eq. 6).

$$MD = EPD + \left|\frac{REL - L}{\sin \alpha}\right| \tag{6}$$

where MD is the maximum possible distance that the truck system can economically operate, 2.3 km for example [9], EPD is the ex-pit truck haulage distance (from the ramp exit point to the primary ex-pit crusher), REL is regarded as the level from which the ramping system exit the pit, L is the level we are looking for and α is the gradient of ramping system. Note that depending on the topography, the level might be in some point above the level that ramp exits the pit.

Pushback Restriction

Since the IPCC systems require a significant capital up-front at the order of a few hundreds of millions of dollars, investment for the IPCC system earlier than payback period is not recommended. This time can be equivalently considered as the time that the first pushback is depleted. Therefore, we can also remove the first pushback from the search space.

Required Space Restriction

Since PBSs consist of a block or a combination of blocks, the space that each PBS can lend for installing the crusher and its accessories is limited to the area of those blocks that the PBS is made of. Some of the PBSs which have been remained in the solution space do not have the adequate space and therefore must be barred from further search.

Radius of Influence Restriction

It is worth noting that the yearly deepening rate in open pit mines is limited to a few benches. As well, the historical data regarding the application of the in-pit crushing systems have shown that the systems are not to be relocated more than once a year. It shows that each candidate location has a vertical radius of influence that effectively affects the mining operation. Hence, we thought a unique form of elimination procedure by which those candidate locations between the two consecutive candidate points with a distance of vertical radius of influence are removed from the search space. In that regard, in each slice, the two candidate locations holding the maximum and minimum depths are kept anyway.

Frozen Economic Values Restriction

Now, those PBSs remained are to be exposed to another restriction. In this step, there might be existed some PBSs that are still in the same level but different slices. We are going to keep just one candidate point in each level no matter in which slice or azimuth it is. For this purpose, we used the concept of locked/frozen values beneath each candidate location. The candidate with the lowest frozen value beneath would be kept and the remaining would be removed. In order to associate a priority to a candidate location, we need to consider its impact on the extraction of other blocks in the future. One such measure proposed by [10, 11] is the block look ahead value. The concept of look ahead value has been practiced mainly for scheduling the sequences of extraction of the blocks.

(a) (b)

Fig. 6. The concept of look ahead value, all the red blocks have the yellow one as a look ahead, (a) 2D view, (b) 3D view

From the practical view, there are some blocks requiring a particular block be removed before they can be extracted (Fig. 6). Actually, all those below it cannot be mined until this key block is mined. Following a stable slope, the set of all blocks that a single block avoids from mining constitutes a cone with that key single block at its apex. In other words, for each block inside the pit, there is a cone downward which is confined by the pit limits and includes all the blocks that have that key block as the look ahead. Since candidate locations are composed of some adjacent blocks, each candidate location prohibits mining all the blocks inside its downward cone. Therefore, there exists a positional weight for each candidate location which is the sum of locked values beneath. Those candidate points with the lowest positional weights have the highest chance among others to be selected as candidates. The best candidates are those who have no value or zero values underneath. Mathematically, the concept can be explained by Eq. 7.

$$\underset{j \in S}{\arg\min} \left\{ \sum_{b \in B_j} BEV_b \right\} \tag{7}$$

Where j is the candidate location, S is the set of candidate locations at the same level, BEV_b is the economic value of block b, B_j is the set of all blocks inside the downward cone of candidate location j confined by the pit boundaries.

4 Validation and Results

We applied the described search algorithm to Sungun open pit copper mine in Iran. The mine is located in the northwest of the country in a dense mountainous and forest area. Figure 7 shows the locations of the main facilities around the pit limit. Waste materials have been planned to be dumped at the same level as the extraction level in the north and northwest of the pit. In this way, the haulage distances for the waste materials are kept constant about one kilometer through the mine life. Therefore, it seems that there is no room for an in-pit waste crushing system. In this study, we are just going to identify the candidate locations for an in-pit ore crushing system. Suppose that a high angle conveyor is to be used for carrying the material from inside the pit to the ground. Due to the topography, location of the ramping system, locations of the ex-pit primary crusher and the main conveyor linking the crusher to the concentrator plant in the south part of the pit, we are able to select a particular part in the southeast of the pit as the target azimuth domain for locating the in-pit crusher. We set this target azimuth domain between 100 and 150°. Therefore, this target domain is our first restriction in the search algorithm or the general rule described in Sect. 2

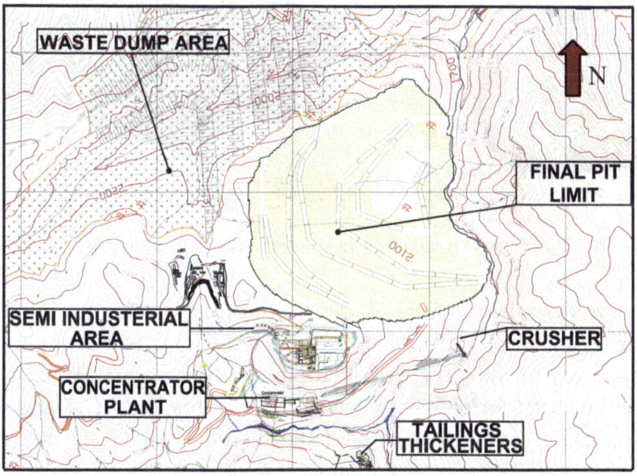

Fig. 7. Plan view of the pit and the locations of the main facilities around the pit

We firstly determined the Ultimate Pit Limit (UPL) and some nested pits based on parameterization of the price factor. The ultimate pit consists of 79038 fixed-size blocks (25 m × 25 m × 12 m). Total rock and ore inside the pit was calculated to be 1219 and 469 million tones, respectively. Figure 8a shows the ore and waste tonnages within the 87 generated nested pits. The automated pushback selection proposed by [12] was used to select the pushbacks. The technique aims to minimize the absolute deviation of tonnage within the pushbacks from a target reference tonnage. This reference value can be calculated by dividing the whole rock tonnage inside the pit by the total number of pushbacks required. The number of pushback was set to four.

Figure 8b shows the rock and ore tonnages within each selected pushback and also the corresponding average grade values. Also, Fig. 9 shows plan and 3D views of the pushbacks 1–4 specified by cyan, pink, red and blue colors, respectively.

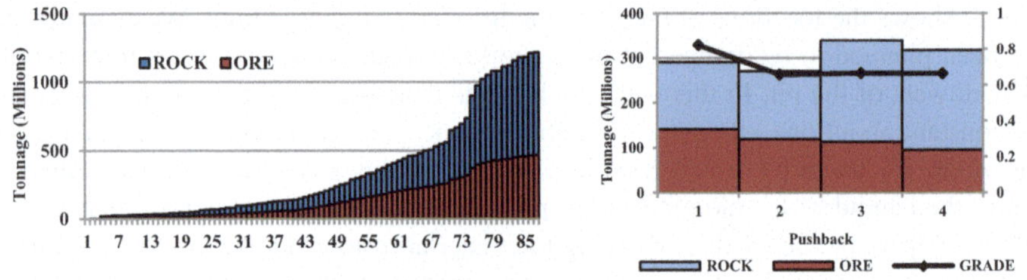

(a) (b)

Fig. 8. (a) Rock and ore tonnage within the nested pits, (b) Characteristic of the automatically selected pushbacks

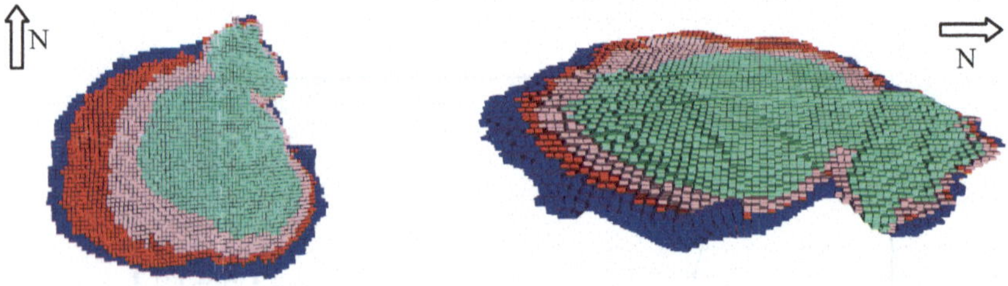

(a) (b)

Fig. 9. Automatically selected pushbacks, (a) plan view, (b) 3D view

The procedure described in Sect. 3 was used to separate the pit into equally sized slices. In order to divide the pit limit into slices, equal azimuth domains of 30° were selected. In this way, the pit was subdivided into 12 distinct equally sized slices. The x and y coordinations of the origin point were also defined as 8724 and 4807 m. The point is located in the centroid of area at the lowest level inside the pit. Figure 10 indicates both plan and 3D views of the pit limit divided by slices. The slices have been differentiated by colors.

(a) (b)

Fig. 10. Slice determination, (a) plan view, (b) 3D view

Considering 65 benches, a total number of 3120 PBSs (4 × 65 × 12) or (number of pushback × number of bench × number of slice) would be expected. However, due to topography, pit and pushbacks' shapes, just 2063 PBSs were generated. According to the target azimuth domains described before and the defined slice size, we selected the fourth and the fifth slices in our search procedure (the numbering format for the slices is based on azimuth increment). The results indicated that 281 PBSs were within these slices. Our first specific rule was the depth restriction. Since the ramp exits the pit at level 1987, the search algorithm was confined to those depths below level 1800. Hence, in this way, 122 PBSs were remained in the subsequent search space. Now, the second specific rule i.e., pushback restriction was applied to the algorithm. We did not let the candidate points inside the first pushback to be kept in our search space. In this step, 21 more PBSs were removed from the search space. Assuming six blocks can provide sufficient space for the system; the third specific rule removed three more PBSs from the search space. Finally, the two other specific rules could reduce the total number of candidate points to only 23 pcs. It is worth noting that to avoid calculating the BEVs which are subjected to volatile economic parameters and also to reduce the calculation time, the total amount of frozen metal contents below a given candidate location was used instead of frozen values.

Figure 11 graphs how the search algorithm works in different steps and how the PBSs are removed from the search space in each slice. The last column in Fig. 11 indicates the selected candidate points. The colors represent the pushbacks.

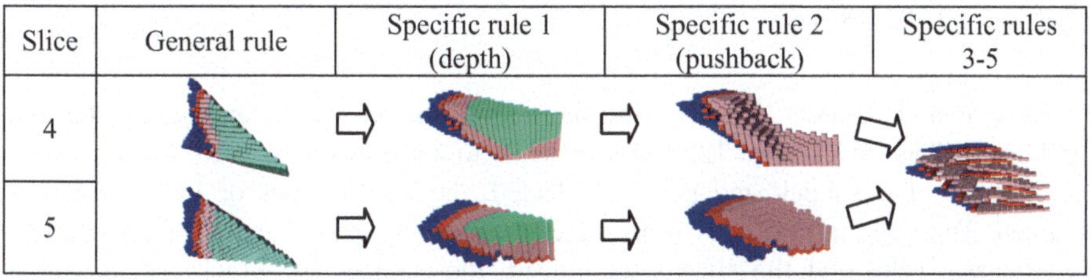

Fig. 11. A graphical representation of steps in the proposed search algorithm

5 Discussion

In the previous section, the performance of the proposed search algorithm was shown by applying it in a real open pit mine in Iran. It was shown that instead of searching all the points inside the pit; we must limit the search procedure to a specific part of the pit. It was called "the general rule". Then, the search procedure was confined by some specific rules. As well, it was shown that how the total number of candidate points were reduced from 281 to only 23 points. Here, we tested the algorithm by changing the input requirements. In the first, the new aggregation procedure was tested by changing the number of pushbacks and slices. The number of pushbacks and slices were changed from 3 to 6 and from 10 to 18, respectively. Totally, sixteen distinct scenarios were built. Table 1 summarizes the results of these scenarios. Figure 12 also shows the characteristics of the pushbacks when 3, 5 and 6 pushbacks are considered. The characteristics associated with the case with four pushbacks were shown before in Fig. 8. Moreover, Fig. 13 shows the 3D views of the pit divided into 20, 24 and 36° slices.

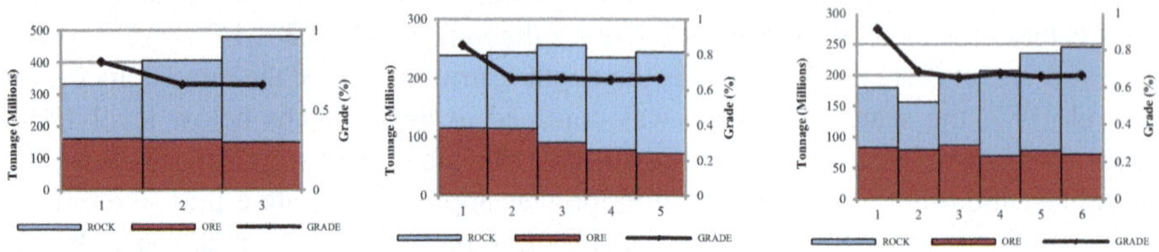

(a) (b) (c)

Fig. 12. Characteristics of the pushbacks, (a) three pushbacks, (b) five pushbacks, (c) six pushbacks

(a) (b) (c)

Fig. 13. 3D views of the UPL associated with different slice sizes, (a) 36° (10 slices), (b) 24° (15 slices), (c) 20° (18 slices)

As shown in Table 1, by changing the number of pushbacks and slices, the total number of PBSs varies from 1332 in scenario 4 to 4236 in scenario 13. Logically, the less number of pushback and slice is selected, the less number of PBSs would be obtained. Then, the target slices were selected according to the target azimuth domain (i.e., general rule) and the slice size. Indeed, those slices by which the maximum possible target azimuth domain is covered were selected. For example, in the first scenario, the sixth, seventh and eighth slices include an azimuth domain between 100 and 160°. Similarly, in the second scenario, the slices 5 and 6 cover an azimuth domain of 96 to 144°. The total numbers of candidate points in different pushbacks and slices have also been brought in the seventh and eighth columns of Table 1. Remember that in all cases, the same specific rules as those described in the previous section were imposed. As shown in the last column of the table, for each pushback size, as the slice size increases the total number of candidate points decreases. In addition, for each slice size, as the number of pushbacks increases the total number of candidate points increases.

Similar to the number of pushbacks and slices, the total number of candidate points can be controlled by the specific rules. For instance, the more space is required for the system, the less number of candidate points would be found. There are the same situations for other specific rules. Anyway, the specific rules must be well suited to each unique operation.

As one of the most critical parameters in the aggregation procedure, the origin point should be determined. Generally, any point inside the pit limit can be an origin point. To illustrate the differences that each origin point may impose into the aggregation procedure, we tested two more origin points. Remember that in this step, we selected the same setting as those used in the base case presented in the Sect. 4. Table 2 shows the results associated with these two origin points. Figure 14 also shows the plan and 3D views for the pit divided by slices based on the new origin points.

Table 1. Summary of the results for different origin points

Origin point scenario	Coordinates of origin	No of PBS	No of candidate location
1	X = 8365 Y = 5175	1495	15
2	X = 8099 Y = 4557	1251	10

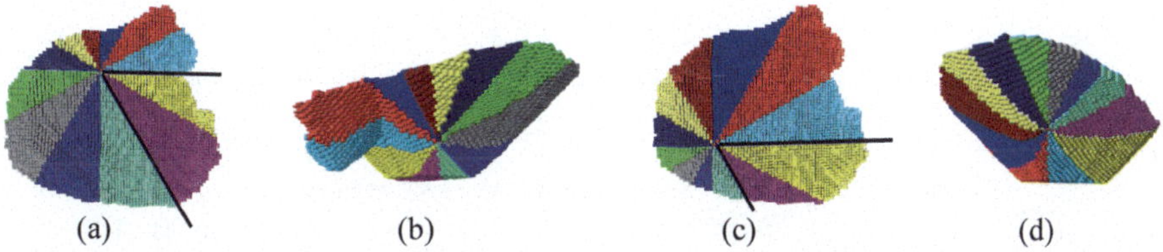

Fig. 14. Plan and 3D views of pit divided by slices, (a) plan view associated with origin point 1, (b) 3D view associated with origin point 1, (c) plan view associated with origin point 2, (d) 3D view associated with origin point 2

As shown in the 3D views, both origin points intersect the pit boundary in a level above the pit floor. Hence, some parts of the target slices (slices 4 and 5) colored yellow and pink and specified by continued black lines in Fig. 14a are in the opposite direction of the pit (see Fig. 14b). Therefore, mistakenly some candidate points are defined in other part of the pit. On the other hand, for the second origin point, almost the entire fifth target slice is located above the depth restriction we defined in our search algorithm (see Fig. 14d). Totally, in some slices, the candidate points are not formed and in some others (i.e., fourth slice) the unshaped candidate location are defined. For these reasons, the origin point must be selected with more obsessions for both aggregation procedure and search algorithm.

Moreover, as shown in Fig. 2, there are a series of open pits wherein a primary production schedule indicates that some parts of the pit inside the target range of azimuths also reaches the final pit boundary. In these cases, instead of eliminating the first pushback from the search space, it must be involved in the search.

In addition to those specific rules defined in Sect. 3, any candidate location requires being as reliable as possible from both safety and geotechnical perspectives. If not, the rock quality can be improved subject to spending money for stabilizing the ground.

Some indexes of rock quality measurements such as RMR (Rock Mass Rating) and RQD (Rock Quality Designation) can be used to rank the candidate locations in order to find the best possible geotechnicaly accepted candidate points. It is also practically important for a candidate location to be in a safe distance vertically from unstable locations.

Table 2. Summary of the sensitivity analysis on the number of pushback and slice

Scenario	No of pushback	Slice size (degree)	No of slice	No of PBS	Target slices	Number of candidate points in pushback						Number of candidate points in target slice					Number of candidate location
						1	2	3	4	5	6	4	5	6	7	8	
1	3	20	18	2275	6, 7, 8	–	8	5	–	–	–	–	–	4	6	3	13
2		24	15	1927	5, 6	–	5	5	–	–	–	–	5	5	–	–	10
3		30	12	1569	4, 5	–	5	5	–	–	–	5	5	–	–	–	10
4		36	10	1332	4	–	5	5	–	–	–	10	–	–	–	–	10
5	4	20	18	2986	6, 7, 8	–	9	10	5	–	–	–	–	11	8	5	24
6		24	15	2530	5, 6	–	9	9	5	–	–	–	10	13	–	–	23
7		30	12	2063	4, 5	–	9	9	5	–	–	12	11	–	–	–	23
8		36	10	1755	4	–	5	5	5	–	–	15	–	–	–	–	15
9	5	20	18	3657	6, 7, 8	–	10	8	9	5	–	–	–	10	10	12	32
10		24	15	3103	5, 6	–	8	5	6	5	–	–	14	10	–	–	24
11		30	12	2537	4, 5	–	8	5	5	5	–	10	13	–	–	–	23
12		36	10	2165	4	–	5	5	5	5	–	20	–	–	–	–	20
13	6	20	18	4236	6, 7, 8	–	7	5	8	9	5	–	–	13	10	11	34
14		24	15	3597	5, 6	–	7	5	5	6	5	–	15	13	–	–	28
15		30	12	2942	4, 5	–	5	5	5	5	5	10	15	–	–	–	25
16		36	10	2512	4	–	4	5	5	5	5	24	–	–	–	–	24

Moreover, the shape and dimensions of the candidate locations, as well as the adjacency of the blocks are important when a candidate location is defined. In this paper, these issues were not considered. Future researches can also consider both geotechnical and shape restrictions.

6 Conclusion

Integration of IPCC plan and open pit mine production plan provides an opportunity to consider both block values and value created due to the position of in-pit crusher, simultaneously. An unmanned position does not permit extracting valuable blocks underneath, nor does it produce further value for the IPCC systems. As well, truly definition of the candidate location for the crusher help reduces the size of the subsequent combined production and IPCC planning. This paper introduces a search algorithm confined by a series of general and specific rules for defining the candidate locations for installing a semi-mobile IPCC system in open pit mines. First, a new aggregation procedure was provided to cluster the blocks located in the same azimuth domains. Both aggregation and search algorithms were implemented in an open pit copper mine in Iran and their abilities were tested. The results showed that the proposed procedure can be a useful tool for semi-mobile IPCC planning problem. The number of candidate locations can be even reduced to a few practical locations.

References

1. Osanloo, M., Paricheh, M.: In-pit crushing and conveying technology in open-pit mining operations: a literature review and research agenda. Int. J. Min. Reclam. Environ. 1–28 (2019)
2. Nehring, M., Knights, P.F., Kizil, M.S., Hay, E.: A comparison of strategic mine planning approaches for in-pit crushing and conveying, and truck/shovel systems. Int. J. Min. Sci. Technol. **28**, 205–214 (2018)
3. Samavati, M., Essam, D.L., Nehring, M., Sarker, R.: Open-pit mine production planning and scheduling: a research agenda. In: Sarker, R., Abbass, H.A., Dunstall, S., Kilby, P., Davis, R., Young, L. (eds.) Data and Decision Sciences in Action, pp. 221–226. Springer, Switzerland (2017)
4. Paricheh, M., Osanloo, M., Rahmanpour, M.: A heuristic approach for in-pit crusher and conveyor system's time and location problem in large open-pit mining. Int. J. Min. Reclam. Environ. **32**, 1–21 (2018)
5. Samavati, M.: New long-term production scheduling methodologies for open-pit mines. Ph.D. thesis, University of New South Wales, Canberra (2017)
6. Arabani, A.B., Farahani, R.Z.: Facility location dynamics: an overview of classifications and applications. Comput. Ind. Eng. **62**(1), 408–420 (2012)
7. Elkington, T., Durham, R.: Integrated open pit pushback selection and production capacity optimization. J. Min. Sci. **47**(2), 177–190 (2011)
8. Rezakhah, M., Newman, A.: Open pit mine planning with degradation due to stockpiling. Comput. Oper. Res. (2018, in press)
9. Bucyrus-Erie Company: Surface Mining Supervisory Training Program, Shovel/Truck. South Milwaukee, Wisconsin (1979)
10. Tolwtnski, B., Underwood, R.: A scheduling algorithm for open pit mines. IMA J. Manag. Math. **7**(3), 247–270 (1996)
11. Gershon, M.: Heuristic approaches for mine planning and production scheduling. Int. J. Min. Geol. Eng. **5**, 1–13 (1987)
12. Jélvez, E., Morales, N., Askari-Nasab, H.: A new model for automated pushback selection. Comput. Oper. Res. (2018, in press)

An Application of an Open Pit Mine Production Scheduling Model with Grade Engineering

Karo Fathollahzadeh[1(✉)], Mehmet Cigla[1], Elham Mardaneh[2], and Waqar Asad[1]

[1] Mining Engineering and Metallurgical Engineering, WA School of Mines: Minerals, Energy and Chemical Engineering, Curtin University, Kalgoorlie, Australia
karo.fathollahzadeh@postgrad.curtin.edu.au
[2] School of Electrical Engineering, Computing and Mathematical Sciences, Curtin University, Perth, Australia

Abstract. Grade Engineering (GE) refers to a range of pre-concentration techniques and operating methods to improve mineral feed grades by early rejecting low uneconomic materials prior to energy and cost intensive processing. While inclusion of a GE plant within the mineral value chain may assist the development of realistic strategic mine plans, the existing models that offer solution to the open pit production scheduling problem ignore the important and relatively new GE techniques. This paper presents an overview of a new mathematical formulation that optimizes annual production scheduling of an open pit mining operation with Grade Engineering subject to a set of operational and technical constraints. The computational results show that the proposed method outperforms than the existing models without GE plant.

Keywords: Grade Engineering · Open pit · Production scheduling · Solution methods

1 Introduction

Open pit mine production scheduling is a decision making process that determines the extraction sequence of which blocks should be mined and when they should be mined. Once mined, the decision is made on where the blocks should be sent e.g. to mining stockpile, processing plants, waste dump, grade engineering plants, grade engineering stockpiles or heap leach. From processing plant and heap leach, the production is obtained to drive profit by selling to the commodity market. In production scheduling, mine planners partitioned the ore-body into thousands of equal-size cubic shapes (blocks???) to make the extraction process more tractable. Each block in the block model has a specific information such as material type (i.e. oxide, transition, fresh) and grade items (i.e. Au, Cu) (Redwood and Scott 2016a, b). By assigning economic attributes (processing cost, metal price, re-handling cost and mining cost) to each block, the resource model is converted into the economic block model so that each block has a specific economic value ($) (Gaupp 2008).

E. Topal (Ed.): MPES 2019, SSGG, pp. 26–32, 2020.
https://doi.org/10.1007/978-3-030-33954-8_3

Grade Engineering is a concept introduced by Cooperative Research Centre for Optimising Resource Extraction (CRC ORE) to refer a span range of preconcentration techniques and operational methods within the mineral value chain to improve feed grades by rejecting undesirable material (Carrasco et al. 2016a, b, 2017). Different preconcentration techniques within grade engineering plant such as bulk sorting, differential blasting (size deportment) and screening (preferential grade) have been developed to improve head grades by removing low grade material prior to moving to processing plant (e.g., flotation, milling) (Carrasco et al. 2016b). Figure 1 shows the mineral supply chain of an open pit mining operation that incorporates GE plant. There are only a few available studies (Montiel and Dimitrakopoulos 2015; Redwood and Scott 2016a, b; Espejel 2017) have proposed mathematical framework of open pit mine production scheduling with multiple processing streams and operations. Montiel and Dimitrakopoulos (2015) presented model of open pit mine production scheduling with multiple processing streams, however, the variation of recoveries for each specific block and the energy consumption was ignored. Later, Redwood and Scott (2016a, b) employed integration of GE within mineral supply chain. In addition, Espejel (2017) shared the framework of open pit mine production scheduling with GE. The proposed model by (Espejel 2017) used revised version of Lane Theory to define cut-off grade strategy as well as considering grade-tonnage distribution (Redwood and Scott 2016b).

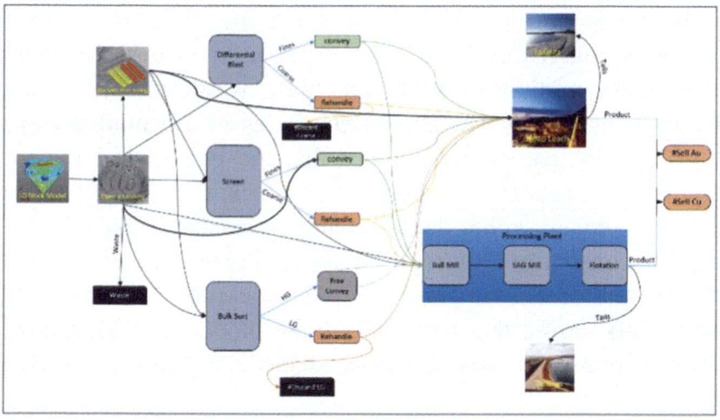

Fig. 1. A framework of an open pit mining operation with GE techniques.

To incorporate the multiple simulated realizations of the GE plant into the traditional open pit mine production scheduling, the mathematical programming is employed so that GE attributes are presented as an additional geological input and GE plant as an additional material destination. This study aims to develop a new mathematical framework of open pit mine production scheduling with GE as well as introducing different pre-processing strategies to improve efficiency of proposed framework by considering different possible destinations (based on content of block) for each block within the mineral value chain.

2 Mathematical Framework

The proposed framework uses the formulation developed by Fu et al. (2019) that aims at optimizing the production schedules (sequencing of extraction) and processing streams (sending material to a specific destination) simultaneously within the mineral value chain. The main difference in the proposed framework is the incorporating of GE plant (bulk sorting, screening and differential blasting) to the traditional formulation of open pit mine production scheduling (OPMPS). The presented framework not only optimizes the network flow of material within the mineral supply chain, but also helps to establish a tractable relationship between mine and processing stages. The proposed mathematical framework aims to maximize the discount Net Present Value (NPV) of future cash flow over planning horizon life of mine (LOM) subject to a wide range of constraints ranging from operational constraints to processing constraints. Objective function and constraints are expressed as follows:

- **The objective function** maximizes the NPV of the open pit mine which contains movement of materials from the pit to the mining grade bins, the pit to the processing plant and heap leach, the pit to the waste dump, the pit to GE and then to the processing plant and heap leach, the pit to GE grade bins, including materials from the mining stockpile to the GE, from the mining stockpile as well as GE stockpile to the processing plant and heap leach.
- **Reserve constraints** ensure that each block is mined only once during LOM.
- **Mining capacity constraints** ensure that the quantity of material which is moving from the pit to the mining stockpile, GE plant, waste dump, GE stockpile, processing plant and heap leach remain within the allowable mining capacity.
- **Processing capacity constraints** ensure that the quantity of material which is moving from the pit, mining stockpile and GE stockpile to the processing plant and heap leach must remain within the allowable processing capacity.
- **Average grade constraints** control the average grade of materials fed to processing plant and heap leach from the pit or from the stockpile.
- **Slope constraints** ensure that for each mining block, all predecessors (overlying blocks) must be mined by creating a cone-shaped extraction at a pre-defined slope angle.
- **Mining stockpile constraints** ensure the inventory balance between the movement of materials from the mine to mining stockpile and the retrieval of material from mining stockpile to processing destinations.
- **GE stockpile balance inventory constraints:** (same mining stockpile)
- **Power and steel consumption constraints** ensure that the quantity of materials fed to the processing plants and heap leach remain within the allowable power and steel available.
- **Fixed cost constraints (mining, processing and GE)** ensure that once mining and processing occurs at each time period, their fixed cost is activated.

3 Pre-processing

Data pre-processing aims to remove excess data before starting of optimization procedure. Not only does it lead to diminish a number of decision variables, but also helps to enhance the contextually relevant information and increase efficiency in terms of space required and processing time (Li et al. 2016). While a wide range of pre-processing techniques exist in the field of production scheduling (Merchan and Maravelias 2016), in this study, we present a set of new pre-processing techniques based on grade content as well as material attributes, which we called "suitability" and two types of suitability are presented as follows:

- Block & mining stockpile (B-Ms)
- Block & GE stockpile (B-GEs)

Tables 1 and 2 show the structure of mining and GE stockpile with pre-defined separate grade bins based on block attributes and grade content. The structure in both mining and GE stockpiles prevents the blending of material except for the grade bin in question. This suitability leads to getting rid of non-linear structure for blending material in both stockpiles, consequently, making model more efficient by decreasing exceed information.

Table 1. Mining stockpile.

Au (g/tonne)	0–0.5																	
Cu (%)	0–0.3						0.15–0.6						0.6–0.9					
Domain	1			2			1			2			1			2		
Rock type	OX	TR	FR	OX	TR	FR	OX	TR	FR	OX	TR	FR	OX	TR	FR	OX	TR	FR
Bin no.	1	2	3	4	5	6	7	8	9	10	11	12	13	14	15	16	17	18

Table 2. GE stockpile.

Au (g/tonne)	0–1																	
Cu (%)	0–0.15						0.15–0.30						0.30–0.40					
Domain	1			2			1			2			1			2		
Rock type	OX	TR	FR	OX	TR	FR	OX	TR	FR	OX	TR	FR	OX	TR	FR	OX	TR	FR
Bill no.	1	2	3	4	5	6	7	8	9	10	11	12	13	14	15	16	17	18

Let's consider a block which has 0.85 g/tonne of Au, 0.27% of Cu within oxide material of domain 1. The variable x_{is} is equal to 1 if grade bin s is suitable for block i, otherwise is zero. Therefore, based on the information in Table 1 and characteristic of each grade bins, the variable x_{is} for block i can be defines as,

$$x_{i7} = 1 \; and \; x_{i1} = x_{i2} = x_{i3} = x_{i5} = \ldots.. = x_{i18} = 0.$$

On the other side, based on attribute information for block i (0.85 g/tonne of Au, 0.27% of Cu within oxide material of domain 1), similar to grade bins in stockpile, just one grade bins in GE stockpile will be suitable. We define y_{is} as equal to 1 if grade bins GE stockpile s is suitable for block b, otherwise is zero. Therefore, based on Table 2 and characteristic of each grade bins, for specific block i,

$$y_{i1} = 1 \text{ and } y_{i2} = y_{i3} = y_{i4} = y_{i5} = \ldots\ldots = y_{i18} = 0.$$

4 Results and Discussion

We apply the proposed framework to a subset of 64, 130 and 543, 5256 mining blocks of copper orebody for a scheduling horizon of 5 years and compare the result of open pit mine production scheduling without GE and with GE in terms of number of decision variables (binary and continuous), number of constraints, computational time, optimality gap and NPV.

The proposed model is implemented in GAMS 27.1.0 and run on a PC with a 2.9 GHz Intel® Core™ i7-7820HQ processor and 32 GB RAM memory. Table 3 shows the value of the input parameters. Results of implementation for both with GE as well as without GE are presented in Tables 4 and 5.

Table 3. Input parameters.

Mining cost ($/tonnage)	1.6
Processing cost ($/tonnage)	9
Price Cu ($/tonnage)	5,500
Price Au ($/gr)	42
Recovery (%)	90
Discount rate (%)	10

Table 4. Numerical results without GE.

Number of blocks	Periods	Continuous variable	Discrete variable	Constraints	CPU time (sec)	GAP (%)	NPV ($)
64	5	4,101	360	1,805	27	0	933,417
130	5	6,236	690	2,231	127	0	4,181,209
543	5	9,874	785	4,231	516	0.06	4,829,723
5256	5	192,456	26,350	41,977	6,256	0.54	79,559,180

Table 5. Numerical results with GE.

Number of blocks	Periods	Continuous variable	Discrete variable	Constraints	CPU time (sec)	GAP (%)	NPV ($)
64	5	22,156	410	13,640	1007	0	1,064,065
130	5	31,271	525	14,821	1279	0.04	4,703,860
543	5	68,861	2,585	25,991	1658	0.12	5,587,990
5256	5	1,350,891	34,505	334,164	14,785	9.67	88,549,361

As shown in Tables 4 and 5, the first column represents number of blocks within the orebody. The second column shows the planning horizon. The next three columns detail the number of continuous and discrete variables as well as constraints. The last three columns provide the required time to solve the model (CPU time), optimality GAP (The gap between best possible objective and best found objective) and the net present value. Regarding the results, in all instances the proposed framework (incorporated with GE) generates consistently higher NPV than without GE framework. For instance, the incorporated GE model generated NPV of $5.5M with 543 blocks at a discount rate of 10% for 5 years life of mine, but without GE model gives $4.8M with the same parameters. Table 6 shows the improvement through with GE Model against without GE Model in terms of NPV. Results confirmed that with GE integrated framework outperforms the without GE and the range of the improvement is [11%–16%].

Table 6. Improvement of with GE framework.

No. blocks	Improvement
64	14% ↑
130	12.50% ↑
543	15.70% ↑
5256	11.30% ↑

5 Conclusion

In this study, we have formulated and implemented a with GE model to integrate GE plant operations with production scheduling of open pit mine and compare the results with and without GE. The GE incorporated framework seeks to maximize the NPV by meeting all the constraints ranging from operational to processing constraints. In order to reduce the size of the problem as well as increase efficiency, we present two types of pre-processing techniques based on material attributes (B-Ms suitability, B-GEs suitability). The numerical results from implementation of the models are presented. The generated NPV in with GE model was about 14% more than that generated by the without GE framework.

Acknowledgements. Cooperative Research Centre for Optimising Resource Extraction (CRC ORE) provided the funding (Project ID Code: P4-007; Curtin University Grant # 58994) for the work contributed in this paper. The collaboration between the authors would not have been possible without the financial support from CRC ORE. CRC ORE is part of the Australian Government's CRC Program, which is made possible through the investment and ongoing support of the Australian Government. The CRC Program supports industry-led collaborations between industry, researchers and the community. We are thankful to the CRC ORE management (specially, Paul Revell, Michael Scott and Luke Keeney) for their valuable collaboration and technical comments.

References

Espejel, C., et al.: Economic benefits and technical complexities of grade engineering® in strategic mine planning of metalliferous projects. In: Proceedings of the International Conference Mine Planning and Equipment Selection (2017)

Carrasco, C., et al.: Unlocking additional value by optimising comminution strategies to process Grade Engineering® streams. Miner. Eng. **103–104**, 2–10 (2017). https://doi.org/10.1016/J.MINENG.2016.07.020

Carrasco, C., Keeney, L., Napier-Munn, T.J.: Methodology to develop a coarse liberation model based on preferential grade by size responses. Miner. Eng. **86**, 149–155 (2016a). https://doi.org/10.1016/j.mineng.2015.12.013

Carrasco, M., et al.: Screening and characterization of amylase and cellulase activities in psychrotolerant yeasts. BMC Microbiol. **16**(1), 21 (2016b). https://doi.org/10.1186/s12866-016-0640-8

Fu, Z., Asad, M.W.A., Topal, E.: A new model for open-pit production and waste-dump scheduling. Eng. Optim. **51**(4), 718–732 (2019). https://doi.org/10.1080/0305215X.2018.1476501

Gaupp, M.: Methods for improving the tractability of the block sequencing problem for open pit mining (2008). http://www.dtic.mil/dtic/tr/fulltext/u2/a486095.pdf. Accessed 8 June 2018

Li, L., et al.: Optimal production scheduling for energy efficiency improvement in biofuel feedstock preprocessing considering work-in-process particle separation. Energy **96**, 474–481 (2016). https://doi.org/10.1016/J.ENERGY.2015.12.063

Merchan, A.F., Maravelias, C.T.: Preprocessing and tightening methods for time-indexed MIP chemical production scheduling models. Comput. Chem. Eng. **84**, 516–535 (2016). https://doi.org/10.1016/j.compchemeng.2015.10.003

Montiel, L., Dimitrakopoulos, R.: Optimizing mining complexes with multiple processing and transportation alternatives: an uncertainty-based approach. Eur. J. Oper. Res. **247**(1), 166–178 (2015). https://doi.org/10.1016/J.EJOR.2015.05.002

Redwood, N., Scott, M.: Application of enterprise optimisation considering grade engineering ® strategies (2016a). https://www.crcore.org.au/images/CRC-ORE/papers/Application-of-Enterprise-Optimisation-with-Grade-Engineering-Strategies-2016.pdf. Accessed 7 May 2019

Redwood, N., Scott, M.: Application of Enterprise Optimisation Considering Grade Engineering ® Strategies (2016b). https://www.crcore.org.au/images/CRC-ORE/papers/Application-of-Enterprise-Optimisation-with-Grade-Engineering-Strategies-2016.pdf. Accessed 23 Aug 2018

Optimised Pit Scheduling Including In-Pit Dumps for Stratified Deposit

Ranajit Das[1(✉)], Erkan Topal[2], and Elham Mardenah[3]

[1] Curtin University and Senior Mining Consultant, Dassault Systemes, Brisbane, Australia
d.ranajit@gmail.com

[2] Faculty of Science and Engineering, Mining Engineering and Metallurgical Engineering, WA School of Mines, Curtin University, Perth, Australia
E.Topal@curtin.edu.au

[3] Faculty of Science and Engineering, Curtin University, Perth, Australia
Elham.Mardaneh@curtin.edu.au

Abstract. In stratified deposits such as coal, lignite, and phosphate, the ratio of waste to ore is often very high and one of the major concerns relates to waste placement to in-pit or external dumps. Unless it is known which waste dump will be used for the extracted material, depending on the availability of the waste dumps (in-pit or external), the haulage cost can largely vary at different times. Hence it is crucial to incorporate the waste placement decisions into the pit scheduling problem for better planning. Ignoring the waste placement and thereby the different waste haulage costs at different times could lead to flawed results. Traditionally, the pit scheduling problem and dump scheduling are studied in isolation with a few exceptions in the literature. Our aim is to fill the research gap in pit scheduling of stratified deposits with in-pit waste dumping and haulage road options. This paper focuses on the integration of pit and dump scheduling including in-pit dumping strategy. This strategy requires a factor of the lag to be considered with the working face. In this paper, we also demonstrate how to maintain a lag space with the dynamically changing mining face and consider waste rock placements in correct dumping (in-pit or external) locations.

1 Introduction

Stratified deposits are normally associated with large volume of waste removal, thereby a significant portion of the cost of mining is in removing the waste. Unless we know which dump the waste will be directed to, depending on the availability of space in the dump (external or in-pit), the cost of haulage will largely differ at different points of time (Li et al. 2013, 2014). The in-pit dump is a factor of the lag to be considered with the working face. While, the haulage cost depends on whether the optimal haul road option has been chosen (Hill et al. 2013). Thus ignoring the proper placement of waste into external and in-pit dumps in a schedule could lead to a flawed result.

The haulage distance for in-pit dumps are often much less compared to transportation to external dumps. Thereby the cost of mining is reduced with increased in-pit

E. Topal (Ed.): MPES 2019, SSGG, pp. 33–41, 2020.
https://doi.org/10.1007/978-3-030-33954-8_4

dumping. Hence, it is preferred to have internal dumping as early as possible in the mine life to be able to do early rehabilitation. Thereby reducing footprint of environmental impact.

Although production scheduling problem has the similar structure with stratified deposit as compared to massive or vein deposit, a few key differentiators can be listed as follow:

Pit Geometry and Layout: For moderately dipping stratified deposits, the base of the layer of ore/coal forms the base of the pit. Unlike, in a non-stratified deposit such as vein type deposit or a massive deposit we may have to mine the footwall to create a stable pit wall. Stratified deposits are large in lateral extent and normally shallow in vertical extent, hence mines like coal are known to extend several kilometres, as shown in Fig. 1 below. Because of this extent, almost all stratified mines try to do backfilling by in-pit dumping as a priority in order to decrease haulage cost and expedite reclamation. It is mostly not the case in vein type or massive deposits.

Fig. 1. Difference in pit geometry of stratified and non-stratified open pit mines using google earth images

Layered Blocks: Other than the pit geometry there is a difference in how blocks are treated in geological modelling and mine planning packages. For metalliferous deposits a block in a block model can be either waste or ore depending on its cut-off grade. Anything above the cut-off grade will be treated as ore and the destination would be a process plant. However, in case of coal deposits each block could consists of multiple coal and waste layers. Further, depending on the thickness of the coal layers and the quality, the layers could be combined with adjacent waste or coal layers while mining and such a model is often referred to as a ROM model with working sections.

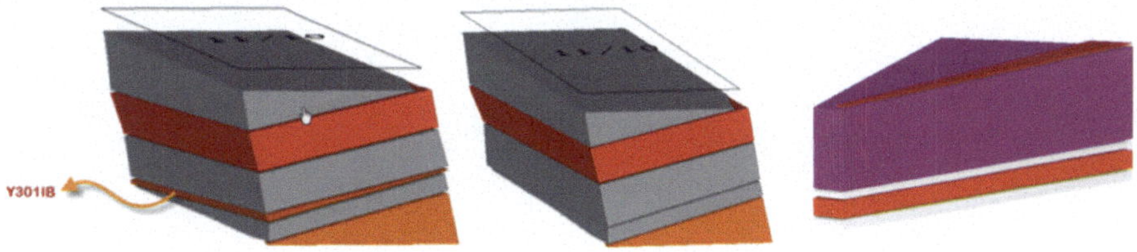

Fig. 2. Typical coal reserves individual block-containing both coal and waste layers

Figure 2 presents the typical blocks in a coal mine. The resource model could contain thin layers such as the Y301IB layer of Inter-burden. While converting the in-situ model to ROM model it is considered as coal and combined with the adjacent coal (Minex Reserves Database tutorial 2018). In order to mine the coal in a block the waste above it has to be removed.

Irregular Shape of the blocks: Another significant difference of a stratified deposit model is the shape of the blocks which are not perfect cubes or cuboids as in non-stratified. As seen in Fig. 2 blocks can have a definite slope angle which follow the ultimate wall or the strip wall angle. The blocks in plan view need not necessarily be a perfect rectangular shape, as commonly known. The block model referred here is only within the pit, where the pit can take any shape laterally. Hence, in order to fit in blocks at the edges or turns in a strip the blocks can be of any shape or size opposed to regular shaped block models in metalliferous deposits.

Pre-designed Pit: The design of these blocks are normally done after the pit optimisation process, hence all blocks in the design are within the optimised pit shell obtained through ultimate pit limit optimisation algorithm such as Lerchs Grossman or other optimisation method. The slope of the pit has already been considered during the design process. It may be observed that the sides of each block also has a slope as shown in Fig. 2. This slope follows the slope specified for the strip or final wall.

Pre-designed Dump: Similar to the pit, the dump is also pre-designed with slopes for each block. Dump blocks are designed and need to be considered as the lag distances are checked from each pit block to corresponding dump blocks. Dump slopes are more important as they are flatter and normally in the range of 35–38°. Hence in the same area the number of strips that can be fitted in the bottom bench may not be same as number of strips in top bench as shown in Fig. 6.

Predecessor Blocks for Mining: Typically 9 blocks above a pit block are considered to be mined in order to mine a block below. However, in a stratified deposit using strip mining method, the pit should be mined as strip by strip. Similarly we also want to proceed the dump strip by strip. Hence the mining or dumping precedence is slightly different. The same 9 blocks are located differently depending on the mining direction. The mining direction being the direction of increase in strip numbers. This is normally important in stratified deposits as they have a dip. The strip numbers normally increase along the dip direction. Mining normally proceeds from the shallower area to the deeper area (Fig. 3).

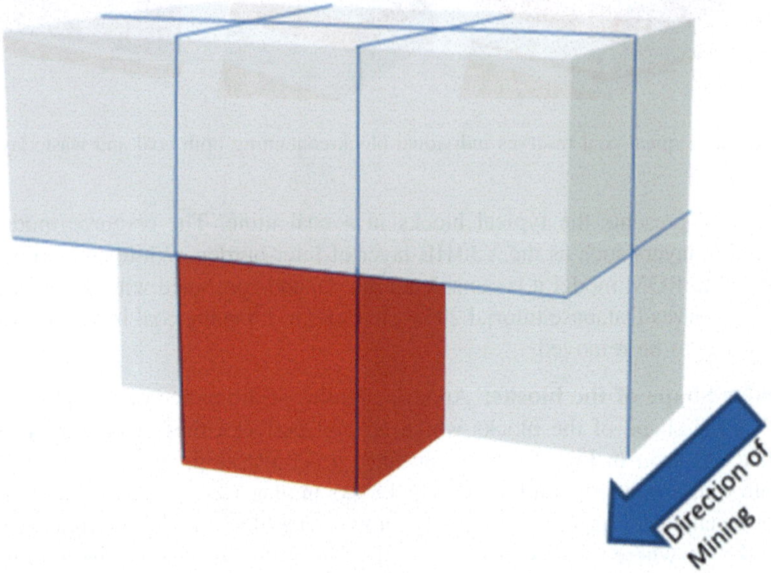

Fig. 3. 9 blocks with respect to block to be mined (red) and direction of mining

2 Methodology for Model

The design of the pit and the dump has been created using a 3D mine planning software. The volumes, tonnes, quality and block coordinates were reported out of the software and tabulated in an Excel workbook. A new mixed integer programming based model has been developed using CPLEX OPL in order to optimise the production schedule for each period and the corresponding dump locations to be used for each block. The mathematical equations considered in this model are similar to the one by (Fu et al. 2018), except certain modifications mentioned below to accommodate stratified nature of the deposit and in-pit dumping with lag distance.

A block here could contain both ore and waste. The dump has been deigned to contain individual blocks. Both pit and dump blocks have been considered to have a location with a coordinate for the centroid. One of the key differentiators of this model is the lag constraint which differentiates it from other models and makes it capable of scheduling for in-pit dumping.

A set of pit blocks have been calculated from each dumping block centroid, within a defined lag distance. The lag constraint checks whether these blocks have been already mined. The dump block can be filled only if the entire set of lagged pit blocks have been mined.

3 Implementation of the Model

The model developed has been tested for a dataset having the components as shown in Fig. 4. It has a single pit with 3 benches and 117 blocks of average size of 50 m 50 m. The coal is transported to either of the 3 stock piles – based on grade or can be directly fed to the coal wash plant. Coal is also fed to the wash plant from the stock pile by rehandling. The values used in the model are describes in Table 1 below.

Table 1. Input parameters for the optimisation model

	Maximum	Minimum	Unit
Wash plant capacity	150,000	120,000	Tonnes/Yr
Mining capacity	750,000	200,000	Cubic m/Yr
Specific energy		15	MJ/kg
		Australian $	
Coal sale price		100.00	per T
Coal mining cost		4.00	per T
Waste mining cost		3.00	per BCM
Waste haulage		0.0001	per BCM meters
Coal wash cost		7.50	per T ROM feed
Coal rehandling cost		0.50	per T ROM feed
Swell factor		1.25	
Discount rate		10%	
Recovery/yield		90%	

The waste is hauled and dumped into two dumps – one external and one in-pit. The in-pit dump maintains a lag distance of 50 m with the pit as both the pit and the in-pit dump progress. The dumps are selected based on the haulage distance and corresponding waste hauling cost. The objective here is to maximise the Net Present Value of the operations considering the above scenario.

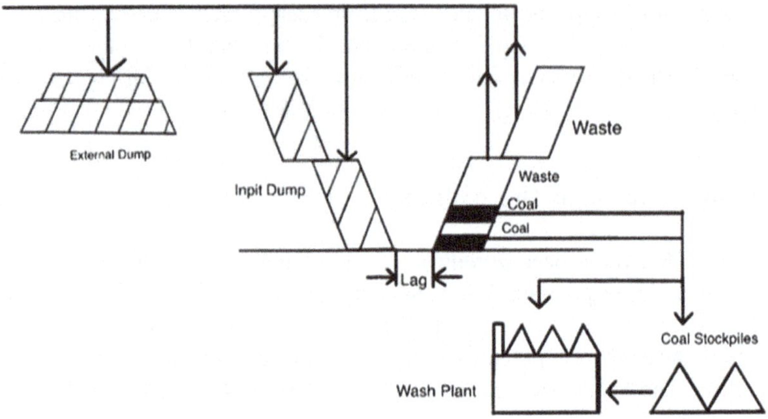

Fig. 4. Material flow schematic diagram

The model has been run on part of a pit for only the first 4 strips which have 117 blocks, in 3 benches as shown in Fig. 5 below. There is also an in-pit dump in the same footprint with 3 benches and 93 blocks as shown in Fig. 6. One ex-pit block has been considered which represents the external dump and has an equivalent capacity. Detailing of the external dump into benches, strips and blocks has been kept out of scope in this model. The pit blocks have multiple layers of coal and waste in them as evident from the colours of the layers in Fig. 5. The lowest bench, not visible in the image is mostly coal. The schedule has been run for 5 periods. The pit and dumps were designed in a 3D Mine planning software and the details including pit and dump block centroid coordinates were exported to excel, which formed an input to the model in CPLEX OPL.

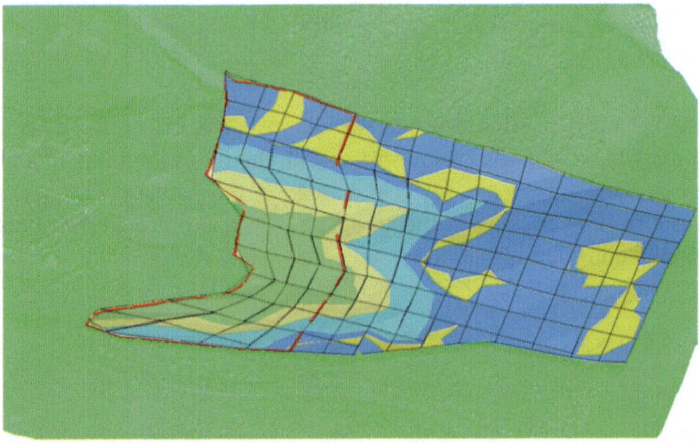

Fig. 5. Complete pit and topography surface (pit coloured by layers of coal and waste). Outline of 4 strips marked in red

Fig. 6. In-pit dump blocks for the first 4 strips considered

The dump has been designed with blocks at 37° as seen in Fig. 6. Due to the slope in the design, although the bottom bench has 4 strips, there is not enough room for 4 strips on the top. The topmost dump bench as visible in Fig. 6 has only 2 strips. A swell factor of 1.2 has been considered for waste placed into dump blocks. Table 2 presents the pit data in excel which contain the block details. Some blocks contain both coal and waste where as some have only waste. There are other tables one of which have coordinates of each block, and one for the dump blocks and their coordinates.

Table 2. A sample format of the pit block details [The complete list has 117 blocks]

Block id	Bench	Strip	Block	CV (Quality)	Coal (Tons)	Coal density	Recovery	Waste vol	Total volume
1	1	1	2	23.76	4435	1.42	90%	39397	45694.7
2	1	1	3	23.91	4938	1.42	90%	31823	38834.96
3	1	1	4	25.39	3677	1.38	90%	18739	23813.26
4	1	1	5	27.04	4058	1.35	90%	20499	25977.3
5	1	1	6	27.88	1509	1.34	90%	36656	38678.06
6	1	1	7					21975	21975
7	1	1	8					17180	17180
8	1	1	9					16952	16952
9	1	1	10					26815	26815
10	1	1	11	27.65	1214	1.41	90%	48640	50351.74
11	1	1	12	27.52	3393	1.42	90%	21513	26331.06
12	1	2	2	24	3730	1.41	90%	38561	43820.3
13	1	2	3	24.04	5090	1.41	90%	34716	41892.9
14	1	2	4	25.49	3925	1.37	90%	19893	25270.25
15	1	2	5	27.04	4773	1.34	90%	22614	29009.82
16	1	2	6	27.88	1096	1.33	90%	23877	25334.68
17	1	2	7					16052	16052
18	1	2	8					11026	11026
19	1	2	9					14525	14525
20	1	2	10					26365	26365

4 Analysis of Results

The results from the run in CPLEX have been tabulated in the format in Table 2 in order to calculate the NPV outside the CPLEX model.

Table 3. Results obtained from model for waste placement from pit to dump blocks for each period

Year	Waste volume	Dump		Coal to plant	Coal to stock pile	Coal from stockpile
		External	In-pit			
1	620,545	615,033	5,512	0	129,455	129,455
2	338,440	193,434	145,006	–	135,349	135,349
3	565,689	536,594	29,095	–	120,000	120,000
4	195,525	–	195,525	0	120,000	120,000
5	236,763	–	236,763	–	120,000	120,000

As can be seen from Table 3, waste has been directed to both external and in-pit dumps. This has been decided based on the shortest distance (lowest cost) or based on blocks eligible for dumping in the in-pit dump in order to maintain the lag distance with the pit blocks being mined. In the early periods when mostly the upper waste is being mined there is not enough room for starting the in-pit dump. Hence the waste is mostly sent to the external dump block. The external dump has been assumed to contain a volume equivalent to a realistic dump outside the pit. From the 4th year onwards there is enough room in the in-pit dump hence all waste is sent there.

Coal could be sent to either wash plant or to stock pile from the pit, in this case all coal went to the stock pile for blending. All coal sent to the stock piles were drawn out by rehandling for feeding to the wash plant.

A sale price for the coal was assumed as \$100 per tonne, and a 90% yield at the coal wash plant. For ease of calculation of NPV each period has been assumed to be an year. The NPV of the current schedule at 10% discount is \$24.91 million, under the given assumptions, as shown in Table 4.

Table 4. Cash flow and NPV calculated from the schedule

Year	Waste mining cost	Waste haulage cost		Coal mining			Coal wash cost	Rehandling cost	Revenue	Margin
		Ext	In-pit	Mining	Haulage to plant	Haulage to stock				
1	1,861,635	3,427,326	2,361	517,820	0	936,014	970,913	64,728	11,650,954	4,806,171
2	1,015,320	643,051	324,154	541,396	–	1,059,358	1,015,117	67,674	12,181,406	8,574,694
3	1,697,068	2,221,300	50,211	480,000	–	667,071	900,000	60,000	10,800,000	5,391,421
4	586,575	–	1,225,708	480,000	0	609,126	900,000	60,000	10,800,000	7,547,716
5	710,289	–	1,806,868	480,000	–	454,355	900,000	60,000	10,800,000	6,842,843
									NPV	\$24,910,484

5 Conclusions

The ability to decide the optimal destination of dump blocks into in-pit and external dumps along with the optimal mining sequence can add significant value to mining operations. This value has remained untapped in several operations globally. Not only does it give the opportunity to maximise value but also allows for quicker backfilling the in-pit dumps and making them available for rehabilitation. Thereby decreasing the footprint of the mine progressively. This research presents the integration of pit and dump scheduling which includes in-pit dumping strategy for stratified deposit. The paper also presents how to maintain a lag space for mining and in-pit filling with the dynamically changing mining face.

6 Future Research

Stratified deposits occur in a sequence, often referred to as the seam sequence. In the current research, the blocks has been considered to have 3 dimensions (i, j, k) which are bench, strip and block. In future research, it is proposed to work with i, j, k and layer name. Where the coal and waste layer in a block are treated separately. This will ensure that the coal and the waste layers are removed in the right sequence for blocks containing multiple layers of coal and waste. At the moment although the model forces all waste in the available block to be removed if the coal is removed. However, it would be more appropriate to have them removed in a particular sequence, giving the flexibility to have a few layers in a block mined in a particular period and remaining in another period.

It is also proposed to work on the development of different strategies and models to obtain faster solution for the larger datasets.

References

Li, Y., Topal, E., Ramazan, S.: Optimising the long-term mine waste dump progression and truck hour schedule in a large-scale open pit mine using mixed integer programming. In: Orebody Modelling and Strategic Mine Planning (SMP 2014), 24 November 2014. The Australasian Institute of Mining and Metallurgy, Perth (2014)

Minex Reserves Database Tutorial, version 6.5.2. Dassault Systemes GEOVIA Minex (2018)

Hill, S., Bates, L., Cooper, A., Elkington, T.: An automated approach to generating haul roads. A Snowden Document, Canadian Institute of Mining Metallurgy and Petroleum (2013)

Li, Y., Topal, E., Williams, D.: Waste rock dumping optimisation using mixed integer programming (MIP). Int. J. Min. Reclam. Environ. **27**(6), 425–436 (2013). https://doi.org/10.1080/17480930.2013.794513

Fu, Z., Asad, M.W.A., Topal, E.: A new model for open-pit production and waste-dump scheduling. Eng. Optim. (2018). https://doi.org/10.1080/0305215X.2018.1476501

A Simulation Model for Estimation of Mine Haulage Fleet Productivity

S. Upadhyay[1], M. Tabesh[2], M. Badiozamani[3](✉), and H. Askari-Nasab[3]

[1] Canadian Natural Resources Limited, Calgary, AB, Canada
[2] Teck Resources Limited, Vancouver, BC, Canada
[3] School of Mining and Petroleum Engineering, University of Alberta, Edmonton, AB, Canada
mbadiozaman@ualberta.ca

Abstract. A good estimation of haulage fleet productivity is a pivotal input to determine the mine fleet requirement. In open pit mining, the two major factors affecting the fleet productivity are the planned tonnage to be moved, that is based on the long-term production schedule, and the time it takes to move the dirt tonnage which depends on the mine road network. The haulage speed, haulage time and the truck load tonnages are all probabilistic values due to the uncertainty involved in real mining operations, and simulation technique is a powerful tool to capture this uncertainty. In this paper, a simulation framework is presented to estimate the productivity of haulage fleet for open pit mining operations with truck-and-shovel system. The Productivity KPI is the Tonne per Gross Operating Hour (TPGOH) for combinations of Dig and Dump locations. Historical data are used to fit probability distributions for haulage cycle components, and the mine road network and long-term production schedule are the main inputs to the model. The developed model is verified and validated through implementing it on a real case from Alberta oil sands open pit operations.

Keywords: Fleet productivity · Open pit mining · Monte Carlo simulation

1 Introduction

Haulage costs as the main part of oil sands truck-and-shovel operations costs contribute for over 50% of open pit mining costs [1–4], which shows the importance of haulage system efficiency in overall profitability of open pit mining operations. Determining the right size of the haulage fleet will optimize the overall fleet utilization and hence, improves the efficiency. Typically, the deterministic mathematical techniques and the "match Factor" [5–7] have been used in Equipment Selection and sizing Problem (ESP) modeling. The other well-recognized mining fleet productivity measure is the Tonne Per Gross Operating Hour (TPGOH), showing how much material is moved per an hour of truck activity. As the two critical values to calculate the TPGOH - the truck load tonnage and the cycle time- are both probabilistic, simulation techniques can be used in TPGOH estimation.

E. Topal (Ed.): MPES 2019, SSGG, pp. 42–50, 2020.
https://doi.org/10.1007/978-3-030-33954-8_5

Different simulation techniques are used extensively in mining operations optimization due to their ability in capturing the inherent uncertainty of mining operations. Some applications of discrete event simulation include scenario analysis of open pit mining operations [8, 9], operational truck dispatching [7, 10, 11], performance analysis of autonomous haulage system (AHS), and so on. Hybrid solution methodologies such as simulation optimization models have been used in a number of applications, such as reliability assessment [12] and short-term mine planning [13]. Monte Carlo simulation covers the second group of simulation techniques, in which the random or pseudo-random numbers are repetitively and independently sampled from a distribution function. Following the early implementations of Monte Carlo simulation in mining optimization [14], it has been used in a number of mining applications such as ultimate pit limit optimization [15] and risk assessment of extraction sequencing [16]. [17] include more comprehensive reviews on the applications of simulation analysis in surface mining short-term planning and mining fleet management systems, respectively.

Improving the productivity of mining fleet has been extensively addressed in the literature. The topics include analyzing productivity measures, different methods of allocating shovels to the mining faces and allocating trucks to the shovels, dispatch optimization and equipment selection. Some of the related works include optimization of truck shovel allocation [18], using simulation for fleet size optimization [19], and decision making model for fleet selection using discrete event simulation [20].

This paper presents a Monte Carlo simulation model to calculate the productivity of mining fleet through measuring of TPGOH. The model takes three inputs, as; (1) historical data of the mining operations including loading, haulage and dumping data records, (2) the production schedule that determines the extraction sequence of polygons with their spatial coordinates and related attributes such as tonnage and grade of different elements, and (3) the mine's road network that introduces all the paths among all the dig locations (polygons) and dup locations such as crushers, stockpiles and waste dumps. A historical data analysis and modeling has been carried out to understand the performance of the mining system and fit meaningful probability distribution functions for different activities of the truck-shovel operations cycle. The proposed model is verified through implementing for small datasets, and finally validated through comparing its results against the historical records of a real oil sands mine. Based on the historical data records, the proposed simulation model estimates the TPGOH more precisely than the mine's current model.

2 Methodology

A Monte Carlo simulation model is developed to estimate the TPGOH as the fleet productivity measure, that can be used later for truck requirement estimation. It takes the production schedule and the mine road network as the initial inputs. At the first step, data preparation, the inputs are compiled and the coordinates of dig and dump locations, and the shortest path between each pair of them is calculated. At the second step, the model takes the fitted probability distributions for variables and samples from them to determine the required data for TPGOH calculation, as the load tonnage and the haulage times. Empirical distributions are used to model all the independent variables. The model iterates over variable sampling to the point that all the scheduled tonnage is moved from sources (dig points) to destinations (dump point). Finally at the third step, the TPGOH is calculated using all the required data, and the statistics of different estimated variables will be generated, together with the corresponding plots. The algorithm is illustrated schematically in Fig. 1.

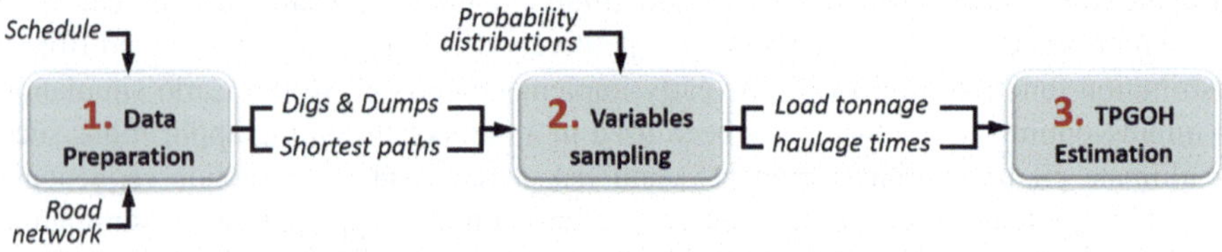

Fig. 1. The simulation algorithm.

The variables involved in TPGOH calculation are classified into two groups of independent and dependent variables. The independent variables do not depend on the production schedule and are also independent of time, such as Loading Time, Dumping Time, Spot Time and the Measured tonnage, which is tonnage of the material loaded by the shovel on trucks per trip. On the other hand, the dependent variables are the ones that depend on the time, such as full and empty haulage times, that are estimated by adjusting the sampled velocities for segments of road network, using the rimpull characteristics of the trucks.

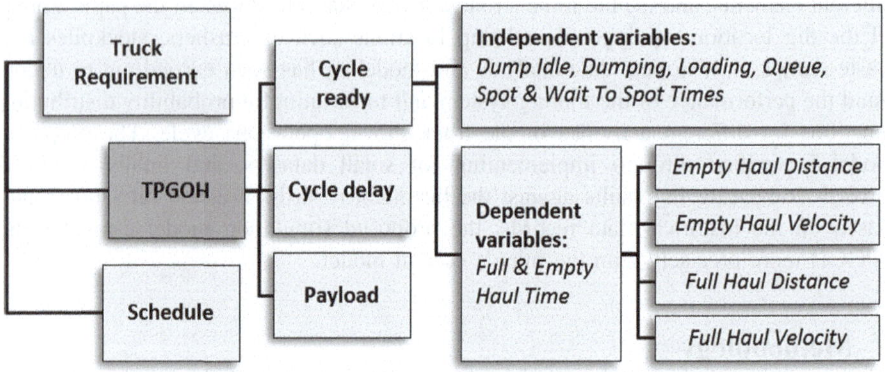

Fig. 2. Fleet requirement and productivity calculation.

The classification and definition of variables are presented in Fig. 2. Empty and Full haul velocity sampling distributions are the adjusted fitted empirical distributions for average empty and full haul velocities. The TPGOH is calculated based on the payload and total cycle time, as payload divided by the total cycle time. Definitions of independent and dependent variables are presented in Table 1.

Table 1. Definition of independent and dependent variables.

Independent variables	
Dump idle (min)	Idle time of truck in dumping area
Dumping time (min)	dumping time of truck at dump
Loading time (min)	Loading time of truck by shovel
Queue time (min)	Time a truck waits in queue at dump
Spot time (min)	Spotting time of truck at shovel
Wait to spot (min)	waiting time of truck for spotting
Cycle delay (min)	Total delay time in the cycle
Measured tonnage (tonne)	Measured load tonnage at the truck
Empty haul velocity (km/h)	Empty truck speed
Full haul velocity (km/h)	Full truck speed
Dependent variables	
Empty haul time (min)	Empty truck travel time
Full haul time (min)	Full truck travel time
Cycle ready (min)	Total time truck is in operative state
TPGOH	Tonne per gross operating hours

Adjusting the Velocities

Empty and Full Haul travel times (TT) are calculated as the summation of the travel times on single road segment of the path, as in (1). $Dist_i$ and v_i are the distance of segment i and the velocity of truck at that segment, respectively. The velocity on each road segment, vi, is calculated by dividing the sampled velocity divided by the velocity adjustment factor as in (2). The velocity adjustment factor, f, adjusts the sampled velocity according to the nominal rimpull velocity considering the total resistance (rolling resistance of 5%, plus the gradient of the segments) and the flat haul rimpull speed for Cat 797B trucks, as Eq. (3). Finally, the Empty and Full Haul times are calculated using Eq. (4).

$$TT = \sum_i \frac{Dist_i}{v_i} \tag{1}$$

$$v_i = \frac{v^{sampled}}{f_i} \tag{2}$$

$$f_i = \frac{v^{rimpull\_FlatHaul}}{v_i^{rimpull}} \tag{3}$$

$$TT = \sum_i f_i \times \frac{Dist_i}{v^{sampled}} \tag{4}$$

3 Case Study

The historically recorded data of an oil sands open pit mining operations in Northern Alberta has been used as the case study to validate the performance of the model. There are three main inputs to the model, as probability distribution functions, the production schedule and the road network. Some initial criteria are considered in early filtering of the provided data records, such as; (1) only the records with CAT797 or related variation of it are accepted as equipment type, (2) only the records with Haulage distance in the range of 0.5 to 11.9 km are considered, and (3) records that generate a TPGOH within the range of 200 and 1500 are accepted. These filters have resulted in a schedule with 67 combinations of Dig and Dump Locations with greater than 5000 tonnes of moved material, for a quarter including three months of January, February and March. The total tonnage of material moved for the selected schedule is 8.3 MT. The mine road network for the selected period is created using three asbuilt surfaces, by velocity map data provided and visually recognition of the roads on the asbuilts. As the pit and dumps remained changing during the quarter, roads were created to reflect an average layout.

4 Discussion of Results

The statistics of recorded and simulated data and the difference between them are presented in Table 2. The selected variables, i.e. the Loading Time and the Full Haul Velocity are two examples representing the independent and dependent variables, respectively. The values are normalized for both recorded and simulated data and include 21,951 records for simulated and 22,045 records for historical data.

Table 2. Comparison of normalized recorded vs. simulated stats.

	Loading time (min)			Full haul speed (kmph)		
	Rec.	Sim.	Diff(%)	Rec.	Sim.	Diff(%)
Min	0.314	0.314	0%	0.072	0.072	1%
Max	1	1	0%	1	1	0%
Mean	0.553	0.553	0%	0.431	0.437	1%
Median	0.527	0.544	3%	0.426	0.429	1%
St. Dev.	0.142	0.143	0%	0.097	0.12	23%
Sum.	12,198	12,140	0%	9,510	9,588	1%
25 Perctl.	0.441	0.439	0%	0.368	0.357	−3%
75 Perctl.	0.644	0.65	1%	0.485	0.506	4%

The statistics of Loading Time is presented as an example of independent variables. Table 2 shows the perfect match between recorded and simulated data's mean, standard deviation and summation for Loading Time. There is only 3% difference in median, meaning that the simulated data are very slightly skewed to the right comparing to the recorded data. The match between the histograms is visually shown in Fig. 3, where the histograms of simulated and recorded data are illustrated. Also the QQ plot at the right, that visually measures the similarity of recorded versus simulated histograms quantile-by-quantile, approves the good match between the two data sets.

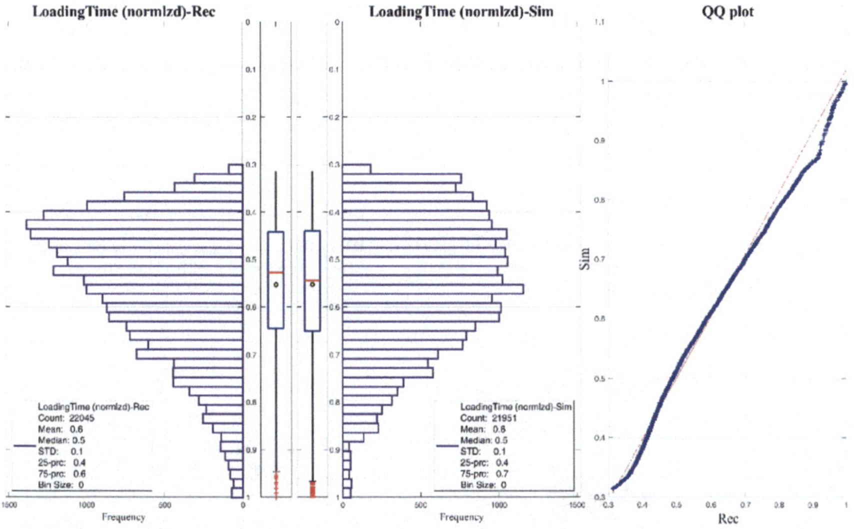

Fig. 3. Histograms of recorded and simulated data and QQ plot for Loading Time.

The statistics for the normalized Full Haul Speed are also presented in Table 2. The comparative results confirm that the model is closely matching the recorded Full Haul Velocity, except the variance of the simulated data and the historically recorded data.

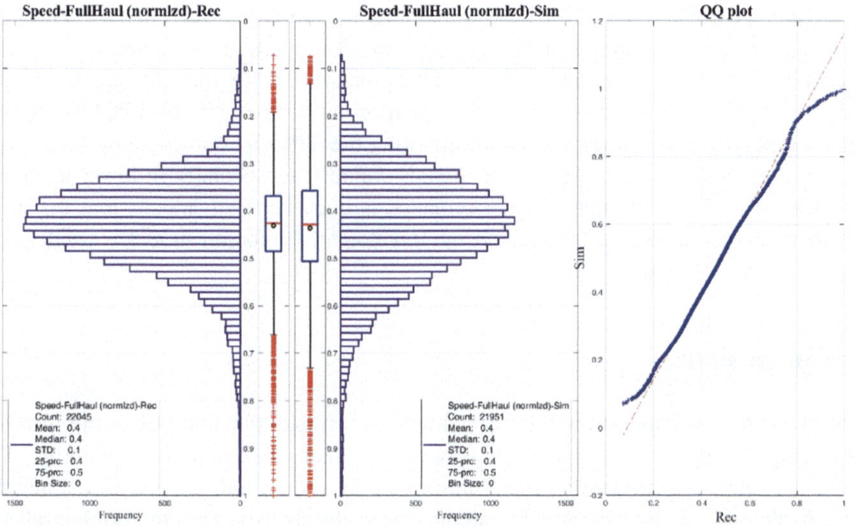

Fig. 4. Histograms of recorded and simulated data and QQ plot for Full Haul Velocity.

This can be seen in Fig. 4. The mean, median and standard deviation as the important KPIs for all independent and most of dependent variables are closely matching that shows an acceptable performance of the model to mimic the behavior of the mining system. For Empty and Full Haul Distances, the summation is the statistic that should be mainly checked, which shows an –11% and a –4% difference for Empty and Full Haul Distances, respectively. That is mainly because of the dispatch logic considered in this model to determine the destination of empty trucks, that is not necessarily aligned with the operational dispatch decisions on site. Gathering more information about mine dispatch rules and tuning the model's dispatch logic accordingly will improve precision of the simulated empty and full haul distance estimations.

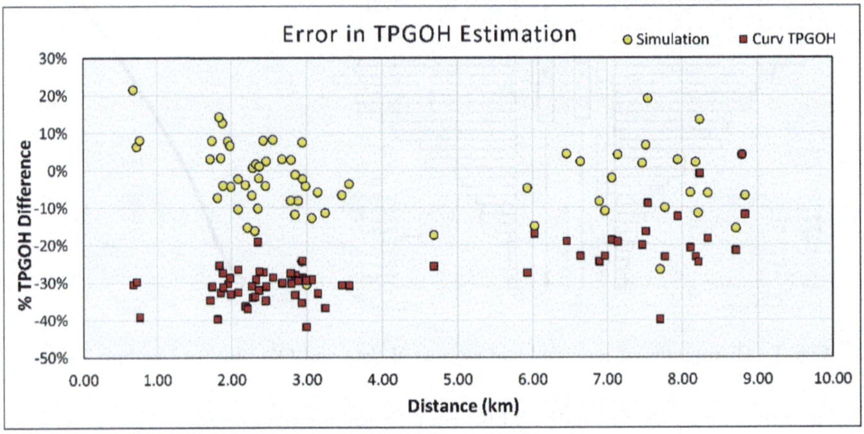

Fig. 5. TPGOH estimation error comparison.

Figure 5 compares the error of two methods in estimating the TPGOH. The light circles show the percentage error of the simulation method in estimating the real calculated TPGOH (based on the historical data records), while the dark squares are the error associated with the current mine estimation curve method for TPGOH. As the figure shows, the simulation model is estimating the TPGOH considerably better than the curve method with error values scattered around 0%. Specially in lower distances (less than 4 km) the simulation's performance is even better. It means that the truck requirement will be estimated better if the developed simulation model is used as the basis for TPGOH estimation.

5 Conclusions

A Monte Carlo simulation model is developed to estimate the TPGOH as the measure for mining fleet productivity, which will be used in fleet size requirement planning. The model gets the fitted probability distribution functions, the production schedule and the mine road network, and iteratively samples from the distributions to simulate all the tonnage movement trips between dig and dump locations through the road network. The variables are classified into two groups, as the variables independent from time and schedule, and variables that may change based on the schedule and road network profile. The statistics and histograms of simulation results are compared against the historical data that shows a good match between and validates the performance of the developed model. The TPGOH estimation error is compared for the simulation model and the current curve method being used in the mine, showing the better performance of the simulation model. However, the model can be improved by implementing a more precise dispatch logic, to better match the summation of full and empty haulage distances. This is what motivates the next research steps.

References

1. Curry, J.A., Ismay, M.J., Jameson, G.J.: Mine operating costs and the potential impacts of energy and grinding. Miner. Eng. **56**, 70–80 (2014)
2. Oraee, K., Goodarzi, A.: General approach to distribute waste rocks between dump sites in open cast mines. In: Sixteenth International Symposium on Mine Planning and Equipment Selection (MPES 2007), pp. 701–712. Reading Matrix Inc. (2007)
3. Akbari, A.D., Osanloo, M., Shirazi, M.A.: Minable reserve estimation while determining ultimate pit limits (UPL) under price uncertainty by real option approach (ROA). Arch. Min. Sci. **54**, 321–339 (2009)
4. de Cunha, R.E., Lima, H.M., de Tomi, G.: New approach for reduction of diesel consumption by comparing different mining haulage configurations. J. Environ. Manag. **172**, 177–185 (2016)
5. Fisonga, M., Mutambo, V.: Optimization of the fleet per shovel productivity in surface mining: case study of Chilanga Cement, Lusaka Zambia. Cogent Eng. **4**, 1386852 (2017)
6. Pasch, O., Uludag, S.: Optimization of the load-and-haul operation at an opencast colliery. J. S. Afr. Inst. Min. Metall. **118**, 449–456 (2018)
7. Chaowasakoo, P., Seppälä, H., Koivo, H., Zhou, Q.: Improving fleet management in mines: the benefit of heterogeneous match factor. Eur. J. Oper. Res. **261**, 1052–1065 (2017)
8. Amin, I., Adil, M., Rehman, S.U., Ahmad, I.: Simulation of truck-shovel operations using simio. Int. J. Econ. Environ. Geol. **8**, 55–58 (2019)
9. Zeng, W.: A simulation model for truck-shovel operation (2018)
10. Ozdemir, B., Kumral, M.: Simulation-based optimization of truck-shovel material handling systems in multi-pit surface mines. Simul. Model. Pract. Theory **95**, 36–48 (2019)
11. Shishvan, M., Benndorf, J.: Simulation-based optimization approach for material dispatching in continuous mining systems. Eur. J. Oper. Res. **275**, 1108–1125 (2019)
12. Yuriy, G., Vayenas, N.: Discrete-event simulation of mine equipment systems combined with a reliability assessment model based on genetic algorithms. Int. J. Min. Reclam. Environ. **22**, 70–83 (2008)
13. Upadhyay, S.P., Askari-Nasab, H.: Simulation and optimization approach for uncertainty-based short-term planning in open pit mines. Int. J. Min. Sci. Technol. **28**, 153–166 (2018)
14. Rist, K.: The solution of a transportation problem by use of a Monte Carlo technique. In: Proceedings of the 1st International Symposium on Computer Application in Mining (APCOM-I), p. L2. Tucson University of Arizona (1961)
15. Deutsch, M., González, E., Williams, M.: Using simulation to quantify uncertainty in ultimate-pit limits and inform infrastructure placement. Min. Eng. **67**, 49–55 (2015)

16. Kumral, M., Sari, Y.A.: Simulation-based mine extraction sequencing with chance constrained risk tolerance. Simulation **93**, 527–539 (2017)
17. Blom, M., Pearce, A.R., Stuckey, P.J.: Short-term planning for open pit mines: a review. Int. J. Min. Reclam. Environ. **33**, 318–339 (2019). https://doi.org/10.1080/17480930.2018.1448248
18. White, J.W., Olson, J.P., Vohnout, S.I.: On improving truck/shovel productivity in open pit mines. CIM Bull. **86**, 43–49 (1993)
19. Suglo, R.S., Al-Hassan, S.: Use of simulation techniques in determining the fleet requirements of an open pit mine. Ghana Min. J. **9** (2007)
20. Ortiz, C.E.A., Curi, A., Campos, P.H.: The use of simulation in fleet selection and equipment sizing in mining. In: Mine Planning and Equipment Selection, pp. 869–877. Springer (2014)

Solving Complex Mine Optimisation Problems Using Blend Vectoring and Multi-objective Production Scheduling

Daniel Htwe(✉)

Orelogy, Perth, WA 6005, Australia
daniel.htwee@gmail.com

Abstract. Developing a large scale open pit mining project constitutes a complex process where decisions made at the planning level can affect the overall project economics and value in the scale of billions of dollars. In this paper, it will be demonstrated how the project reserve, and expected economic value can be increased substantially depending on the ability within both the optimisation and scheduling process in accounting for material process classification based on a blend vectored resource. This will be one of the first examples within the literature where the differences in ability to handle this fundamental concept is quantified as comparative analysis is provided to conventional approaches. These are applied within a practical implementation for a Western Australian iron ore case study. The elementary methodology of traditional cut-off optimisation to determine 'ore' is evaluated in comparison to results generated from applying metaheuristic blend vector techniques and derived mixed integer linear programming formulations. It will be shown how generating the results based on these techniques is computationally efficient in practice, with minimal engineering time consumed to generate solutions despite datasets generally being very large. Furthermore, post-schedule level results obtained from conventional scheduling and multi-objective genetic algorithms are provided. These explicitly demonstrate how the overall project value possesses a strong dependence on the mine planning conditions imposed by different approaches. A point of view on the open pit mine production schedule optimisation problem is also given that has allowed the development of high quality solutions in practical instances.

Keywords: Mixed integer linear programming · Iron-ore blending · Optimisation · Production scheduling · Operations research · Mining engineering

1 Introduction

1.1 The Role of Mine Planning During Feasibility Studies

During the development of a feasibility study, maximising the profitable value of a resource is of paramount importance for mining engineers at all stages of the planning phase. Whether at the resource level where an upper bound is required on saleable product contained within a full block model, at determination of the ultimate pit

E. Topal (Ed.): MPES 2019, SSGG, pp. 51–66, 2020.
https://doi.org/10.1007/978-3-030-33954-8_6

delineation or generating a practical production schedule that underpins the bankability of the study, there exists many opportunities where this critical goal can be realised. Net present value (NPV), total undiscounted cash flow (UCF) or total ore tonnes are the general criterion upon which strategic plans are compared and measured. The realised value on these parameters for any given project are primarily driven by the decisions generated from the optimisation strategy. In the open pit mining space, there has been significant research for many applications, see [7] for the pit limit and [10] for the production scheduling problems. The current practices within both the academic literature and in the industry however, often fail to either truly maximise value, generate good quality solutions or account for essential requirements such as a blended product when given practical problem instances. This is especially true for iron ore projects which have the special property that realised price is not a linear function of material composition as generally is the case for generic metalliferous mining. It is observed that for many proposed methodologies, project economics are generally not capable of being maintained throughout all stages of the planning process and iterative multiple solutions within a specified timeframe is usually not possible. This is due to the computational difficulty that arises due to the inherent algorithmic complexity in the problems of blend vectoring, pit limit optimisation and life-of-mine production scheduling. In the case of mine scheduling, there are very few or even no known instances of practical sized problems being solved to truly full optimality by mixed integer non-linear programming.

1.2 Our Contribution

Within this paper, methodologies that have been utilised to obtain optimal solutions (with respect to their formulated mathematical models) to occurring sub-problems throughout the mine planning process during feasibility studies will be demonstrated. Their strength lies in the relative ease of implementation, use and algorithmic tractability as compared to commonly proposed approaches within the academic literature as well as the ability to unlock the intrinsic value of any given project; from the writer's experience, they will often be performed on very large datasets multiple times during very short time windows as is the case and requirement during a feasibility study. It is by comparison, impossible using the traditional alternatives to the optimisation techniques discussed to obtain solutions that are both practical and maximise value despite the significant amount of research. Generic heuristic approaches lead to sub-optimality, mathematical programming to intractability and simplifications of problems to impracticability. In Sect. 2 we present the first set of results sourced from an equipment height study that demonstrate the difference in ability of a heuristic approach and that of a mixed integer linear program for determining an optimal combination of ore blocks within a resource or pit that yields a maximised potentially convertible inventory at a blend specification. In Sect. 3 we show the importance of simultaneously solving the ultimate pit limit and blending problems as opposed to conventional pit optimisation that requires utilising graph-based algorithms such as that of Lerchs-Grossmann. In Sect. 4, we discuss the Orelogy™ developed strategy contained within Maptek's Evolution software to the production scheduling problem and demonstrate the difference in ability and realised value of an approach that is able to account for a blend vectored resource. In Sect. 5, we provide an optimisation view point where near-optimal solutions (in regards to block sequencing models) for the scheduling problem have been obtained from integration of a mixed integer linear program with the mining sequence from multi-objective evolutionary algorithms. The previously discussed concepts are applied to an anonymous Western Australian Iron Ore project and an Australian nickel laterite project. The projects consist of multiple open pits with varying material compositions and a targeted blend strategy that largely prioritises a singular blend specification as the main product though with residual secondary by-products. They are proposed large-scale operations that utilise correspondingly large equipment and possesses a multitude of complex mining constraints inherent with projects of their caliber and scale. The results acquired from the optimisation methodologies proposed, have demonstrated the technical and financial acceptability for various strategic options proposed during the development of these projects.

2 Blend Vectoring a Resource

2.1 The Iron Ore Optimisation Problem

Optimisation for iron-ore projects differ to that of generic metalliferous projects as ore is not linearly priced on metal content but on a price per ton basis for material that meets a certain product specification. This special characteristic means that revenue is not a continuous function of the original material composition. Heavy price discounting is usually incurred when a narrow band of product specifications in terms of key analytes are deviated from even marginally. There also exists an upper limit on realised prices for a given product grade interval (i.e. no metal credits). The specification is usually comprised of a combination of multiple element grades; our metal analyte is usually iron while the contaminant analytes can include silica, alumina, phosphorous and more. The systematic approach utilised throughout the entire mine planning process to determine the definition of ore must take into account this complexity in order to fully utilise the reserve.

Blend vectoring is an approach used to maximise the conversion of resource to saleable blended product. A blend vectored model utilises variable rather than fixed product/contaminant grade criteria that trades off material with varying qualities to end up with the maximum yield of product at the required specification. The results of blend vectoring a resource is usually the minor loss of high risk ore that is high in both contaminant grades and low in metal grade. This loss is off-set by a significant increase in ore that has poor (in regards to specification) quality in only one grade analyte leading to an overall maximised resource. Figure 1 shows the difference between the traditional fixed cut-off methodology using quadrant analytes to define ore that meets a certain product specification and that of a blend vector approach that uses a dynamic cut-off criterion. The blend vector is performed across the entire population of available material and results in the ore contained within section a to be discarded in favor of the ore in sections b.

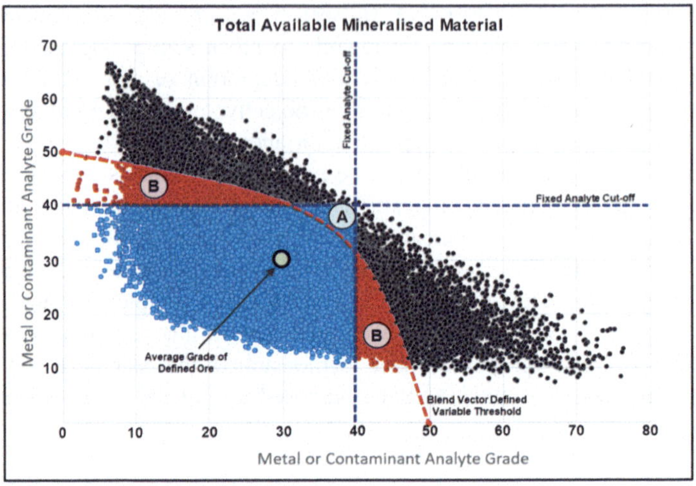

Fig. 1. Blend vector ore definition

2.2 The Knapsack Formulation

The problem of obtaining a maximised resource for a given set of different blend targets and product specifications is the equivalent of a multiple, multi-dimensional knapsack problem that is a generalisation of the standard 0–1 knapsack. The mixed integer linear programming formulation is shown in Eqs. (1) to (5) and is exactly the same as the standard form with the exception that we apply an additional weighting function on the decision variables based on each possible blend specification. Typically, this is a combination of either the price or penalty of each product specification, a mass yield regression or implied value based on desirability of a particular blend. P is the set of priced element analytes, K is the set of possible blends, C is the set of contaminant analytes, B is the set of all blocks under consideration, Q is the tonnage weighting and g is the grade. For a given project, the number of blends can range from 1–4 and the number of contaminants (|C|) between 2–9. Although the number of binary variables that usually result from typical problem instances tend to be in the millions, in practice, it is very easy to obtain optimal solutions in an efficient manner. In addition to modern computing, this is due to the fact that many of the cardinality constraints corresponding to different contaminants described by Eqs. (2) and (3) can often be dominated due to the contaminant and metal grade analyte co-efficients being proportional to each other. As the incidence vector of the optimal solution is not bounded by these constraints, we can often ignore them. Finding dominated constraints during the linear programming stage is important as any of the heuristic methods are by definition incapable of returning an optimal solution under the conditions imposed by tolerance vectors. It is also observed that in terms of the weighting applied to each of the different blends, the target is always maximising the highest ranked product spec over lower ranked alternatives. This is done regardless of assumed price and overall cashflow in certain cases, as the overall salability of higher quality product is more easily achievable and reduces the possibility of excessively downgrading high-grade ore. When this target is taken into account, the problem can then be easily reduced to solving a series of decomposed m-dimensional knapsack problems. Although the problem would still require utilising an algorithm that is technically exponential in running time such as branch-and-cut, it is only exponential in the number of grade/contaminant constraints rather than the millions of decision variables if we are to believe proposition 1 in [13] holds. Arguably, under these conditions the overall complexity is equivalent to that of a fixed parameter tractable algorithm as we only ever branch on a small, fixed total number of basic variables.

$$Maximize\ Z = \sum_{i \in B}^{i=1} \sum_{k \in K}^{k=1} \gamma_k \left[Q_i\right] x_{ik} \tag{1}$$

Subject to:

$$\frac{\sum_{i \in B}^{i=1} x_{ik} Q_i g_{ip}}{\sum_{i \in B}^{i=1} x_{ik} Q_i} \geq R_{pk}^{min} \quad \forall\, p \in P, k \in K \tag{2}$$

$$\frac{\sum_{i \in B}^{i=1} x_{ik} Q_i g_{ic}}{\sum_{i \in B}^{i=1} x_{ik} Q_i} \leq R_{ck}^{max} \quad \forall\, c \in C, k \in K \tag{3}$$

$$\sum_{k \in K}^{k=1} x_{ik} \leq 1 \quad \forall\, i \in B \tag{4}$$

$$x_{ik} \in \{0, 1\}^{|B| * |K|} \tag{5}$$

While it is possible to solve the blend vector problem to full integer optimality by branch and cut, the methodology that has been adopted in practice is different and a relatively simple one; we assume first that each block is not an elementary selective mining unit, and then take the solution to the linear programming relaxation as an intermediate optimal solution. In the LP relaxation each decision variable corresponds to block proportions. As the LP relaxation is a tight upper bound on the "true" integral optimum with respect to the formulated model, we then apply an approximate heuristic rounding procedure that rounds up fractional blocks in a way that ensures that the objective value is at least as high as that of the LP relaxation while only violating the blending constraints by a small fraction ε. The number of blocks rounded is often extremely small due to the underlying completely integral matrix that the reserve constraints form and small number of cardinality constraints, especially in proportion to the total number of blocks within a certain blend. The corresponding super-optimal solution is then guaranteed at least better than the original integer optimum in terms of objective value with any violated grades for each blend specification only exceeding the target by a very small percentage. The resulting blend grades obtained by this methodology for the results presented did not exceed more than 0.0001%. The reasoning for this method is obvious; the small violations are insignificant when considering the uncertainty of data, the fact that these violations can be thinned over a multi-period schedule and the level of accuracy we require. The strength of the process lies in its computational complexity; A branch-and-bound tree is never required to be built and solutions can be obtained in polynomial time. The results themselves do not differ from the true optimal integer solutions by more than a few blocks.

2.3 Finding the Optimal Resource

The following results in Fig. 2 have originated from a mining bench height/equipment study. The data used comprised of multiple block models at different bench heights and lateral mining extents that have had product yield regressions applied to determine product grades at block level resolution. Each blend vector optimised across 7 models, over a million blocks and 4 grade/contaminant targets within designed pit limits simultaneously. The results from the MIP, are compared to a python implementation of the Hill-climbing approach of [4] to generate a blend vector; in this approach, a linear or non-linear regressive equation is used that acts as a dynamic cut-off to repetitively test decision variables within a dataset until the desired objective is maximised. The analysis was undertaken using a linear method as it has been found that using a polynomial function does not result in any substantial or beneficial difference in resulting solutions. The product tonnes from a cut-off approach has not been graphed as they were originally too low and resulted in the project being uneconomical.

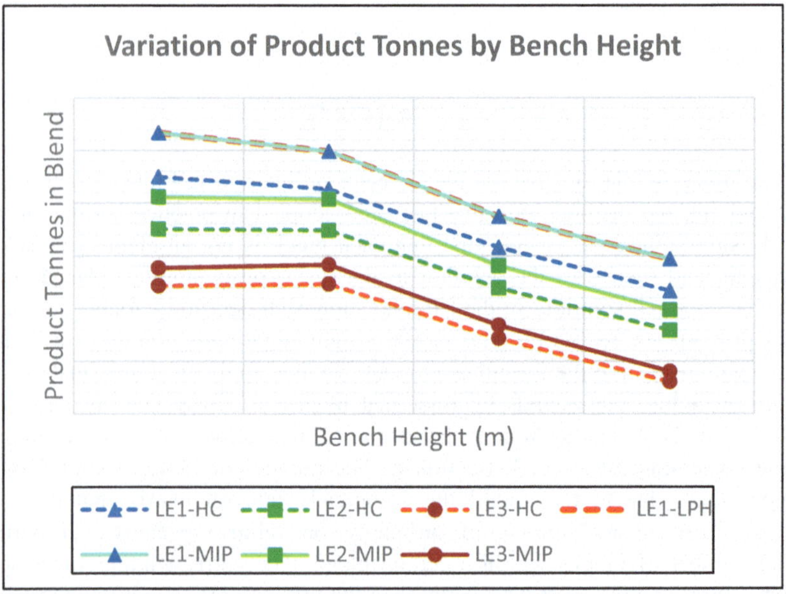

Fig. 2. Product tonnes at different block dimensions

The mixed integer programming approach was able to generate an increase in the total tonnes of product by up to 3% more across the different selective mining units. At glance, this does not appear significant but cashflow wise, it is a swing of over a billion dollars and consideration must be made as to the optimisation being constrained to within designed pit boundaries; material that would otherwise be 'waste' incurring only stripping costs has been converted to potential ore that generates revenue. An important observation is the ability of the MIP to generate a blend resource, if the targeted objective is fixed, at least as good as that of the heuristic but at larger bench sizes (i.e. with a more diluted resource). This possesses the implication that should we wish to maintain blend continuity at the fixed objective, we would be able to do so while also minimising production costs (less drilling costs due to larger blasting patterns, cost-effective practices of utilising larger equipment) and increasing productivity due to a larger bench size without a trade-off in total reserve.

3 Ultimate Pit Limit Optimisation

3.1 Optimal Design of Open Pit Mines

The primary objective of open pit design is creating practical pits that maximise profit. The ultimate pit limit problem in the context of a blending project introduces mining precedence constraints and stripping trade-off determination that must be additionally be accounted for within an optimisation model. The model outlined in Eqs. (1) to (5) is only suitable for application on a set of ore blocks under consideration usually either at the resource estimation stage where we are evaluating the entirety of a deposit for potential product or post-design where only the possible ore within the pit boundaries are decision variables. The solution gives an upper bound on the maximum total product/ore that is achievable given the blending targets and the material composition of the available resource. The model Eqs. (6) to (11) is the formulation that is required to be solved in order to simultaneously optimise both the undiscounted cashflow value/profit in terms of an ultimate pit limit with consideration of stripping precedence and the blending targets on products. The formulation has been re-arranged to a "by"-formulation with generalised upper bounds for the blending constraints. It is an extension of the integer programming form of the Lerchs-Grossmann (LG) defined pit limit problem that is much stronger, as conventional LG does not integrate processing cut-off and ore definition by destination directly during the optimisation process; these decisions are usually pre-defined and is the core weakness of a graph-based approach. Equation (6) is the objective function that is the total accumulation of cashflow, (7) and (8) are the blending constraints on product specification, (9) are the mining precedence constraints as in traditional LG and (10) enforces the "by" structure of the model.

$$Maximize\ Z = \sum_{i \in B}^{i=1} \sum_{k \in K}^{k=1} \gamma_k [Q_i] (x_{ik} - x_{ik-1}) \tag{6}$$

Subject to:

$$\sum_{i \in B}^{i=1} x_{ik} Q_i \left(R_{pk}^{min} - g_{ip} \right) \leq 0 \quad \forall p \in P, k \in K \tag{7}$$

$$\sum_{i \in B}^{i=1} x_{ik} Q_i \left(g_{ic} - R_{ck}^{max} \right) \leq 0 \quad \forall c \in C, k \in K \tag{8}$$

$$x_{ik} \leq x_{i'k} \quad \forall i \in B, k \in K, i' \in B' \tag{9}$$

$$x_{ik-1} \leq x_{ik} \quad \forall i \in B, k \in K \tag{10}$$

$$x_{ik} \in \{0, 1\}^{|B| * |K|} \tag{11}$$

This formulation is of course, strongly NP-hard in computational complexity as it is a precedence-constrained knapsack problem. However, the formulation yields special structure as the number of constraints are usually only large in the precedence and reserve constraints which comprise a totally uni-modular sub-matrix. Mining variables are not needed binary specification. The lagrangian dual obtained by relaxing the cardinality/resource constraints is a maximum flow/minimum cut problem that can be solved in polynomial time using a variety of efficient algorithms such as [5, 6]. This has been known since [12]. This behaviour is readily exploited in many column-generation style algorithms such as [1] to solve linear programming relaxations of much larger sized problems of the same form. There also exists many pre-processing, probing and problem size reduction strategies that may be employed to help yield solutions more efficiently than if we were to straight up optimise over raw data. Discussion on various pre-processing strategies for an equivalently structured MIP that assist in providing solutions such as providing a strong initial integer feasible solution and pre-eliminating blocks with certain uneconomic profitability are outlined in [8]. When solving LP or lagrangian relaxations of the pit limit problem, attempting to utilise a rounding heuristic/algorithm to achieve a solution is generally not desirable as was the case in blend vectoring due to the structural nature of the variables. As the formulation amounts to a single period equivalent problem of the scheduling problem, an integer solution can generally be obtained from utilising an efficient branch and cut algorithm combined with strategies that improve the models' tractability. For example, A methodology to obtain an IIFS to the MILP of solving the ultimate pit limit problem under the conditions imposed by the blending constraints can be efficiently found by essentially repeated application of the individual sub-problems of blend vectoring and UP. In this approach, a blocks definition of ore is decided by the blend vector rather than a cash flow or cut-off calculation. Since neither of the two subproblems are 'difficult', this approach can find solutions efficiently however will not always return an optimal solution in terms of cashflow due to its disjoint nature.

3.2 Assessing the Strategies for Pit Shell Generation

We present results obtained from application of 3 strategies to the pit limit optimisation problem; solving the mixed integer program, utilising a blend vector with the LG algorithm contained within the industry standard Whittle software and traditional cut-off optimisation. See Fig. 3. The optimisation was performed using 3 separate pre-processed, block models simultaneously with the economic parameters modified (from real numbers). As is demonstrated, adopting the traditional strategies as practiced commonly in industry leads to not only sub-optimal solutions that deviate from the optimal cashflow by at least ~15% for the LG+BV case and ~50% for the LG+CO case but also fail to effectively adhere to a target grade. An interesting observation that has been found is that for the case where we utilised a blend vector and LG strategy, the resulting 'total product' was in fact higher than that of the MIP. This is expected as when the definition of ore is simultaneously being optimised along with the pit delineation, the cashflow is maximised based on a key property; ore reserves with less incremental strip and individual processing cost being chosen for blending as opposed to sub-optimal material that might lead to an overall larger resource but result in a decreased UCF. The MIP driven strategy was able to reduce the overall mining footprint and cost substantially while only losing a very small portion of total reserve.

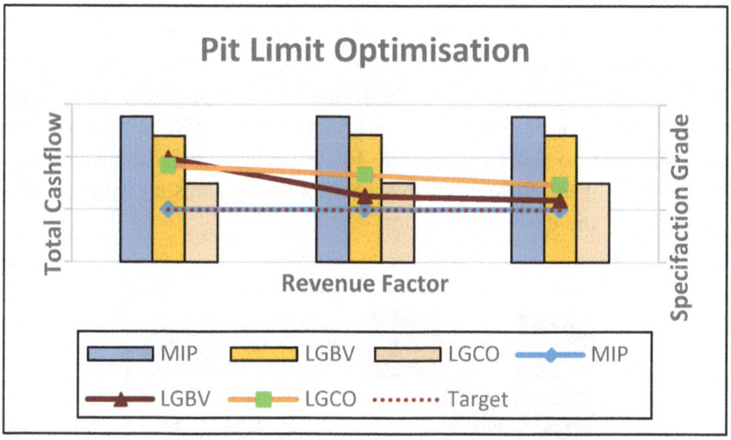

Fig. 3. Cashflow and target grades of pit optimisations

4 Open-Pit Mine Production Scheduling

4.1 Scheduling in Practice

The scheduling problem is significantly more computationally difficult and complex than the design or resource maximisation problems. While, there has been significant research within this area, most examples within the literature either fail to account for important concepts that are required for maintaining the projects economics such as blending or are incapable of providing an efficient solution within a specified timeframe that maintains optimality. This is attributed to the inherent inclusion of constraints required to model practical mining conditions and time-based resource allocation. The additional cardinality constraints required for production tonnages further violate the unimodularity of the di-graph structure and attempting to model activities of a non-linear nature such as stockpiling demands that linearity assumptions are not made on their behaviour. There also exists an enormous range of mining constraints that cannot be easily modelled with an integer programming formulation such as minimum operating widths, cutback widths, minimising excavator relocation, stage lag, bench turnover, etc.

Attempting to generate solutions based on classical optimisation techniques such as a branch and cut algorithm, rounding a linear programming relaxation or conventional heuristics generally requires making significant sacrifices either on quality or computational efficiency. Large scale optimisation models are commonly intractable and cannot easily model the previously mentioned constraints. It can be shown easily using extremely simple problem examples with small scale models that all integer programming scheduling models based on block sequencing in the literature from [3–9] fail under truly practical mining conditions. See Fig. 4; an optimal solution to a MIP will return the top sequence however if our mining width is 2 then the true practical optimum is the bottom sequence. While the mining width could be modelled with additional binary variables, this will cause increased complexity that can easily result in insolvability for any real sized problem instance. There also exist additional objectives that cannot be mathematically quantified such as providing allowance of tails/waste backfilling to minimise environmental footprints and compatibility with a conveyance system to minimise truck haulage. These objectives often arise in the case where there exists minimal access to flat land suitable for building waste dumps/external storage facilities in the case of the former and where the mining areas are spread across a very large project area in the case of the latter.

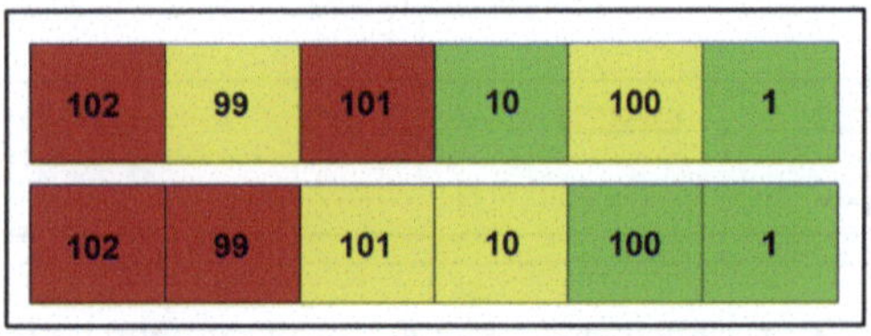

Fig. 4. A simple sequencing problem

It is possible to attempt to generate solutions by an integer programming approach however given specified time constraints and desired optimality it is often not reliable in the context of a feasibility study. In practice, attempting such an approach will often result in a solution being impossible to obtain due to practical instances possessing millions of decision variables and constraints. The problem sizes and intractable nature leads to even linear programming relaxations being unobtainable in some cases. The application of metaheuristics is an approach well suited to handle the complexity that arises in the scheduling problem and have seen heightened interest in the past few decades. Evolutionary algorithms for example, possess a significant edge due to their ability in handling large-dimension, multimodal, multi-objective, non-convex and discontinuous problems [11]. The scheduling methodology we employ here utilises implementations of multi-objective evolutionary algorithms for the mine production scheduling problem that are contained within Maptek Evolution originally developed by Orelogy Consulting and a particle swarm post-schedule blend optimiser. The capability of taking in the definition of ore based on a blend vectored resource is one of the primary strengths of Evolution as it remains possibly the only available package in the industry that can do so.

4.2 Genetic Algorithms

The general structure of a genetic algorithm (GA) is composed of an initial step where a randomized population of candidate solutions is spawned after which, various operators are recursively applied over several generations (iterative steps of the algorithm). These operators fall under three main categories; reproduction, cross-over and mutation. Possible candidate solutions are encoded as real-valued vectors that contain the necessary information that is required to assign an appropriate fitness value; a weighted measure of quality for an individual solution. In the mine scheduling problem, the genetic information contained in these real valued vectors include the extraction sequence. The advantage of using GA's is that the search-operators such as crossover and mutation can be designed in such a way that feasibility is maintained w.r.t precedence and other mining constraints. In classical approaches these mining constraints are the sole cause of intractability due to the very large number of precedence arcs required to model intermediate pit structures. The similarities to conventional mathematical programming are obvious; incidence vectors to chromosomes, fitness value to an objective function, spawning to initial integer feasible/basic solutions and generations to simplex pivots.

The role of the selection operator involves the elimination of candidate solutions within a population with weak fitness values as well as the promotion and duplication of ones with strong fitness values while maintaining a constant population size. It is responsible for the emphasis of solutions that have the genetic potential (building blocks), with the help of the other operators, to eventually assemble the optimal building blocks to produce an optimum solution. This behaviour is equivalent to natural selection in biological science. The cross-over operator is the primary search algorithm responsible for exploring the search space by combining two or more existing solutions to produce new solutions. Binary tournaments are played and the winners of these are combined by the cross-over operator to produce new solutions which contains building blocks from all the winners. In the cross-over function, two parent strings are chosen randomly to breed, producing descendants that are a hereditary combination of both. This is done by exchanging some portion of the phenotype of one parent with the other. The mutation operator promotes diversity within the population and also acts as a secondary search heuristic by locally modifying an individual solution. It operates under a small permutation probability proportional to the number of encoded variables.

The three operators are straightforward and work together over a number of generations, to emphasize good quality solutions and eliminate poor ones. In Evolution, the genetic algorithms are highly specialised for the open-pit scheduling problem. For commercial reasons, the specific and detailed characteristics of the methodology cannot be disclosed. Reproduction is done using a tournament selection approach. Cross-over operations are conducted in a way that ensures that derived solutions adhere to the complex mining constraints implicitly over the course of optimisation. Crossover locations are chosen within periods to ensure that all periods have ample opportunity to evolve.

4.3 The Evolution Application

Block level production scheduling was performed on 7 different block models constrained to designed pit boundaries simultaneously with a total of hundreds of thousands of mining blocks and over 15 scheduling periods. The three schedule grade profiles displayed in Fig. 5 were pulled from different stages of feasibility studies but essentially possessed the same input data and required objectives; to maintain a production schedule feeding the plant at a specified process rate while simultaneously adhering to blending and mining constraints. The competitor case (CPT-BFS) utilises a supposedly mixed integer programming-based cut-off approach to attempt to maximise cashflow. The value obtained from the Evolution approaches (OR-SC1 and OR-SC2) is obvious; revenue generated is maximised due to the metal grade target being achieved owed to the ability of being able to, not only input a blend vectored resource ore definition, but adhere to it throughout the full life of mine schedule. Utilising the forward-looking capability of evolutionary algorithms, a marketable blend is sustained and the potential upside of blend vectoring is captured. Targeting a higher grade was a strategic option only possible due to blending capability. For CPT-BFS, the scheduling periods during which the grade exceeds the target are essentially lost potential as high-grade portions of the reserve are priced at a grade lower than that of the material average. The scheduling periods where the target grade is failed to be adhered to will result in either heavy price discounting on the product or unsaleability leading to an extreme loss in revenue.

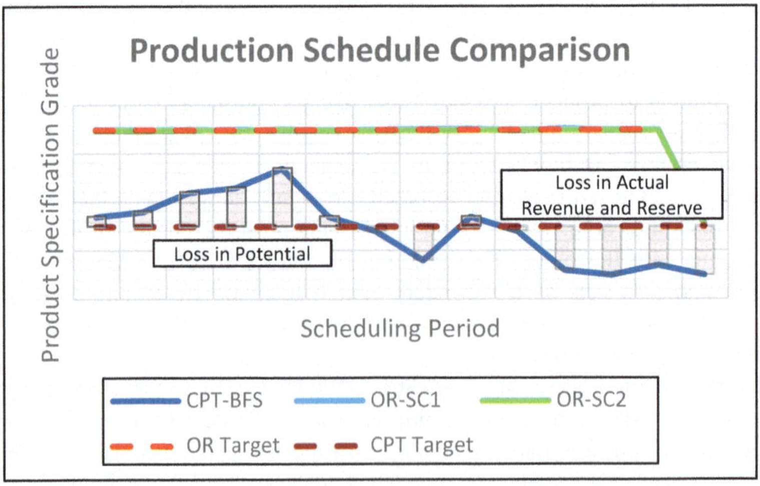

Fig. 5. Meeting the blend targets during scheduling

Often, it is the case that a blended resource does not retain a high conversion yield (rate) as in the case of conventional schedule optimisation practice within industry however our experience has demonstrated that this is not the case with Evolution where perfect or near perfect conversion of a stated resource is met. Adherence of the grade blend target objectives results in potential of the resource not being destroyed and price discounting not occurring. If we are to believe for example, that anything below the targeted specification grade is unsalable than the realised life of mine and overall reserve is effectively half that of the potentially recoverable inventory in the competitor case. Importantly, the mining sequence in regards to previously mentioned physical constraints are met in the Evolution schedules in conjunction with the economic and total material movement objectives. The adopted strategy also allowed for additional objectives to adhered to such as backfilling pits, compatibility with a conveying system and minimising stockpile peaks until very late periods. See Fig. 6 for the schedule visualisation of one of the pits within the project.

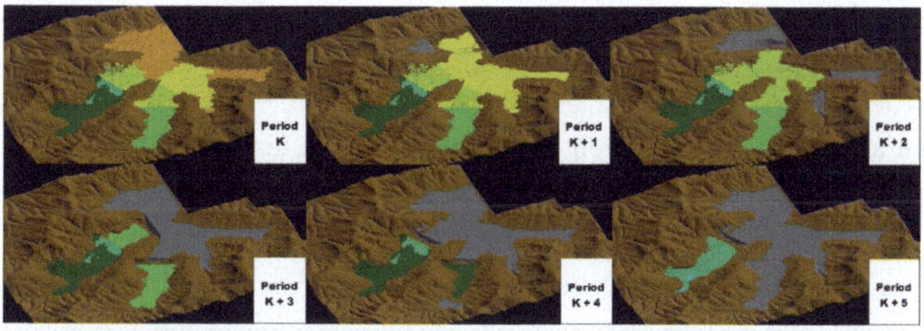

Fig. 6. Mine production schedule physical development

4.4 Post-mining Sequence Process Scheduling

The blending problem at the processing level is inherently non-linear due to the certain heavy usage of multiple stockpile processes and the associated allocation problem as well as the possibility of obtaining pricing credits that scale with grade in some projects similar to that of generic metal mining. Variable processing costs for different material types and therefore blocks also presents an opportunity for maximising value as costs can potentially be deferred. The non-linearity of the problem would usually mean automatic preclusion of conventional approaches such as linear programming. However, as observed by multiple researchers [2], the overall value within a schedule is often largely determined by processing level decisions which generates a perfect opportunity for iterative post-mine sequence improvement by either a linear program or metaheuristic local search approach on an otherwise complete production schedule. The reasoning is obvious; given a near finalized schedule, selectively modifying it by solving the easier reclaim scheduling sub-problem to full optimality can only provide an improvement in the worst case. A population-based approach using particle swarm optimisation (PSO) is one such method that has been applied by Orelogy. We present results from using a mixed-integer programming-based implementation that has been written to integrate with the scheduling process performed by applying Evolution. In this approach, the full scheduling problem is solved by Evolution-Origin at block level resolution after which the mining sequence is seeded and the processing optimisation performed by the MIP. In this methodology, schedule solutions with stronger mining fitness objectives within the pareto front are chosen for seeding over those with better processing objectives.

The approach we use for where mine stockpiling is the only intermediate stage during the processing cycle is simple; we first assume that there are an infinite number of possible stockpiles and therefore the blocks can be selectively reclaimed at the same individual material composition that they were originally extracted at. Under these conditions, mass balance constraints or stockpile variables are not required within the mathematical programming formulation. There is only inclusion of extra time discounting on the value of blocks in the objective function based on costs associated with reclaiming material from a stockpile using a loader. The resulting problem is a multi-knapsack problem similar to that of Eqs. (1) to (5) that can be solved easily using either linear programming or even a greedy algorithm to full optimality. In the case of additional process complexity, there is additional modelling that is required which will be discussed in a future paper. The resulting solution provides the necessary information we require to accumulate blocks into a realistic number of stockpiles based on their processing period. The objective value obviously does not change when the blocks are allocated to the different stockpiles and is at least as high as if we were to have directly modelled the accumulated stockpiles. This is equivalent to fully optimal cut-off optimisation.

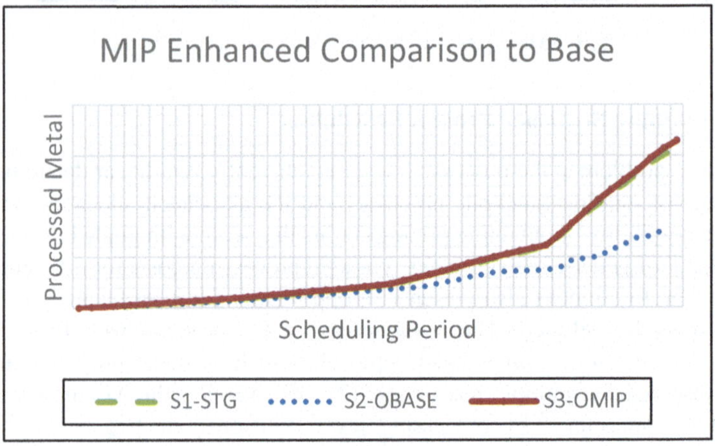

Fig. 7. An improvement to NPV maximisation for origin

When scheduling in Evolution, we obtain solutions with near optimal value and full adherence to complex mining constraints when cashflow is capped by the project economics so thus we instead present results for a metalliferous mining project. Figure 7 demonstrates the application of the approach outlined above on a complex nickel laterite mine comprising dozens of pits, millions of blocks and a processing set-up involving blocks possessing multiple parcels due to a beneficiation process and multiple intermediate stockpiling options. Evolution strategy was used to guide the Origin mining sequence. The graph spans the first 40 periods of a 60+ period schedule; Series 1 is the cumulative processed metal within an indicative schedule with maximum discounted value based on strategic optimisation using bench aggregation. Series 2 then shows for a particular production schedule produced from Origin; the resulting cumulative metal processing based on adopting the same mining sequence and TMM as that indicated by the strategic optimisation but with a proper block-by-block extraction sequence suitable for practical mining. Series 3 then shows how the production sequence from Origin is then used to fully optimise the processing decisions at selective block resolution using mixed integer programming to maximise the metal processed/cash flow and obtain a near optimal value for NPV. The NPV difference between the baseline Origin schedule and the one enhanced by MIP is in excess of 10%.

5 Conclusion

The mine planning process during a feasibility study is comprised of many optimisation problems that possess inherent complexities that must be accounted for. Failure to do so results in a significant difference between the achieved value of a project and what is possible as demonstrated by the results provided within this study. The tools to obtain optimal or near optimal solutions with respect to the mathematical models of mine planning problems exist and have been used throughout multiple feasibility studies by Orelogy to maximise the realised value.

References

1. Bienstock, D., Zuckerberg, M.: A new LP algorithm for precedence constrained production scheduling. http://www.optimization-online.org/DB_HTML/2009/08/2380.html
2. Boland, N., Dumitrescu, I., Froyland, G., Gleixner, A.M.: LP-based disaggregation approaches to solving the open pit mining production scheduling problem with block processing selectivity. Comput. Oper. Res. **36**(4), 1064–1089 (2009)
3. Chicoisne, R., Espinoza, D., Goycoolea, M., Moreno, E., Rubio, E.: A new algorithm for the open-pit mine production scheduling problem. Oper. Res. **60**(3), 517–528 (2012)
4. Everett, J.E.: Planning an iron ore mine: from exploration data to informed mining decisions. Issues Inform. Sci. Inform. Technol. **10**(1), 145–162 (2013)
5. Goldberg, A., Tarjan, R.: A new approach to the maximum-flow problem. J. Assoc. Comput. Mach. **35**(4), 921–940 (1988)
6. Hochbaum, D.: A new-old algorithm for minimum-cut and maximum-flow in closure graphs. Networks **37**(4), 171–193 (2001)
7. Hochbaum, D., Chen, A.: Performance analysis and best implementations of old and new algorithms for the open-pit mining problem. Oper. Res. **48**(6), 894–914 (2000)

8. Htwe, D., Asad, M.W.A.: Performance evaluation and an implementation of mixed integer linear programming for open pit optimisation. Unpublished Paper (2015). https://www.slideshare.net/slideshow/embed_code/key/15gqbYxx1GUxbP
9. Johnson, T.B.: Optimum open pit mine production scheduling. Doctoral dissertation, University of California, Berkeley, California (1968)
10. Lambert, W.B., Brickey, A., Newman, A.M., Eurek, K.: Open-pit block sequencing formulations: a tutorial. Interfaces **44**(2), 127–142 (2014)
11. Myburgh, C., Deb, K.: Evolutionary algorithms in large-scale open pit mine scheduling. In: Proceedings of the 12th Annual Conference on Genetic and Evolutionary Computation (GECCO 2010), Portland, Oregon, pp. 1155–1162. ACM (2010)
12. Picard, J.C.: Maximal closure of a graph and applications to combinatorial problems. Manag. Sci. **22**(11), 1268–1272 (1976)
13. Puchinger, J., Raidl, G., Pferschy, U.: The multidimensional knapsack problem: structure and algorithms. INFORMS J. Comput. **22**(2), 250–265 (2010)

Effective Methods to Reduce Grade Variability in Iron Ore Mine Operations

Oscar Parra Trouchina(✉) and Erkan Topal

Western Australia School of Mines, Curtin University, Perth, Australia
oscar.parratrouchina@postgrad.curtin.edu.au

Abstract. This paper explores possible strategies to meet iron ore customer requirements (reduce variability and keep Fe% and contaminants such as Al_2O_3% and P% within the acceptable tolerance) by optimizing the mine plan and scheduling strategy in the rehandling process. The paper aims to obtain more homogenous iron ore grade by redesigning current rehandling strategy and creating selective Run of Mine (ROM) with defined Fe grade and contaminants tolerances in the supply chain process. A new Linear Programming Model was developed and sample data from an Iron Ore operation in Western Australia was used to conduct this analysis. The outcomes indicate that it is possible to achieve better grade results for Fe% and Al_2O_3% keeping required tonnage targets. Furthermore, the results reflect Low-Grade materials could play a more important role in the weekly schedule and that grade contents could balance alterations of High-Grade products. Although there are extra costs due to rehandling activities, these costs could be offset by selling more consistent products. It is an invitation to mining companies to change current strategy and focus more on the downstream and customer requirements in terms of quality product instead of just quantity.

Keywords: Grade variability · Iron ore · Rehandling · Linear programming

1 Introduction

Grade control happens at all the stages of the iron ore supply chain process. Planning, scheduling and execution processes play a relevant role together in the aim of keeping the grades under control. Previous research has explored alternatives to mitigate the grade variability impact by using better planning and scheduling tactics. However, as the presence of the mineral deposits is not homogeneous, mine plans are prepared and executed based on expected values, leaving a failure factor that needs to be fixed during the execution. The mine execution may follow the plan, but the plan might not represent the best results in terms of quality and quantity contents. Thus, a strategy which combines not just planning and scheduling against execution, but also different approaches in the rehandling process is required.

When controlling grades, the mine operations face a dilemma - Should mine sites continue limiting number of stockpiles and prioritizing material movement to the crusher, Low-Grade and waste dump (and put the focus on the process improvement of

E. Topal (Ed.): MPES 2019, SSGG, pp. 67–74, 2020.
https://doi.org/10.1007/978-3-030-33954-8_7

the mine plan and scheduling)? Or instead, should the mine site use stockpiles as an effective grade control strategy? (Fig. 1).

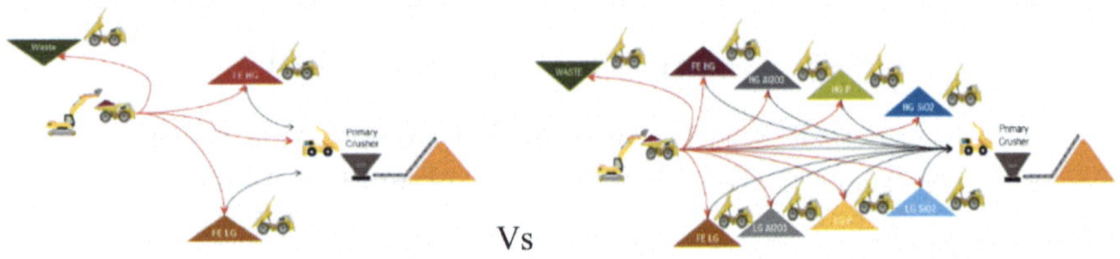

Fig. 1. Control grade in the mine process (pre-crushing) using limited stockpiles vs. using several stockpiles with different grades for Fe% and contaminants.

According to Parra Trouchina (2018), currently mining operations are driven by the required tonnage target. A significant focus and pressure are put on reaching those targets. Thus, processing plants receive all type of materials and downstream processes (after crushing) are in charge of making enough blends to hit the grade market requirements. However, as the process is reactive, often there is no time or space to blend the material and achieve the grades, meaning that the companies need to offer lower quality products or pay penalties. A different management mindset should consider the primary crusher as the first main customer in the supply chain. The crushing process should define which material (tonnage and grade) is needed daily (reversing the analysis) – based on the tonnage and grade mine plan targets. Fingers with different Fe% and contaminant tolerances could be used to join the fresh ex-pit material. This practice should ensure that, if available in the pit, the material should meet at least the grade requirements. If tonnage is not met (due to lack of required grade in the pit), then the planning team would have to ask for the remaining tonnage from another mine site or pit.

Several researchers have investigated alternatives to mitigate grade variations along the iron ore production process (Mai et al. 2019). However, due to the different constraints and lay out of the operations, it is difficult to find one single solution or recommendation that addresses the grade control issues in iron ore operations.

In practice, pre-crusher stockpiles are mainly used as buffers between the mine and primary crushing processes. According to Everett et al. (2015) the use of pre-crusher stockpiles to store ore and buffer short-term fluctuations in production processes is generally well recognised and accepted. Notably, Everett et al. (2015) state a gap in current practices when indicating that the potential to reduce short-term grade variation of ore entering the crusher is rarely recognised and generally poorly understood. Furthermore, Everett et al. (2015) proposes that design options for pre-crusher stockpiling should consider the four competing roles of storage; buffering, blending, and grade control, to produce predictable and uniform crusher feed grades. The selection of alternative grade allocation methods requires careful consideration as decisions at this early stage of the production process have been shown to affect shipping and customers (Everett et al. 2015). Everett (2010) asserts the execution of the daily crusher schedule, in a well-designed system supported by an effective pre-crushing stockpiling strategy is a relevant step in grade control in the short-term variability. For Everett, "pre-crusher stockpiles are relevant in smoothing out short term variability, where short-term implies variations within a time period comparable to the time interval over which the stockpile is built" (Jupp et al. 2013, p. 244).

Whilst these studies have demonstrated that, if conducted correctly, stockpiling is a good alternative to reduce grade in the pre-crusher processing, the question remains which type of stockpiles are used currently in the industry and what are their features. Using the amount of variation in reclaimed material Robinson (2004) highlights many useful predictive models. He established four different scenarios when building stockpiles: (i) Perfectly mixed material; the limitation of this model is that it does not provide any useful prediction of the amount of variation that may happen in the reclaimed material. (ii) Sampling theory, which determines the minimum variation that could possibly be achieved in the case that all the material in a stockpile were combined. (iii) Bed-blending theory, which defines the variation that would occur if the geometry of stacking and reclaiming were perfect (meaning that each parcel of stacked material was equally represented in each parcel of reclaimed material), (iv) Stockpiles with uniform stacked layers. From his predictive models, Robinson (2004) concludes that blending stockpiles is often effective when reducing variation between large output blocks, however, they often produce a variable grade output with a cycle that follows the cyclical behavior of the reclaimer.

Even though these techniques exist, the operations often find difficulties in reducing grade variability control which affects the quality of the product offered. Based on this outcome, this paper explores how efficient the use of stockpiles could be when trying to improve grades. The paper aims to demonstrate that with the use of stockpiles, contents of Fe% and contaminants like Al_2O_3% and P% in material outcomes are in tolerance, keeping the same required tonnages. The analysis has been modelled using a Linear Programming technique.

2 Description of Methods and Mathematical Model

A quantitative research was carried out supported by the development of a Linear Programming model (LP). Samples from a current process were used and several experiments under different constraints were performed. According to Chapra et al. (1998), linear programming is an optimization approach that deals with meeting a desired objective such as maximizing profit or minimizing cost in the presence of constraints such as limited resources. The mathematical formulation for this grade blending constraints is defined as:

Definition of decision variables X_{it}, where X_{it} represents the tonnes from different material type i, (i = 1, 2, 3, 4, 5, 6, 7, 8 and 9) in time period t.

Then, the objective function is:

$$Max\ Z = \sum_{t=1}^{n} \cdot \sum_{i=1}^{n} C_{it} X_{it} \tag{1}$$

Where the coefficient $C_{i,t}$, represents the percentage of material taken from every type of material i in time period t.

Subject to:

For grade blending constraints

$$\sum\nolimits_{m=1|m\in i}^{n} (g_{mt} - G_{min}).O_{it}.X_{it} \geq 0 \quad \forall t \tag{2}$$

$$\sum\nolimits_{im=1|m\in i}^{n} (g_{mt} - G_{max}).O_{it}.X_{it} \leq 0 \quad \forall t \tag{3}$$

Where G_{min} and G_{max} represent the minimum and maximum (tolerance) Fe%, Al_2O_3% and P% grade value for the final product. g_{mt} represents the average grade of material such as Fe%, Al_2O_3% and P% available in time period t. O_{it} represents the ore tonnage i available in time period t. Finally, non-negativity constraints:

$$X_{it} \geq 0 \tag{4}$$

3 Model Implementation and Results Analysis

Sample data from an Iron Ore operation in Australia was used for 5 weeks' time intervals. Each week the material available for each type of material (stockpile) is represented as the material non-used from the previous week (already sitting in the stockpile) + the fresh material coming from the mine. Three types of experiments were modelled:

Experiment 1: Provide the maximum Saleable Ore Product (SOP) capacity by keeping grades in target.
Experiment 2: Keeping current Saleable Ore Product (SOP) and the grades using all available material types including low grade materials.
Experiment 3: Keeping current Saleable Ore Product (SOP) and the grades but forcing the model to use 30% of the SOP coming from low grade material types.

In this case study, the iron ore operation produces materials from one pit. These materials are categorized as: HG (high Fe% content), LG (low Fe% content), mixed (material with high Fe% but some level higher level of contaminants) and waste (low Fe% contents). Table 1 shows current grade ranges:

Table 1. Current grade ranges of the study case.

Material	Min Fe%	Max Fe%	Min Al_2O_3%	Max Al_2O_3%	Min P%	Max P%
HG	60.3	64.1	1.41	3.04	0.113	0.166
Mix	59.2	62.8	2.20	4.43	0.148	0.204
LG	54.7	58.6	2.50	6.45	0.141	0.211
Waste	43.3	51.1	1.83	10.37	0.098	0.143

Depending on the features and the requirements of the operation, the materials coming from the pit go to different destinations. Table 2 presents the current plan for the tonnages and the destinations of these materials.

Table 2. Scheduled material distribution during the five week periods.

From	To	Tonnage	Material type
Mine	Crushers	3,575,848	HG material
Mine	ROM Blend (SP)	991,658	Blend material (1 SP)
Mine	ROM HG (SP)	567,381	HG material
Stockpile (ROM)	Crushers	567,378	HG material
Mine	Dump LG	1,163,121	LG material
Mine	Dump waste	1,668,574	Waste material

As can be seen from Table 2, the total tonnages of the material is 8,533,960 tonnes for 5 weeks of time period and total material crushed (scheduled) is 4,143,226 tonnes. The current mine schedule proposes to utilize the following sources:

- Directly from the mine: ~86%. From the Stockpiles (ROM): ~14%
- Directly from the mine mixed and LG materials: <1%. From Dump LG: <1%

The initial assumption is that the current model (with only one ROM) does not offer enough flexibility to the operation to reduce the variability and control the grades. By creating more stockpiles (during pre-crushing) with defined Fe% and contaminants tolerances, is it possible to get a more homogenous iron ore with controlled contaminants in the planning is the main aim to answer in this research.

Table 3. Results of the experiments – linear programming model.

Type of material	Feature	Current schedule	Exp. 1: Max. SOP	Exp. 2: Improve current schedule	Exp. 3: Alternative
HG	H Fe%	From Mine	From Mine	From Mine	From Mine
HG1	H Fe%	From SP	From SP	From SP	From SP
LG1	L Fe%				
LG2	L Fe%		From SP	From SP	From SP
LG3	L Fe%			From SP	From SP
LG4	L Fe%			From SP	From SP
Mix1	Mid Fe%	From SP	From SP	From SP	From SP
Mix2	Mid Fe%	From SP	From SP	From SP	From SP
Mix3	Mid Fe%	From SP	From SP	From SP	From SP

After solving the models, Table 3 presents how the proposed low-grade stockpiles were used in comparison to the original schedule.

As a reference current total crusher capacity is 960,000 Tonnes/week.

Figure 2 shows the SOP results of the 3 experiments. Notice that the variation in experiment 1 was expected as the model was asked to deliver the maximum SOP capacity:

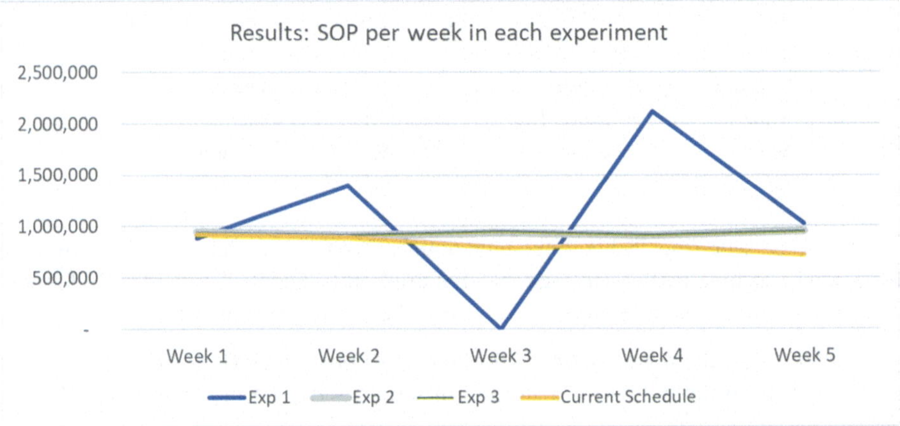

Fig. 2. Results for SOP for experiments 1, 2 and 3 and current schedule.

In Experiment 1, the objective was to understand the maximum SOP available for the targeted grade. In week 3, the stockpile is exhausted and the new (fresh) material is not able to meet the grade requirements. Higher P% was an outcome compared to the original requirement.

In Experiment 2, the objective was to improve the current schedule. The model was asked to keep the same target (as current schedule) for SOP and grades. The model used stockpiles like LG2, LG4, Mix1, Mix2 and Mix3 (rather than only HG and HG1 as in current schedule). Grades and tonnage were smooth and consistent in the periods.

In Experiment 3, the model was forced to find 30% of the SOP from Low-Grade and mixed stockpiles. Tonnage and grade targets were met in a consistent transition between weeks.

Regarding the grade results, as seen in Figs. 3 and 4 below, positive outcomes came for Fe% and Al_2O_3%. Regarding the P%, the results were above current schedule but still within the tolerance of the site, see Fig. 5.

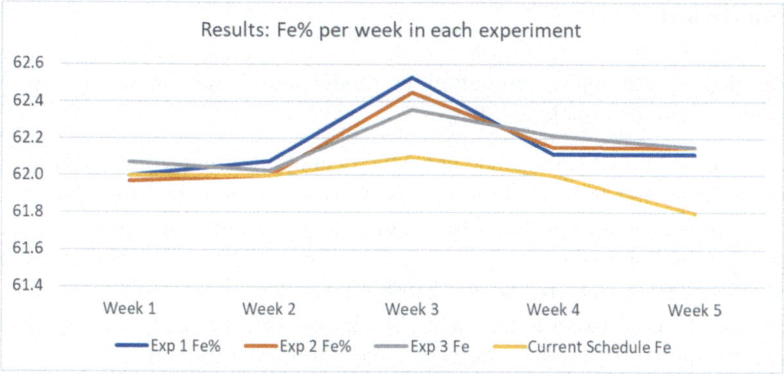

Fig. 3. Results for Fe% in experiments 1, 2 and 3 and current schedule.

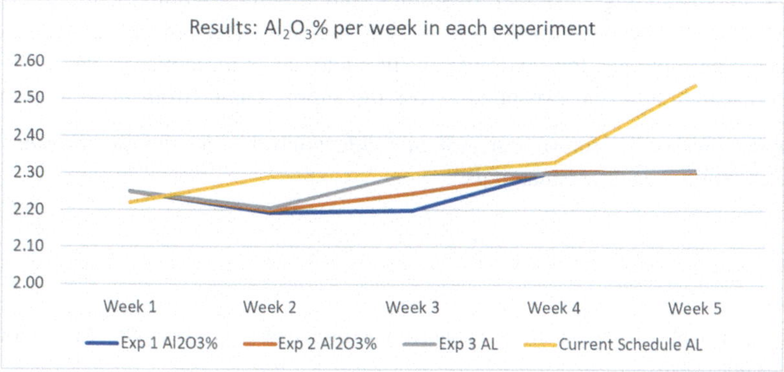

Fig. 4. Results for Al_2O_3% in experiments 1, 2 and 3 and current schedule.

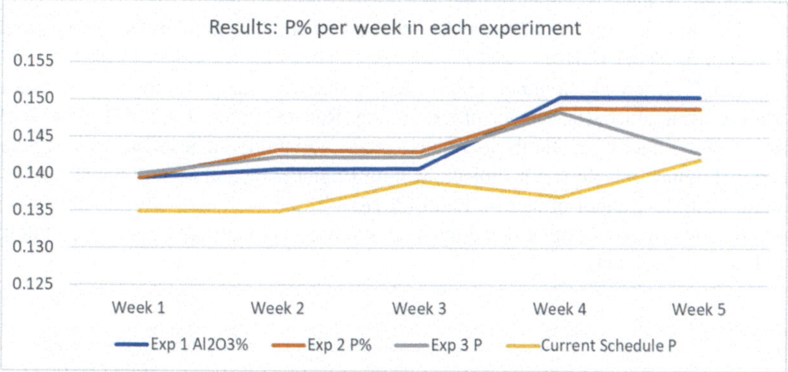

Fig. 5. Results for P% in experiments 1, 2 and 3 and current schedule.

4 Conclusions

After developing the linear programming model and based on the results of the experiments, it can be concluded that:

- The use of additional Low Grade and Mixed stockpiles in the pre-crushing process could be an option to control the variability of Fe%, Al_2O_3% as well as P%.
- The 3 experiments took material from LG's stockpiles, demonstrating that there are alternatives to take advantage of Low-Grade materials by combining them with good quality HG products. As the distribution of the materials was different along the weeks in the 3 experiments, it is difficult to conclude that there are irrelevant stockpiles (however, some of them provide minor contributions).
- The Linear Programming model is a simple and practical method to consider when solving decision problems. Once the mathematical steps are formulated several scenarios can be tested quickly.
- After understanding customer quantities and qualities requirements, a different mindset could be developed by choosing the plant (bottleneck) as the decision driver. The mine has the flexibility in time and space to deliver as required.
- This research could be expanded across the supply chain (from mine to ship).

New opportunities for future research in grade control strategies are opened using stockpiles and lower grade materials.

References

Chapra, S., Canale, R.: Numerical Methods for Engineers, pp. 375–398. Mc Graw Hill, New York (1998)

Jupp, K., Howard, T.J., Everett, J.E.: Role of pre-crusher stockpiling for grade control in iron ore mining. Appl. Earth Sci. **122**(4), 242–255 (2013)

Everett, J.: Simulation modeling of an iron ore operation to enable informed planning. Interdisc. J. Inf. Knowl. Manag. **5**, 101–114 (2010)

Everett, J., Howard, T., Jupp, K.: Pre-crusher stockpile modelling to minimise grade variability. TOS Forum **5**, 35–41 (2015)

Mai, N., Topal, E., Erten, O., Sommerville, B.: Risk reduction strategy for the iron ore production scheduling using stochastic integer programming. Resour. Policy J. **62**, 571–579 (2019)

Trouchina, O.P.: Effective methods to reduce grade variability in iron ore mine operations. Master of Engineering Research Project, Mining Engineering and Metallurgical Engineering, Western Australia School of Mines, Perth (2018)

Robinson, G.: How much would a blending stockpile reduce variation? Chemometr. Intell. Lab. Syst. **1**, 121–133 (2004)

Open Pit Mine Scheduling Model Considering Blending and Stockpiling

Mojtaba Rezakhah[1(✉)] and Eduardo Moreno[2]

[1] Tarbiat Modares University, Tehran, Iran
m.rezakhah@modares.ac.ir
[2] Faculty of Engineering and Sciences, Universidad Adolfo Ibanez, Santiago, Chile

Abstract. Open pit mine production scheduling (OPMPS) is a decision problem which maximizes net present value (NPV) by determining the extraction time and destination of each block of ore and/or waste in a deposit. Stockpiles can be used to maintain low-grade ore for future processing, to store extracted material until processing capacity is available, or to blend material based on single or multiple block characteristics (i.e., metal grade and/or contaminant).

We modify an existing integer-linear program to maximize NPV and provide a schedule and stockpiling strategy for an operational open pit mine, in which the stockpile is used to blend materials based on multiple block characteristics. We compare the schedule of $(\hat{p}^{la})$ with that produced by (p^{ns}) which does not consider stockpiling, and with $(\tilde{p}^{la})$, which controls only the metal content in the stockpile and ignores the contaminant level at the mill and in the stockpile. We show that our model improves the NPV of the project while satisfying operational constraints.

Keywords: Stockpiling · Linear and integer programming · Mine planning · Open pit mining

1 Introduction

The OPMPS problem is commonly formulated as an integer program with binary variables representing if and when each block is extracted. Some researchers present nonlinear-integer models to solve open pit mine production scheduling with stockpiling (OPMPS+S) problems. In their models, the authors assume that the material mixes homogeneously in the stockpile and that the grade of material leaving the stockpile is equal to the average grade of all the material within the stockpile, but these models are di cult to solve. Here, we use a linear-integer model to approximate the (OPMPS+S) problem, and instead of computing the average grade in the stockpile, we force the stockpile to have an average grade above a specific limit. We compare the output of this model in terms of a schedule and stockpiling strategy against that provided by state-of-the-art software.

We develop a production schedule for an operational open pit mine with two metal types and a contaminant. Our data show that a considerable number of blocks contain both high metal grade and a high level of contaminant. The processing plant requires us

E. Topal (Ed.): MPES 2019, SSGG, pp. 75–82, 2020.
https://doi.org/10.1007/978-3-030-33954-8_8

to keep the contaminant level of material below a specific limit. Therefore, blending material in the stockpile based on the metal's grade and contaminant level may result in a higher NPV. In other words, blocks with high metal content and a high contaminant level should be mixed with other material to satisfy the plant requirement; without a good blending strategy, we might lose the value of the aforementioned blocks by sending them to waste. Existing models either do not consider stockpiling as part of OPMPS problem, or just control one grade in the stockpile.

1.1 Integer-Linear and Nonlinear Models Considering a Stockpile

Linear and mixed integer programming models have been used significantly to optimize open pit production scheduling by focusing on the extraction sequence. Akaike and Dagdelen (1999) propose a model for long-term mine planning considering a stockpile. The authors use a graph theory-based method which considers an in nite number of stockpiles, meaning that every block has its associated stockpile, i.e., there is no blending in the stockpile and the material grade in a stockpile is the same as that of the associated block. Hoerger et al. (1999) describe a mixed integer-linear programming model to optimize mine scheduling for Newmont Mining Corporation's Nevada operations. The authors consider multiple stockpiles, each of which has a specific grade range. When the material is removed from the stockpile, its grade is considered to be the minimum of the associated grade range.

However, a more realistic assumption is that material in the stockpile is mixed homogeneously. In other words, the characteristics (e.g., grade) of the material change by entering the stockpile and should be treated as variables. Bley et al. (2012) propose two different non-linear integer models for stockpiling and assume that the grade of the material removed from the stockpile is the weighted average of the material inside the stockpile. Due to the non-linearity of this model, it cannot be used for solving real-sized instances of the problem. Recently, Moreno et al. (2017) propose different linear integer models to consider stockpiling in open pit mine scheduling, compare the objective function values of these models, and suggest that their (p^{la}) model provides more accurate solutions in which material with a single grade is mixed in the stockpile. However, a deposit usually contains more than one metal and sometimes a contaminant. The use of stockpiling to blend extracted material in these cases has not been treated in previous linear and mixed integer programming models.

2 Data

Our initial data set is taken from an operational open pit mine in Southeast Asia and consists of 30,100 blocks, four phases, 56 benches, and 16 time periods. A phase corresponds to a sub-region of the pit, and can be obtained by employing the Lerchs-Grossmann (1965) algorithm. A bench is a ledge that forms a single level of operation to extract both mineral and waste materials. Figure 1 shows the pit, which includes four phases differentiated by shading.

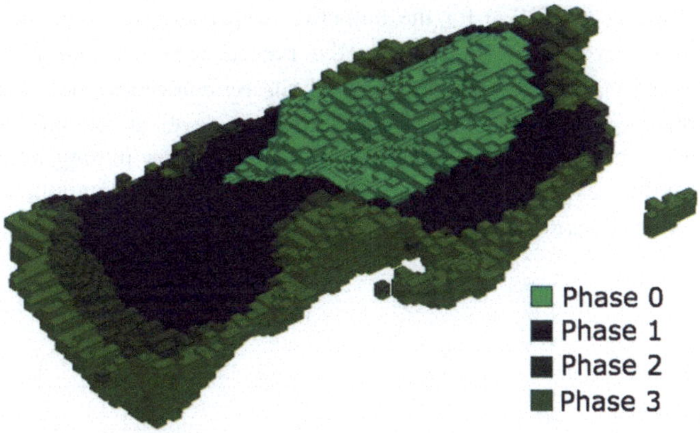

Fig. 1. The pit contains four phases, differentiated by shading. The center light green area is Phase 0, the black region is Phase 1, the dark green region is Phase 2, and the green perimeter region is Phase 3.

Each block may contain two metals, gold and copper, whose selling prices are assumed as \$20 per gram and \$2.5 per pound, respectively. Copper and gold recovery rates are 60% and 80%, respectively. The mining and processing costs are \$2.5 and \$7.5 per ton, respectively. For the sake of simplicity, we represent the copper and gold contained in each block as a copper-equivalent by jointly considering their selling prices, mining and processing costs, and their recovery rates at the mill.

Blocks also contain arsenic as a contaminant, which should be limited at the mill to 150 parts per million (ppm). In other words, the average grade of arsenic processed in each period should not exceed 150 ppm. To maximize NPV, we need to process the blocks containing high copper-equivalent grade as soon as possible, but there is a considerable number of blocks whose material includes both high-copper-equivalent grade and a high arsenic level. Here, material blending is very important because, without a good blending strategy, we might send some blocks with high-copper-equivalent grade to the waste dump.

There are 56 benches in the mine and the height of each bench is 12 m. The aforementioned blocks are contained in four predefined phases, computed using the Whittle commercial mining software, resulting in 220 phase-benches to schedule. To extract any block in a given phase-bench, the predecessor phase-benches must be mined completely. As a phase-bench is mined, equal proportions must be taken from each block. Mining capacity is 116.8 million tons per year and processing capacity is 67.89 million tons per year. The cash ow discount factor is 10% per year.

3 An Optimization Model Considering Stockpiles and Blending Constraints

Moreno et al. (2017) propose a linear-integer model, (Pla), which is tractable and provides an objective function value that is very close to the nonlinear-integer model proposed by Bley et al. (2012) for the numerical experiments presented. We modify (p^{la}) in such a way that it (i) enforces the precedences between phase-benches, (ii) homogeneously mixes material in the stockpile by controlling both metal content and contaminant level, (iii) satisfies the contaminant level at the mill by blending material from the mine and the stockpile, and (iv) considers mining and processing resource constraints explicitly. The rst subsection introduces notation, and the following subsections provide the mathematical formulation.

3.1 Notation

Indices and Sets:

$b \in \mathcal{B}$: blocks; $1, \ldots, B$
$n \in \mathcal{N}$: phase-benches; $1, \ldots, N$
$b \in \tilde{\mathcal{B}}_n$: all blocks in a phase-bench n that must be mined together
$\hat{n} \in \hat{\mathcal{N}}_n$: all phase-benches that must be mined directly before phase-bench n
$r \in \mathcal{R}$: resources $\{1 = \text{mine},\ 2 = \text{mill}\}$
$t \in \mathcal{T}$: time periods; $1, \ldots, T$

Parameters:

δ_t: discount factor for time period t (fraction)
C^m: mining cost per ton of material ($/ton)
C^p: processing cost per ton of material ($/ton)
P: profit generated per ton of metal ($/ton)
W_b: tonnage of block b (ton)
M_b: metal obtained by completely processing block b (ton)
C^h: rehandling cost per ton of material ($/ton)
L_b: grade of copper-equivalent in block b
$\bar{L}$: average grade of copper-equivalent in the stockpile
G_b: grade of contaminant in block b (ppm)
$\bar{G}$: average grade of contaminant in the stockpile (ppm)
$\hat{G}$: contaminant limit at the mill (ppm)
R_{rt}: maximum amount of resource r available in time t (tons/yr)

Decision Variables:

y^m_{bt}: fraction of block b mined in time period t
y^p_{bt}: fraction of block b mined in time period t and sent to the mill
y^s_{bt}: fraction of block b mined in time period t and sent to the stockpile
y^w_{bt}: fraction of block b mined in time period t and sent to waste
x_{nt}: 1 if all blocks in $\tilde{\mathcal{B}}_n$ have finished being mined by time t; 0 otherwise
i^p_t: tonnage of ore sent from the stockpile to the mill in time period t
i^s_t: tonnage of ore remaining in the stockpile at the end of time period t

3.2 L-Average Bound Model $\left(\hat{p}^{la}\right)$

Here, we use a modified version of $\left(p^{la}\right)$ to explore the effect of blending material while controlling different grades in the stockpile. This model requires the blocks that enter the stockpile to have an average metal grade of at least $\bar{L}$ and an average contaminant grade of at most $\bar{G}$. The model is as follows:

$$\left(\hat{\mathcal{P}}^{la}\right) : \max \sum_{t \in \mathcal{T}} \delta_t \left[P\left(\sum_{b \in \mathcal{B}} M_b y_{bt}^p + \bar{L} i_t^p \right) - C^p \left(\sum_{b \in \mathcal{B}} W_b y_{bt}^p + i_t^p \right) - C^m \left(\sum_{b \in \mathcal{B}} W_b y_{bt}^m \right) - C^h i_t^p \right] \tag{1}$$

$$\sum_{t \in \mathcal{T}} y_{bt}^m \leq 1 \quad \forall b \in \mathcal{B} \tag{2}$$

$$y_{bt}^p + y_{bt}^w + y_{bt}^s = y_{bt}^m \quad \forall b \in \mathcal{B},\ t \in \mathcal{T} \tag{3}$$

$$x_{nt} \leq \sum_{t' \leq t} y_{bt'}^m \quad \forall b \in \tilde{\mathcal{B}}_n, n \in \mathcal{N},\ t \in \mathcal{T} \tag{4}$$

$$\sum_{t' \leq t} y_{bt'}^m \leq x_{n't} \quad \forall n \in \mathcal{N},\ b \in \tilde{\mathcal{B}}_n,\ n' \in \hat{\mathcal{N}}_n,\ t \in \mathcal{T} \tag{5}$$

$$\sum_{b \in \mathcal{B}} W_b y_{bt}^m \leq R_{rt} \quad \forall t \in \mathcal{T},\ r \in \mathcal{R} : r = 1 \tag{6}$$

$$\sum_{b \in \mathcal{B}} W_b y_{bt}^p + i_t^s \leq R_{rt} \quad \forall t \in \mathcal{T},\ r \in \mathcal{R} : r = 2 \tag{7}$$

$$\sum_{b \in \mathcal{B}} G_b W_b y_{bt}^p + \bar{G} i_t^s \leq \hat{G} \left(\sum_{b \in \mathcal{B}} W_b y_{bt}^p + i_t^s \right) \quad \forall t \in \mathcal{T} \tag{8}$$

$$i_t^p \leq i_{t-1}^s \quad \forall t \in \mathcal{T} : t \geq 2 \tag{9}$$

$$i_t^s = \begin{cases} \sum_{b \in \mathcal{B}} W_b y_{bt}^s & t = 1 \\ i_{t-1}^s - i_t^p + \sum_{b \in \mathcal{B}} W_b y_{bt}^s & t \in \mathcal{T} : t \geq 2 \end{cases} \tag{10}$$

$$\sum_{b \in \mathcal{B}} \sum_{t' \leq t} L_b W_b y_{bt'}^s \geq \bar{L} \sum_{b \in \mathcal{B}} \sum_{t' \leq t} W_b y_{bt'}^s \quad \forall t \in \mathcal{T} \tag{11}$$

$$\sum_{b \in \mathcal{B}} \sum_{t' \leq t} G_b W_b y_{bt'}^s \leq \bar{G} \sum_{b \in \mathcal{B}} \sum_{t' \leq t} W_b y_{bt'}^s \quad \forall t \in \mathcal{T}. \tag{12}$$

$$0 \leq y_{bt}^{m}, y_{bt}^{p}, y_{bt}^{w}, y_{bt}^{s} \leq 1; \; x_{nt} \in \{0, 1\} \quad \forall b \in \mathcal{B}, \; n \in \mathcal{N}, \; t \in \mathcal{T} \tag{13}$$

The objective function is the sum of the revenues of blocks sent to the mill from the mine or the stockpile, minus the sum of the processing, mining and rehandling costs. To represent NPV, all terms are multiplied by a discount rate associated with each time period t.

Constraint (2) ensures that each block is not extracted more than once. Constraint (3) establishes that the amount of material sent to different destinations is equal to the amount of extracted material. Constraint (4) forces all of the blocks in each phase-bench to be mined by time period t if the binary variable associated to that phase-bench is set to one. Constraint (5) enforces mining precedence constraints by ensuring that all phase-bench predecessors are completely mined before the successor phase-bench.

Constraints (6) and (7) represent extraction and processing restrictions, respectively. Constraint (8) requires that the average arsenic level (ppm) of material at the mill in each time period should not exceed G. Constraint (9) ensures that the amount of material sent from the stockpile to the mill in time period t is at most the amount of material in the stockpile in time period t 1. Constraint (10) enforces inventory balance for an initial time period and a general time period t, ensuring that the amount of material in the stockpile during time period t is equal to that of the previous period plus or minus any material added or subtracted from the stockpile and sent to the mill, respectively. Constraint (11) ensures that blocks in the stockpile in any time period have an average metal grade of at least $\bar{L}$, and Constraint (12) guarantees that blocks in the stockpile in any time period have an average contaminant level of at most $\bar{G}$.

4 Results

We analyze the schedules obtained by solving $(\hat{p}^{la})$, $(\tilde{p}^{la})$ and (p^{ns}). Specially: we compare the tonnage of extracted and processed material, the average grades of the material sent to the different destinations, and the resulting NPV from these schedules.

Figure 2a compares the extraction schedule of the $(\hat{p}^{la})$, $(\tilde{p}^{la})$ and (p^{ns}) models in each time period. Although the amount of extracted material is nearly the same for all cases,

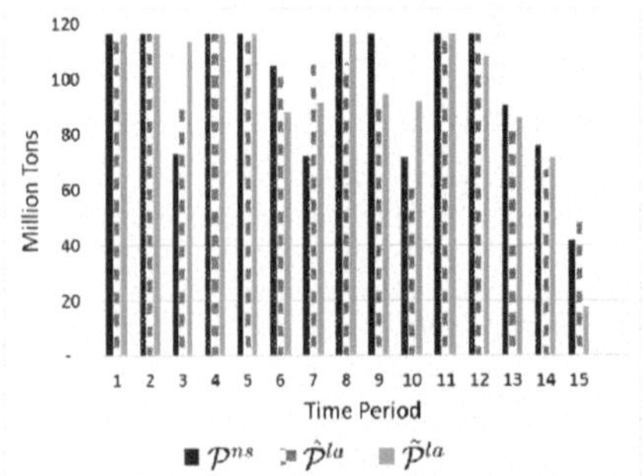

(a) Extracted tonnage from each of the three models

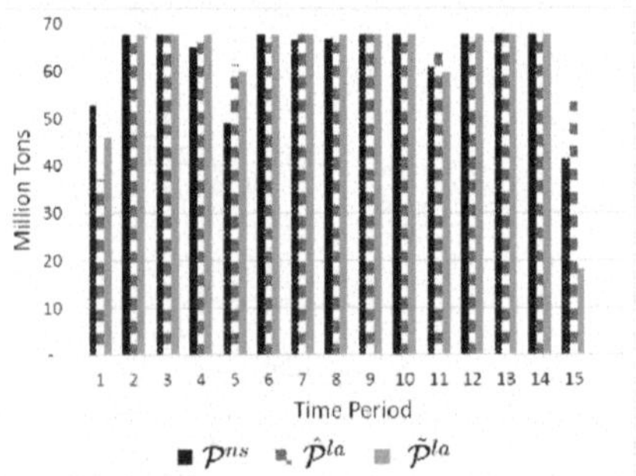

(b) Processed tonnage from each of the three models

Fig. 2. Extracted and processed tonnage comparison between the three models over the life of the mine.

Fig. 2b shows that the $(\tilde{p}^{la})$ model extracts more material than the other models early in the mine life, and in the last time period it only uses 15% of the mining capacity.

Figure 2b compares the processing schedule of three models in each time period. The discount makes it preferable to obtain pro t sooner rather than later, which is effected by the efficient use of processing capacity. The figure shows that the (p^{ns}) model has a better utilization of the mill in the rst time period, but over the life of the mine, $(\hat{p}^{la})$ processes 3.4% and 1.7% more material than the $(\tilde{p}^{la})$ and (p^{ns}) respectively. This is due to the better blending strategy of (P) which allows this model to process more material by blending it in the stockpile.

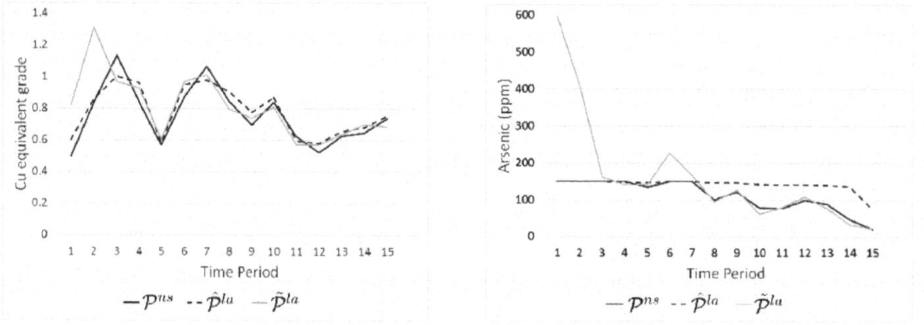

(a) Average copper-equivalent grade

(b) Average arsenic level

Fig. 3. Comparison of average grades at the mill between the three models.

Figure 3 compares the average copper-equivalent and average arsenic per ton processed at the mill between these three models in each time period. The high level of arsenic in the mine makes its limit at the plant a binding constraint for at least the rst seven time periods, requiring a good blending strategy to satisfy this requirement. Figure 3b shows that the arsenic level at the mill fluctuates over the life of the mine for $(\tilde{p}^{la})$ and violates the arsenic limit constraint at the mill. Since we do not impose this constraint in this model, the corresponding schedule is operationally infeasible. On the other hand, the mill arsenic level for $(\hat{p}^{la})$ is about the same over the life of the mine. This figure shows that the amount of arsenic at the mill decreases for (p^{ns}) over the life of the mine, but considering that $(\hat{p}^{la})$ processes more material than (p^{ns}) while not violating the arsenic limit constraint, $(\hat{p}^{la})$ has a better blending strategy. Figure 3a demonstrates that $(\tilde{p}^{la})$ processes ore with a higher copper-equivalent grade than the other models over the life of the mine but violates the arsenic limit at the mill.

5 Conclusion

We use a modified version of (p^{la}) presented by Moreno et al. (2017) to provide long-term planning for an operational open pit mine and compare its schedule with no stockpiling option model. By comparing the schedule from $(\hat{p}^{la})$ to those of $(\tilde{p}^{la})$ and

(p^{ns}), we detect that $(\hat{p}^{la})$ provides a schedule with higher NPV than (p^{ns}). The difference between the NPV of $(\hat{p}^{la})$ and (p^{ns}) represents the value addition of the stockpile. The difference between the NPV of $(\tilde{p}^{la})$ and $(\hat{p}^{la})$ represents overestimation of NPV caused by not adequately accounting for contaminants. We show that $(\hat{p}^{la})$ produces a blending strategy in the stockpile that controls more than one grade. Here, we recognize that $(\hat{p}^{la})$ uses processing plant capacity more efficiently.

References

Akaike, A., Dagdelen, K.: A strategic production scheduling method for an open pit mine. In: Proceedings of the 28th International Symposium on Application of Computers and Operations Research in the Mineral Industry (APCOM 1999), pp. 729–738. Colorado School of Mines (1999)

Bley, A., Boland, N., Froyland, G., Zuckerberg, M.: Solving mixed integer nonlinear programming problems for mine production planning with stockpiling. Optimization Online (2012). http://www.optimization-online.org/DB_HTML/2012/11/3674.html

Hoerger, S., Seymour, F., Homan, L.: Mine planning at newmont's Nevada operations. Min. Eng. **51**(10), 26–30 (1999)

Lerchs, H., Grossmann, I.: Optimum design of open-pit mines. Can. Min. Metall. Bull. **LXVIII**, 17–24 (1965)

Moreno, E., Rezakhah, M., Newman, A., Ferreira, F.: Linear models for stockpiling in open-pit mine production scheduling problems. Eur. J. Oper. Res. **260**(1), 212–221 (2017). https://doi.org/10.1016/j.ejor.2016.12.014

Impact of Geological Uncertainty at Different Stages of the Open-Pit Mine Production Planning Process

Enrique Jélvez[1(✉)], Nelson Morales[1], and Julián M. Ortíz[2]

[1] Delphos Mine Planning Laboratory, Advanced Mining Technology Center and Department of Mining Engineering, Universidad de Chile, Av. Tupper 2007, 837-0451 Santiago, Chile
enrique.jelvez@amtc.cl

[2] The Robert M. Buchan Department of Mining, Queen's University, Goodwin Hall 25 Union Street, Kingston, ON K7L 3N6, Canada

Abstract. This work addresses the long-term open-pit mine production planning. The solution for this problem indicates how and when the ore reserves will be extracted in order to maximize the value of the mining business, generating a promise that commits the mine production over time. Usually, due to the complexity of the problem, the planning process is divided into stages, generating three related problems that are sequentially solved to obtain a tentative production plan, that is: (i) determination of the final pit, which consists of delimiting the subregion of the mine where the extraction will be carried out; (ii) pushbacks selection, that corresponds to a partition of the final pit that allows to guide the sequence of extraction and to control the design; and finally, (iii) temporary production scheduling, which defines when the different zones will be extracted and which of them will be processed.

One of the disadvantages of the traditional methodology is the geological uncertainty is not taken into account, despite the great impact it can have on the production objectives. In this work we show some approaches to incorporate the uncertainty by means of conditional simulations to different stages of production planning, evaluating their impact. The results show that, on one hand, it is possible to increase the expected value of the business, and on the other hand, to reduce the risk of failure to meet production targets, allowing to generate more robust plans. In the case study presented, the results show that it is possible to obtain a discounted expected value increase of 2% and an uncertainty total cost decrease of 69% with respect to the usual methodology, which does not consider the geological uncertainty. Therefore, better decisions could be made in the long-term open pit mine production planning.

Keywords: Open-pit mine production scheduling · Geological uncertainty · Stochastic mixed integer programming · Conditional simulation

E. Topal (Ed.): MPES 2019, SSGG, pp. 83–91, 2020.
https://doi.org/10.1007/978-3-030-33954-8_9

1 Introduction

Mine planning is the discipline of mine engineering that conjugates geological resources with the market to delineate the best productive business for the mining company, generating a mining plan, which defines how and when the mining reserves will be extracted, allowing to quantify the human, economic and technical resource, generating the company's business plan. The productive promise generated in the mining plan is expressed through a production plan and is supported by a schedule, which compromises production over time. In order to generate a production plan, the orebody is discretized in blocks and the planning horizon is discretized in time periods. To generate a production plan, the planning process is divided into three stages, which are solved sequentially: (i) ultimate pit limit problem, which delimits the extraction area, (ii) pushback selection, which splits the pit limit into volumes that meet certain operational requirements, and (iii) temporary production scheduling, which assigns to each period which blocks must be extracted, respecting limits on operational resources consumption and maximizing the net present value (NPV).

Since the 1960s the methodology based on the Lerchs and Grossmann algorithm [1] has been the basis for scheduling the production of open pit mines. However, traditional methods do not consider the uncertainty associated with input parameters, such as ore grades or metal prices, which can lead to large deviations from production targets. In recent years some authors have recognized the importance of taking into account multiple sources of uncertainty and including them in the optimization of the mine production scheduling process, for example [2] have shown the consequences of considering a single block model as an input to the scheduling process, which is described as a non-linear transfer function. This is a critical issue for mine planning because the inability to quantify the impact of uncertainty on the performance of the downstream processing operations is a key reason why mining companies are often unable to meet production targets and financial forecasts [3].

Geological uncertainty represents the level of ignorance of mineralogical characterization, in particular, of the different types of material and their respective concentrations (grades), as well as of the extent and position of geological units. Since the estimates are continuous interpolations of discretely obtained data, the models do not capture the real variability of the deposit, the origin of which lies in the inherent variability of the deposit and the variability of errors in data collection, preparation and analysis. In recent years, the use of tools such as geostatistical simulation has shown better results, since they reproduce the real spatial variability of the regionalized variable. In particular, the use of conditional simulations (see, for example, [4]) has made it possible to incorporate this type of uncertainty in planning process. It has been widely reported (see [2, 5–8] and references cited there) that geology is one of the sources of uncertainty that contributes most to the differences between planned and operational solutions.

This paper assesses the impact of uncertainty in each of the three stages of the open-pit mine production planning process: (i) ultimate pit limit, (ii) pushback selection, and (iii) temporary production scheduling. Although geological uncertainty is not considered in the traditional methodology, most of the efforts made to incorporate it using optimization models have only been made within the scheduling stage, defining it within a deterministic ultimate pit, inside which there exists a set of scenarios that model uncertainty (see [9–11]). Moreover, these results do not even consider the selection of pushbacks as a stage within the process, going directly from a deterministic definition of ultimate pit limit to scheduling under uncertainty, which can generate impractical results. In this work we propose and evaluate a methodology that incorporates geological uncertainty in the three mentioned stages. For each one of them, the risk associated to the ignorance of the mineralized zone is considered.

2 Methodology

The methodology will be based on a sequential development for the long-term open-pit production planning process by considering the geological uncertainty represented by a series of $|R|$ conditional simulations of the block model B. This sequential methodology includes developing strategies for each of the three stages: (i) definition of the ultimate pit limit, (ii) pushback selection, and (iii) temporary production scheduling. The objective will be to successively incorporate the uncertainty along stages, in order to assess how the uncertainty impacts on the total process. The following cases will be considered:

- **Base Case:** it corresponds to the process that does not consider uncertainty at any stage.
- **Case 1:** it follows the approach of most of the published works, that is, incorporating uncertainty in the scheduling only (stage (iii)).
- **Case 2:** it starts from a deterministic ultimate pit limit, but then incorporates the geological scenarios for selecting pushbacks and scheduling, (stages (ii) and (iii)).
- **Case 3:** it incorporates uncertainty in the three stages of the planning process.

For each case, an expected NPV will be obtained, which can be used to see the effect of incorporating uncertainty in the different stages, for example, through the value of the information, which is interpreted as the cost of ignoring uncertainty in decision-making.

2.1 Ultimate Pit Limit Problem

The ultimate pit is found considering geological uncertainty by means of conditional simulations of metal grade, based on the development made by [12], who propose a multi-objective optimization model that seeks a balance between maximizing the expected benefit and minimizing the risk, expressed in terms of the Conditional Value at Risk (CVaR) [13].

It is assumed that there is a value v_{br} for each block b and each grade simulation r. A binary variable $x_b = 1$ is defined if block b belongs to the ultimate pit, and zero if not. The precedence relations are coded by means of a set A of pairs of arcs (a, b), where to extract block a, block b must first be extracted. Taking a level of confidence $\delta \in (0, 1]$ the ultimate pit limit problem under uncertainty is defined as:

$$\text{(P1) Max } \left(\frac{1}{R} \sum_{b \in B} \sum_{r \in R} v_{br} x_b - \left(\alpha + \frac{1}{|R|(1-\delta)} \sum_{r \in R} z_r \right) \right) \tag{1}$$

$$\text{s.t. } x_a \leq x_b \quad \forall (a, b) \in A \tag{2}$$

$$z_r \geq - \sum_{b \in B} v_{br} x_b - \alpha \quad \forall r \in R \tag{3}$$

$$z_r \geq 0 \quad \forall r \in R \tag{4}$$

$$x_b \in \{0, 1\} \quad \forall b \in B \tag{5}$$

where α and $\left(\alpha + \frac{1}{|R|(1-\delta)} \sum_{r \in R} Z_r \right)$ correspond to the discrete approximations of Value at Risk (VaR) and CVaR, respectively. It is important to note that an optimal solution $\boldsymbol{x}^*$ to the problem (P1) determines which blocks belong to the ultimate pit so that the expected value is maximized throughout all simulations, while minimizing the risk of loss measured by the CVaR. The deterministic version of this problem (P1D) considers a single representation of grade distribution such as kriging or e-type [3, 7].

2.2 Pushback Selection Problem

Let B the ultimate pit obtained from Sect. 2.1. Applying the methodology of Lerchs and Grossmann [1] and scaling the metal price in the valorization by a series of n revenue factors $0 < \lambda_1 < \cdots < \lambda_n = 1$, we have a value for each block b, each realization r and each revenue factor λ_i. Taking the average value over all realizations, we have an expected value v_b^i of each block b associated to the revenue factor λ_i. To obtain the stochastic nested pits we solve n (P1D) problems, one for every revenue factor, similar to [14] for the ultimate pit limit.

To select pushbacks from the set of stochastic nested pits, the formulation given in [15] is used, where an optimization model chooses the best pushback candidates on the basis of clearly defined criteria, for example, the minimizing the gap problem [16], so that the resulting pushbacks have the minimum difference between them in ore and waste tonnages.

For the deterministic case, the same process is repeated, but the nested pits generated are not stochastic as [14], but a single valuation is generated v_b^i for each revenue factor i, based on kriging or e-type model, for block b. The pushback selection is done by using the same model as the stochastic case [15] for a fair comparison of pushback selection.

2.3 Production Scheduling Problem

The production scheduling model is based on a mixed integer program by considering the following features:

- **Multiperiod:** The temporal dimension is included into the scheduling model, allowing to the model to decide which is the best extraction period for each block, considering a discount rate.
- **Multidestination:** The destination of each block is decided in the model, therefore a variable cutoff grade is implemented.
- **Explicitly considers geological scenarios:** The information given by the conditional simulations is directly an input to the model in one-run to be considered simultaneously, maximizing the expected discounted value and minimizing the deviations of the production targets along the set of realizations. Each deviated tonnage is penalized by a unitary deviation cost in the objective function. Taking an average on the realizations we penalize the expected NPV with a total cost due to uncertainty in the objective function.
- **Pushback design:** The schedule respects the sequence imposed by the pushbacks. A min/max lead is imposed between bench-phases.
- **Operational resource consumption constraints:** Upper and lower limits are imposed on the resource consumption such as mining and processing capacities.
- **Blending constraints:** The quality of the material sent to the plant should be controlled for an optimum processing performance, such as impurities (arsenic in copper ore, or silica in iron ore).

For a detailed description of the mathematical model see [8, Section 5.3].

2.4 Case Study

The block model B corresponds to a porphyry copper deposit, known as BM. The model consists of $|B| = 407{,}179$ blocks of $10 \times 10 \times 10$ m. Each block has information about spatial coordinates, density, $|R| = 50$ realizations of copper grade, and E-Type copper grade. The parameters to generate the economic block valuation are: Price 2.5 ($/lb), Metallurgical recovery 85%, Mining cost 3.2 ($/ton), Processing cost 9.0 ($/ton), and Selling cost 0.4 ($/lb). Regarding to slope control, slope angles of 45° were considered and no additional consideration is taken over the geomechanic in the pit walls. An ultimate pit limit is computed with a $\delta = 95\%$ level of confidence.

For the generation of nested pits the same economic valuation is considered, but scaling the metal price by a series of 90 revenue factors given by $\lambda_i = i/90$ with $i = 1, \ldots, 90$.

Finally, the additional parameters of the scheduling stage are: horizon planning 22 (year), discount rate 10%, 2 destinations (plant and waste dump), maximum mining capacity 13 (Mton/year), minimum/maximum processing capacity 6/7 (Mton/year), minimum head grade of 0.5% per year. A maximum depth of 8 benches (max lead) is allowed. The cost of under and overproduction of ore is set at 18.5 ($/ton) and the cost of underproduction of metal is set at 39 ($/ton).

3 Results

Since it is not possible to show in detail the results of tonnages, grades, operative geometries of sequences, and economic values of each stage and each case, the results will be focused individually in each stage, without considering uncertainty (deterministic) and considering uncertainty (stochastic). By sequentially doing all deterministic cases, Base Case is generated. Likewise, all stochastic stages generate the Case 3. Finally, the results of generating intermediate cases (Cases 1 and 2) are shown in Table 1 to evaluate the impact of each stage in the open-pit mine production planning process.

The ultimate pit limit for stochastic and deterministic approaches are shown in Fig. 1. Differences can be seen at the pit bottoms. Stochastic pit presents 5.3% more mineral, obtaining a larger pit when compared to deterministic approach. The economic value is almost the same, in fact, stochastic pit presents a 0.4% higher value, but the risk (CVaR) is 15.7% lower.

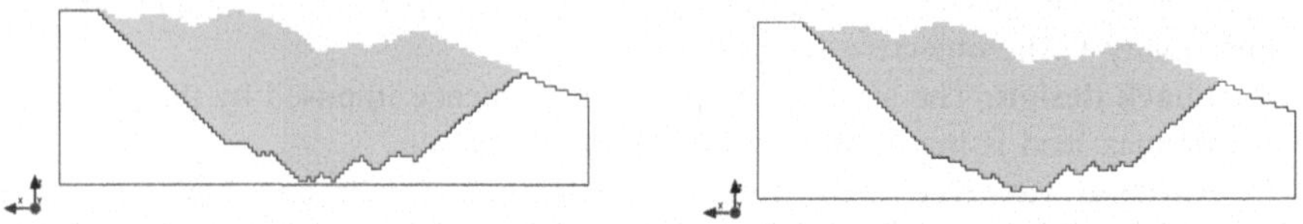

(a) N section view - 5.030m. Stochastic pit limit (b) N section view - 5.030m. Determ. pit limit

Fig. 1. Section views of alternatives for stochastic and deterministic ultimate pit limits

Figure 2 presents the main results of pushback selection. The tonnages of both approaches (stochastic and deterministic) are very similar. The same is true for the average grade per pushback. This implies that the how the uncertainty was incorporated at this stage did not have much impact on the results.

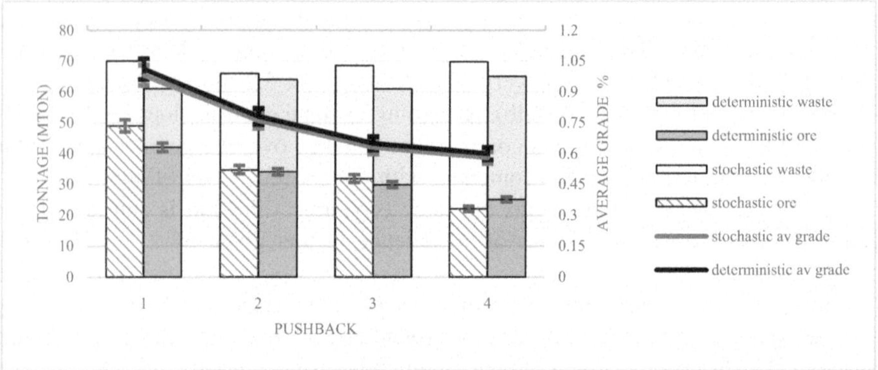

Fig. 2. Pushback selection results. Average tonnages and grades per pushback (Case 3 and Base Case). Error bars show percentiles 5th and 95th (P5 and P95).

Figure 3 shows the results of the final stage, production scheduling. As usual, the results of the Best case and Case 3 are shown, both their extraction geometries and comparative production plan, where the bars represent the percentiles P5 and P95, both in tonnage (ore and waste) and head grade. The main differences are observed in the last periods of the planning horizon, especially considering that the stochastic ultimate pit limit presents greater ore tonnages: The life of the mine is extended by one year more than deterministic case. Stochastic scheduling has 3.8 (Mton) of underproduction ore and the deterministic scheduling has 15.9 (Mton).

Based on Table 1 results, we can compute the value of the information if uncertainty is incorporated in each stage: (i) Best case v/s Case 1: 10.7 (M$); Case 1 v/s Case 2: 2.4 (M$); Case 2 v/s Case 3: 6.0 (M$). Therefore, the total value of the information in this case is 19.1 (M$).

Table 1. Results of comparison between each approach respect to Base Case.

Approach	Expected NPV (M$)	Relative variation %	Total cost due to uncertainty (M$)	Relative variation %
Base Case	904,7	–	86,5	–
Case 1	915,4	+0,9	41,8	−51,7
Case 2	917,8	+1,1	34,2	−60,5
Case 3	923,8	**+2,1**	26,7	**−69,1**

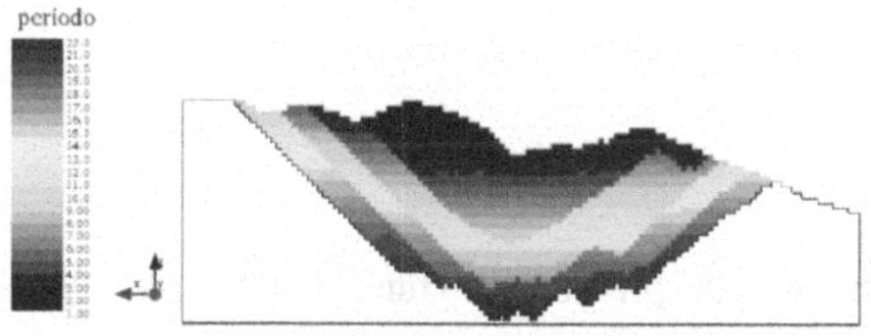

(a) N section view - 5.030m. Case 3

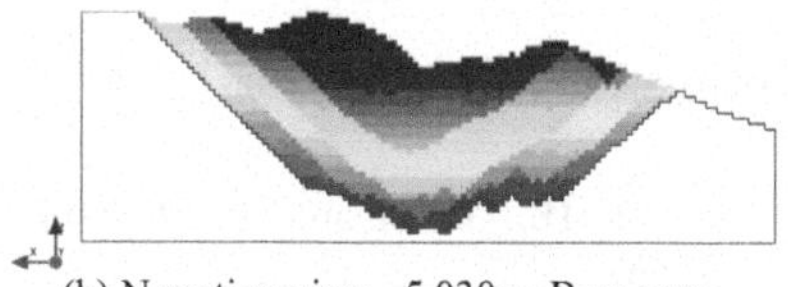

(b) N section view - 5.030m. Base case

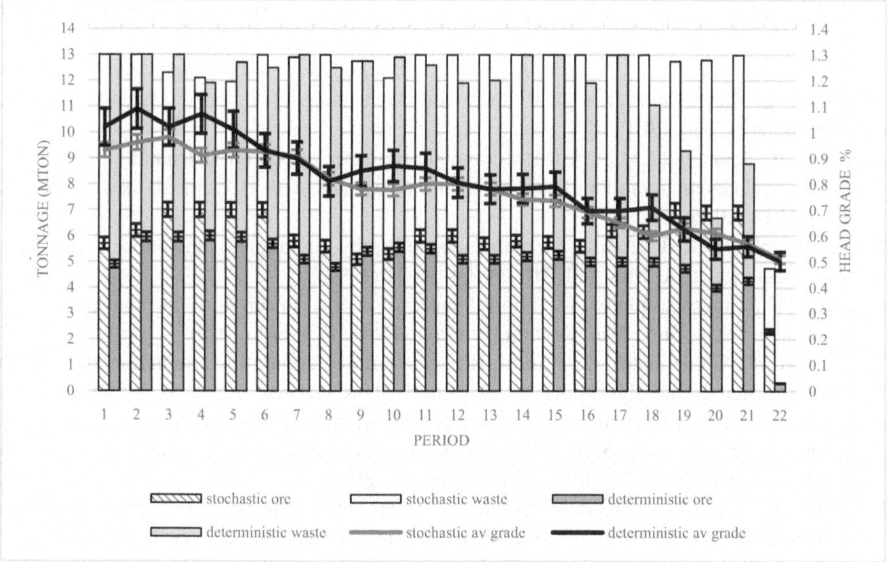

(c) Production scheduling: stochastic and deterministic

Fig. 3. Scheduling results. Stochastic (Case 3) and deterministic (Base case)

4 Conclusions

In this work we study the impact of incorporating geological uncertainty in each of the three stages that are sequentially solved to generate a production plan in strategic open-pit mines: (i) ultimate pit limit, (ii) pushback selection, and (iii) temporary production scheduling. Four cases were generated for evaluation, depending what stage the uncertainty is introduced: Base case, without uncertainty; Case 1, uncertainty included only in stage (iii), Case 2, uncertainty included in stages (ii) and (iii), and Case 3, where uncertainty is considered in all stages.

Results show that incorporating uncertainty helps to reduce risk of losses due to noncompliance with production targets. Likewise, the three stages contribute in different proportions to the total value of the information, in this case study: ultimate pit limit contributes 31.4%, pushback selection contributes 12.6%, and production scheduling contributes 56%. This is very important because it helps to identify where to concentrate efforts and up to what amounts it is justified to pay for more information to reduce this uncertainty, e.g. more drillings in specific areas of the mine.

Acknowledgements. Enrique Jélvez and Nelson Morales were supported by CONICYT/PIA Project AFB180004 – Advanced Mining Technology Center – Universidad de Chile.

References

1. Lerchs, H., Grossmann, I.: Optimal design of open-pit mines. Trans. C.I.M. **68**, 17–24 (1965)
2. Dimitrakopoulos, R., Farrelly, C., Godoy, M.: Moving forward from traditional optimization: grade uncertainty and risk effects in open-pit design. Min. Technol. **111**(1), 82–88 (2002)
3. Morales, N., Seguel, S., Cáceres, A., Jélvez, E., Alarcón, M.: Incorporation of geometallurgical attributes and geological uncertainty into long-term open-pit mine planning. Minerals **9**(2), 108 (2019)
4. Emery, X., Lantuéjoul, C.: TBSIM: a computer program for conditional simulation of three-dimensional Gaussian random fields via the turning bands method. Comput. Geosci. **32**(10), 1615–1628 (2006)
5. Ravenscroft, P.: Risk analysis for mine scheduling by conditional simulation. Trans. Inst. Min. Metall. Sect. A. Min. Ind. **101**, A104–A108 (1992)
6. Smith, M., Dimitrakopoulos, R.: The influence of deposit uncertainty on mine production scheduling. Int. J. Surf. Min. Reclam. Environ. **13**(4), 173–178 (1999)
7. Osanloo, M., Gholamnejad, J., Karimi, B.: Long-term open pit mine production planning: a review of models and algorithms. Int. J. Min. Reclam. Environ. **22**(1), 3–35 (2008)
8. Jélvez, E.: Metodología multietapa para la planificación de la producción de largo plazo en minas a rajo abierto bajo incertidumbre geológica. Ph.D. thesis, Departamento de Ingeniería de Minas, Universidad de Chile, Santiago, pp. 1–189 (2017)
9. Godoy, M.: The efficient management of geological risk in long-term production scheduling of open pit mines. Ph.D. thesis, University of Queensland, Brisbane, pp. 1–256 (2003)
10. Dimitrakopoulos, R., Ramazan, S.: Stochastic integer programming for optimising long term production schedules of open pit mines: methods, application and value of stochastic solutions. Min. Technol. **117**(4), 155–160 (2008)

11. Mai, N.L., Topal, E., Erten, O., Sommerville, B.: A new risk-based optimisation method for the iron ore production scheduling using stochastic integer programming. Resour. Policy **62**, 571–579 (2018)
12. Jélvez, E., Morales, N., Ortíz, J.M.: Stochastic ultimate pit limit: an efficient Frontier analysis under geological uncertainty (submitted)
13. Rockafellar, R., Uryasev, S.: Optimization of conditional value-at-risk. J. Risk **2**, 21–42 (2000)
14. Marcotte, D., Caron, J.: Ultimate open pit stochastic optimization. Comput. Geosci. **51**, 238–246 (2013)
15. Jélvez, E., Morales, N., Askari-Nasab, H.: A new model for automated pushback selection. Comput. Oper. Res. (2018, in press)
16. Meagher, C., Dimitrakopoulos, R., Avis, D.: Optimized open pit mine design, pushbacks and the gap problem: a review. J. Min. Sci. **50**(3), 508–526 (2014)

Review of Mathematical Models Applied in Open-Pit Mining

J. Githiria(✉)

School of Mines and Engineering, Taita Taveta University, Voi, Kenya
j.muchiri07@ttu.ac.ke

Abstract. Mining is the art of extraction of material from the earth surface for profit. The main objective of mining companies is to maximize net present value (NPV) while considering health, safety and environmental issues. Revolution in the mining industry has seen mathematical modelling being the key component in improving the mining process. Mining models has been proven as the next frontier that needs all the focus it can get. There are three mining stages namely: mining or extraction, processing and refining involved in this exercise. The major optimisation problems found in surface mining operations are ultimate pit limit problem, cut-off grade optimisation and open pit production scheduling. In order to maximise the output from the three mining stages, there is need to apply mathematical equations and models to optimise the whole process. There are extensive research studies since 1960s on development of mathematical models optimising mining operations. However, the mining industry has not fully appreciated and applied all these optimisation models. This paper highlights the strides made in development of mathematical models in the mining industry and their application. It discusses ways in which the mining industry can increase the application of these models to improve the output generated from the mining projects.

Keywords: Open-pit mining · Mine planning · Optimisation · Mathematical model · Mining 4.0 · Operation Research · Net present value

1 Introduction

Mining has reached a stage where mine planners have embraced optimisation models when planning their operations. Mine planning for any mine operation (surface or underground) is aimed at maximising the value realized from the ore body. Consequently, most mines use computer-programming tools to optimise the ore extraction processes. The tools are used minimise costs and maximise profits from the whole mining system. This paper aims to highlight deterministic and stochastic mathematical models developed and applied in the mining industry.

There are extensive research studies done on the development of complex mathematical models optimising mining operations, but they are not fully appreciated and applied in the mining industry. The application of mathematical models in daily mining operations remains a mystery to mining engineers in most mines due to the complexity in their computation processes. Since the 1960s, major development of mathematical

E. Topal (Ed.): MPES 2019, SSGG, pp. 92–102, 2020.
https://doi.org/10.1007/978-3-030-33954-8_10

models applied have been developed to solve mining problems in both surface and underground mining. Generally, these mining-related problems involve exploration, mining, processing and refining. It is worth noting that exploration has not been fully developed as a problem, however, it is imperative to develop algorithms to determine how and where to explore. Currently, optimisation problems are categorised into surface and underground mining problems with three major stages: mining, processing and refining. Most optimisation problems in surface mining operations are found when undertaking block economic evaluation, cut-off grade optimisation, pit design and production scheduling. On the other hand, when optimising underground mining operations the major concern is optimising development, stope outlining, and ore production scheduling in a single run.

Mining companies must advocate and develop global strategic mine plans that optimise both surface and underground operations to manage any changes that may affect the operations according to their profiles of operations. In order for mine plans to incorporate such changes, optimisation methods using Operation Research (OR) techniques need to be adopted and applied (Musingwini 2016). Operation Research (OR) is a mathematical science commonly applied to optimisation problems that are concerned with improving the performance of a real-world system. It maximises profit, performance, or yield or minimises loss, risk, or cost of a real-world objective (Musingwini 2016). Some of the techniques applied in the mining industry to solve optimisation problems are linear programming techniques, heuristic methods, dynamic programming and stochastic programming.

There are five main categories of OR models namely: optimisation models, simulation models, network models, multi-criteria decision-making (MCDM) model and global optimisation model (Musingwini 2016). Developing an algorithm is the first step when coming up with a mathematical model to optimise mining operations both in surface and underground mining. An algorithm is a systematic procedure to accomplish a specific task. It is the idea behind any computer program. Application of programming languages such as C++ and Python are used to solve algorithm design problems.

1.1 Relevance of Application of Mathematical Models in Open-Pit Mining

Several algorithms have been used for over 30 years for open pit mine design such as LG 3D algorithm, Lane's cut-off grade theory, among others. They are well known and have been implemented in commercial software such as Whittle, MaxiPit, Maptek™, Minemax, Minesight, GEOVIA Surpac™, Talpac, Arena, MINVEST software package (Dowd and Xu 2000), and LINGO software (Yasrebi et al. 2015). Some mathematical models have been developed based on the specific mining operation such OptiPit (Dagdelen and Kahawata 2008), Cut-off Grade Optimiser (Githiria et al. 2016) and NPVMining (Githiria and Musingwini 2019).

Lately, there has been an increase in the development of mathematical optimisation models in solving stochastic optimisation problems in the mining industry. For instance, network flow algorithms such push-relabel (Goldfarb and Chen 1997) and pseudoflow (Hochbaum and Chen 2000) could theoretically solve the pit optimisation problem more efficiently and some have been implemented commercially like

MineMax which uses push-relabel algorithm. Lane's cut-off grade approach has been incorporated in several commercial software such as GEOVIA Whittle™. Myburgh et al. (2014) applied modern heuristics in an evolutionary algorithm to come up with a mine planning software package, Maptek™ Evolution, which maximises net present value through cut-off grade optimisation. This paper shows the relevance of mathematical modelling and application of computer-aided application in optimising open-pit mining operations to provide an optimal strategic mine plan.

2 Mathematical Models Applied in Open-Pit Mining

The development and implementation of mathematical formulations to solve complex optimisation problems in mine planning has been progressive, especially in open-pit mining. Algorithms developed in surface mining involve production scheduling which embodies several major aspects that affect the overall mine output. These include (i) cut-off grade optimisation, (ii) block economic evaluation, (iii) slope design and requirements, (iv) pit design and optimisation and (v) pushback design.

An ideal mining system constitutes of a mine, ore processing stream and metal refinery, which are optimally planned to avoid losses (Githiria et al. 2016). Modelling of these optimisation problems related to mining operations follows the same thought-process. They follow the same conventional mining layout as shown in Fig. 1. The changes arising along the process depend on the nature of the problem, if surface or underground mining related.

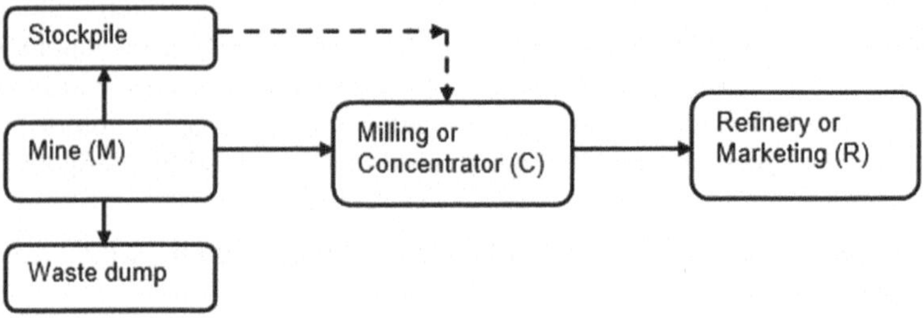

Fig. 1. Layout of an ideal mining system (Githiria and Musingwini 2019)

2.1 Cut-off Grade Calculation and Optimisation

Cut-off grade is the grade that is normally used to distinguish between ore and waste within a given ore body. It also distinguishes between various ore types before processing takes place for different metallurgical processing options. The development of cut-off grade models can be traced back to the 1960s in the published work of Henning (1963), Johnson (1969), Lane (1964, 1988), among others. Henning (1963) created a framework of relationships for cut-off grade calculation with varying objectives. Lane (1964, 1988) came up with a 3D algorithm in calculating cut-off grades. Lane's deterministic theory accounts for economic and geological parameters and production capacities in the calculation of cut-off grades. This theory uses mathematical derivations to get six cut-off grades, which are then sorted using a sorting algorithm to get an optimum cut-off grade. Three of the cut-off grades are based upon costs, price and mining, processing and refining capacities with each independently constraining throughput. The other three are balancing cut-off grades that are determined by assuming that two of the three stages are operating in unison at their capacity limits. They are based on the grade, and capacities of the production stages. The rate of production in a mining system is determined by the production capacities (Githiria and Musingwini 2018). This cut-off grade approach has been modified in several studies by incorporating different mining scenarios. Research studies done by Asad (2002), Dagdelen and Kawahata (2008), Whittle and Wharton (1995), Gholamnejad (2009) and Githiria et al. (2016) applied and modified Lane's deterministic approach to incorporate certain aspects in a mining operation.

Stochastic Approach on Cut-off Grade Optimisation. Several studies (Dowd 1976; Cetin and Dowd 2016; Asad and Dimitrakopoulos 2013; Li et al. 2012; Myburgh et al. 2014) have applied stochastic and/or dynamic programming approaches in the calculation of cut-off grades. The models applied the dynamic aspect of certain parameters affecting daily mining operations such as price and costs.

Lane's deterministic approach uses mathematical formulation in selecting cut-off grades by maximising net present value (NPV) subject to mining, processing and refining constraints. In modifying this approach, there is need to account for the uncertainty in metal price and grade-tonnage distribution. Given a set of equally probable grade-tonnage curves (w), the model develops the cut-off grade policy by determining cut-off grade (G) from periods 1 to T. The objective function of this cut-off grade optimisation model is represented as in Eq. 1:

$$Maximise\,NPV = \sum\nolimits_{i=1}^{N} \frac{P_{wi}}{(1+d)^{i}} \tag{1}$$

However, the cash flow equation (P_{wi}), as represented in Eq. 2, changes to incorporate the uncertainty of both the economic and geological parameters.

$$Cash\,flow, P_{wi} = \left(\sum\nolimits_{i=1}^{N} s_i - r_i\right) * Qr_{wi} - c_i * Qc_{wi} - m_i * Qm_{wi} - f_i \tag{2}$$

Subject to: ($Qm_{wi} \leq M$ for $i = 1\ldots T$), ($Qc_{wi} \leq C$ for $i = 1\ldots T$), and ($Qr_{wi} \leq R$ for $i = 1\ldots T$).

Maximising the NPV is the objective function and is represented as Eq. 1 where annual profit (P_{wi}) in Eq. 2 is consequently discounted to produce the NPV. Decision variables/parameters used in the model are divided into economic and geological parameters. Economic parameters include price of metal (s_i), mining cost (m_i), processing cost (c_i), refining/selling/marketing cost (r_i), fixed/period cost (f_i), recovery (y) and discount rate (d).

The geological and operational parameters are defined by the grade-tonnage distribution inside the ultimate pit limit or pushbacks. They include optimum cut-off grade (G), average grade (g), quantity of material corresponding to a specific grade (q), total quantity of material inside the pit or pushback (Q), quantity of material mined (Q_{mi}), quantity of material processed (Q_{ci}) and quantity of metal refined (Q_{ri}). M, C and R represent mine, concentrator and refining capacities, respectively.

The constraints that bound the decision variables in this study consist of equalities and inequalities represented as follows: ($Q_{mi} \leq M$ for i = 1…T), ($Q_{ci} \leq C$ for i = 1…T), and ($Q_{ri} \leq R$ for i = 1…T). The variables are governed by a set of non-negativity restrictions: $Q_1, Q_2 \ldots Q_n \geq 0$ where $Q_1, Q_2 \ldots Q_n \in [0, 1]$.

Some studies have applied modern heuristics and integrated them with mathematical techniques to develop cut-off grade optimisation models that maximise NPV. Such models employ evolutionary algorithms in developing the cut-off grade model that caters for the dynamic nature and complexity of mining parameters (Myburgh et al. 2014). The algorithms stochastically produce high-quality results quickly through several iterations.

From the literature, it is evident that stochastic cut-off grade models generate higher NPV than deterministic models. For instance, the cut-off grade models compared in a study undertaken by Githiria and Musingwini (2019) showed that stochastic approaches generate higher NPVs than deterministic and conventional methods. NPVMining (Githiria and Musingwini 2019) and Maptek Evolution® produced approximately 30% higher NPV than OptiPit® (Dagdelen and Kawahata 2008), Cut-off Grade Optimiser (Githiria et al. 2018) and break-even cut-off grade model (Table 1).

Table 1. Comparison of the mine life, cash flow and NPV from the cut-off grade models (Githiria and Musingwini 2019)

	Deterministic cut-off grade approaches			Stochastic cut-off grade approaches	
	Break-even cut-off grade model	Heuristic cut-off grade model	Cut-off grade optimiser	Maptek evolution®	NPVMining
Mine life (years)	35	18	10	10	9.12
Cashflow (US$ million)	863	885.60	825.44	760.10	887.09
NPV (US$ million)	163.42	347.08	435.52	413.84	467.17

2.2 Block Economic Evaluation

The basis of calculation for block economic evaluation is the cut-off grade and net value of material being mined. Net value of material is considered as a function of grade, whereby, the grade at which net value is equal to zero is breakeven cut-off grade. Block economic values vary when the material mined has either single or multiple material destinations along the conventional layout of an ideal mining system.

The block economic value (BEV) formula in Eq. 3 is used to represent the revenue generated after mining each block. The cost of sending the waste material to the waste dump is considered in the mining cost.

$$BEV = T_oGRP - T_oC_p - TC_m \tag{3}$$

where:

BEV—is block value, \$, T_o—is tonne of ore in the block, G—is grade, unit/tonne, R—is recovery, P—is unit price, \$/unit, C_p—is processing cost, \$/tonne, T—is total amount of rock (ore and waste) in the block, C_m—is mining cost, \$/tonne.

Block economic value calculations generally apply deterministic conventional approaches based on geological, technical and economic parameters. The ore grade or metal content of each block and economic variables such as metal prices and operating costs are known with certainty. The block economic values are calculated at the present time like a static parameter and is assumed to be constant throughout the mine life. This formula is incorporated in most of the computer-aided applications such as Whittle 4X (an open-pit optimisation software) used in mining industry.

There are various BEV formulae depending on the specifications of the mine. Ataee-pour (2005) introduced a linear programming approach for determination of block values in underground metalliferous mines. There are different factors that influence the economic value of a block which are the location of blocks, when the blocks will be mined, and mining method applied. Location of the blocks represent the distance of the mineable blocks to the processing plant and waste dump site. The time value of money which is specified by when the blocks are mined depends on the revenue and operational costs discounted by a factor that increase over time. The economic value of a block is not the same in both surface and underground mining. The mining method selected, especially in underground mining, affects the block values due to different mining activities involved in the operation.

It is evident from the calculations that it is tedious, time consuming and prone to mistakes to calculate the BEVs for a large mining operation. Mathematical models have been developed to ease the calculation process and make it accurate especially when uncertainty in the grade-tonnage distribution is incorporated. The output from the calculation is useful for feasibility studies and preliminary resource appraisal.

Uncertainty on the Block Economic Values. When calculating BEV, the grade of the ore is commonly considered to be static which is not the case. The grade-tonnage distribution of the orebody varies thus the need to incorporate develop models that will cater for such uncertainty. Henry et al. (2005), Dimitrakopoulos and Abdel Sabour (2007) and other researchers have undertaken several studies on the effect of both geological and market uncertainty on BEVs. BEV is dynamic in nature because metal price and operating costs are dynamic over time. The value of each block must be determined according to their mining time along the mining sequence. The block economic value under uncertainty is calculated using a binomial tree to solve Eq. 3 (Dehghani and Ataee-pour 2012).

$$BEV_n = T_o\left(G_{nw}R_nP_{ni} - C_{pnj}\right) - TC_{mnj} \tag{4}$$

where:

BEV—is block value, \$, T—is block tonnage, To—is tonne of ore in the block, n—is the period number, R—is the recovery, P—is unit price, \$/unit, C_p—is processing cost, \$/tonne, C_m—is mining cost, \$/tonne, G_{nw}—is the block grade in a set of equally probable grade-tonnage curves (w), P_{ni}—is the ith price in the nth period in metal price variations, C_{pnj}—is the jth processing cost in the nth period in total cost variations, C_{mnj}—is the jth mining cost in the nth period in total cost variations.

2.3 Pit Design and Optimisation

Ultimate pit is the pit with the highest profit value defining the size and shape of the mine after depleting all the economic resources. The main factors considered in designing a feasible pit outline are the slope, block and bench requirements. There is no block that can be added or removed to/from a feasible pit outline that will increase the value without violating the slope constraints. Other mining operational parameters that are determined when designing the ultimate pit limit include width, length and depth of mined pit, haul roads, location of waste dump, stripping ratio, mine life, minable ore tonnage, waste tonnage and production scheduling.

Some of the ultimate pit limit methods are floating or moving cone (Carlson et al. 1966), other versions of Floating cone (Floating cone II and III) (Wright 1999; Khalokakaie 2006; Elahizeyni et al. 2011), Lerch-Grossman (LG) approach (Lerch and Grossman 1965), Pseudoflow network model (Hochbaum and Chen 2000), Korobov algorithm (David et al. 1974), dynamic programming methods (Roman 1974; Koenigsberg 1982; Yamatomi et al. 1995) and maximum flow algorithm (Johnson 1969), Zhao and Kim algorithm (Zhao and Kim 1992), Korobov algorithm (David et al. 1974), variations of Lerch-Grossman algorithm using graph theory, and other methods/models based on linear programming.

Floating cone is one of the common pit limit calculation methods that entails a series of interlocking frustum shaped removal increments. This method works on an economical block model of the deposit whereby a cone is constructed with sides oriented parallel to the pit slope angles for each positive (ore) block. The value of the cone is then determined by summing the values of blocks enclosed within it. If the value of the cone is positive, all blocks within the cone are mined. This process starts from top to bottom level searching for positive blocks until no positive cones remain in the block model (Carlson et al. 1966). It can be easily computerized and thus is easy to understand by engineers coupled with a simple computational algorithm.

The Lerch-Grossman (L-G) algorithm obtains the ultimate pit limits and pushback regions, and then obtains scenarios of subsequent production scheduling that should be followed to maximise the mine value with emphasis on maintaining consistent production. Mining software programmes such as Whittle 4X have incorporated the algorithm. L-G 2D algorithm is based on dynamic programming while the 3D version of the algorithm is based on graph theory.

Recently, the algorithms developed based on network flow are more efficient than the LG algorithm with examples of the network flow algorithms that include push-relabel and pseudoflow algorithms. These algorithms are based on the LG algorithm and incorporate the lowest label and highest label methods. The highest label methods are consistently faster than the lowest label methods and the generic LG methods. They can reduce the solution time by increasing the speed of calculation from 2 to 50 times faster than the LG methods. They are theoretically much faster for larger problems. Hochbaum and Chen (2000) generalised the LG algorithm to a pseudoflow network model which proved to be theoretically more efficient than the LG algorithm since it did not allow selection of constraints in its calculations. Pseudoflow models give new life to the Lerchs-Grossmann pit optimisation (Muir 2004). Due to the uncertainty in the operational and economic conditions in the mining industry there is a need to adopt stochastic optimisation approaches in the calculation of the best pit limits.

2.4 Production Scheduling

Production scheduling is the basis of preparing and controlling the mine's development and production with an objective of maximizing the net present value of future cash flows. It determines the quantities of ore that can be mined, processed and refined in any given period under specified production capacities, mining and milling capacities. There are two different ways of scheduling the mine's operations. It is either done by pushback/cutback/phase design or block-by-block production schedule. The constraints considered when designing the schedule are mining, processing and refining capacities and slope or block precedence. It is an iterative process considering all these constraints and the labour and equipment requirements.

When mining operation starts, the schedule considers input and output generated during the development stage, preproduction period and production period. Waste mined in the preproduction period is accounted for in the calculations. There is need to maintain a sustainable supply of ore to the processing facilities throughout the life of operation. The iterative process involved in production scheduling has been simplified by the use of mathematical models that identifies the optimal sequence of extraction and chooses the best combination of blocks to mined at a particular period.

Milawa NPV, Milawa Balanced and Fixed lead algorithms are some of the algorithms applied in production scheduling using Whittle 4X software. They implement mathematical models that combine pit shells to generate a pushback design. Using Whittle 4X along with calculated mining parameters, the periodic production and mine value can be predetermined thereby eliminating guesswork and its subsequent effects on mining results (Whittle 1999).

Optimisation tools help in tackling mining problems to generate strategic and global solutions in production scheduling such as blending problem, product mix problem, transportation problem and flow capacity problem.

3 Conclusion

Most of the mine planning methods used in the mining industry are implemented on the basis of estimated models and parameters, without being influenced by the uncertainty present in the values of these inputs. Several approaches take this uncertainty into account when generating solutions, with the aim of forming mine plans that are robust under different realisations of the underlying uncertainty. In general, the design and planning tasks for mining deposits using optimisation tools involves geological modelling, pit limit design, optimum cut-off grade selection, production scheduling and valuation.

There are major insights that can be drawn from the mathematical models discussed and compared in this study. The mathematical models designed to undertake block economic evaluation, cut-off grade optimisation, pit design and production scheduling should take into account the uncertainty of the mining parameters to improve the output (NPV). Uncertainty in mining is generated from quantitative factors such as grade-tonnage distribution, price fluctuations and variations of operational costs. These factors are numerical outcomes from which a decision can be measured when undertaking financial analysis in a mining project. These parameters fluctuate with time hence the need to develop stochastic approaches that will account for uncertainty. Qualitative factors such as political stability, technological advancement and social impacts are decision outcomes that cannot be measured but should be considered when undertaking risk and decision analysis.

From the work done on cut-off grade optimisation by Githiria and Musingwini (2019), it is evident that the NPV generated from most of the developed stochastic models have improved the NPV by 25% on average. Osanloo et al. (2008) and Samavati et al. (2018) reviewed of major production scheduling models and algorithms in open-pit mining and demonstrated the advantages and disadvantages of the models. The stochastic models generate maximum NPV and schedules blocks with discounted uncertainty of the mining variables such as commodity price, mining costs, processing cost, investment required, grade, tonnages. The algorithms converge quickly to optimal solution compared to deterministic approaches. The self-adaptiveness of such stochastic algorithms contributes greatly to its ability to successfully address complex real-world problems.

References

Asad, M.W.A., Dimitrakopoulos, R.: A heuristic approach to stochastic cut-off grade optimisation for open pit mining complexes with multiple processing streams. Resour. Policy **38**, 591–597 (2013)

Asad, M.W.A.: Development of generalized cut-off grade optimisation algorithm for open pit mining operations. J. Eng. Appl. Sci. **21**(2), 119–127 (2002)

Ataee-pour, M.: A linear model for determination of block economic values. In: The 19th International Mining Congress and Fair of Turkey, İzmir, pp. 289–294 (2005)

Carlson, T.R., Erickson, J.D., O'Brain, D.T., Pana, M.T.: Computer techniques in mine planning. Min. Eng. **18**(5), 53–56 (1966)

Cetin, E., Dowd, P.A.: Multiple cut-off grade optimization by genetic algorithms and comparison with grid search method and dynamic programming. J. South Afr. Inst. Min. Metall. **116**(7), 681–688 (2016)

Dagdelen, K., Kawahata, K.: Value creation through strategic mine planning and cut-off grade optimization. Min. Eng. **60**(1), 39–45 (2008)

David, M., Dowd, P., Korobov, S.: Forecasting departure from planning in open pit design and grade control. In: Proceedings of the 12th Symposium on the Application of Computers and Operations Research in the Mineral Industries (APCOM), vol. 2, pp. F131–F142. Colorado School of Mines, Golden (1974)

Dehghani, H., Ataee-pour, M.: The role of economic uncertainty on the block economic value – a new valuation approach. Arch. Min. Sci. **57**(4), 991–1014 (2012)

Dimitrakopoulos, R.G., Abdel Sabour, S.A.: Evaluating mine plans under uncertainty: can the real options make a difference? Resour. Policy **32**(3), 116–125 (2007)

Dowd, P.A., Xu, C.: MINVEST User's Manual, p. 272. Computer Aided Mine Design Centre, University of Leeds (2000)

Dowd, P.A.: Application of dynamic and stochastic programming to optimise cut-off grades and production rates. Trans. Inst. Min. Metall. Sect. A. Min. Technol. **85**(1), 22–31 (1976)

Elahizeyni, E., Kakaie, R., Yousefi, A.: A new algorithm for optimum open pit design: floating cone method III. J. Min. Environ. **2**(2), 118–125 (2011)

Gholamnejad, J.: Incorporation of rehabilitation cost into the optimum cut-off grade determination. J. South Afr. Inst. Min. Metall. **108**(2), 89–94 (2009)

Githiria, J., Musingwini, C.: Comparison of cut-off grade models in mine planning for improved value creation based on NPV. In: Proceedings of the 6th Regional Conference of the Society of Mining Professors (SOMP), Johannesburg, pp. 347–362 (2018)

Githiria, J., Musingwini, C.: A stochastic cut-off grade optimisation model to incorporate uncertainty for improved project value. J. South Afr. Inst. Min. Metall. **119**(3), 1–12 (2019)

Githiria, J., Muriuki, J., Musingwini, C.: Development of a computer-aided application using Lane's algorithm to optimise cut-off grade. J. South Afr. Inst. Min. Metall. **116**(11), 1027–1035 (2016)

Goldfarb, D., Chen, W.: On strongly polynomial dual simplex algorithms for the maximum flow problem. Spec. Issue Math. Program. B **78**(2), 159–168 (1997)

Henning, U.: Calculation of cut-off grade. Can. Min. J. **84**(3), 54–57 (1963)

Henry, E., Marcotte, D., Samis, M.: Valuing a mine as a portfolio of European call options: the effect of geological uncertainty and implications for strategic planning. Geostat. Banff. Quant. Geol. Geostat. **14**, 501–510 (2005)

Hochbaum, D.S., Chen, A.: Performance analysis and best implementations of old and new algorithms for the open-pit mining problem. Oper. Res. **48**(6), 894–914 (2000)

Johnson, T.B.: Optimum open-pit mine production scheduling. In: Weiss A. (ed.) A Decade of Digital Computing in the Mineral Industry, pp. 539–562. AIME, New York (1969)

Khalokakaie, R.: Optimum open pit design with modified moving cone II methods. J. Eng. Tehran Univ. **4**(3), 297–307 (2006)

Koenigsberg, E.: The optimum contours of an open pit mine: an application of dynamic programming. In: Proceedings of the 17th Symposium on the Application of Computers and Operations Research in the Mineral Industries (APCOM), pp. 247–287. AIME, New York (1982)

Lane, K.F.: Choosing the optimum cut-off grade. Colorado Sch. Min. Q. **59**(4), 811–829 (1964)

Lane, K.F.: The Economic Definition of Ore: Cut-off Grade in Theory and Practice, p. 149. Mining Journal Books, London (1988)

Lerchs, H., Grossmann, I.F.: Optimum design of open pit mines. Trans. CIM **68**, 17–24 (1965). Joint CORS and ORSA Conference, Montreal, May

Li, S., Yang, C., Lu, C.: Cut-off grade optimization using stochastic programming in open-pit mining. In: IEEE, pp. 66–69 (2012). https://doi.org/10.1109/icicse.2012.24

Muir, D.C.W.: Pseudoflow, new life for Lerchs-Grossmann pit optimisation. Orebody Model. Strateg. Mine Plan. Spectr. Ser. **14**, 97–104 (2004)

Musingwini, C.: Presidential address: optimization in underground mine planning – developments and opportunities. J. South Afr. Inst. Min. Metall. **116**(9), 1–12 (2016)

Myburgh, C.A., Deb, K., Craig, S.: Applying modern heuristics to maximising net present value through cut-off grade optimisation. In: Proceedings of Orebody Modelling and Strategic Mine Planning Symposium, Melbourne, pp. 155–164 (2014)

Osanloo, M., Gholamnejad, J., Karimi, B.: Long-term open pit mine production planning: a review of models and algorithms. Int. J. Min. Reclam. Environ. **22**(1), 3–35 (2008). https://doi.org/10.1080/17480930601118947

Roman, R.J.: The role of time value of money in determining an open pit mining sequence and pit limits. In: Proceedings of the 12th International Symposium on Application of Computers and Operation Research in the Minerals Industries (APCOM), pp. 72–85 (1974)

Samavati, M., Essam, D., Nehring, M., Sarker, R.: Open-pit mine production planning and scheduling: a research agenda. In: Data and Decision Sciences in Action, pp. 221–226 (2018)

Whittle, J.: A decade of open pit mine planning and optimisation – the craft of turning algorithms into packages. In: Proceedings APCOM 1999 Computer Applications in the Minerals Industries 28th International Symposium, Colorado School of Mines, Golden, pp. 15–24 (1999)

Whittle, J., Wharton, C.: Optimising cut-offs over time. In: Proceedings of the 25th International Symposium on the Application of Computers and Mathematics in the Mineral Industries, Brisbane, pp. 261–265 (1995)

Wright, E.A.: MOVING CONE II - a simple algorithm for optimum pit limits design. In: Proceedings of the 28th Symposium on the Application of Computers and Operations Research in the Mineral Industries (APCOM), pp. 367–374 (1999)

Yamatomi, J., Mogi, G., Akaike, A., Yamaguchi, U.: Selective extraction dynamic cone algorithm for three-dimensional open pit designs. In: Proceedings of the 25th Symposium on the application of computers and operations research in the mineral industries (APCOM), Brisbane, pp. 267–274 (1995)

Yasrebi, A.B., Wetherelt, A., Foster, P., Kennedy, G., Ahangaran, D.K., Afzal, P., Asadi, A.: Determination of optimised cut-off grade utilising non-linear programming. Arab. J. Geosci. **8**(10), 8963–8967 (2015)

Zhao, H., Kim, Y.C.: A new optimum pit limit design algorithm. In: Proceedings of the 23rd International Symposium on the Application of Computers and Operations Research in the Mineral Industries, pp. 423–434 (1992)

Mine Planning and Optimisation Techniques Applied in an Iron Ore Mine

Moore Theresa Malisa and Bekir Genc(✉)

University of the Witwatersrand, Johannesburg 2050, South Africa
bekir.genc@wits.ac.za

Abstract. The surface mine planning and optimisation techniques such as block modelling, pit optimisation, scheduling, stockpiling and reconciliation that are and have been applied at an existing iron ore open pit operation are discussed in this report. The material of economic interest is separated as ON grade, OFF grade and Waste. There is a Mineral Resource Management (MRM) department that includes geology, rock engineering and planning; however, survey is excluded from the MRM structure, which is outside the traditional MRM hierarchy. Mine planning begins with a block model which is created through the Datamine software using information from exploration drilling. The block model is used in the whittle software to obtain an ultimate pit outline. The outline is used to create an ultimate pit design with the recommended geotechnical parameters. The mining block model is further used in the mine planning process. Life of mine plans, 12-month rolling plans and six-week plans are done. Stockpiling is included in the plans and there is a strategy in place to reduce the stockpiles. The material movement is done through load and haul using trucks and shovels. Reconciliation was considered and it was found that there was not a proper reconciliation process for the iron ore mine.

Keywords: Optimisation · MRM · Plan · Block model

1 Introduction

Mine planning and optimization techniques are an important part of a mine's operation. The iron ore mine is in the Northern Cape in South Africa and it has been in production for approximately ten years. The iron ore produced is transported through trains to Saldhana which is approximately 861 km away from the mine. The in-situ ore types are laminated ore, conglomerate ore and detrital ore. The major ore bearing mineral is Hematite (Fe_2O_3) which constitutes 93% of the orebody mass. The material mined is separated into ON grade ore, OFF grade ore and Waste. ON grade is the ore material with Fe greater or equal to 63.5% and OFF grade is the ore material with Fe less than 63.5% and greater or equal to 54%. There is a required ON and OFF grade split, that is, of the total ore tonnes produced or mined, 30% should be ON grade ore and 70% should be OFF grade ore, although it can change depending on the requirements of the iron ore mine for a certain period. Table 1 indicates the grades of the ore that is mined in detail including contaminants. Material with Fe grade of less than 48% is regarded as waste which is sent to the waste dump. ON grade and OFF grade ore are sent for further

E. Topal (Ed.): MPES 2019, SSGG, pp. 103–110, 2020.
https://doi.org/10.1007/978-3-030-33954-8_11

processing to produce the product to be sold to clients. After the processing, the ore is separated into three product stockpiles: Lumpy, fines and Direct Reduction (DR) Lump, (Table 1). The three product stockpiles are then loaded onto trains using stacker reclaimers at the load-out station and then transported to Saldhana.

Table 1. Cut-off grades and product sizes

Cut-off grades	Fe (%)	Al_2O3 (%)	K_2O (%)	Mn (%)
ON grade	≥63.5	≤2.6	≤0.4	<2
ONOFF grade	≥63.5	>2.6	>0.4	<2
OFF grade	<63.5 and ≥54			<2
Low grade	<54 and ≥48			

Lumpy Stockpile: -32mm to +6mm
DR Lump Stockpile: -18mm to +6mm
Fines Stockpile: -6mm to +0.2mm

2 Surface Mine Planning Optimisation Techniques

2.1 Mineral Resource Management (MRM) Structure

MRM is a part of a mining activity that ensures integration between different mining activities. The purpose of MRM is to create a plan that ensures that a mineral resource is extracted to its optimum throughout the mine value chain. The iron ore mine has a dedicated MRM department and it comprises of planning, geology and rock engineering. Unlike the traditional MRM hierarchy, the MRM structure of the iron ore mine does not include surveying, instead surveying forms part of technical services. The structure that is currently in place works in an integrated manner to ensure that the goals of the operations are met. Geology is responsible for exploration drilling, infill drilling, blasthole sampling and the creation of block models that depict the orebody. The block models are then sent to the planning section for long, medium and short-term planning. Rock engineering ensures that the plans produced result in a safe working environment and to analyse if there aren't other foreseeable geotechnical issues that might disrupt the plan. The block models are then provided to the planning section for long, medium and short-term planning.

2.2 Block Modelling

A block model is a 3D representation of an orebody. It typically has sizes x, y and z. The block model sizes are usually related to the smallest selective mining unit (SMU). SMU is the smallest volume of material that an excavator can be able to extract (Darling 2011). A block model is created from information obtained from exploration drilling. For the iron ore mine, the initial exploration drilling was done on a 100 m $\times$ 100 m grid in order to cover the major outcrop. The orebody was found to be complex; therefore, the 100 m $\times$ 100 m grid was then reduced to a 50 m $\times$ 50 m grid in order to obtain sufficient information below the ground surface. The 50 m $\times$ 50 m grid assisted in increasing the level of confidence and the definition of the orebody was clearer, therefore allowing for geological block models which categorizes resources into measured, indicated and inferred to be created. The resources then were converted to reserves based on modifying factors such as mining losses, geological losses, dilution, economic factors, marketing and legal. After applying the modifying factors and reserves were obtained, mining block models of sizes 6.25 × 6.25 × 10 m which can be used for planning purposes were created in Datamine software. The mining block model has the attribute that separates the ore material into ON grade, OFF grade and waste which are the mineable material of economic interest. Figure 1 shows a reduced mining block model for the iron ore mine.

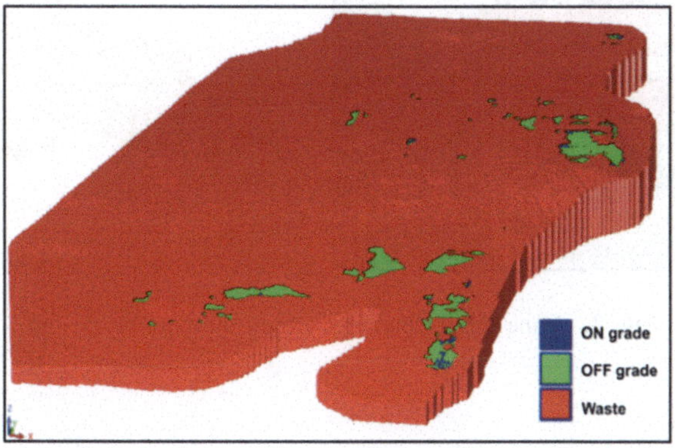

Fig. 1. Mining block model for the iron ore mine

2.3 Pit Optimisation and Mine Design

Pit optimisation using the block model along with technical and economical parameters for the iron ore mine was done through the Whittle software in order to obtain the ultimate pit. The whittle optimisation process produces a set of nested pit shells or pit outlines. The pit outlines are then compared with one another to see which pit outline produces the best Net Present Value (NPV). The pit outline that produces the higher NPV is taken as the ultimate optimum pit outline. From the outline, a mine design can be created using a set of mine design parameters that are obtained through a geotechnical process that analyses the characteristics of the rock. The design parameters include the bench height, berm width, bench angle, overall slope angle, ramp width, ramp gradient and a safe mining distance from the top bench to the bottom bench. Figure 2 indicates the actual mine design parameters that were used to create the ultimate pit design. "*The ultimate pit is the maximum valued pit possible that obeys slope and physical constraints*" (Meagher et al. 2014, p. 508). The ultimate pit design for the iron ore mine was created in Surpac and is currently being used as a reference in the long term, medium term and short-term planning. The ultimate pit is approximately 4.3 km in length and 1 km wide. The size of the final pit limit requires pushbacks. Pushbacks are manageable small pits with their own volumes inside the ultimate pit.

Pushbacks have their own accesses and allow for a manageable short-term plan (Meagher et al. 2014). Pushback designs are currently lacking for the iron ore mine. This causes a challenge in terms of knowing where to start mining in order to achieve a higher production and reach the ore with high grades in order to maximise the NPV in the earlier mine life. Production and the ease of doing short-term and medium-term plans can be potentially improved by using pushbacks for this iron ore mine.

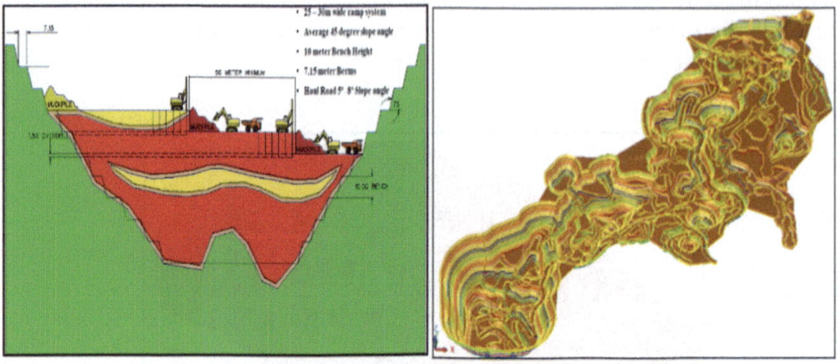

Fig. 2. Mine design parameters and ultimate pit-design

2.4 Scheduling

Scheduling indicates in which sequence material should be extracted and how much tonnes can be practically achieved annually, monthly or weekly. The purpose of scheduling is to maximise the NPV while taking into considerations constraints such as mill throughput, tonnes that can be extracted per period, stockpile constraints, blending constraints and logistic constraints (School of Mining Engineering 2018). The iron ore mine before the beginning of 2018 only had schedules that ranged from Life of Mine (LOM) plans, five-year plan and short term (eight weeks). The schedules are created using the Geovia MineSched software for both the long-term and short-term plans. From 2018, a process has been put in place that considers a 12-month rolling plan. In the 12-month rolling plan, the first three months' schedule is done in weeks in order to show enough details of what blast blocks are needed to achieve the required ON, OFF and waste targets, the infrastructure required and the short-term challenges that the mine might come across. The remaining nine months is an overview of how the mine scheduling advanced and the potential tonnes that can be produced and where temporary and permanent ramps need to be placed. The eight-week short-term plans have currently been reduced to six-week plans. The five-year schedule is based on the LOM plan while the 12-month rolling plan is based on the five-year schedule and the six-week plan is based on the 12-month rolling plan. Figure 3 shows some of the outputs obtained from the 12-month rolling plan. The output include the monthly survey (contours), the areas where mining will take place, which week, the amount of tonnes to be produced, the current hauling roads and the critical tasks in that specific month.

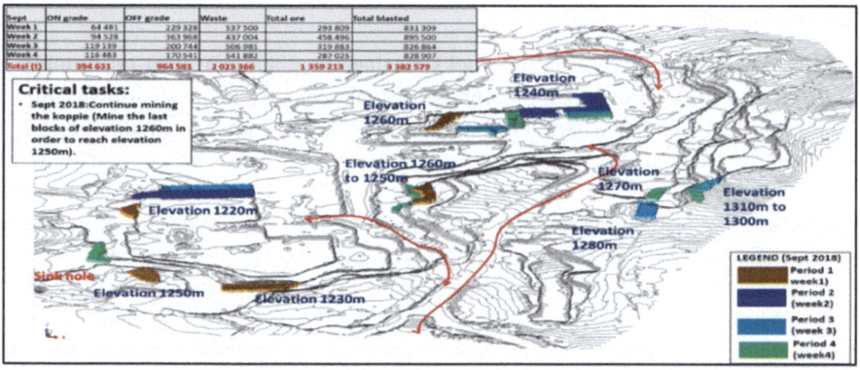

Fig. 3. Outputs from the 12 months rolling plan

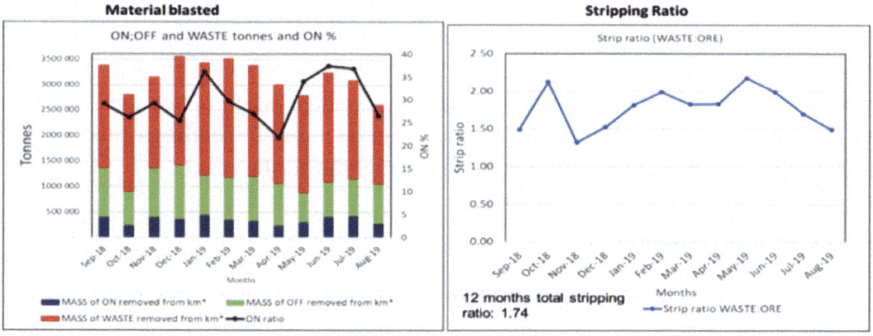

Fig. 4. Production profile and strip ratio from the 12 months rolling

The five-year schedule indicate the average stripping ratios that is required in order to make sure that there is enough waste stripping to expose the ore. From the 12-month rolling plan, stripping ratios and summarised tonnes per month are calculated and shown as profiles for better interpretation and understanding. Figure 4 indicates the production profile and strip ratios obtained from the 12-month rolling plan. The challenges with scheduling is the integration of the 12-month rolling plan with the six-week plan since it is still a new process and the process of reporting out is different from what was already in place in the planning process. Another challenge is reaching ON grade ore targets in some scheduling periods since the orebody cannot be changed and most of the time, the OFF grade ore is required to be mined first in order to expose the ON grade ore. In some cases, the OFF grade can be found to be ON grade after blast hole sampling and final predictions. Infill drilling should be done in order to increase the level of confidence. Another challenge that disturbs the scheduling is related to geotechnical issues. The iron ore mine has two sinkholes which have been discovered inside the pit. The sinkholes cause a risk because they can cause subsidence thus resulting in failure of the pit and damage to equipment. Sinkholes are delaying scheduling since areas closer to the sinkholes must be avoided, therefore they cannot be scheduled resulting in the delay of waste stripping and therefore, delaying ore exposure.

2.5 Stockpiling

Stockpiles are used to create buffer stocks so that the plant can keep on running on periods of low production due to scheduled maintenance, mine shutdowns due to holidays and bad weather where the pit is inaccessible. Stockpiles can also be used to stock marginal ore so that the marginal ore is not treated as waste. The marginal ore can be used in times of high commodity prices. Stockpiles can also be used for blending purposes whereby low-grade ore can be blended with high grade ore to reach a target average grade ore (School of Mining Engineering 2018). The iron ore mine currently has an abundance of OFF grade ore stockpiles due to the prioritising of ON grade ore at the plant. Stockpiling is included in the five-year and 12-month rolling plans. If the ore that is mined is more than what the plant can process, it is then sent to the stockpiles. The OFF grade ore stockpiles are growing at a rapid rate and there is a strategy in the 12-month rolling plan to reduce the amount of OFF grade ore mined in the pit so that priority should be moving material from the stockpile to the plant, therefore reducing the stockpiles. The current stockpile balance is approximately 1 800 000t of OFF grade ore. It is currently not on record on how long the stockpiles have been there, what Fe% grade and contaminants are on the stockpiles and if the stockpiled ore can be taken to the plant and be usable. OFF grade ore tends to weather when is not used for a long time. The solution to the OFF grade stockpile can be through blending of the stockpile material with high-grade material allowing throughput to the plant.

2.6 Equipment Selection

The iron ore mine is an open pit operation whereby the material movement method is load and haul. The load and haul is done through shovels, Front End Loaders (FEL) and haul trucks. The material is drilled, blasted and then loaded into haul trucks. The haul trucks then transport the ore to the crusher and waste to the waste dumps. The ore after it has been crushed is sent to the plant through conveyor belts. There are three hydraulic front shovels and one FEL. The shovels are then matched with rigid haul trucks. The rigid haul trucks are a mixture of the CAT 789 (180t payload) and the Komatsu 860E (280t payload). The drilling, loading and hauling equipment were selected with a consideration of the planned production rate. When the equipment is utilised to its full capacity, the production targets can be comfortably reached. The iron ore mine also uses contractors for cleaning purposes. The contractors use 40t payload Articulated Dump Trucks (ADT).

2.7 Quality/Grade Control

At the iron ore mine, the Run of Mine (ROM) is taken through a grade control process to check if the ROM meets the product specifications in respect to the physical properties and grade. The process incorporates the use of real-time online analytical analysis in order to trace and record the quality of the products; enhance grade control and blending capabilities, and to implement immediate corrective measures. The iron ore mine uses a geoscan which uses a multi-detector array and spectrometer electronics combined with gamma rays into a spectrum to determine the iron composition.

The geoscans are placed on the conveyor belts that transport ore from the crusher to the plant. The geoscans check the quality of the ON and OFF grade ore before it is taken for further processing. Figure 5 is a picture of a geoscan and the targets the geoscan is supposed to measure.

ON grade Geoscan target

$Fe \geq 64.5\%$

$SiO_2 \leq 7\%$

$Al_2O_3 \leq 2.3\%$

$K_2O \leq 0.48\%$

OFF grade Geoscan target

$Fe \geq 60\%$

$SiO_2 \leq 12\%$

$Al_2O_3 \leq 5\%$

$K_2O \leq 0.64\%$

Fig. 5. Geoscan and the targets required to be measured by the geoscan

The weighed average target specifications measured by the geoscan for ON grade ore should have a Fe% of greater or equal to 64.5% while for OFF grade ore the Fe% should be greater or equal to 60%. The ON grade cut-off grade in Table 1 is of Fe% of greater or equal to 63.5% and the weighted average ON grade in Fig. 5 is of Fe% of greater or equal to 64.5%. If the weighed average on the geoscan is less than 64.5%, the material will be considered as OFF grade. Care must be taken in ensuring that the material that is taken to the plant is classified accordingly in order to not confuse ON grade ore with OFF grade ore. There are major consequences if the ON grade is treated as OFF grade and vice versa. The OFF grade is not treated the same as ON grade and the economic benefits between the two are different. ON grade ore processing is only crushing and screening. OFF grade ore is taken through crushing, screening and further processing through a jig plant. If ON grade is treated as OFF grade, there is an increase in cost due to the extra processing that is required in the jig plant.

2.8 Reconciliation

Reconciliation is comparing an estimate with a measurement. The estimate is typically a mineral resource model or mineral reserve model. The measurement is usually survey information or the production reports. The purpose of reconciliation is to measure what was achieved against the target of an operation; confirm the grade and tonnages; ensure that the mineral asset is evaluated accurately and to obtain key performance indicators for grade control predictions (Morley 2003). Currently, the reconciliation process in the iron ore mine is lacking, therefore, it is difficult to keep accountability into the areas that have been mined with what was planned. Currently, the reconciliation does not account for spatial reconciliation whereby it can be seen visually the areas that have been planned and what was mined. Therefore, improvements need to be implemented into the overall reconciliation process.

3 Conclusion and Recommendations

Planning and optimisation techniques are an important part of a mining operation to ensure that targets are achieved. The planning techniques that are applied in the iron mine include MRM structure, block modelling, pit optimisation, scheduling, stockpiling and grade control. The iron ore mine has an MRM structure that includes rock engineering, planning and geology, however survey is not included into the MRM structure. Mining block models are created using Datamine software and are used in the planning process. Pit optimisation to obtain an ultimate pit is created through the Whittle software. The ultimate pit design for the iron ore mine covers a large area and there could be potential improvement in the production through the use pushbacks. Long term plans, 12-month rolling plans and six-week plans are currently in place and the scheduling is done through Geovia MineSched. Stockpiling is used to create buffer stocks and for blending purposes. Load and haul through truck and shovels is used for material movement in the iron ore mine. Geoscans are used for grade control. The reconciliation process for the iron ore mine is currently lacking. It is recommended that survey forms parts of the MRM organisational structure as recommended by the traditional MRM hierarchy. This has the potential to improve integration and the flow of information throughout the mine planning process. Pit optimisation with the aim of creating optimised pushback designs is recommended. This will ensure concentrated mining and creates a manageable short-term and medium-term planning. Drilling and mapping of the current OFF grade stockpiles is recommended in order to understand what is exactly on the stockpiles and if the OFF grade ore can produce a usable product after it has gone through processing. A reconciliation process should be put in place so that mineral assets can be tracked from resource estimation until the load-out station where the ore has already been converted to a saleable product. The reconciliation will assist in analysing if targets are met and if mining is taking place where it should.

References

Darling, P.: SME Mining Engineering Handbook, 3rd edn. Society for Mining, Metallurgy, and Exploration, Inc. (SME) (2011)

Meagher, C., Dimitrakopoulos, R., Avis, D.: Optimized open pit mine design, pushbacks and the gap problem—a review. J. Min. Sci. **50**(3), 508–526 (2014). http://link.springer.com/10.1134/S1062739114030132. Accessed 13 Sept 2018

Morley, C.: Beyond reconciliation: a proactive approach to using mining data (2003). https://www.researchgate.net/publication/228799954_Beyond_reconciliation_a_proactive_approach_to_using_mining_data. Accessed 13 Sept 2018

School of Mining Engineering: Planning and optimisation of surface mines. University of the Witwatersrand, Johannesburg, South Africa (2018)

Modeling Long-Term Production Scheduling Problem and Its Solution Using a Bat Meta-heuristic Method

Ehsan Moosavi(✉)

Department of Petroleum and Mining Engineering, South Tehran Branch, Islamic Azad University, Tehran, Iran
Se.Moosavi@yahoo.com, se_moosavi@azad.ac.ir

Abstract. The long-term production scheduling problem, in which a schedule must obey physical constraints and operational constraints between pairs of activities, is one of the most studied scheduling problems. An important variation of this problem is to find a schedule which maximizes the net present value. Since production scheduling problems are NP-hard, there is a need of improving scheduling methodologies to get good solutions. At present, the research on scheduling theory has been paid an overwhelming attention, and has made great progress. However, it is still not mature enough. Among them, the research on complexity of the scheduling problem has become a branch of applied mathematics with a strong engineering background. The mathematical model of production scheduling problem is established, and a novel improved bat algorithm (BA) is proposed. For the purpose of expressing the relationship effectively between the process and the bat population, a new method of encoding strategy based on dual flexibility degree is proposed. For the purpose of overcoming the shortcomings of the fixed parameters in the bat algorithm, the value of the inertia weight was adjusted, and a linear decreasing inertia weight strategy was proposed. This shows that the proposed algorithm is more excellent in solving the production scheduling problem, and it is an efficient scheduling algorithm.

Keywords: Long-term production scheduling · Expected economic loss · Open pit mine · Bat algorithm

1 Introduction

Production scheduling is a process of allocating limited ore/waste blocks to operations (activities) over mine life. Long-term production scheduling (LTPS) is a combination of mixed integer and linear optimization problem. A LTPS process plays a vital role for the economic use of mining sequencing in mine planning. LTPS schedules the block extraction sequences in accordance to the economic parameters in which include all constraints of open pit mines.

In recent years, with the need of optimization problems in reality, all kinds of bioinspired optimization algorithms or swarm intelligence optimization algorithms have been proposed, such as the genetic algorithm (GA) [1–4], ant colony optimization

E. Topal (Ed.): MPES 2019, SSGG, pp. 111–119, 2020.
https://doi.org/10.1007/978-3-030-33954-8_13

(ACO) [5, 6], particle swarm optimization (PSO) [7], simulated annealing (SA) [8–10], dynamic programming (DP) [11–15] and artificial intelligence (AI) [16–18]. These bionic intelligent algorithms are random search methods which mimic natural biological systems [19]; entirely instinct depends on the organism itself, by the unconscious optimization behavior to optimize its survival to adapt to the environment. Compared with the traditional optimal algorithms, they do not depend on the characteristic of strict mathematical optimization problem itself and have the characteristic of strong parallelism. Each individual has self-organization and evolution; as a result, some high dimensional optimization problems are superior to the traditional methods.

The bat algorithm (BA) is a new metaheuristic method which was proposed by Yang in 2010 [20]. The capability of echolocation of microbats is fascinating as these bats can find their prey and discriminate different types of insects even in complete darkness Inspired by the echolocative behavior of bats, this algorithm carries the search process using artificial bats as search agents mimicking the natural pulse loudness and emission rate of real bats. When these bats are chasing prey, they tend to decrease the loudness and increase the rate of pulse emission. The BA has a good optimization performance in low dimensional case. The BA has been employed which is more effective and capable of solving linear optimization problems faster and with better accuracy in detecting the global best solution [21].

With the proposed method, the total generation cost can be remarkably decreased while considering various constraints reflecting operational mining. The rest of the paper is organized as follows. In Sect. 2, mathematical problem formulation of LTPS problems is defined. The proposed model considering the constraints as soft constraints is shown Sect. 3. The proposed BA is discussed in Sect. 4.

2 LTPS Model for Open Pit Mines

In the LTPS problem, the extraction of each block in each period is dependent upon the definition of cut-off grade. The cut-off grade dictates the block classification. Obviously, the choice of type in choosing the different types of processing is directly related to the selection of block classification in mining sequence, and is effected by special recovery, costs and special block economic a value, as well. Consequently, the misclassification of a block results in a wrong block economic value calculation. In this regard, Richmond [22] defines a function as "economic loss", which is used to distinguish between ore and waste block. The actual economic loss associated with each type of processing d ($d = 1, 2, \ldots, D$ ordered from lowest grade to highest grade) is the potential value less than the recovered value, which can be calculated using the Eq. (1).

$$L_{ijk}^{d} = \left[\left(P - C_{sel}\right) \cdot \overline{\alpha}_{ijk} \cdot R^{d} - C_{p}^{d} - C_{m}^{d}\right] - \left[\left(P - C_{sel}\right) \cdot \overline{\alpha}_{ijk} \cdot R^{d'} - C_{p}^{d'} - C_{m}^{d'}\right] \quad (1)$$

Where:

P: unit selling price of the metal,
C_{sel}: unit selling cost of the metal,
ijk: is the block identification number,

$\overline{\alpha}_{ijk}$: average grade of block *ijk*,
R^d: total metal recovery of material if processed as type *d*,
C_p^d: unit processing cost of the material if processed as type *d*,
C_m^d: unit mining cost of the material if processed as type *d*,
d: correct processing type for block *ijk*,
d': chosen processing type for block *ijk*.

Taking from the Eq. (1), if the correct processing type for block *ijk* is chosen, then economic loss is zero. In practice, the ore block is unknown and it will be represented by a cumulative distribution function. A more realistic method is to use the conditional simulation techniques, which allows the generation of a number of equally probable realizations of block grades. Taken this into consideration the expected economic loss (EEL) for each alternative d is calculated using the probabilities distribution and average grade for each type processing, generated from independent realization as described in Eq. (2).

$$EEL_{ijk}^d = \sum_{d=1}^{D} \left[P_{ijk}^d | O \right] \cdot L_{ijk}^d \tag{2}$$

Where:
$P_{ijk}^d|O$: probabilities distribution of block *ijk* if processed as type *d*.
The optimal processing type for block *ijk* is that d for which the *EEL* is minimized, i.e.:

$$L(opt)_{ijk} = Min \quad [EEL_{ijk}^d] \tag{3}$$

2.1 Formulation as an Integer Programming Problem

In light of the definitions and assumptions described above, the mathematical programming model of the mining sequence in terms of integer decision variables imposes the period the particular block is extracted and its destination is determined correctly. In fact, this model can simultaneously optimize the block extraction sequence and cut-off grade strategy. As mentioned earlier, the objective function of the model can be represented mathematically as the following [23]:

$$Minimize \quad Z = \sum_{ijk \in \Gamma} \sum_{t}^{T} \frac{L(opt)_{ijk}^t \cdot b_{ijk}^t}{(1+r)^t} \tag{4}$$

The objective function in Eq. (4) does not explicitly maximize net present value (NPV), but rather optimizes feasible extraction sequencing and ensures a desired cut-off grade. The reason is that feasible extraction sequences and the amount of ore having the desired quality to be sent to the mill need to be prioritized. So, the mentioned objective function indirectly leads to a maximum NPV that is optimal. Otherwise, the generated NPV would only be optimal in the theory but not in mining practice. However, economic loss for the suitable ore block which has the desired properties has been integrated in the present model, to maximize the chances of delivering to the mill, the amount and quality of ore required during mining operations. EEL minimization and feasible sequences; result in maximum NPV. The proposed model in Eq. (4) contains a series of constraints as follow:

$$\sum_{ijk\in\Gamma}\left(\bar{\alpha}_{ijk}-U^t_{\bar{\alpha}}\right)\cdot Q^o_{ijk}\cdot b^t_{ijk}\le 0\,\forall t\in T \tag{5}$$

$$\sum_{ijk\in\Gamma}\left(\bar{\alpha}_{ijk}-L^t_{\bar{\alpha}}\right)\cdot Q^o_{ijk}\cdot b^t_{ijk}\le 0\,\forall t\in T \tag{6}$$

$$\sum_{t=1}^{T} b^t_{ijk}=1\,\forall\, ijk\in\Gamma \tag{7}$$

$$\sum_{ijk\in\Gamma}\left(Q^o_{ijk}\cdot b^t_{ijk}\right)\le U^t_o\,\forall t\in T \tag{8}$$

$$\sum_{ijk\in\Gamma}\left(Q^o_{ijk}\cdot b^t_{ijk}\right)\ge L^t_o\,\forall t\in T \tag{9}$$

$$\sum_{ijk\in\Gamma}\left(Q^o_{ijk}+Q^w_{ijk}\right)\cdot b^t_{ijk}\le U^t_{w\&o}\,\forall t\in T \tag{10}$$

$$\sum_{ijk\in\Gamma}\left(Q^o_{ijk}+Q^w_{ijk}\right)\cdot b^t_{ijk}\ge L^t_{w\&o}\,\forall t\in T \tag{11}$$

$$b^t_k-\sum_{y}^{Y}\sum_{r=1}^{t} b^r_y\le 0\,\forall t\in T,\forall k\in K \tag{12}$$

Where:

$L(opt)^t_{ijk}$: the optimal processing type for block *ijk* in period *t*,
T: the total number of scheduling periods,
t: is the scheduling periods index, $t = 1, 2, \ldots, T$,
Γ: the total number of blocks to be scheduled,
r: the discount rate in each period,
b^t_{ijk}: the binary variable equal to: $\begin{cases} 1 & \textit{if block ijk is extracted in period t} \\ 0 & \textit{Otherwise} \end{cases}$
Q^o_{ijk}: the ore tonnage in block *ijk*,
Q^w_{ijk}: the waste tonnage in block *ijk*,
$\bar{\alpha}_{ijk}$: the average grade of block *ijk*,
$U^t_{\bar{\alpha}}$: the upper bound average grade of material sent to the mill in period *t*,
$L^t_{\bar{\alpha}}$: the lower bound average grade of material sent to the mill in period *t*,

U_o^t: the upper bound total tons of ore processed in period t,
L_o^t: the lower bound total tons of ore processed in period t,
$U_{w\&o}^t$: the upper bound total amount of material (waste and ore) to be mined in period t,
$L_{w\&o}^t$: the lower bound total amount of material (waste and ore) to be mined in period t,
Y: the total number of blocks overlaying block k,
K: the index of a block considered for extraction in period t,
y: the counter for the Y overlaying blocks.

Constraints (5), (6) limit the average grade of the material sent to the mill to a certain value. Constraint (7) enforces that a block is removed in one period only. Constraints (8), (9) ensure that the milling capacities hold. These upper and lower bounds are necessary to secure a smooth feed of ore. Constraints (10), (11) relate the production capacity for each period. These upper and lower bounds are total amount of material (ore and waste) to be mined in each period. Constraint (12) is the wall slope restriction on the basis of Y constraints for each block per period.

3 Proposed Model as Multi Objective Optimization Problem

The objective function is composed of two parts. The first part is for the minimization of the expected economic loss of the production schedule, and the second part represents the minimization of the expected recourse costs whenever the stochastic constraints (Eqs. 8 to 11) are violated, similar to the one proposed in Ramazan and Dimitrakopoulos [24]. In this section, the multi-objective function is presented in accordance with the research topic.

Subject to:

$$Min\ Z_{mo} = \sum_{t}^{T}\left[Z + \frac{1}{S}\sum_{s=1}^{S}\left(M_{U_o}^t\lambda_{SU_o}^t + M_{L_o}^t\lambda_{SL_o}^t + M_{U_{w\&o}}^t\lambda_{SU_{w\&o}}^t + M_{L_{w\&o}}^t\lambda_{SL_{w\&o}}^t\right)\right] \tag{13}$$

$$\sum_{ijk\in\Gamma}\left(Q_{ijk}^o \cdot b_{ijk}^t\right) - \lambda_{SU_o}^t \le U_o^t\,\forall t\in T \tag{14}$$

$$\sum_{ijk\in\Gamma}\left(Q_{ijk}^o \cdot b_{ijk}^t\right) - \lambda_{SU_o}^t \ge L_o^t\,\forall t\in T \tag{15}$$

$$\sum_{ijk\in\Gamma}\left(Q_{ijk}^o + Q_{ijk}^w\right)\cdot b_{ijk}^t - \lambda_{SU_{w\&o}}^t \le U_{w\&o}^t\,\forall t\in T \tag{16}$$

$$\sum_{ijk\in\Gamma}\left(Q_{ijk}^o + Q_{ijk}^w\right)\cdot b_{ijk}^t - \lambda_{SL_{w\&o}}^t \le L_{w\&o}^t\,\forall t\in T \tag{17}$$

Where:

S: Total number of simulations of the orebody,
s: is the scheduling periods index, s $= 1, 2, \ldots, S$,
$M^t_{U_o}$, $M^t_{L_o}$: are the discounted unit costs for shortage or surplus ore produced in period t respectively,
$M^t_{U_{w\&o}}$, $M^t_{L_{w\&o}}$: are the discounted unit costs for shortage or surplus metal produced in period t respectively,
$\lambda^t_{sU_o}$, $\lambda^t_{sL_o}$: are the second-stage scenario-dependent continuous variables representing the shortage or surplus amount of ore produced in period t if scenario s occurs, respectively
$\lambda^t_{sU_{w\&o}}$, $\lambda^t_{sL_{w\&o}}$: are the second-stage scenario-dependent continuous variables representing the shortage or surplus amount of metal produced in period t if scenario s occurs, respectively. The second-stage (recourse) decision variables are dependent on the outcome of the orebody realizations and of the first-stage decision variables $i.e.\ b^t_{ijk}$.

The upper and lower processing capacity constraints (Eqs. 14, 15) and the upper and lower metal production constraints (Eqs. 16 and 17) represent the scenario-dependent stochastic constraints. These constraints are modelled as soft constraints where the violation is M large method (penalized) in the objective function. The constraints as Eqs. 5 to 7 and 12 are included in the model.

4 Bat Algorithm (BA)

Bats are fascinating animals. They are the only mammals with wings and they also have advanced capability of echolocation. In the BA, for the ith bats of swarm having position x_i (solution), velocity v_i, and frequency f_i, each bat will move toward the current best position (solution), and its position, velocity, and frequency are updated during the course of iteration as follows:

$$f_i = f_{min} + (f_{max} - f_{min})\beta$$
$$v_i^t = v_i^{t-1} + \left(x_i^{t-1} - v_g^{t-1}\right) f_i \tag{18}$$
$$x_i^t = x_i^{t-1} + v_i^t$$

Where β is a random number of a uniform distribution in [0, 1] and x_i^{t-1} represents the current global best solution (position) after comparing all the solutions (positions) among all the n bats. These equations can guarantee the exploration ability of the BA. For the local search, when a solution is selected among the current best solutions, a new candidate solution can be generated as Eq. (19).

$$x_{new} = x_{old} + \varepsilon \bar{A}^t \tag{19}$$

Where ε is a random number in [− 1, 1] and directs new solution apart from or close to the current best solution. Here, $\bar{A}^t$ is mean value of all bats of different loudness A_i. When finding prey, a bat will gradually decrease the loudness and increase the rate of pulse emission in order to track its prey to capture it. The loudness and pulse emission rate update accordingly as the iterations proceed as shown in Eq. (20).

$$A_i^t = \alpha A_i^{t-1} \tag{20}$$

$$r_i^t = r_i^0(1 - \exp(-\gamma t))$$

Where α and γ are constants. In fact, the α parameter controls the convergence of BA and therefore plays a similar role as the cooling factor in the simulated annealing algorithm [25]. For simplicity, we set $\alpha = \gamma = 0.9$ in our simulations.

4.1 Pseudo Code of the Bat Algorithm

The basic steps of the BA can be summarized as the pseudo code shown in Fig. 1.

```
Begin
Step 1.  Initialize parameters such as bat population x_i, velocity v_i, loudness A(i), pulse rates R(i) and
         so on
Step 2.  Define pulse frequency F_i
Step 3.  Calculated fitness value fit_min and the current global optimal solution x*
Step 4.  While (t < Max number iterations)
Step 5.    For (i = 1 to pop size)
Step 6.      Generate new solutions by adjusting frequency, and updating velocities and locations
             via Eq.18
Step 7.      If (rand>R(i))
Step 8.        Select a solution among the optimal solution set
Step 9.        Generate a new solution via Eq.19
Step 10.     end if
Step 11.     Calculated fitness value fit_new
Step 12.     If (fit_new<fit(i)&rand< A(i))
Step 13.       Accept the new solutions
Step 14.       Update A(i) and R(i)
Step 15.     end if
Step 16.   end for
Step 17.   Find the current optimal fitness value fit_new and the current optimal x*
Step 18. end while
End
```

Fig. 1. Bat algorithm (BA) pseudo code.

5 Conclusion

In the present work, a BA based approach is developed and implemented successfully to solve the LTPS problem. It is observed from the simulation results that the fitness value obtained from the proposed BA method after satisfying all the constraints of all the systems, are better than the earlier best reported results. It is also found that the proposed method has superior features, including stable convergence characteristic and avoids premature convergence. Moreover, it can be observed that the proposed approach appears to be a robust and reliable optimization algorithm for solving small and large-scale LTPS problems. The computational analysis shows that the proposed method is capable of reaching the optimum solution in less computational time than that required for other algorithms. Therefore, it may finally be concluded that the proposed BA has great potential to generate near optimal solution in less computational time than several other solution approaches presented in existing literature.

References

1. Denby, B., Schofield, D.: Open-pit design and scheduling by use of genetic algorithms. Trans. Inst. Min. Metall. Section A: Min. Technol. **103**, A21–A26 (1994)
2. Denby, B., Schofield, D.: Genetic algorithms for open pit scheduling-extension into 3-dimensions. In: Proceedings of the Mine Planning and Equipment Selection Conference, MPES 1996, pp. 177–185, Sao Paulo, Brazil (1996)
3. Denby, B., Schofield, D., Surme, T.: Genetic algorithms for flexible scheduling of open pit operations. In: Proceedings of the 27th Application of Computers and Operations Research in the Mineral Industry (APCOM) 1998, London, UK, pp. 473–483 (1998)
4. Zhang, M.: Combination Genetic algorithms and topological sort to optimize open-pit mine plans. In: Proceedings of 15th Conference on Mine Planning and Equipment Selection, Torino, Italy (2006)
5. Sattarvand, J., Niemann-Delius, C.: Long-term open-pit planning by ant colony optimization. Institut für Bergbaukunde III, RWTH Aachen University (2009)
6. Sattarvand, J., Niemann-Delius, C.: A new metaheuristic algorithm for long-term open-pit production planning. Arch. Min. Sci. **58**(1), 107–118 (2013)
7. Khan, A.: Long-term production scheduling of open pit mines using particle swarm and bat algorithms under grade uncertainty. J. South Afr. Inst. Min. Metall. **118**, 361–368 (2018)
8. Boucher, A., Dimitrakopoulos, R.: Multivariate block-support simulation of the yandi iron ore deposit. Math. Geosci. **44**(4), 449–468 (2012)
9. Kumral, M., Dowd, P.A.: Short-term mine production scheduling for industrial minerals using multi-objective simulated annealing. In: Proceedings of the 30th Application of Computers and Operations Research in the Mineral Industry (APCOM) 2002, Fairbanks, Alaska, USA, pp. 731–742 (2002)
10. Kumral, M., Dowd, P.A: A simulated annealing approach to mine production scheduling. J. Oper. Res. Soc. **56**(8), 922–930 (2005)
11. Roman, R.J.: The role of time value of money in determining an open pit mining sequence and pit limits. In: Proceedings of 12th Symposium on Application of Computers and Operations Research in the Mineral Industry (APCOM) (1974)
12. Dowd, P.A., Onur, A.H.: Optimizing open pit design and sequencing. In: Proceedings of 23rd Symposium on Application of Computers and Operations Research in the Mineral Industry (APCOM) (1992)
13. Tolwinski, B., Underwood, R.: A scheduling algorithm for open pit mines. IMA J. Math. Appl. Bus. Ind. **7**(3), 247–270 (1996)
14. Tolwinski, B.: Scheduling production for open pit mines. In: Proceedings of 27th International Symposium on Application of Computers and Operations Research in the Mineral Industry (APCOM), pp. 651–662 (1998)
15. Wang, Q., Gu, X., Chu, D.: A dynamic optimization method for determining cut-off grades in underground mines. Mineral Resources Management (2008)

16. Tolwinski, B., Underwood, R.: An algorithm to estimate the optimal evolution of an open pit mine. In: Proceedings of 23rd Symposium of Application of Computers and Operations Research in the Mineral Industry (APCOM), pp. 399–409 (1992)
17. Elevli, B.: Open pit mine design and extraction sequencing by use of OR and AI concept. Int. J. Surf. Min. Reclam. Environ. **9**(4), 149–453 (1995)
18. Ibrahimov, M., Mohais, A., Schellenberg, S., Michalewicz, Z.: Scheduling in iron ore open-pit mining. Int. J. Adv. Manufac. Technol. **72**, 1021–1037 (2014)
19. Talbi, E.: Metaheuristics: From Design to Implementation. Wiley, Chichester (2009)
20. Yang, X.S: A new metaheuristic bat-inspired algorithm. In: Nature Inspired Cooperative Strategies for Optimization, vol. 284, pp. 65–74 (2010)
21. Yang, X.: Nature-Inspired Metaheuristic Algorithms. Luniver Press, London (2010)
22. Richmond, A.J.: Maximum profitability with minimum risk and effort. Application of Computers and Operations Research in the Mineral Industry, pp. 45–50 (2001)
23. Moosavi, E., Gholamnejad, J.: Optimal extraction sequence modeling for open pit mining operation considering the dynamic cutoff grade. J. Min. Sci. **52**(5), 956–964 (2016)
24. Ramazan, S., Dimitrakopoulos, R.: Stochastic optimization of long-term production scheduling for open pit mines with a new integer programming formulation. In: Proceedings of Orebody Modelling and Strategic Mine Planning: Uncertainty and Risk Management Models. Australasian Institute of Mining and Metallurgy, Melbourne, pp. 385–392 (2004)
25. Afrabandpey, H., Ghaffari, M., Mirzaei, A., Safayani, M.: A novel bat algorithm based on chaos for optimization tasks. In Proceedings of the Iranian Conference on Intelligent Systems (ICIS 2014), pp. 1–6. IEEE, Bam, Iran (2014)

Comparison of Methods to Define the Final Pit - A Case Study

Ana Luiza Medeiros Moreira, Bárbara Isabela da Silva Campos, Pedro Henrique Alves Campos(✉), Viviane da Silva Borges Barbosa, and Pedro Benedito Casagrande

Universidade Federal de Minas Gerais,
Antônio Carlos Av. 6627, Belo Horizonte, MG, Brazil
pedrocampos@demin.ufmg.br

Abstract. Elaborating a proper mine-planning is a crucial stage to define the success of a mining project. In the long-term it is desired to define the final pit and the mine sequencing, and this is the objective of the present paper, in which different methods were compared to get to the best viable economic result. Out of a database from a known copper deposit, two main methods were used to describe the behaviour of the mineral grade. The methods used were the Geostatistical model, using Ordinary Kriging, and the Inverse Distance Squared (IDS). Each one of them defined three pits using, respectively, Floating Cones algorithm, Lerchs-Grossmann (LG) algorithm and Direct Block Scheduling (DBS). The tests and their analysis showed that there are great variations on the results provided from each methodology. For IDS methodology, the economic difference between Floating Cone and the LG algorithm is about 62%, while the Kriging method resulted on a difference of 50%. In both cases, LG algorithm returned higher results. The comparison between scheduling methods shows that for IDS, DBS returned an economic result 17% higher than the Best Case and 23% for the Worst Case. Considering the Kriging method, the difference is about 16% higher for DBS in comparison to the Best Case and 22% for the Worst Case. All the results were obtained using specific software for each case.

Keywords: Mine planning · Final pit · Scheduling

1 Introduction

The objective of Strategic Mine Planning is to define the final pit geometry in order to obtain the highest possible profit, improving the extraction capacity with minimum costs. The main methods used to determine the final pit are the Floating Cone technique and Lerchs-Grossmann algorithm. The first one considers physical and geomechanical characteristics, not always returning a maximum pit value. On the other hand, Lerchs-Grossmann algorithm presents an optimum pit, since it aims to maximize its financial return [1].

Once final pit is determined, it is necessary to define a production chart in long, medium and short term. This phase will be responsible for defining the sequence in which each block will be removed from the pit, considering different restrictions [1].

E. Topal (Ed.): MPES 2019, SSGG, pp. 120–127, 2020.
https://doi.org/10.1007/978-3-030-33954-8_14

To obtain this scheduling some methodologies can be used, such as the traditional scheduling or Direct Block Scheduling (DBS). The traditional method is developed starting from the pit benches or the defined pushbacks, which will lead the way to mine the blocks, respecting these restrictions. DBS does not consider restrictions, determining directly which blocks will be mined and the best moment to do it.

The methods to evaluate a mine deposit are classified as follows: classic methods, statistical methods and geostatistical methods. In this paper two methodologies will be used to estimate a mine deposit: Kriging and Inverse Distance Squared (IDS). The first one is a geostatistical method that determines values in places that were not sampled using experimental data and their spatial correlation. On the other hand, IDS is a classic linear method, where the weight attributed for a sample is inversely proportional to the squared distance between the estimated point and the sample [2].

This paper is a case study about a copper mine, and it can contribute as a reference for similar cases about long-term mine planning.

2 Methods

2.1 The Deposit

The chosen mine deposit for analysis in this paper was Radomiro Tomic, which belongs to Corporación Nacional del Cobre de Chile (CODELCO). It is a copper deposit located in Chile and its main minerals are chrysocolla and atacamite. Main data of this deposit were obtained from [3] and [4].

2.2 Compositing

The composition was made based on the bench size starting from the highest elevation found among the holes. The maximum number of intervals was used for more precise results. Table 1 shows the data used to define the composites.

Table 1. Data for calculate of the composite

Data	Value
Highest bench elevation (m)	3013
Composite interval (m)	15
Number of interval	1000

2.3 Variogram

The data used to elaborate the Variograms are in Table 2 and its parameters in Table 3.

The model presented good approximation to experimental points, regardless the direction being analysed. This shows that the deposit, in general, have a good spatial correlation of the grades and an isotropic behaviour.

Table 2. Data for elaboration of variogram

Data	Value
Number of variogram	4
Horizontal direction of first variogram	0
Horizontal direction increment size	25
Horizontal half window (°)	45
Variogram cell size	50
Number of cells	18
Vertical bench height (m)	15

Table 3. Parameters of variogram

Data	Value
Model	Spherical
Nugget (C_0)	0,1
Sill (C)	0,12
Range (m)	290

2.4 Block Model

The block model was defined using the data on Table 4.

Table 4. Data for elaboration of block model

Data	X	Y	Z
Key coordinates (m)	720	260	3013
Block size (m)	50	50	15
Number of blocks (m)	45	30	15

2.5 Kriging Estimative and Inverse Distance Square (IDS)

The parameters of Table 5 were used to find the Inverse Distance Squared (IDS) and Kriging, and the maximum search distance was defined as 200 m in both cases. For Kriging, it was necessary to apply the parameters obtained from the variograms, such as nugget, sill and range (Fig. 1).

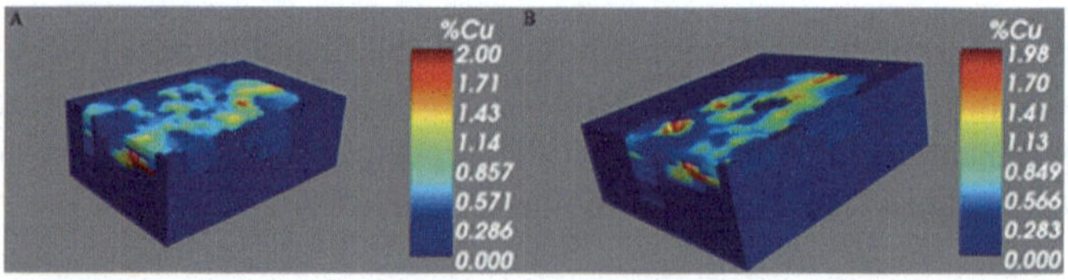

Fig. 1. Final pit solid defined from Floating Cone technique using (A) IDS and (B) Kriging method.

2.6 Blocks Economic Value

The data used to evaluate the blocks are in Table 5.

Table 5. Parameters for valuation of blocks

Data	Value
Cut-off grade (%)	0,2
Density (t/m^3)	2,59
Average grade (%)	0,5
Mine and haul ($/t)	−0,60
Increment mine and haul ($/level)	−0,10
Administrative ($/t)	−0,24
Mill ($/t)	−1,20
Recovery (%)	78
Refine cost ($/lb)	−0,24
Mine price ($/lb)	3,00

To describe the economic values of each block, it was made an evaluation of the revenue and the costs associated with the extraction, becoming possible to define the economic block value. Benefit Function is defined as:

$$F_{benefit} = \text{Revenue} - \text{Cost} \tag{1}$$

$$F_{benifit} = R * T * P - \left(C_l + C_p + C_{G\&A}\right) \tag{2}$$

Where, R is the global recovery of the plant, T is the block grade, P is the ore price, C_l is the mine cost, C_p is the processing cost and $C_{G\&A}$ represents general and administrative costs [5].

2.7 Floating Cone Pits

Out of the block model, the Floating Cone technique was applied and the results are shown in Fig. 2 below.

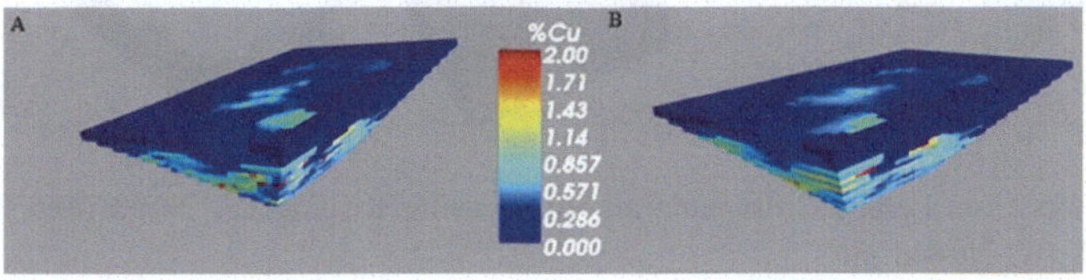

Fig. 2. Final pit solid defined from Floating Cone technique using (A) IDS and (B) Kriging method.

2.8 Lerchs-Grossmann Pits

To define the pits using Lerchs-Grossmann algorithm it is necessary to define the maximum value of each block using the following equation:

$$V = \text{máx}\{0{,}01 \times R \times G \times T \times F \times (r_f \times P - C_s) - (C_m + C_p) \times T; -C_m \times T\} \quad (3)$$

Where, V is the output block value, R is the recovery, G is the grade, T is the block tonnage, F is a conversion factor, r_f is the revenue factor, P is the ore price, C_s is the sell cost, C_m is the mine cost and C_p is the processing cost.

The data used to calculate these values are basically the ones in Table 5, with some additional parameters shown in Table 6.

Table 6. Lerchs-Grossmann parameters

Data	Value
Tonnage (t)	97125,00
Conversion factor (lb/t)	2204,60
Revenue factor interval	0 to 1

The block tonnage is calculated by the multiplication of the block dimensions and its density, and the revenue factor ranges from 0 to 1 using a step of 0.2 (Figs. 3 and 4).

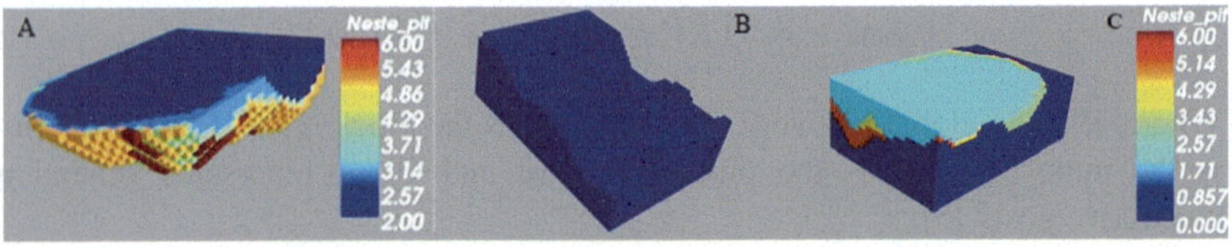

Fig. 3. Results from Lerchs-Grossmann algorithm using IDS. (A) Solid from the final pit. (B) Final pit. (C) Nested pits.

Fig. 4. Results from Lerchs-Grossmann algorithm using Kriging. (A) Solid from the final pit. (B) Final pit. (C) Nested pits.

2.9 Traditional Scheduling

The traditional scheduling was made for the Worst and the Best Case scenarios. In both cases, it is necessary to define the maximum production capacity of the mine, which is about 30.000.000 tons per year. In the Worst Case, the scheduling is based on mining each bench in sequence, so the revenue factor used was 1 to include all the pit blocks. On the other hand, the Best Case uses the pushbacks to develop the scheduling, so it was necessary to include all the defined revenue factors, so all the nested pits were included (Figs. 5 and 6).

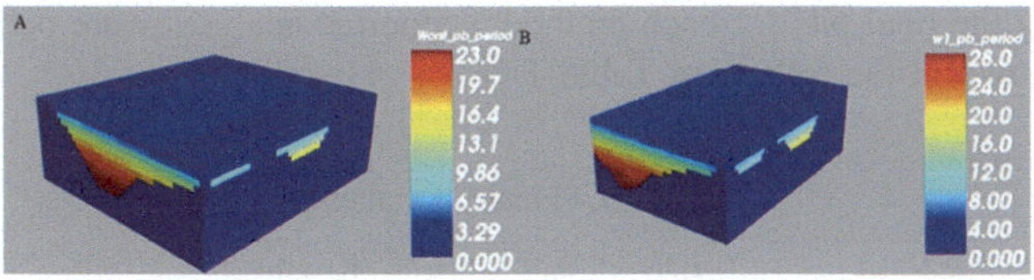

Fig. 5. Worst case scheduling using (A) Kriging and (B) IDS.

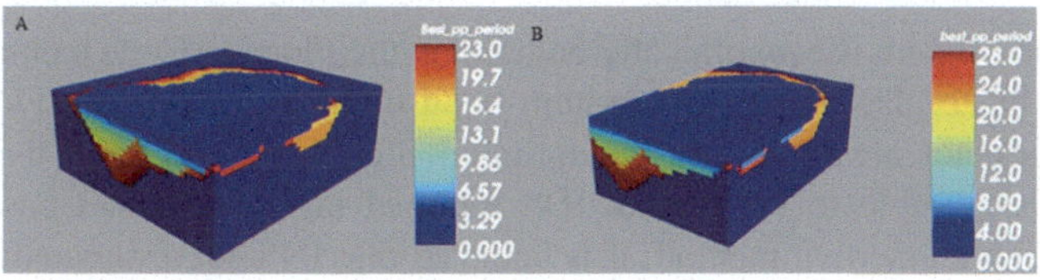

Fig. 6. Best case scheduling using (A) Kriging and (B) IDS.

2.10 Direct Block Scheduling

DBS was processed using the data obtained from the application of the Lerchs-Grossmann algorithm. Most of the parameters used to run the DBS were defined before, such as tonnage, prices and costs, annual production and nested pits. The discount rate equals 0.1 (Figs. 7 and 8).

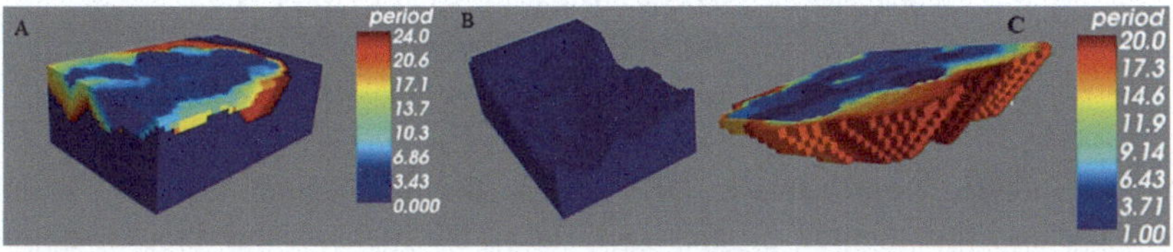

Fig. 7. DBS using IDS method.

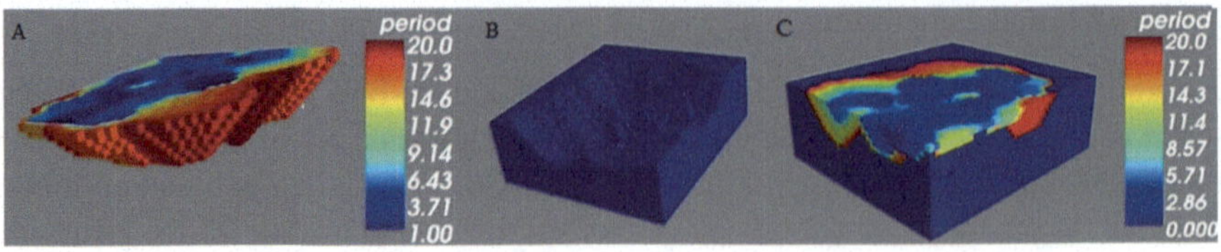

Fig. 8. DBS using kriging method.

3 Results and Discussion

After defining the final pits for each method, economic analyses were performed. The results are shown in the following Table 7.

Table 7. Final pit values

	IDS	Kriging
Floating Cone	$ 5.444.299.745,80	$ 5.443.953.850,67
Lerchs-Grossmann	$ 14.502.476.462,90	$ 10.997.376.339,90

Lerchs-Grossmann is economically better than the Floating Cone technique, which is explained by the fact that the algorithm is an optimizer, while the Floating Cone technique does not always return an optimum final pit. Comparing the estimation methods, it is possible to notice that IDS returned higher values. Considering that Kriging is a more assertive method, it is believed that IDS overestimated the deposit.

After the scheduling process it was possible to compare the Net Present Values (NPVs) of each situation. The traditional scheduling returned a life of 28 years for the IDS method and 23 years for Kriging, while DBS defined a horizon time of 24 years for IDS and 20 years for Kriging. The Table 8 shows the NPVs for the estimation methods used.

Table 8. Net Present Value for each pit defined from Lerchs-Grossmann algorithm using traditional scheduling and DBS.

	IDS	Kriging
Best case	$ 6.404.454.419,75	$ 5.554.184.127,23
Worst case	$ 5.948.697.084,39	$ 5.122.852.738,78
DBS	$ 7.761.455.399,67	$ 6.578.561.883,03

As it is noticed from Table 8, DBS returns the best economic result, which is justified because this method progressively takes the higher value blocks, mining a great volume of ore in the first years with minimum waste being extracted, while the final years present an opposite situation. The Best and the Worst Cases have a variable production of ore and waste along the years, not selecting a block based on its value, but in pushbacks and benches, respectively.

4 Conclusion

The comparison between the estimation methods showed that Kriging returns more assertive results than IDS method. That could be explained because Kriging considers the spatial correlation between the samples. Analysing the obtained results, it is believed that IDS overestimated the mineral reserve, which could be seen from the bigger size of the pits calculated by this method. In both methods to define the final pit, IDS returned higher economic values.

Lerchs-Grossmann algorithm presented better financial results than the Floating Cone technique, with a significant difference: for IDS, the results were 62% higher and, for Kriging, 50% higher. That happens because the algorithm is an optimizer, which means it always chases the best economic solutions for the project.

The best scheduling chart was obtained from DBS, since it aims to optimize the block mining without considering a geometric restriction. Traditional scheduling presented good results in both Best and Worst Cases and, besides there were some differences in the NPVs obtained by these two methods, they were not as relevant when compared to DBS method. For the IDS method, the difference was about 17% higher for DBS in comparison to the Best Case and 23% to the Worst Case. Considering the Kriging method, this difference was, respectively, 16% and 22%.

References

1. Flores, B.A.: Planejamento de Lavra Estratégico e Tático de Morro da Mina - Conselheiro Lafaiete/MG. 2008. 150 f. Dissertação de Pós-Graduação em Engenharia Mineral. Departamento de Engenharia de Minas. Universidade Federal de Ouro Preto, Ouro Preto
2. Curi, A.: Minas a Céu Aberto: Planejamento de Lavra. Oficina de textos, São Paulo (2014)
3. Codelco. Divísion Radomiro Tomic. Operaciones. https://www.codelco.com/division-radomiro-tomic/prontus_codelco/2016-02-25/163906.html. Accessed 23 Mar 2019
4. Hustrulid, W., Kuchta, M., Martin, R.: Open Pit Mine Planning & Design, vol. 2, 3rd edn. CRC Press/Balkema, Leiden (2013)
5. Peroni, R.: Planejamento de Lavra a Céu Aberto. Curso de Lavra a Céu Aberto, 29 p. Departamento de Engenharia de Minas, Universidade Federal do Rio Grande do Sul, Porto Alegre (2007)

Advances in Underground Mine Planning Design and Computer Application

A Simulation Study of Underground Coal Mining Logistics and Roadway Development Performance

Ernest Baafi(✉), Senevi Kiridena, and Dalin Cai

Faculty of Engineering and Information Sciences, University of Wollongong, Wollongong, NSW 2522, Australia
ebaafi@uow.edu.au

Abstract. The rate of underground coalmine roadway development is normally affected by multiple factors including the support cycle and operational delays. A simulation study of underground coal development rates together with associated roof support operations showed potential for a 30% performance increase utilising current support technology and faster coal transportation by the shuttle car to the boot end. In the case of the material supply requirements at the development face, the simulation results showed that the delivery rate of the material supply system has only a minor effect on the roadway development rate, regardless of either the duration or frequency of the supply operation.

Keywords: Simulation · Roadway development · Logistics

1 Introduction

Most Australian coalmines strive to achieve higher roadway development rates; a rate of seven (7) meters per operational hour (MPOH) is common. These mines all face challenges of around the timely supply of materials to the roadway development face, and more broadly inadequate logistical systems. A study conducted by the Australian Coal Association Research Program (Gibson 2015) concluded that amongst the major factors that affect the efficiency/ability of the underground logistical systems are the extensive traveling distances, limited transportation provisions and poor mine floor conditions for mobile equipment.

Generally underground mining logistics, constitutes integrated systems consisting of different forms of material handling along with materials/personnel/equipment movement, information flow and activity/operations control sub systems (Feng et al. 2010). Any attempt to optimise these sub systems separately, without considering the often dynamic and complex interactions might not yield the desired results. Simulation tools have been used to visualise and understand the dynamics of underground mining logistics systems, as well as evaluating them for improvements. Feng et al. (2010) highlighted the need for studying what they called the 'production logistics system' to improve underground coal mining operations from both the economic and safety perspectives. To this end, they advocated a comprehensive approach, which included the development of a simulation model using the software tool WITNESS to identify

E. Topal (Ed.): MPES 2019, SSGG, pp. 131–138, 2020.
https://doi.org/10.1007/978-3-030-33954-8_16

bottlenecks in an underground coalmine logistical system. Feng et al. (2010) focused on modelling and analysing the whole coal transportation system, i.e. from the shearer cutting coal at the longwall face through to the coal transportation to the surface. They established a relationship between the shearer cutting speed and coal output, and also the relationship between the speed of each run-time and the working face productivity.

2 Discrete Event Simulation (DES)

Page Jr (1993) defines simulation as "The use of a mathematical/logical model as an experimental vehicle to answer questions about a referent system" and concludes, "arriving at the correct decision is the overriding objective of simulation". According to Shannon (1998) simulation is "a process of designing a model of a real system and conducting experiments with this model for the purpose of understanding the behaviour of the system and evaluating various strategies for the operation of the system". Nance (1993) identified the six characteristics of a discrete event simulation language including:

- Generation of random numbers to represent uncertainty.
- Process transformers to permit other than uniform random varieties to be used.
- A list processing capability so that objects can be created, manipulated and deleted.
- Statistical analysis routines to provide a descriptive summary of model behaviour.
- Report generation to provide a presentation of potentially large amounts of data in an effective way for decision making.
- A timing executive or time flow mechanism.

Inside a simulation model are logic statements and tangible elements (or entities) e.g. shuttle cars and a continuous miner. The logical relationships link different entities together and define the overall behaviour of the model. Another key part of a simulation model is the simulation engine or executive that controls the time advance. A central clock is commonly used to keep track of time, so the executive controls the logical relationships between entities and advances the clock to a new time to provide the dynamic behaviour of the model. Two other elements that are vital to any simulation software are the random number generator and results collation and reporting. The random number generator provides random behaviour typical of the real world.

3 Discrete Event Simulation (DES) Model

Figure 1 shows the typical working area of a discrete simulation model, MINESIM, developed by Cai (2015) to simulate the roadway development activities using a general-purpose simulation package, FlexSim. MINESIM was configured and validated using data collected from various operating Australian underground coalmines. A simulation model MINESIML, is a refined version of MINESIM with the capability of handling material supplies to the face.

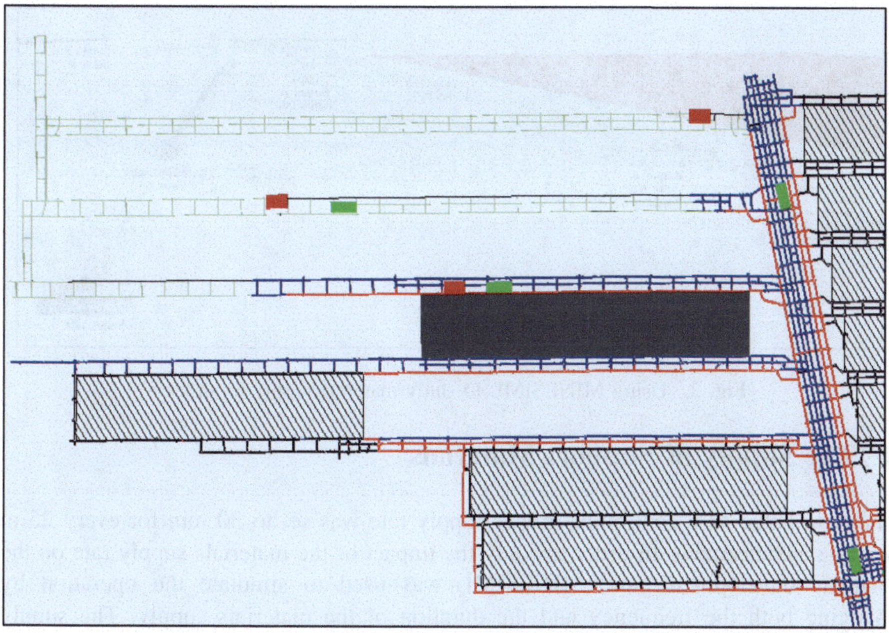

Fig. 1. An underground working area of MINESIM DES model

Sensitivity analysis was performed using MINESIML (Fig. 2) to identify bottlenecks/constraints, and highlight the production potential of the logistical component of the roadway development in the following ways:

i. identifying the deficiencies of the existing logistics system, e.g. process bottlenecks.
ii. evaluating alternative process improvement initiatives, e.g. addition of resources.
iii. assessing the impact of adopting new technology or alternative transport/infrastructure systems, e.g. monorails, and
iv. examining the interplay between the logistical system capacity and future expansion or growth strategies, e.g. production capacity, infrastructure upgrade and mine expansion.

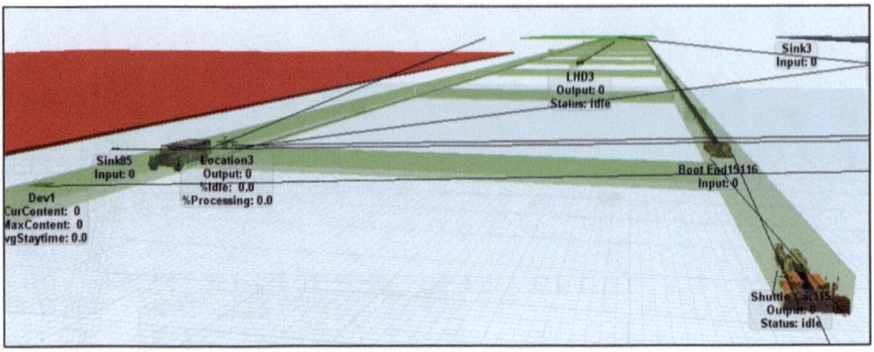

Fig. 2. Using MINESIML to study material supply to an area

4 The Supply of Support Materials

The base value of a continuous miner supply rate was set to 30 min for every 25 m heading advancement. In order to study the impact of the materials supply rate on the roadway development rate, MINESIML was used to simulate the operation by changing both the frequency and the duration of the materials supply. The supply frequency was set to a series of intervals, 15 m, 20 m, 25 m, 30 m, 35 m and 40 m of heading developments; while the duration was set to range from 10 min to 40 min. The simulation was performed for the entire set of roadway development operations, which include cutting, loading, supporting, miner supply, extending ventilation tubes, cutting the breakaway and stone dusting. The idea was to simplify the process as much as possible in order to better understand the influence of the material supply to the development panel. The base model of the continuous miner was set with the following parameters without panel advance delays, breakdowns and shift schedules:

- pillar size: 125 m × 71.8 m (66.6 m centre to centre)
- overdrive: 25 m
- boot end to cut through: 20 m
- cut depth: 0.5 m
- cut time: 1.617 min (87 s cutting time + 10 s positioning time)
- time before cutting: 0.5 min
- time after cutting: 0.5 min
- support time: 17.0 min/m
- Shuttle car (SC) discharge time: 1.6 min (86 s +10 s positioning time)
- SC speed: 96 m/min
- cut breakaway time: 300 min
- extend vent tube: 2 min every 4 m
- stone dusting: 60 min every 50 m or every 24 h which comes first
- miner supply: 30 min every 25 m

The simulated results are shown in Fig. 3. As can be seen from Fig. 3, the duration of material supply has a linear relationship with the roadway development rate across all supply intervals. Specifically, the slope ratio increases if materials supply is more frequent. However, with four times the duration change, the development rate only changes from about 3.33% at the least frequent supply to about 8.33% at the most frequent supply, which means that the material supplement duration only has a minor influence on the roadway development rate.

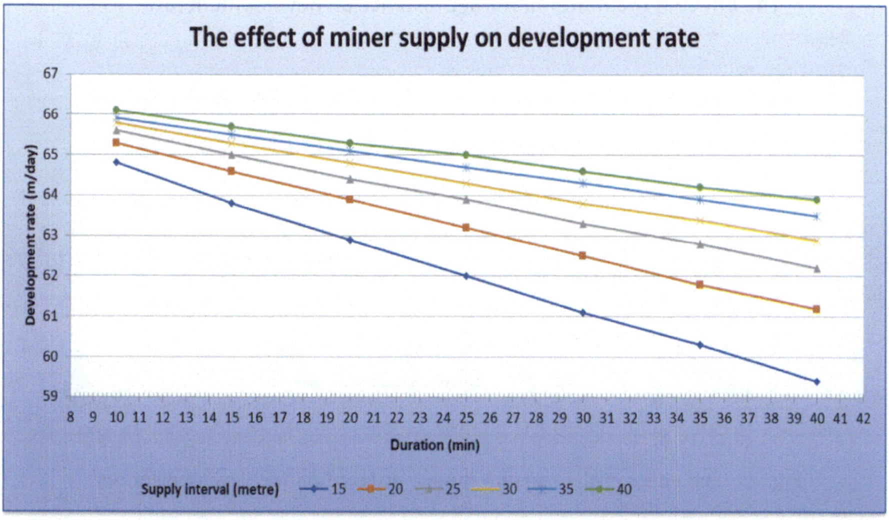

Fig. 3. The effects of miner supply on development rate

A further simulation was performed to study how the distance from the material storage location to the continuous miner affected the roadway development rate. The following simulation assumptions were made:

- the materials were transported by a LHD with a travel speed of 80 m/min both loaded and empty;
- the total time to load the LHD and discharge materials from the LHD is 10 min fixed;
- the miner stops when material supply is called for and when the LHD starts to load, the miner resumes work when the LHD leaves the panel working zone;
- the frequency of the material supply is at 25 m intervals.
- the two continuous miners are supplied by two LHDs without interference or interaction;
- the distance is measured from the last cut through of the tail heading to the material storage location. Therefore, the distance from CM007 is 66.6 m longer than that from CM008. In this study, the distances were set to range from 50 m to 550 m at an interval of 100 m.

The simulated results are depicted in Fig. 4. As can be seen from Fig. 4, the material storage distance has a linear relationship with the roadway development rate. However, the effect is minor, where the change is only 0.7 m per day (1%) with about

500 m difference over the total distance. For every 100 m decrease in distance between the material storage and the development heading, there is only about a 0.2% improvement in the development rate.

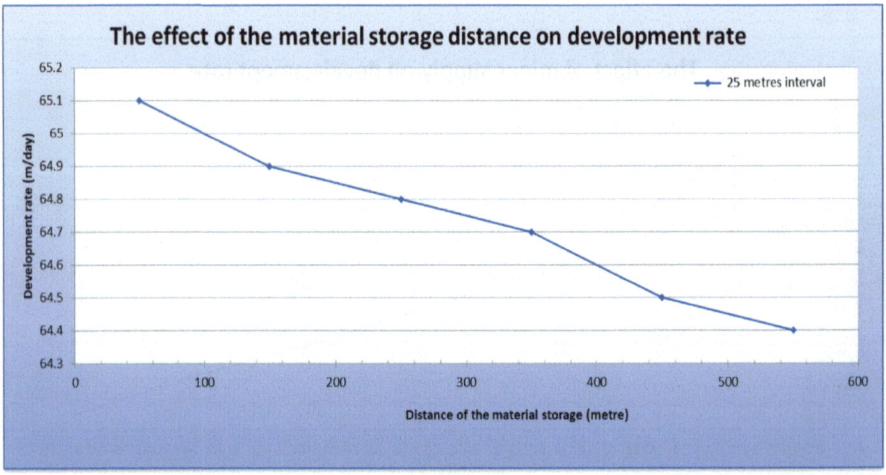

Fig. 4. The effect of the material storage distance on the development rate

From the simulation and analysis of the material supply to the roadway development headings, it can be concluded that material supply has only a minor effect on the roadway development rate, with respect to the duration, frequency of the material supply operation and the distance to the material storage. However, as observed in actual practice, logistics does have a noticeable effect on the roadway development performance. As can be seen from the simulation, the effect does not come from the material supply of the development panel, nor the operation, the equipment, or the location of the material storage hub. Therefore, the actual logistics effect may come from other parts of the operation, such as the communication and scheduling of the material supply from the surface of the mine to the panel material storage hub or machine breakdown, which usually causes a long-time delay.

5 The Effect of Downtime and Scheduling Delay of Material Supply on the Development

Based on the analysis above, a long delay away from the development face might be the major cause of the material supply constraint, rather than the supply interactions at the working face, which have a minor effect. A further study was done by integrating random and scheduled long downtimes to the material supply process using the base value of a 10 min fixed delay plus a travel delay when the materials are stored 50 m away from the heading. The following model assumptions were made:

- Case 1: uniformly distributed delays from 2 h to 4 h and 3 h to 6 h for every 40 m developed

- Case 2: uniformly distributed delays from 2 h to 4 h and 3 h to 6 h happening randomly every 8 h to 16 h which means a delay per every one to two shifts (about every 3–5 shifts with actual breakdowns and scheduled down time which is about 60%–65% of total calendar time);
- Case 3: integrating delays of both Case 1 and Case 2.

The simulation results are as shown in Fig. 5. As can be seen from Fig. 5, the integrated delays have a very large effect on the roadway development rate. An average two hours delay for every 40 m advancement can cut down the total development rate by 15%, while a 22% development rate drop would be caused by an average two hours delay at a frequency of about every 12 h. The effect of both type of delays caused a 39% drop in the development rate when compared with the base model. With a one-hour incremental delay applied to both types of delay a 51% decrease in the development rate was simulated. Similarly, an average three hours delay for every 40 m advancement can cut down the total development rate by 20%, while a 33% development rate drop would be caused by an average three-hour delay at the frequency of about every 12 h. The simulation results supported the opinion that the random long-time delay of logistics causes the major development problem.

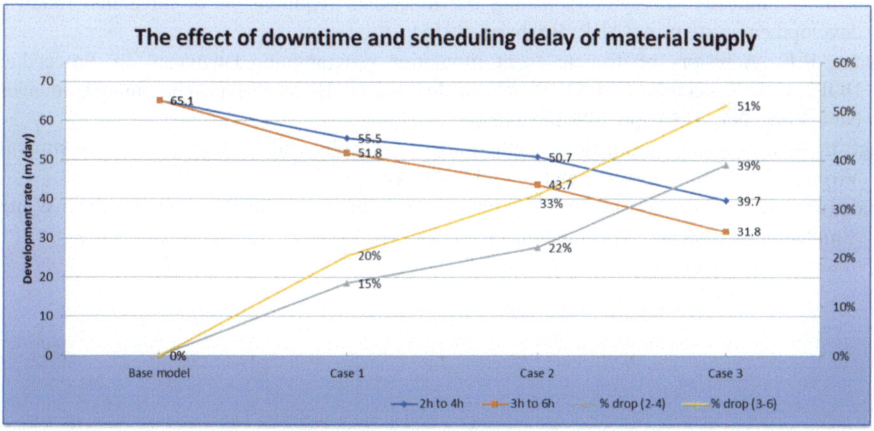

Fig. 5. The effect of downtime and scheduling delay of material supply

The two major logistical activities in the roadway development were the coal transport out of the working face and material supply to the working face. The simulation results show that the major logistical problem was not caused by coal transport from the working face or material supply to the working face when the operation is support constrained. More specifically, if the support operation can be improved to 12 min/m from the current support operation at 17 min/m, the overall performance can be improved by 20%. If the 12 min/m support operation can be achieved, the performance can further be improved by another 10%, i.e. 24 m advance rate to more than 26 m a day advance utilising a faster shuttle car for coal transport to the boot end.

6 Concluding Comments

A validated underground coal mining FLEXIM model focusing on roadway development and relevant logistical systems have been developed and utilized for this study. In the case of material supply requirements at the development face, the simulation results showed that the material supply system only has a minor effect on the roadway development rate, regardless of either the duration or frequency of the supply operation. Further simulation studies demonstrated that the distance between the material storage hub and the development face has a linear relationship with the roadway development rate.

References

Cai, D.: Using FlexSim to simulate the complex strategies of longwall mining production systems. Ph.D. Thesis, University of Wollongong, 260 p. (2015)

Feng, X., Zhao, Z., Sun, J.: Establish of coal production logistics simulation system based on WITNESS. In: Proceedings of International Conference on Electrical and Control Engineering, Wuhan, China, pp. 915–918. IEEE (2010)

Gibson, G.: Industry survey- establishing the functional requirements of integrated roadway development system. ACARP, Brisbane (2015)

Nance, R.E.: A history of discrete event simulation programming languages. In: Proceeding HOPL-II The second ACM SIGPLAN conference on History of programming languages, New York, NY, USA, pp 149–175 (1993)

Page Jr, E.H.: Simulation modelling methodology: principles and etiology of decision support. Virginia Polytechnic Institute and State University (1993)

Shannon, R.E.: Introduction to the art and science of simulation. In: Proceedings of the Winter Simulation Conference. Washington, DC, USA, pp. 7–14 (1998)

Optimization and Sequencing a Semiautomated Ramp Design in Underground Mining: A Case Study

Sergio Montané[(✉)], Pierre Nancel-Penard, and Nelson Morales

DELPHOS Mine Planning Laboratory, AMTC and Department of Mining Engineering, University of Chile, Santiago, Chile
smontane@delphoslab.cl

Abstract. Access ramp design in underground mining is an important challenge to mining engineers, however, there are few optimizing methodologies that may guide the process. These methodologies consider only very simple costs when they made the economic evaluation of the project, because they design access first and then they complete the extraction schedule.

This work presents a methodology to generate ramp designs in underground mining, their development plan and the extraction of ore plan, and then apply it in a gold and silver mine extracted by Bench & Fill method. We consider two models to approach the problem, which are combined. A first model determines a ramp design that aims to minimizes both CAPEX and OPEX without considering time. The second model considers a given ramp and maximizes the net present value (NPV), i.e., it schedules the construction and production over time. Both approaches work on the block model and rely on mathematical programming for optimization. The methodology uses both models iteratively to find a design with high NPV.

We apply the methodology to a case study consisting of three zones of a Bench & Fill mine, for which different scenarios are compared, including an initial design developed using traditional design methods. The results show that the methodology improves the NPV in comparison with initial designs. Interestingly enough, the result also shows there is not necessarily a direct correlation between NPV and costs, that is, the designs with best NPV are not necessarily those with the lowest costs.

Keywords: Ramp design · Scheduling · Underground mining

1 Introduction

In the mining industry, there is constant research in methodologies which can improve processes and get the best feasible economic indicators. However, in underground mining, there is little research around access ramp design and tasks scheduling.

[1] and [5] identified the opportunity to develop algorithms for access design in an underground mine to maximize NPV, so a genetic form to solve this kind of problems is proposed. A workspace is created to develop a tool to obtain an optimum access ramp design to the mine. [5] proposed the discounted junction point algorithm (DJPA)

E. Topal (Ed.): MPES 2019, SSGG, pp. 139–145, 2020.
https://doi.org/10.1007/978-3-030-33954-8_17

which must find the location of junction which was formulated with using Steiner networks. The proposed model assumes that ramp links are straight lines, so the proposed theory and algorithm can only be applied directly to underground mines in which terminal points are in a near-horizontal plane to satisfy the gradient constraint, i.e., they cannot be used in mines with several production levels.

On the other hand, [4] determined that the lack of available software to assist the production scheduling in underground mines has implied that this task continues being a hand-made procedure, which it includes also an intensive and complex use of spreadsheets. Thus, there is no doubt that this is a highly tedious and time demanding process where a feasible solution can be found, but there is a minimum possibility to obtain the optimum solution, in some sense.

[3] proposed a heuristic with the optimization model capable of obtaining operative ramp designs in a little time, which uses some operational and economical inputs. The input parameters focused on development cost and operational cost of ramps because the objective is to minimize the total cost to obtain a design. This work does not consider temporal factor so the evolution of mine development and operation are not included, because of this the decision is set by the minimum cost.

In this paper, we propose a methodology from the work made in [3] to assist ramp design in underground mines while economic benefit of the project is maximized, and then we obtain a ramp design with development and production schedule.

2 Methodology

The proposed methodology in this paper utilizes two tools to work: a heuristic developed to obtain preliminary ramp design while it minimizes costs, and UDESS, a scheduling tool that maximizes NPV of the project.

The heuristic mentioned uses several design parameters to achieve a preliminary operational ramp and some cost values according to the case to minimize the total cost of configuration. This model returns a ramps and crosscuts configuration from inputs like connection points location to access to production levels, start point and endpoint and tonnage to extract.

[2] UDESS is a scheduling tool which needs three inputs: activities, precedences, and constraints. Activities are the tasks that we want to manage in a schedule; the precedences are the relations among activities that set the requirements that need each one to start to develop it; the constraints set limits in using of some resources that are necessary to carry out the activities.

First, we must determine the project in which we are going to work, so a block model is necessary to have a spatial reference of the mine and it allows to define the available zone to design ramp and crosscuts. Then, the inputs of heuristic are set according to each case: gradient, curvature radius, development cost, operational cost, ventilation cost, production levels (quantity, location, and tonnage to extract from them).

When those inputs are got into, the heuristic is executed to obtain a preliminary ramp and crosscuts configuration to refine. With this design, we set activities, precedences, and constraints. Activities are the segments of ramp, crosscuts and production sectors from where ore is extracted. Precedences set the sequence in which the tunnels must be developed and ore from levels must be extracted, and they define relationships among production levels like the direction of exploitation (for example, from bottom to top). Constraints restrict the resource consumption to carry out the tasks, like available hours of equipment, maximum extraction capacity per period, etc.

Later, we use UDESS to obtain a development plan of ramp and crosscuts, and ore extraction plan from production levels and NPV of the project. With those results, we adjust the development, haulage and ventilation costs value with the time value according to the plan, so the previous costs are changed. We run the heuristic again with the new values and changes in the obtained design are analyzed. If the configuration of ramp and crosscuts vary, we carry on the next steps, otherwise, we have a final design, because the possible changes that it can have will be very mild. This is an iterative process with using heuristic and scheduling tools to obtain the Gantt chart of the project with its NPV (Fig. 1).

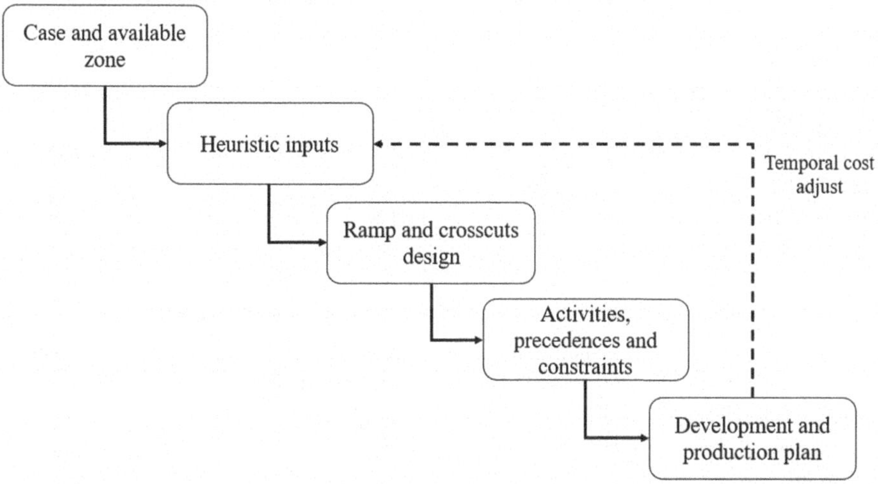

Fig. 1. Proposed methodology steps.

3 Case Study

We used a gold and silver mine which contemplated Bench & Fill method to extract the ore. The data was provided by an engineering mine company who made a pre-feasibility study, which included from geostatistical study until final economic analysis.

The mine has three exploitation zones: Y East mine, Y West lower mine, and Y West higher mine. The first zone had 26 production levels, but the design had to reach an access to 13 main drift because in these levels the drift system was implemented, which allows to access to two adjacent levels, so the number of developed tunnels can be reduced. The second zone there were 11 drifts to access; therefore, there were 22 production levels. Finally, the third zone did not have the drift system; there were 15 production levels to access directly.

The extraction method considers accessing to the mine by a shaft from surface to the bottom of each mine zone and starts to develop the ramp and crosscuts from there until the higher production level. Thus, as the development is advancing to higher production levels, the journeys are increasingly longer.

From the case study, we took two ramps and crosscuts designs base case to apply the proposed methodology. The first one is a simulation of real design and the second one is a design which seeks to minimize total cost associated without considering the time.

Tables 1 and 2 show the design parameters obtained when we used methodology with three zones of the case study. Each zone was considered independently from the others.

Table 1. Comparison of design parameter results in simulation case.

	Scheduling of design obtained by simulation	Application of methodology in design obtained by simulation	Variation
Length developed [m]	8,178	8,891	+8.7%
Development cost [USD/ton]	1.22	1.29	+5.7%
Haulage cost [USD/ton]	1.01	1.01	−0.2%
NPV [MUSD]	1,008	1,019	+1.1%

Table 2. Comparison of design parameter results in alternative case.

	Scheduling of alternative design	Application of methodology in alternative design	Variation
Length developed [m]	9,978	10,285	+3.1%
Development cost [USD/ton]	1.41	1.49	+5.6%
Haulage cost [USD/ton]	1.24	1.30	+4.4%
NPV [MUSD]	994	1,011	+1.6%

We can note that in both scenarios the use of methodology allows obtaining a better NPV of the project. In general, length developed, development cost and haulage cost increase, because the methodology makes a trade-off between length and NPV. It is a problematic situation because the increasing total cost sometimes is a difficult scenario when the investors are not willing to risk. But in other scenarios, the investors are willing to risk, and they prefer to pay more expenses in early periods and obtain better NPV of the project. This decision depends on the aim of the company and politics in the state where the project is.

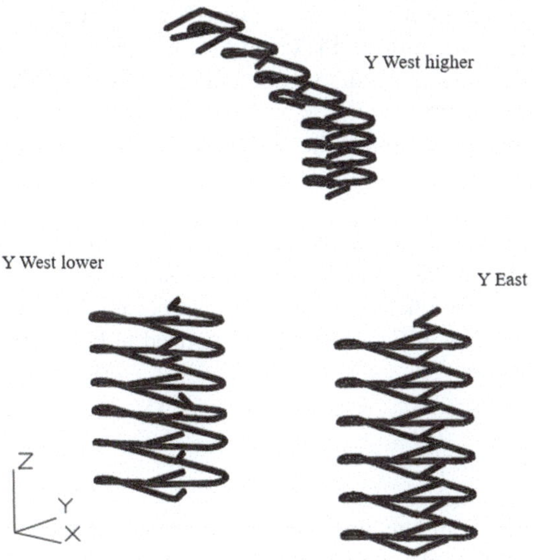

Fig. 2. Ramp and crosscuts configuration.

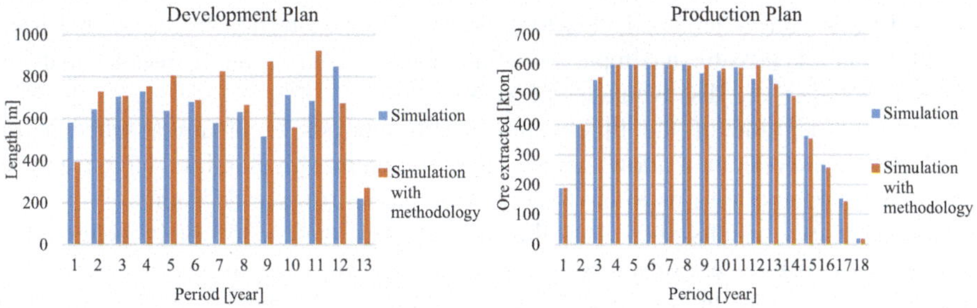

a. Simulation case.

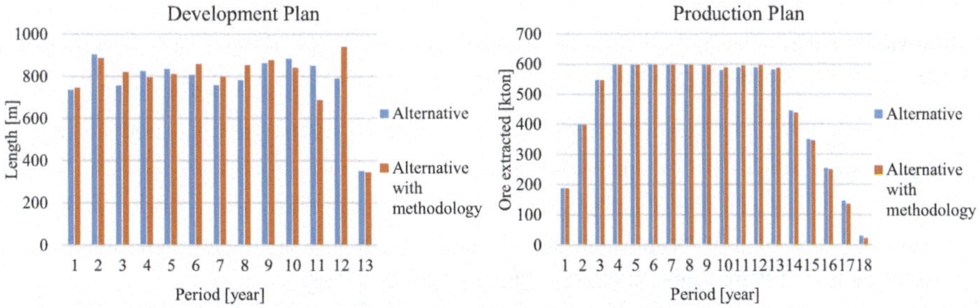

b. Alternative case.

Fig. 3. Development and production plans.

An interesting observation is the second case because in the static methodology it was a good solution to minimize OPEX and CAPEX, but when we consider temporality in this analysis, the sequencing shows a different behavior because of the trade-off to obtain a better benefit.

In Fig. 2 we can note that when the design is developing upper levels, distance is increasing, and crosscuts are longer. This explains the different development plan in both cases.

Figure 3 shows the evolution in the development of ramp and crosscuts period by period. When we applied the methodology, the ramp and crosscuts configuration was different, so it allowed the change production plan because it was possible to extract more ore in previous periods. This tendency is noticed in both case and it can explain the better NPV of them.

4 Conclusions

The designs obtained with the proposed methodology in both cases are operative because they accomplish all operational constraints to allow equipment work and generate a possible plan to perform.

This methodology is capable to assist in designing and scheduling of the access development for the mine and shows its impact on production performance to maximize NPV. On the other hand, this tool allows to obtain results in reasonable time execution, so it is possible a comparison of alternative scenarios and makes an analysis of the advantages and disadvantages of each one.

The incorporation of the time value in this analysis to obtain a schedule of development and operation allows achieving a solution that gives a time reference to be carried out, which allows us to have more information to make decisions.

In both cases, results show that the methodology prefers to obtain better NPV despite increasing meters developed. It is an interesting analysis because while the benefit is higher, sometimes companies are not willing to pay more costs. It would be a good analysis to make in future, changing the objective function to minimize cost while a minimum NPV is required as a constraint to improve the base case.

Acknowledgments. The authors would like to thank CONICYT Basal Grant AFB180004 - Advanced Mining Technology Center.

References

1. Brazil, M., Grossman, P., Rubinstein, J.H., Thomas, D.: Improving underground mine access layouts using software tools. Interfaces **44**(2), 195–203 (2013)
2. Espejo, N., Nancel-Penard, P.: Modelos matemáticos UDESS. Technical Inform (2017)
3. Montané, S., Nancel-Penard, P., Morales, N.: Methodology to optimize and sequence the semiautomated ramp design in underground mining. In: Proceedings of the 27th International Symposium on Mine Planning and Equipment Selection-MPES 2018, pp. 143–151. Springer, Cham (2019)

4. Nehring, M., Topal, E.: Production schedule optimization in underground hard rock mining using mixed integer programming. In: Proceedings-Project Evaluation, pp. 169–175. The Australasian Institute of Mining and Metallurgy, Melbourne (2007)
5. Sirinanda, K.G., Brazil, M., Grossman, P.A., Rubinstein, J.H., Thomas, D.A.: Strategic underground mine access design to maximize the net present value. In: Advances in Applied Strategic Mine Planning, pp. 607–624. Springer, Cham (2018)

Developing a Tool for Automatic Mine Scheduling

Kateryna Mishchenko[1(✉)], Max Åstrand[1], Mats Molander[1], Rickard Lindkvist[1], and Torbjörn Viklund[2]

[1] ABB Corporate Research, Västerås, Sweden
{kateryna.mishchenko,max.astrand,mats.a.molander, rickard.lindkvist}@se.abb.com
[2] New Boliden, Boliden, Sweden
torbjorn.wiklund@boliden.com

Abstract. Presented is an automated short-term scheduling for underground mines which is an important part of the overall mining process.

The presented algorithm is a constructive heuristics applied for scheduling of the production cycles of a cut-and-fill mine. The goal of the heuristics is to minimize the finishing times for each task. To achieve this, the tasks as scheduled as early as possible.

The automated schedules were validated at the New Boliden mine both in offline and online tests and now the algorithm is going through later steps of product development.

The main result is the creation of high quality schedules, outperforming existing manual ones in terms of both computational time and quality. The auto scheduler was tested and compared to manual scheduling. The results show a 10% increase in ore production and produce always feasible schedules with respect to all operational limitations. Thereat, the automated algorithms take minutes versus hours for manual scheduling.

Keywords: Short-term scheduling · Heuristics · Cut-and-fill

1 Introduction

Underground mining which is a process of extracting of minerals under the surface giving a profit, is divided into 5 stages, see [3]. Profitability of the process depends on the performance of these stages, where one of them is due to short-term scheduling including the activities that need to be conducted for reaching a certain production goal. For each activity, a set of necessary resources must be allocated and the time for accomplishing each operation needs to be specified.

Since underground mining process includes some specific features that make manual scheduling cumbersome, complex and time consuming, mining companies seek to automate the scheduling using algorithms. By introducing these algorithms, one can create standardized and high-quality schedules within seconds or minutes instead of hours or days as is the case of manual scheduling.

Algorithms for scheduling depend on the types of mines, underground mining methods and nature of material excavated, see [1, 3, 4]. This area is relatively new due

E. Topal (Ed.): MPES 2019, SSGG, pp. 146–153, 2020.
https://doi.org/10.1007/978-3-030-33954-8_18

to more complex nature of underground operations. The models generally of mixed-integer type, and the solution algorithms are either exact methods or heuristics.

Due to computational complexity of the scheduling problems, heuristics is the most common approach used for scheduling. These methods are fast and could be quite efficient for delivering good enough solutions. They are even more valuable in cases when it is impossible to solve the problem in any other way. Nevertheless, the issue with a heuristic approach could be the optimality of the schedule obtained.

This paper summarizes the main findings after many years of research, prototyping, and testing of systems for closed-loop automatic scheduling for underground mining. This work disseminates a long collaboration between the technology supplier ABB (Asea Brown Boveri) and the mining company New Boliden.

2 Constructive Heuristic for Automated Scheduling

This paper deals with short-term heuristic scheduling for cut-and-fill. This mining method is suitable for steep ore bodies that require high selectivity [1]. In a cut-and-fill mine there are many active locations, i.e. there are numerous production cycles taking place at a multiple faces using the same fleet of mobile machines. The coordination of the machines and operators by scheduling is non-trivial and it is part of the reason why even in modern mines the machine utilization can be as low as 50% [2].

The approach used in this paper is called by constructive heuristic. These kind of procedures work with a number of unscheduled tasks starting with the task on top of the list and build schedules sequentially one by one. The goal of the heuristic presented is to minimize the finishing times for each task. To achieve this, the algorithm is constructed in a way to schedule tasks as early as all requirements for doing that task are fulfilled subject to some constraints.

2.1 Algorithm for Pilot Heuristic

The algorithm creates short-term scheduling for tasks in the production cycles of an underground mine and implemented in MATLAB. With a production cycle we mean the sequence of repeated tasks executed at production locations in order to excavate ore. This algorithm could be easily extended to handle other production tasks as maintenance or activities related to building infrastructure.

To create a valid schedule, several parameters of the mine needs to be known.

The main input are the unscheduled tasks, information about the tasks usually includes the order of the tasks at each face, which is defined by the production cycle. The tasks scheduled by this algorithm are of two types: the "fixed" tasks that cannot be shifted in time (e.g. safety inspections) or casual tasks from the production cycle e.g. drilling.

Information about the number and type of available and required resources, data about the kind of resources used for each task, as well as competencies of the operators including the list of machines they can operate, need to be known. Finally, input to the scheduling includes some extra information as blast window, shift hours, etc.

The output is a schedule that allocates necessary resources and determines the start and end time of all tasks.

Based on all inputs above, a schedule fulfilling the requirements on the production tasks and resources is constructed. Additionally, all local processes as well as safety rules should be satisfied, too.

Pseudocode for pilot heuristics for scheduling

```
while {unplanned tasks exist}
    Create the set TS of all first unscheduled tasks of
    each drift
  for {all tasks in TS}
    if there are tasks that are not fixed in time then
      Select the task T_NotFixed in TS
        if max number of tasks for this task type has  not
        been reached then
          Select this task as the next task, T_next = T_NotFixed
           Update the start time s_next
        and estimate the end time e_next accordingly
        end
    end
    if there are tasks fixed in time then
      Assign s_fixed to be the start time of the fixed task
      if e_next > s_fixed then
        Reassign T_next to be the fixed task
        Update the start time s_next accordingly
      end
    end
  end {all tasks in TS}

  If the current task cannot be scheduled before the next
  upcoming blast window then
    update the earliest possible start time s_next to the end
    of the blast window
  end

  Determine the set of machines MS that can process the
  selected task T_next
  for {each machine M in MS}
    Determine the set of operators OS that can operate
    machine M
    for {each operator O in OS}
      Calculate performance measures F(O,M) based on the
      combination of machine and operator
    end {loop for operator}
    Schedule machine M and operator O that maximize
    F(O,M) to process T_next at start time s_next
  end {loop for the machines}
end {while}
```

In this algorithm, the performance measures F(O,M) calculates the suitability of assigning machine M and operator O for the selected task. For confidential reasons this function cannot be disclosed.

3 Results and Numerical Evaluation of Heuristic

This section presents the results of the scheduling based on the heuristics described above using real operational data from the New Boliden mine.

In this mine, weekly plans are created manually during each week-end. The staff then tries to follow the plan during the next week, but numerous adjustments are normally required during the week.

To test the proposed algorithm and minimize the interference with the daily work, a framework interfacing the Gantt-system used in the mine for manual scheduling was developed.

3.1 Evaluation of the Heuristic Scheduling

Before presenting numerical results based on comparison of heuristic and manual scheduling, let address the problem of the intractability. As an example, consider the scheduling problem with 100 tasks and 5 machines. Even for this small problem, the number of feasible combinations of machine queues (ignoring the possibility to change execution orders in the queues) would be in the order of 5^{100}. Such a complexity of the problem makes the CPU time of exact methods being of the order of 1052 years. Quite often scheduling problems are of bigger size, which means that it is practically impossible to find a proven optimal solution to these problems.

The upper part of Fig. 1 depicts a part of the manual plan scheduled for the part of the mine and the lower part of the Fig. 1 shows the part of plan generated by heuristic. Both schedules are visualized by Mine Gantt viewer. The complete schedules are not presented to make the figure easier to read.

In order to properly compare schedules presented by Fig. 1, the complete plans with all activities and resources included have to be analyzed. Note that despite that manual schedule ends up earlier, the schedule produced by the automated algorithm is preferred, since the manual schedule violates production constraints.

Evaluation of the schedules is done by comparison of two performance metrics:

1. Finishing times for tasks in each drift obtained by manual and heuristic scheduling;
2. Blasting pace. It is an accumulated number of blasts as a function of time.

Here, but drift it is meant a production location where blast cycle is executed. Only one task at a time can be executed at one drift.

Figures 2 and 3 present the results of testing of one week scheduling after two weeks. This test includes 31 machines, with 4 machines feasible for each task and 42 operators divided into four shift teams included in the schedules. As mentioned, the scheduling was applied for the part of the mine.

Figure 2 presents finish times for the tasks in the drifts scheduled. It includes the manual plan, automated plan generated using the heuristic as well as the plan when one

Fig. 1. Example Mine Gantt View, the manual (top) and heuristic (bottom) schedules

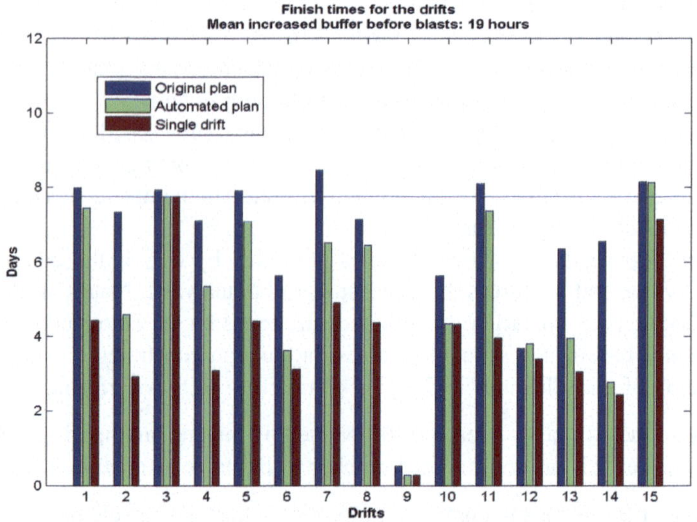

Fig. 2. Comparison of end times of manual and automated schedules after 2 weeks

drift at a time is considered, with all resources devoted only to that drift. This latter case is used as a benchmark to measure the best theoretical end time for all the tasks.

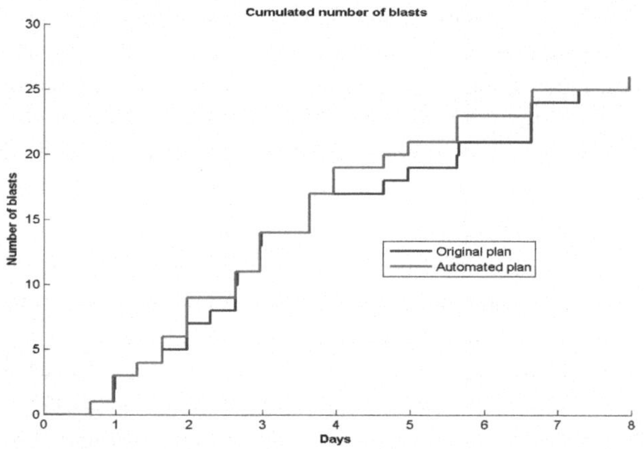

Fig. 3. Comparison of blasting pace for manual and automated schedules after 2 weeks

A blue horizontal line in the figures shows the theoretically earliest time when all tasks in the schedule can be completed. The mean increased margin before the blasts were calculated with a rather impressive result of 19 h after the second week.

Figure 3 shows an increase in the production rate by applying the automated schedule versus manual one.

To compare schedules, a rolling horizon for the schedules must be used, where the state of the mine is continuously fed into the algorithm in a close-loop fashion. Several weeks of production were compared, and the schedules constructed by the heuristic managed to pack the same amount of blast tasks as the manually constructed schedules in 10% less calendar time. Roughly, this results corresponds to 10% increase in amount of ore extracted. Note that those results took minutes to obtain new schedules compared to hours for manual scheduling.

4 Discussions and Conclusions

4.1 Feedback from Test in New Boliden Mine

The automatic scheduling framework was tested in the New Boliden mine. The schedules constructed by the algorithm were verified by mine operators. Before presenting the main findings from the tests and discussions with the mine site, it is important to mention that close collaboration and direct communication between the mine planner at the New Boliden and the ABB developer was a crucial success factor.

After the week of pilot testing of the algorithm, the mine planners noted the following advantages in automating the scheduling process:

- Fast rescheduling: Frequent machine breakdown and activities delay should be addressed by correspondent revision of the current schedule. Introduction of the automated scheduling helps the mine planner to do this re-planning work in fast and efficient way, independently on any external factors.

- Predictable planning: By introducing supportive algorithms, the scheduling process becomes more predictable since the schedules are created under a common set of business rules, rather than by the individual expertise of the mine planner.
- High quality: One strength of the automatic scheduling is that it always constructs feasible schedules, which need not be the case in manual scheduling. By helping the mine planner to instantly construct accurate short-term schedules, a lot of mine planner's time can be used to focus on other important tasks.

It was also noted that for full benefit of automating scheduling, reviewing the whole planning process, routines and interactions between short and long-term planning, mine engineering, as well as maintenance and service, is also required.

At the same time, when working on formalizing the short-term scheduling process, it becomes evident where standard practices are missing. Thus, the process of formalizing the business rules is an important step towards operational standardization. All this leads to another important point, which is the quality of the data such as task durations, process descriptions, and machine sets. Thus, the quality of the data is seen as a prerequisite to possibility to generate efficient and stable schedules.

Talking about the potential improvements of the algorithm, the following point should be mentioned:

- The algorithm creates a schedule based on default duration of activities and it was obvious that the planning process and its parameters need to be reviewed before the automatic scheduling product goes in full operation.
- Adjusting the algorithm is done by several parameters and it is not always obvious for the operators how to tune these parameters to reach a certain behavior.
- Even though algorithms are important, the usability of the solution is equally important.

All these findings were acknowledged and remedied in final product development.

4.2 Conclusions and Future Work

This paper addresses automating the short-term scheduling process in underground mining. In daily operations, scheduling, rescheduling, and analyzing the short and long-term effects of a schedule, is a complex and time consuming task. This is especially true for unexperienced mine operators or in stressful situations requiring fast actions. In case of automatic scheduling, any mine operator can do fast and efficient scheduling without risk of making human errors. Which in turn, helps to continuously follow up e.g. weekly production KPI's.

In this work we have introduced and analyzed a constructive procedure for creating short-term schedules in a fast and efficient way. The automatic scheduling algorithm presented in this paper is going through later steps of product development at the moment, and it is being implemented as a part of the EU-funded project Sustainable and Intelligent Mining Systems (SIMS). In continuation, it would be interesting to expand the scope to include optimization taking into account down-stream processes such as transportation or even the state of the concentrator plant. Moreover, a tighter integration of the scheduling model with the mine maps would give a significant contribution into development of new generation of automated underground mine scheduling algorithms. In this case the expansion of new and existing mining areas can be easily integrated into scheduling, additionally other auxiliary parameters like fueling could be included in an easy way.

Finally, the quality of the schedule can be improved by "what if" (sensitivity) analysis, since sensitivity of the schedule with respect to a change in number of resources gives an important information about the potential bottlenecks of the schedule. It helps mine operators to choose the most suitable schedule as well as re-plan the current schedule according to the changes in operation conditions.

References

1. Hustrulid, W.A., Bullock, R.L: Underground Mining Methods: Engineering Fundamentals and International Case Studies. SME, Littleton (2001)
2. Gustafson, A., et al.: Development of a markov model for production performance optimisation. Application for semi-automatic and manual LHD machines in underground mines. Int. J. Min. Reclam. Environ. **28**(5), 342–355 (2014)
3. Newman, A.M., et al.: A review of operations research in mine planning. Interfaces **40**(3), 222–245 (2010)
4. O'Sullivan, D., et al.: Optimization-based heuristics for underground mine scheduling. Eur. J. Oper. Res. **241**, 248–259 (2015)

Resource Modelling and Geometallurgical Planning

Mineral Resource Classification Based on Uncertainty Measures in Geological Domains

Nasser Madani(✉)

School of Mining and Geosciences, Nazarbayev University,
Nur-Sultan 010000, Kazakhstan
nasser.madani@nu.edu.kz

Abstract. Mineral resource classification is of paramount importance for mining industry. The main challenge for this, however, is related to the geostatistical modeling approach, in which there is no unique algorithm for such a significant act. The deterministic approaches such as kriging, indeed is not proper, because of its smoothing effect and ignoring the proportional effect that lead to possible misinterpretation of kriging variance. As an alternative, stochastic simulation based on modeling the continuous variable can be employed. Besides of legitimate criticism against this approach, it is still usable for mineral resource classification. One of the dispute is related to setting parameters and choosing the optimum Gaussian simulation algorithm. In this study, an alternative is proposed in reliance on stochastic modeling of categorical variables rather than continuous variables such as estimation domains and rock types. The algorithm is founded on probability assumption, in which definition of thresholds for different categories can be manipulated with reference to opinion of the competent person as defined in JORC code.

Keywords: Mineral resource classification · Plurigaussian simulation · Lithology domaining · JORC code

1 Introduction

Mineral resource classification is necessary for public reporting and internal company assessments, financial institutions etc. [1]. This approach is based on the level of confidence inferred from geologically modeled block. Based on JORC code, a mineral resource can be classified into "Measured", "Indicated" and "Inferred", depending on the level of confidence (www.jorc.org). The modeling process for deriving the corresponding category, can be either deterministic or stochastic. The conventional approach of stochastic block categorization is mostly based on geostatistical simulation of continuous variable such as grade of interest. Although this approach is trustworthy; however, the choice of proper geostatistical simulation algorithm and setting the relevant parameters may be challenging [2]. In this study, an alternative of mineral resource classification is proposed on the basis of stochastic modeling of categorical variables (e.g., lithologies, rock types) rather than continuous variables through a probabilistic paradigm. The results are illustrated through an actual case study.

E. Topal (Ed.): MPES 2019, SSGG, pp. 157–164, 2020.
https://doi.org/10.1007/978-3-030-33954-8_19

2 Methodology

The proposed methodology is based on quantification of geological uncertainty. In this technique, underlying categorical values are available through the boreholes and one needs to calculate the uncertainty of each rock unit at unsampled locations. This step can be implemented by any stochastic paradigms such as multiple-point statistics, sequential indicator simulation and plurigaussian simulation [3], depending on the complexity of geological formation and their contact relationships. Plurigaussian simulation as an extension of the truncated Gaussian simulation is more capable of handling the complex contacts relationship among the geo-domains. In this context, the allowed and forbidden contacts can be injected into the modeling process. In a nutshell, it has found wide acceptance between the practitioners for modeling the petroleum reservoirs. Once the simulated categories are available at targeted locations, the uncertainty or probability of finding that rock unit different form others can be straightforwardly computed for each node. Through the probability maps, a model (one unique map) can be constructed by selecting, for each grid node, the most probable rock type domain. The value of this map is varying between 0 and 1, useful for quantification of uncertainty at target nodes. The high amount of this measure (close to 1) indicates that one is certain about the simulated value for that location irrespective of the type of simulated rock unit and low amount of this probability map (close to 0) implies that one is uncertain about the availability of that simulated rock unit irrespective of the simulated category. This interpretation explicitly pronounce the level of availability of the information such as sampled locations and boreholes, for which the confidence in simulated rock unit can be defined in each node. The proposed approach in this paper, takes into account this most probable map to classify the mineral resources in each block into measured, indicated and inferred. This can be realized through break points in the graph of probability plot where the global distribution of most probable values are illustrated. The advantage of this approach is that, the selection of the thresholds of interest for the purpose of classification depends on the opinion of competent person and should be derived manually. The procedure of the proposed algorithm is:

1. Exploratory data analysis of rock units through borehole data
2. Selection of optimum approach for probabilistically domaining of each rock unit taking into account the complexity
3. Computation of most probable map through probability map of each rock unit
4. Inference of thresholds for mineral resource classification in the probability plot of most probable values
5. Classification of each block based on the derived thresholds and underlying breaking points

This algorithm is illustrated through an actual case study.

3 Actual Case Study

In order to show the performance of the proposed approach for mineral resource classification, a dataset from a porphyry copper deposit in Chile is now employed. The data consist of 2,376 composites from an exploratory borehole campaign [4]. The composites are 12 m long with semi-regular sampling pattern. Each composite is assigned a geological domain which is related to lithology type. The rock codes were originally six, however, they were grouped into three main types (Fig. 1):

(a) Granodiorite: this is host rock where breccia intruded. It is mostly located in eastern and southern parts of the deposit.
(b) Tourmaline breccia: This breccia has granodiorite clasts with minerals correspond to tourmaline and sulphides such as chalcopyrite, pyrite, molybdenite, and some bornite. Its emplacement is related to the main alteration-mineralization event. This rock has the highest mean grade and is centrally located in the deposit.
(c) Other breccias: They are organized by three different breccia types and outcrop in the western and southern areas of the deposit. Their emplacement is simultaneous or more recent than the intrusion of tourmaline breccia, relocating and diluting it.

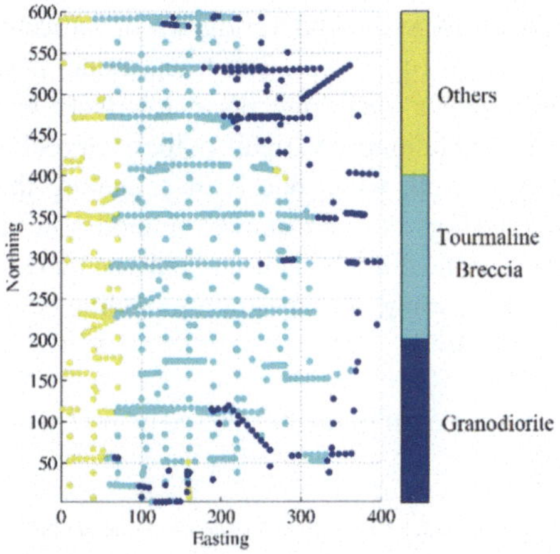

Fig. 1. Location map (horizontal plane) of drill hole samples showing the local distribution of each rock type

Based on exploratory data analysis, roughly 15%, 69% and 17% are the proportion of region that are dominated by Granodiorite, Tourmaline breccia and other breccia, respectively. Based upon the geological interpretation and local distribution of these three rock types, Granodiorite is in direct contact with Tourmaline Breccia and Tourmaline Breccia is in contact with other breccia. Therefore, there is a forbidden contact between Granodiorite and other breccia. This implies that there is a restriction in contact relationship and justify applying the truncated Gaussian simulation approach for probabilistically domaining the rock types. To do so, the flag of interest is illustrated in Fig. 2. As can be seen, there exist one Gaussian random field for modeling purposes.

For the theory of truncated Gaussian simulation, the interested readers are referred to the reference therein [3, 5, 6].

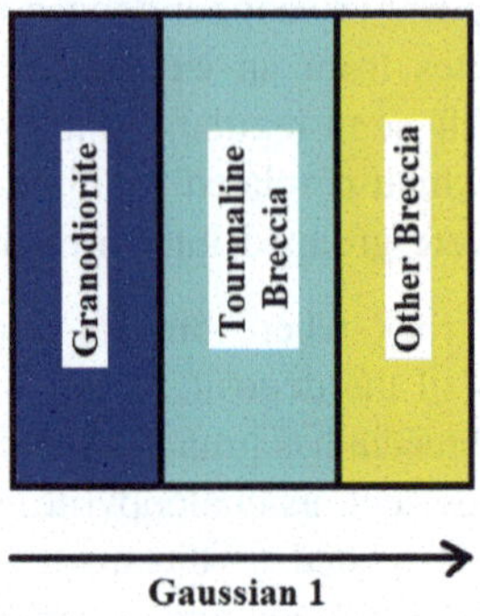

Fig. 2. The flag expressing the contact relationship among the rock types

The next step, following inference of flag and truncation rule, is to compute the theoretical model of variogram for the Gaussian sought. This can be carried out through an iterative manner between experimental variogram of indicator data and theoretical variogram of Gaussian data [3, 6] (Fig. 3). The fitted model is spherical isotopic with the range of 300 m and show a very long range relatively to the dimension of the drilling grid, for which a non-stationary variability for each rock type may persist. This also can be corroborated as well, when one is investigating the local distribution of each rock type in the region (Fig. 1). For instance, Granodiorite can be just found in the right side of the area whereas the other breccia can only be met along the left side of the region. Therefore, such a long range of variogram modeling is apparently evident. The underlying formulae is as follow:

$$\gamma(h) = Spherical(300\,\mathrm{m}, 300\,\mathrm{m}, 300\,\mathrm{m})$$

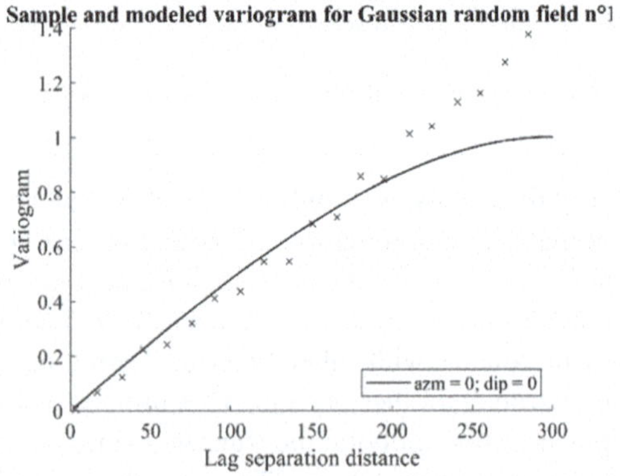

Fig. 3. Fitted theoretical variogram for Gaussian random field. Experimental variogram of indicators and fitted theoretical variogram of Gaussian random filed are illustrated by black crosses points and solid black line, respectively.

Once the variogram model of Gaussian values is derived, the categorical data can be converted to Gaussian values corresponding their truncation rule as presented in Fig. 2 by Gibbs sampler. For this purpose, two thresholds defining each rock type can be derived as −1.0407 and 0.9826 for the first and second threshold. The first threshold is only taking into account the first rock type, Granodiorite and the second one introduces the proportion of Granodiorite and Tourmaline breccia together. The inverse of proportions can be inferred from normal standard Gaussian cumulative density function. After this conversion, the Gaussian values at sample locations can be considered as conditioning data for multi-Gaussian conditional simulation. This step can be implemented through any Gaussian simulation approach, however, in this study; we use turning bands simulation for producing the realizations. Based upon the truncation rule in Fig. 2, the Gaussian maps can be truncated to the categorical maps, so that each grid node, show the possible occurrence of the rock types (Fig. 4). Following the proposed approach in this study, we continue working on the calculation of probability of finding each rock type at either grid nodes. This gives an insight about the probable area for searching that specific rock type. The probability map for each rock type is illustrated in Fig. 5.

The probability maps allow calculating the most probable maps at each block location. For this, it is necessary to identify the maximum probable value at each block over the probability of rock types. For instance, if the probability of Granodiorite is 0, Tourmaline breccia is 19% and other breccia is 81% for block No. 1, then the most probable value for this block will be 81%, showing the fact that satisfying density of information (i.e., borehole data) of other breccia around this block exist and one is certain about the availability of the surrounding information. In contrast, block No. 2, may show different characteristics, in such a way that the probability for Granodiorite is 48%, Tourmaline breccia is 51% and other breccia is 1%. In this block, the most probable value is 51%, which is quite low relatively to block No. 1. This corroborates that the supporting information surrounding this block is poor or rather far from the conditioning data.

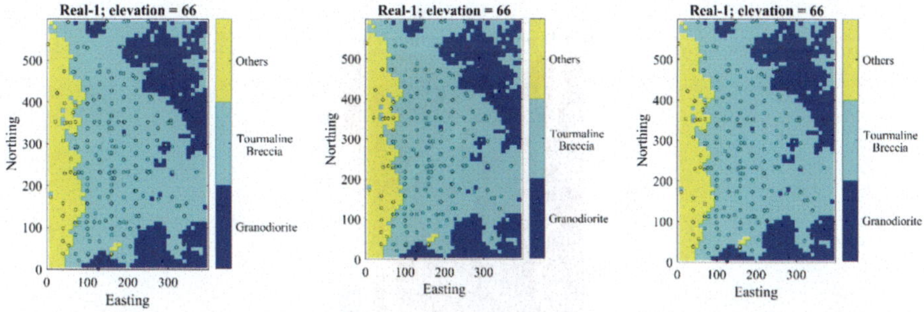

Fig. 4. Three different realizations obtained from truncated Gaussian simulation approach

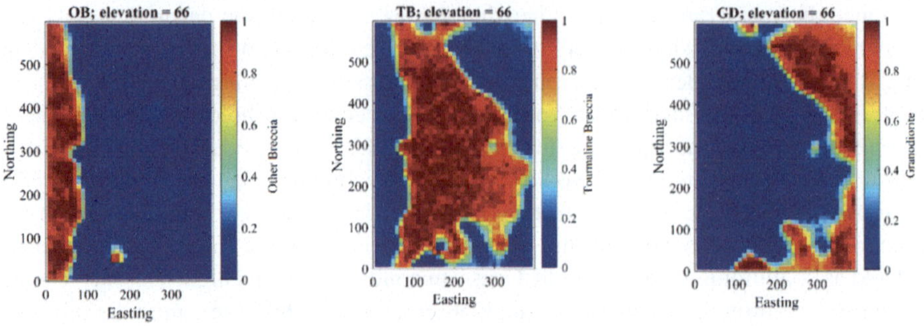

Fig. 5. Probability map for each rock type, GD: Granodiorite, TB: Tourmaline breccia, OB: Other breccia

Figure 6 shows the most probable map. The location close high probability indicates that there are enough information for that block to be simulated and one is confident in the derived value for that block. The location with low probability indicates that one is not confident in the simulated value. As can be seen, comparing to Fig. 4 that show the simulated rock types, the most probable areas are located on the place that the borehole information is available, and the low values manifest themselves through the boundary of the lithologies. Following the proposed approach for mineral resource classification, the next step is to derive the thresholds that can capture three zones based on the level of confidence, indicating measured, indicated and inferred categories. One of the useful tools, for this purpose, however, can be showing the distribution of the underlying variable i.e., most probable values on the probability plot. In principle, this graph is for assessing the distribution whether or not a dataset follows the normal or lognormal distribution. The data is plotted against a theoretical distribution so as to the points should form approximately a straight line. In this graph, the breaking points can show the different populations, for which in the case of resource estimation, can introduce the area of each category (Fig. 7).

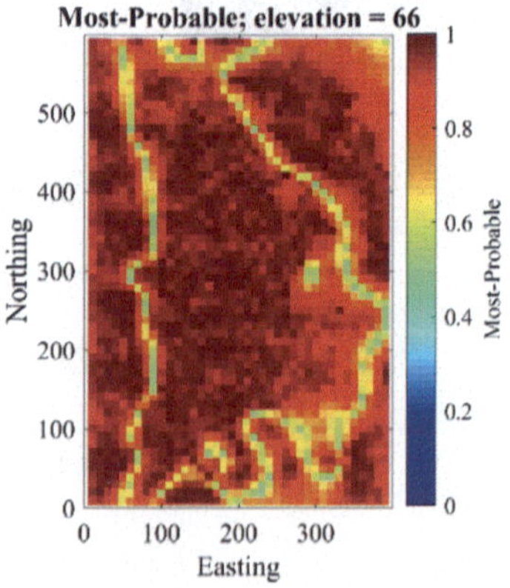

Fig. 6. Most probable map calculated over the probability of each rock type

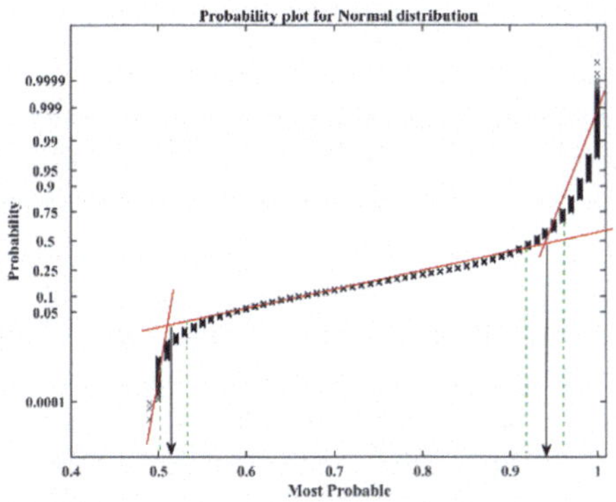

Fig. 7. Probability plot of most probable values, the arrows show the thresholds indicating the different categorization areas

As can be seen in Fig. 7, the breaking points show the thresholds, representing the different populations and categories. Based on this graph, the blocks with most probable value less than 0.52, between 0.53 and 0.94 and above 0.95 can be classified into inferred, indicated and measured, respectively. The dashed green lines over this graph provides this opportunity to identify the underlying thresholds in possible ranges (between two green lines). This can be corroborated in accordance with the opinion of competent person who is responsible for this type of deposit. Once the block are categorized, the continuous variables can be estimated or simulated taking into account the uncertainty quantified through the stochastic modeling of geological domains. Since the category of all the blocks are identified, recovery functions such as tonnage, mean grade above cut-off and metal quantity can be reported in either categories (Fig. 8).

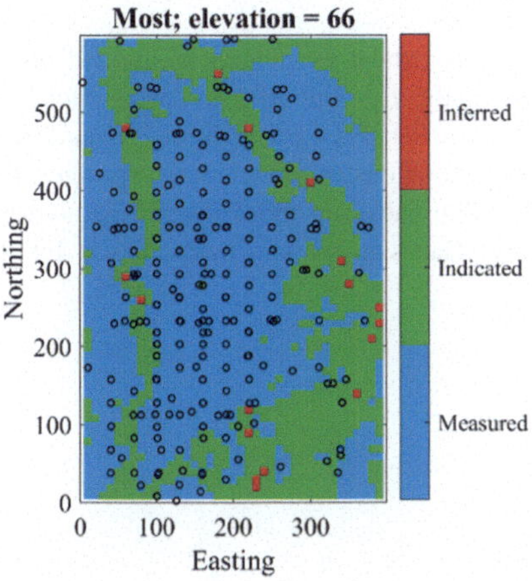

Fig. 8. Categorization based on most probable plot; the area with high density of boreholes

4 Conclusion

Mineral resource classification is important in mining industry and is vital for public reporting of mineral resources and ore reserves based on JORC code. In this study, an algorithm is proposed for this classification based on modeling the geological domains. In this approach, the categorical variables, introducing the estimation domain should be first modeled in a stochastic manner. Second, through producing different scenarios, one can calculate the probability maps for each category. Through these probability measures, the third step, is to calculate the most probable value at each grid node and produce the most probable map. Since, this value is obtained from the probability of each categorical domain, the high values indicate that the simulated values irrespective of the type of simulated rock type is significant and can show that the surrounding information is good enough. In contrast, the low values show that one is less certain about the simulated category and the supporting information for modeling that category in that block may be poor. Based on this concept, different thresholds can be defined by plotting the most probable values in a probability plot and infer the underlying thresholds. The proposed algorithm showed that this method is capable of handling the uncertainty even in the captured thresholds, in which the competent person can easily comment on that and classify the resource sought based on JORC code.

References

1. Battalgazy, N., Madani, N.: Categorization of mineral resources based on different geostatistical simulation algorithms: a case study from an iron ore deposit. Nat. Res. Res. **28**(4), 1329–1351 (2019)
2. Rossi, M.E., Deutsch, C.V.: Mineral Resource Estimation. Springer, Berlin (2014)
3. Madani, N., Emery, X.: Simulation of geo-domains accounting for chronology and contact relationships: application to the Rio Blanco copper deposit. Stoch. Environ. Res. Risk Assess. **29**, 2173–2191 (2015)
4. Ortiz, J.M., Emery, X.: Geostatistical estimation of mineral resources with soft geological boundaries: a comparative study. J. South Afr. Inst. Min. Metall. **106**, 577–584 (2006)
5. Armstrong, M., Galli, A., Beucher, H., Le Loc'h, G., Renard, D., Renard, B., Eschard, R., Geffroy, F.: Plurigaussian simulations in geosciences. Springer, Berlin (2011)
6. Madani, N., Emery, X.: Plurigaussian modeling of geological domains based on the truncation of non-stationary Gaussian random fields. Stoch. Environ. Res. Risk Assess. **31**, 893–913 (2017)

Strategic and Tactical Geometallurgical Application in an Underground High-Grade Narrow-Vein Gold Operation

Simon Dominy[1,2(✉)], Louisa O'Connor[2], Hylke Glass[1], and Saranchimeg Purevgerel[2,3]

[1] Camborne School of Mines, University of Exeter, Cornwall TR10 9FE, UK
s.dominy@e3geomet.com
[2] Western Australian School of Mines, Curtin University, Perth, WA 6102, Australia
[3] MSA Global LLC, Ulaanbaatar, Mongolia

Abstract. Vein gold deposits are often characterised by multiple sub-parallel veins and free-milling coarse gold. Inherent heterogeneity results in grade and process parameter variability, which increases project risk if not quantified and controlled. The geometallurgical approach can be broadly split into two activities: strategic and tactical. The strategic approach focuses on the whole orebody and long-term life-of-mine view, whereas tactical geometallurgy relates to a more short- to medium-term view during mining. The geometallurgical approach requires spatially distributed samples within a deposit to support variability modelling. A variability sampling and testwork protocol was developed to quantify gold grade and recovery. Additional attributes from core logging, mineralogical determination were integrated with grade and recovery data. This contribution presents a case study of strategic and tactical geometallurgical programme application to a narrow-vein deposit. It exemplifies how data can be used to support resource estimation, a pre-feasibility study, trial mining and production. Subsequent to production commencing, a tactical geometallurgical/ore control programme was introduced to optimise mine scheduling and process activities.

Keywords: Geometallurgy · Narrow-vein gold · Variability determination

1 Overview

The prime objective of geometallurgy is to improve the profitability of mines through the use of spatial models of rock properties that have a significant impact on value [1, 2, 5]. It aims to correlate geology and mineralogy with data from testwork and develop a model to predict variability [1, 2, 5]. A key output is 3D geometallurgical mapping, where diverse attributes from core logging, mineralogical/textural determination and small-scale tests are used to resolve grade, process parameter and rock mass variability. The geometallurgical approach emphasises early stage intervention and progression during the project to optimise the mine plan. Geometallurgy can be split into two components: strategic and tactical. The strategic component focuses on the whole orebody and

E. Topal (Ed.): MPES 2019, SSGG, pp. 165–174, 2020.
https://doi.org/10.1007/978-3-030-33954-8_20

long-term life-of-mine view, whereas tactical geometallurgy relates to the short- to medium-term operational view.

This case study presents a geometallurgical approach initiated at the San Antonio project, located in South America [3]. The deposit is characterised by sheeted vein mineralisation, which comprises multiple parallel to sub-parallel veins. Gold is generally coarse, with grade displaying a high nugget effect. A 14,000 m NQ diamond drilling programme was undertaken between 2005–2008 on a 40–100 m grid pattern. In 2009, an Inferred Mineral Resource of 1.6 Mt at 5.2 g/t Au at a bulk-mine cut-off grade of 4 g/t Au was reported.

Recognising the presence of grade heterogeneity and different metallurgical domains, the 2013–2016 programme aimed to quantify variability in grade and metallurgical recovery and inform a pre-feasibility study (PFS). In 2013, a four-stage programme was initiated, including: (a) ore characterisation study; (b) geometallurgical drilling; (c) bulk sampling; and (d) resource estimation; leading to a PFS and trial mining.

The ore characterisation study accessed the mineralisation to undertake testwork in support of a geometallurgical programme design [3]. The four-stage programme aimed to upgrade part of the Inferred resource to the Indicated category to support the first 2–3 years of production. The geometallurgical drilling programme was completed in late 2015. The PFS, trial mining and resource estimate were completed in mid-2017. Production commenced in July 2017, with the current mine plan targeting 125,000 tpa for 80,000 oz Au recovered.

2 Geology and Mineralisation

The deposit comprises a sheeted vein zone (SVZ), which can be traced along strike for 2.5 km and is hosted in a composite granodiorite intrusion. It strikes 075–085° and dips at 80–90 °NW. The resource zone represents 650 m of strike to a depth of 300 m. The SVZ retains a relatively consistent width of 15 m (ranging 12–16.5 m) along the drilled zone (Fig. 1). All veins are dominated by quartz, with locally up to 20% sulphides. Sulphides are typically pyrite with traces of galena and sphalerite. The pyrite contains free-gold generally <100 μm in size.

A central 4–5 m wide core zone (CZ) displays the highest density of veins, strongest alteration and highest grades. Within the CZ, some veins reach 0.4 m in width (Fig. 1). The hangingwall (HW) and footwall (FW) zones are characterised by a lower density of veins, weaker silicification and lower grades. Outside of the SVZ, the granodiorite displays weak sericitic alteration. There is localised strong silicification, which results in quartz-rich zones comprising quartz veins and silicified wall rocks. A characterisation study confirmed the presence of coarse-gold, where sampling "gold liberation diameter" (d_{95Au}) values reach 1,700 μm [3].

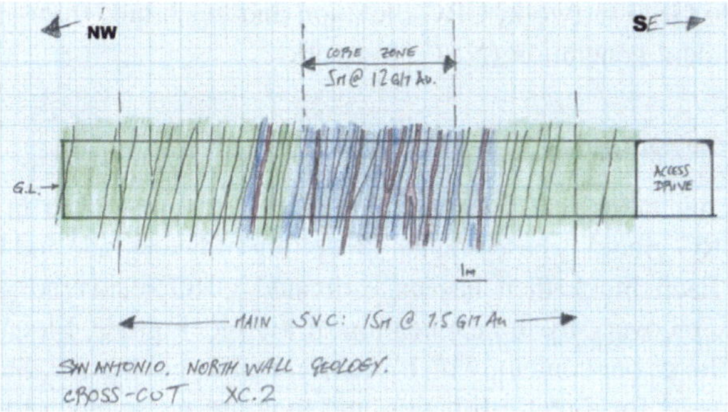

Fig. 1. Geological field sketch of cross-cut #2 through the SVZ NE wall. Green: weak to moderate silicification; and Blue: strong silicification.

3 Strategic Geometallurgical Programme

3.1 Sampling and Testwork Strategy

The 2015 programme included 45 HQ core holes (c. 5,500 m) at a spacing of 20 m by 20 m, and provided information on geology/lithology, geochemistry, rock mass properties, metallurgical recovery and grade. Drilling was initially based on 2 m HQ core, each whole-core sample had a mass of approximately 17 kg. Each core sample was treated as a variability sample [3, 4]. After preparation, each sample was subjected to a single-stage gravity recoverable gold (GRG) test (Table 1) [3, 6]. The gravity concentrate was sized and assayed, and the tails retained for leach, flotation, geochemical and mineralogical analysis. The tails leach result was recombined with the gravity grade to give a sample head grade. After an initial phase, the samples were reduced to 1 m ± 0.25 m to resolve variability across the SVZ and investigate the CZ. Samples continued to be processed in their entirety (Table 1).

Table 1. Summary of sample processing locations and activity.

Sample	On-Site Activity	Off-Site Activity
Geomet core	Dry, crush, pulverise, Knelson GRG, concentrate cleaning and fire assay	Comminution tests, all geochemistry (incl. NAG) and mineralogy, flotation testing and LeachWELL tails
Ore control	Dry, crush, pulverise, Knelson GRG, concentrate cleaning and fire assay	Geochemical analysis (incl. NAG) and LeachWELL tails

Additional data collection related to core logging (e.g. recovery, rock quality designation "RQD" and EQUOtip) and density determination. A series of 100 kg underground panel samples were collected for comminution (Bond ball work and abrasion indices) and flotation (rougher, cleaner and locked cycle) testwork. Thirty core samples were used for Bond ball work index and uniaxial compressive strength testwork, prior to GRG testwork. GRG, flotation and leach tails from all samples were subjected to net acid generation (NAG) testwork.

3.2 Programme Outputs

The strategic geometallurgical programme outputs are summarised in Table 2. GRG recovery (e.g. GRG grade) is based on the modelling of head grade and GRG recovered fraction grade. Each block then has an estimated GRG recovered grade, which is reverted to GRG recovery by the application of the block grade. Flotation recovery is estimated from the pyrite content, which is correlated to gold grade. Sulphur, lead and zinc assays are used for direct estimation and then are recombined to produce a sulphide distribution and density models. Throughout the SVC, density content varies from 2.7 t/m^3 to 3.4 t/m^3 and is controlled by sulphide content.

Table 2. Geometallurgical programme block model inputs and outputs.

Data Input (Units)	Primary Model	Secondary Model(s)
Grade (g/t Au)	Grade	GRG recovery
GRG recovery (%)	GRG recovered grade	
Sulphur (%)	Sulphur	Sulphide Density, Flotation recovery and NAG
Iron (%)	Iron	
Lead (%)	Lead	
Zinc (%)	Zinc	
Density (t/m^3)	Density	–
EQUOtip	Hardness	–

Gold recovery reflects the relative proportions of free gold in quartz versus sulphides. The presence of sulphide-rich vein sets reflects locally less GRG and increased fine gold. These features are reflected by two gold recovery domains within the Indicated resource model. The AUREC.1 domain reflects dominantly free-gold with high GRG potential (e.g. CZ: >50% GRG), whereas domain AUREC.2 reflects the sulphide-hosted gold amenable to flotation.

Comminution relationships were determined through paired uniaxial compressive strength and Bond ball work index samples, which were correlated to point load EQUOtip values. Comminution variability ranged from hard-very hard in quartz-dominated silicified zones to soft kaolinised fault zones. The majority of the mineralisation relates to the hard classification. Three comminution domains are identified; defined as COMM.1-very hard (75% of model); COMM.2-hard (20% of model); and COMM.3-soft-medium (5% of model). Localised soft zones (COMM.3) reflect cross-faulting which results in sheared granodiorite with some kaolinite alteration. The CZ geological domain is dominated by COMM.1.

4 PFS and Trial Mining

4.1 PFS Outcome

During the PFS, consideration was given to two underground mining strategies: (1) a 350,000–400,000 tpa bulk-mine operation based on longhole stoping, and (2) a high-grade small-scale operation based on shrinkage stoping. The conclusion was that a selective operation was the best option allowing for acceptable economics, a small footprint, fast production ramp-up (to 50,000 tpa in 6 months), less capital expenditure and aligned better with local stakeholder expectations. The selective option was considered the more sustainable option.

The selective mining scenario was based on a narrow high-grade zone (HGZ) within the centre of the CZ. It is based on a consistent 2–3 m wide zone comprising a number of 0.1–0.4 m wide veins that run in a parallel to anastomosing fashion. Individual intersection grades across the HGZ vary from 0.01–225 g/t Au. The HGZ contains 60–65% of the grade across the mineralised SVZ. Both mining scenarios related to ore zone domains with different grade-recovery characteristics (Table 3). The HGZ is dominated by recovery domain AUREC.1 (65%), though includes intercalated zones of AUREC.2 (35%).

Table 3. Comparison between ore domains.

Mining Scenario/Ore Type	Minable Width (m)	Grade (g/t Au)	Nugget Effect	>100 μm	d_{95Au} (μm)	GRG	Gold Deportment
Bulk FW, CZ, HW	10–15	6–8	50%	30%	650	15–40%	Free gold, finer gold within pyrite
Selective HGZ	2–3	20–26	60%	55%	1700	50–75%	Free gold Coarse gold, with less fine gold

A key outcome of the HGZ mining strategy, was that production could commence immediately based on a gravity-only circuit. Additional recovery could be gained by the addition of a flotation circuit.

4.2 Trial Mining

As validation for the PFS, the company instigated a trial mining programme based on a shrinkage stope above the existing development (Fig. 2).

The trial stope aimed to extract 2,200 t at 21–23 g/t Au with a modelled GRG recovery of 61%. The actual performance was 1,820 t at a reconciled head grade of 20.5 g/t Au, against an estimate of 22.6 g/t Au. The lower tonnes represent a tight stope

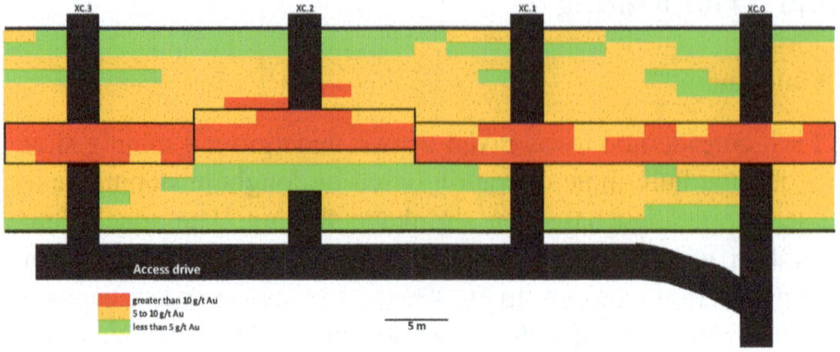

Fig. 2. Trial mining area grade control model (plan view) showing HGZ amenable to narrow vein mining. Trial stope focused on areas defined by black outlines.

width with reduced planned dilution from the wall rocks. As a result, the company commenced production, whilst development was undertaken to access additional levels and stope blocks.

4.3 Mineral Resources and Ore Reserves

The 2017 estimate for the HGZ yielded an Indicated Mineral Resource of 155,000 t at 21.9 g/t Au (6 g/t Au cut-off grade), providing the first 2.5 years of production. An updated estimate (July 2019) yields an Indicated Mineral Resource of 330,000 t at 22.8 g/t Au (6 g/t Au cut-off grade). This supports a Proven Ore Reserve of 265,000 t at 20.5 g/t Au. The estimate was dominated by 20 m by 20 m drilling, augmented by 10 m by 10 m drilling, and development and surface sampling. The updated Inferred Mineral Resource was 475,000 t at 17.9 g/t Au, based on 50–120 m drilling.

As part of the July 2019 resource update, conditional simulation was used to evaluate grade risk within the Indicated resource for the next 12 months of production. Sequential Gaussian simulation was applied. The simulated points (nodes) were regrouped into 5 m (strike) by 5 m (dip) by 0.5 m (width) SMU blocks. The Indicated resource displayed a global precision of $\pm$ 21% at the 80% confidence limits (e.g. 21.9 g/t Au $\pm$ 4.6 g/t Au; range: 17.3–26.5 g/t Au).

4.4 Project Financial Metrics

The original consideration (Q2 2017) was for a 7-year bulk stoping 400,000 tpa operation at a run of mine grade of 6.5 g/t Au. This required around US$70 M in capital expenditure, principally for a new process plant and infrastructure. This yielded a post-tax NPV_{15} of US$55 M and an IRR of 60%. The metrics of a selective high-grade (22–24 g/t Au) 50,000 tpa option were also positive, where a 7-year mine life yielded a post-tax NPV_{15} of US$51 M and an IRR of 108%. The capital expenditure for the smaller operation was US$15 M. The smaller selective operation commenced in July 2017.

A revised model (July 2019) for the 7-year mine life (2017–2019 actuals and 2019–2023 predicted; production rate of 125,000 tpa) yields a post-tax NPV_{10} of US$120 M and an IRR of 80%. The improved NPV relates to increased annual production and gold recovery, and reduced OPEX. It is noted that the last 2.5 years (out of 7-years) of production are based in Inferred resources. The operator has confidence in conversion of Inferred to Indicated resources (nominal conversion of 65–70%) based on on-going drilling and underground development. As a private entity, the operator is not required to report its results publicly.

5 Tactical Geometallurgy Programme

5.1 Ore Control

With commencement of production, a tactical geometallurgical approach was incorporated into the operation. Outputs remain the same as the strategic programme (Table 2). Initial production was 140 t per day, increasing to 350 t per day in early 2019. Ore control is a critical part of tactical geometallurgy, where the focus is on defining and controlling plant feed variability (e.g. grade, recovery and hardness) for forecasting and blending purposes [2].

Development faces (nominally 2.6 m high by 2.5–3 m wide) are saw-cut channel sampled at a length of 1 m ± 0.25 m and used for development control, tracking grade distribution and contribution to a local ore-control model. The saw-cut channels are designed to approximate drill core sample support and minimise bias. In-stope samples are hand chip-channelled from the backs to achieve lengths of 1 m ± 0.25 m.

Short 2–4 m AW (35 mm) core holes are placed into the drive sidewalls every 10–15 m or as required to track the position of the HGZ in relation to the FW and HW mineralisation. Ore control and resource core sampling remains essentially the same, where 1 m ± 0.25 m samples are collected to resolve the narrow mineralisation target.

All core and ore control samples pass through the geometallurgical drilling protocol, including assay for sulphur, lead and zinc. The on-site laboratory runs separate circuits for drill core and ore control samples (Table 1). Quality assurance/quality control (QA/QC) is applied to all samples [3, 4].

5.2 Tactical Geometallurgical Modelling

The medium-term (>6 months) resource model is based on diamond core drilling on a 20 m by 20 m grid augmented by face channel samples (as available) to achieve an Indicated resource. The short-term ore control model (<6 months) is based on core drilling, augmented by face channel samples. Key model outputs across both the short- and medium-term models are: (1) grade; (2) GRG recovery; (3) sulphide content; (4) flotation recovery; (5) density; and (6) hardness (Table 2).

5.3 Short-Term Stope Panel Modelling

The target stope panel size is 60 m by 20 m, including lower development to yield approximately 8,700 t assuming a stope width of 2.5–3 m. Estimation is based on 10 m by 10 m by 1 m (w) blocks using ordinary kriging. For each panel, the model is compiled to yield key variables (Table 2).

In the stope panel example presented in Table 4, the reconciled head grade and tonnage are 108% and 103%, respectively, of the prediction. The increase in grade, recovery and tonnage improved the profit margin of the stope.

Table 4. Example of stope panel ore control (short-term) model and actuals.

Parameter	Panel Estimate	Reconciled Actual
Tonnage	7,360 t	7,555 t
Width	2.25 m	2.30 m
Grade	22.7 g/t Au	24.5 g/t Au
GRG recovery	71%	74%
Flotation recovery	19%	Pre-float circuit
Sulphide	4%	5%
Density	2.97 t/m^3	3.10 t/m^3
Hardness	Hard	Hard
Stope profit	US$1.1 M	US$1.4 M

The block model supports mine design, planning and ore blending. Each estimation block informing the panel is reviewed and applied to final stope design (e.g. smaller stopes will be designed as required and/or low grade areas planned as pillars where possible). The stope panel grade, mineralogy, recovery, hardness and RQD are used to design a final ore control model incorporating pillars and dilution. Additional studies using multiple indicator kriging and conditional simulation are in progress.

Ore from development drives and stope blocks are held in individual stockpiles. Blending of plant feed aims to: optimise head grade and recovery; and reduce the effect of soft material. Blending (if required) is facilitated by feeding different stockpiles to the plant via two primary crushing and feed belt systems. Both jaw crushers feed a fine ore bin prior to passing to the secondary crusher. Additional flexibility is being added to the plant via the capability to switch to either gravity-only (two Knelson units), or gravity (one Knelson unit) plus flotation.

6 Production Results

Production results from 2017 are summarised in Table 5. Like many operations displaying a relatively high-nugget effect, on a monthly basis reconciliation is variable. In this case the critical parameter is gold grade, which varies within ±20% monthly. This is reasonable for an Indicated resource in high-nugget mineralisation. On an annual basis, reconciliation is within ±10% which is an acceptable result indicating that the sampling and estimation process is performing well.

Table 5. 2017-2019 production results. Compares short-term ore control model with actual.

Period	Actual tonnes	Actual grade	Tonnes	Grade	Ounces rec.
Jun-Dec 2017	33,150	23.0	+2%	−3%	+1%
2018	104,800	22.2	+8%	+7%	+7%
Jan-Jul 2019	73,555	24.6	+5%	+4%	+9%
Total	211,505	23.2	+6%	+4%	+7%

The low tonnage variability reflects good ground conditions and mining practice. In-stope sampling allows control of each bench onto the HGZ within a minimum to maximum stope width of 1.5–3 m. The plant is recovering 90–95% of the gold by gravity (c. 70–75%) and flotation.

7 Summary

- San Antonio represents a moderate-grade bulk-mine sheeted-vein deposit, with a high-grade selective-mine core zone. Both styles pose challenges for grade and metallurgical sampling by virtue of their heterogeneous gold distribution.
- The geometallurgical study aimed to link ore characteristics with key operating parameters, principally grade and gold recovery, but also geochemistry and mineralogy, rock mass properties and plant throughput. The outcome was the production of 3D block models to describe variability [1, 2, 5].
- The selection of whole-core variability samples for grade determination and testwork was driven by the need to quantify gold grade and recovery variability. Variability samples allow the distribution of ore types to be modelled in 3D [4].
- The geometallurgical protocol presented was designed to respond to the coarse-gold nature of the mineralisation and to provide maximum data from drill core to support a resource estimate and PFS. The change from 1 m-half core samples to 2 m whole-core assay indicated a 20% increase in grade. The sampling strategy reverted to 1 m whole-core samples in response to the change to selective mining.
- A fit-for-purpose sampling, testwork and assaying programme was designed to support a Mineral Resource reported in accordance with the 2012 JORC Code.
- The PFS concluded that the selective operation was a better option allowing for a smaller mine footprint, fast production ramp-up, less capital expenditure and aligned better with the expectations of stakeholders.
- The tactical geometallurgical programme defines and predicts plant feed variability (e.g. grade, recovery, throughput, etc.) which improves forecasting and blending, and assists with managing the mine schedule, stockpiling and process plant operation.
- As initially a gravity-only circuit, the sulphide content needed to be determined so that NAG potential could be managed. A flotation plant has now been added to the

circuit to recover a further 20–30% of gold and remove sulphides from the plant tails stream and thereby reducing the NAG potential.

Acknowledgements. The authors acknowledge OCX Gold SA who consented to this case study within a Confidentiality Agreement. The opinions expressed are those of the authors and not necessarily those of their given affiliations or the mine owner.

References

1. Bye, A.R.: Case studies demonstrating value from geometallurgy initiatives. In: Proceedings Geometallurgy Conference, pp. 3–90. The AusIMM, Melbourne (2011)
2. Dominy, S.C., O'Connor, L., Parbhakar-Fox, A., Glass, H.J., Purevgerel, S.: Geometallurgy – a route to more resilient mine operations. Minerals **8**(560) (2018). https://doi.org/10.3390/min8120560
3. Dominy, S.C., O'Connor, L., Glass, H.J., Xie, Y.: Geometallurgical study of a gravity recoverable gold orebody. Minerals **8**(186) (2018). https://doi.org/10.3390/min8050186
4. Dominy, S.C., O'Connor, L., Purevgerel, S.: Importance of representative metallurgical sampling and testwork programmes to reduce project risk – a gold case study. Mining Technol. **128** (2019). https://doi.org/10.1080/25726668.2019.1628462
5. Dunham, S., Vann, J.: Geometallurgy, geostatistics and project value – does your block model tell you what you need to know. In: Proceedings Project Evaluation Conference, pp. 189–196. The AusIMM, Melbourne (2007)
6. Laplante, A.R., Spiller, D.E.: Bench-scale and pilot plant test work for gravity circuit design. In: Proceedings Mineral Processing Plant Design, Practice and Control, pp. 160–175. SME, Littleton (2002)

Investigation of Uncertainties in Data Imputation Through Application of Sequential Co-simulation

Dauletkhan Orynbassar(✉) and Nasser Madani

School of Mining and Geosciences, Nazarbayev University,
Nur-Sultan 010000, Kazakhstan
{dauletkhan.orynbassar,nasser.madani}@nu.edu.kz

Abstract. The main principle of mineral resource estimation process is to quantify a mineral grade in target places. However, in many cases, available dataset show heterotopic sampling pattern that makes the current practice of mineral resource estimation challenging. In order to cope this difficulty, one alternative consists of removing the sample locations that only one variable is available. Despite an attractive simplicity of this method, one may mention that such excluding of valuable data may lead to biased results for resource modeling. Another solution for this impediment is using imputation algorithms, for which the data at sampled location is imputed by stochastic techniques. This is beneficial not only because one can keep the co-variate data, but also the uncertainty can be quantified. In order to show the capability of the imputation method, in this study, a homotopic dataset from a limestone deposit located in south of Kazakhstan is selected as the actual case study accompanying with another homotopic dataset from Chile to evaluate the proposed approach of imputation. Then, to investigate uncertainties in data imputation process, a technique called Sequential Gaussian Co-simulation was utilized. For this purpose, homotopic dataset was transformed to heterotopic dataset. This allows performing validation of the proposed technique. Lastly, sensitivity analysis was carried out to observe the effect of amount of missing data. Overall results of imputation techniques showed satisfactory results, and it can be concluded that derived simulated grade values for drillholes can be incorporated to the ore body modeling.

Keywords: Heterotopic sampling · Sequential Gaussian Co-simulation · Data imputation · Variogram analysis

1 Introduction

The mineral resource estimation represents an application of a set of geostatistical techniques for predicting spatial characteristics and quantifying mineral grade of ore deposit at target locations [1, 2]. In multi-element deposits, the bivariate or multivariate relationship among the variables is significant for taking into account their spatial cross dependency, in the corresponding mineral resource estimation [1]. However, this practice encounters numerous challenges with respect to the available data, especially

E. Topal (Ed.): MPES 2019, SSGG, pp. 175–182, 2020.
https://doi.org/10.1007/978-3-030-33954-8_21

heterotopic sampling pattern that highly impacts the confidence level of the resource model. In most of the geostatistical packages, it is necessary that input dataset to be homotopic, i.e. all the variables should be available through sample locations. However, in case of heterotopic sampling, one simple solution, yet inaccurate is to convert this unequal sampling pattern to homotopic. In this regard, one important source of information from one of the variable, is intentionally removed [3]. Another alternative for making the sampling pattern homotopic while reserving the information, corresponds to imputation wherever the missed data of the underlying variable is stochastically simulated through unsampled locations. For this purpose, Sequential Gaussian Co-Simulation (SGCOSIM) can properly fit the purpose of this technique. Moreover, this method does not only produce spatial distribution, but it also reproduces the correlation coefficient between variables [3–6]. In this study, two variants of SGCOSIM, one based on Simple Co-Kriging and another one based on Multi-collocated Co-Kriging [2, 5] are taken into account for imputation of variable of interest at un-sampled locations. Besides the testing of data imputation methods, sensitivity analysis was performed, in order to investigate the accuracy of the proposed approaches with respect to different amount of missing data through two actual case studies, one from Kazakhstan and one from Chile. The results showed that SGCOSIM with Multi-collocated Co-Kriging outperforms the SGCOSIM with Simple Co-Kriging.

2 Methodology

2.1 Sequential Gaussian Co-simulation

The methodology in the present study is Sequential Gaussian Co-Simulation (SGCOSIM), which is an extension of Sequential Gaussian Simulation (SGS) algorithm. Rationale of this algorithm is built on Multi-Gaussian random function model and it represents numerous advantages over other stochastic approaches. Convenient properties and ease in implementation with acceptable representation of spatial distribution makes this method remarkably attractive in mining and petroleum industries [2, 6–8]. One of the exclusive features of SGCOSIM is that besides the reproduction of the original distributions, it also reconstructs the correlation between the variables. Indeed, for using SGCOSIM algorithm, one of the required condition is presence of spatial correlation between variables, regardless of auxiliary variable type or features [2, 6]. In this algorithm, there can be two alternatives for applying the Co-Kriging systems.

2.2 Simple Co-Kriging

The Simple Co-Kriging is extension of Simple Kriging, a traditional geostatistical interpolation technique used for estimating grade values at un-sampled locations. It incorporates all attributes of Simple Kriging with known mean value and variance of the prediction error, and it also takes into consideration the variables' cross spatial correlation for grade prediction purpose [2, 9].

2.3 Multi-Collocated Co-Kriging

The difference between Multi-Collocated Co-Kriging and Simple Co-Kriging is that it uses all available information provided by auxiliary data. The secondary data is assumed to be known not only at primary data locations but also at target nodes. Thereby, besides the cross dependency of variables, the secondary variable's direct covariance assists to estimate the main variable [2, 5, 9–11].

3 Actual Case Studies

In order to implement and examine two aforementioned algorithms, two sets of data are available to carry out geostatistical modeling and apply data imputation techniques. Because of confidentiality, the name of deposits that the datasets belong to, is not disclosed. For this purpose, the original coordinates also are converted to local coordinates.

The first case study which is a gold-copper deposit is located in Chile and the second one, limestone deposit located in southern Kazakhstan. These two datasets can be distinguished according to their bivariate shape of relationship and the level of complexity. For instance, gold-copper deposit consists of 2,376 borehole sample points and cross correlation between the gold and copper shows a linear characteristic (simple), corresponding to ρ value of 0.7133. The representative scatter plot can be observed in Fig. 1, left. Whereas, limestone deposit with 6672 points shows quite complex cross relationship (Fig. 1, right). In second case, the bivariate relationship between two variables, Calcium Oxide (CaO) and Silicon Dioxide (SiO2) get more complicated. These two variables have represented ρ value of −0.5385 (Fig. 1, right). Through the paper, the Complex dataset refers to limestone deposit and Simple dataset refer to gold-copper deposit.

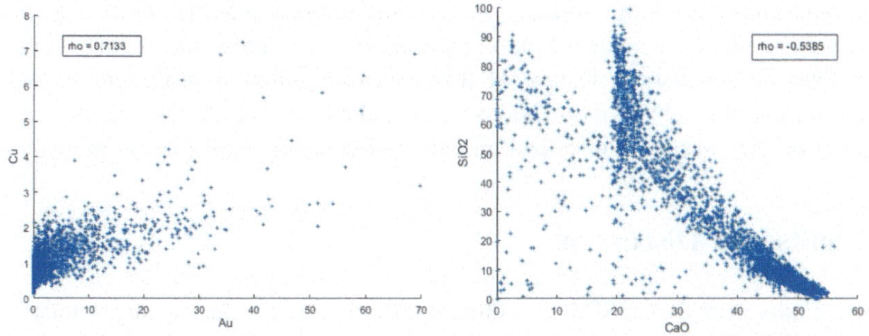

Fig. 1. Scatter plots of variables (left: simple dataset and right: complex dataset)

In order to assess the performance of two algorithms in SGCOSIM, both homotopic datasets are intentionally converted to partially heterotopic sampling pattern, with 5%, 15%, 25% and 50% of missing data. These new datasets then are composed of two groups of homotopic and heterotopic sampling patterns. Since, the original values at those missed locations are known, one can compare the exactitude of the algorithms by comparing the imputed versus original values. For the beginning, 5% of the total number of datasets (100 points in simple dataset, 334 points in complex dataset) was taken out for this purpose.

4 Stochastic Imputation

The imputation was carried out by ISATIS, which is a widely recognized software as the geostatistical package solution. Moreover, this software is quite suitable for our purpose – testing the SGCOSIM algorithms [12].

The initial step for geostatistical modeling is to transform original datasets, both heterotopic as well as homotopic ones into standard normal score distribution through Gaussian Anamorphosis Modeling. Thereby, we need to make sure that the transformed data are within allowed range with mean "0" and standard deviation "1". Then, direct and cross variograms of Gaussian experimental data need to be analyzed. It is worth mentioning that cross variogram plays an important role in co-simulation both in simple and multi-collocated cokriging. Furthermore, experimental cross variogram analysis allows to examine grade spatial distribution attributes like isotropy or anisotropy patterns. Next step is to fit calculated experimental variogram into existing theoretical models such as spherical, exponential, power function or logarithmic function model. To do so, developed variogram models' eigenvalues clearly reflects the spatial variation of the orebody parameters, through the produced characteristics like 'range' or 'sill'. Now moving to main simulation part, let's recall the methodology: SGCOSIM with simple Co-Kriging and multi-collocated Co-Kriging options. The main difference in application of these methods is that multi-collocated Co-Kriging variant requires Gaussian values of auxiliary variable [2, 12]. In both cases, simulation is performed with standard neighborhood parameters, and neighborhood distance was chosen regarding the variogram range index. Furthermore, input files inserted for two algorithms are identical namely homotopic datasets with cross theoretical variogram model. One of the distinct feature of this overall simulation procedure is that for simulation instead of block option we use punctual alternative, thereby we do not have to create grid for macrovariables and all calculation is accomplished on point base.

5 Results and Discussion

Derived results from SGCOSIM algorithm, namely imputed values at target nodes were exported to MATLAB, in order to assess the correctness of the proposed methodologies. Correlation between imputed values and true values was drawn as can be observed in Fig. 2 with imposed mean values over 100 realizations (read line). So the results follow as: for 5% of missing data, SGCOSIM that has used multi-collocated Co-Kriging has produced higher accuracy than conventional simple co-kriging SGCOSIM. Moreover, either approaches has lower results for complex dataset. One of this issue is related to poor correlation between variables and ambiguity of scatter plot's pattern.

As a result, the mean value results for Simple dataset is almost twice higher than the results for Complex dataset.

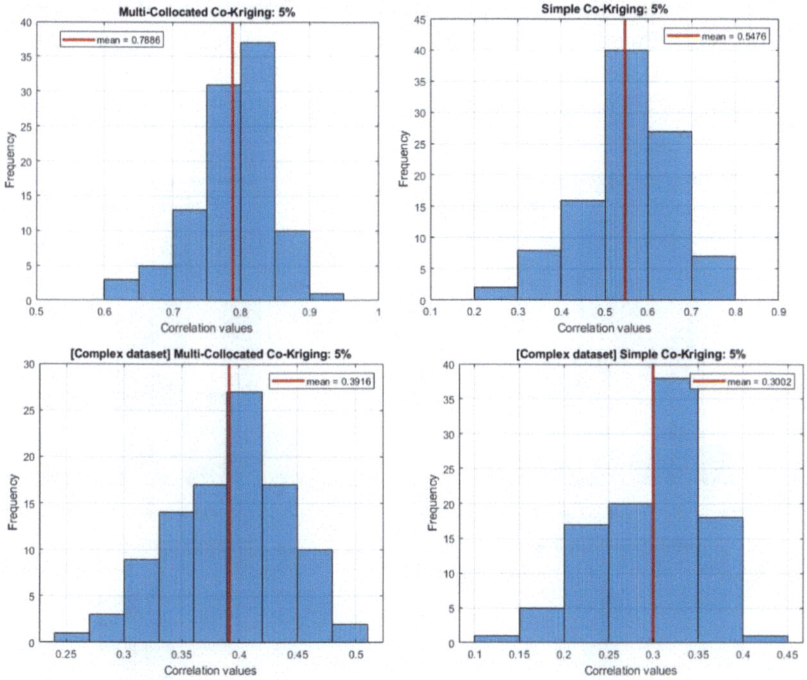

Fig. 2. SGCOSIM Simple Co-Kriging and SGCOSIM Multi-Collocated Co-Kriging imputation results: top – simple case, bottom – complex case

5.1 Sensitivity Analysis

The taken ascending percentages 5%, 15%, 25% and 50% of missing data had explicit impact on a validity of imputed values (see Table 1 & Figs. 3 and 4). Nevertheless, changes or reduction of accuracy is not significant, and the results even for 50% missing dataset are still acceptable. Furthermore, produced results corroborate the fact once, that a presence of adequate correlation between variables as well as consistent spatial distribution of dataset play essential role in data imputation. For instance, from table below, it can be seen that the case where 50% of data is missing (simple dataset) has higher correlation than the case where only 5% of data (complex dataset) needs to be imputed. Lastly, in all cases for sensitivity analysis, it has proved that SGCOSIM considering multi-collocated co-kriging shows better results rather than SGCOSIM with conventional co-kriging system.

Table 1. Correlation Coefficient between true and imputed data

Missing Data %			5%	15%	25%	50%
Variables	Au	Correlation Coef. [SGCOSIM SC]	0.5476	0.4996	0.4825	0.4678
		Correlation Coef. [SGCOSIM MC]	0.7886	0.7068	0.5986	0.6681
	CaO	Correlation Coef. [SGCOSIM SC]	0.3002	0.2591	0.2759	0.2056
		Correlation Coef. [SGCOSIM MC]	0.3916	0.3559	0.3479	0.2558

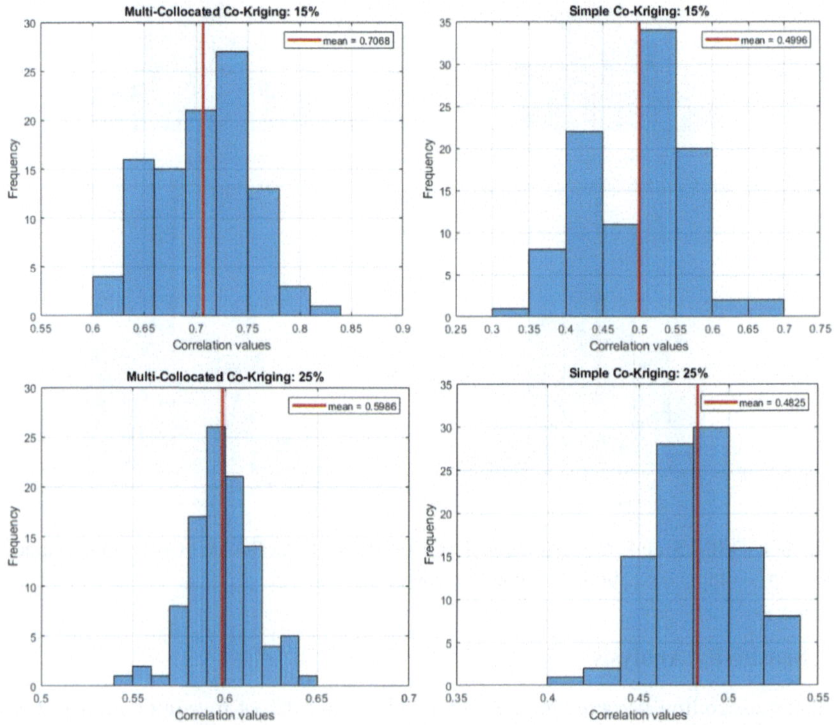

Fig. 3. SGCOSIM Simple Co-Kriging and SGCOSIM Multi-Collocated Co-Kriging imputation results for Simple dataset, red line represents the mean value, for brevity, only the cases with 15% and 25% missing values are shown.

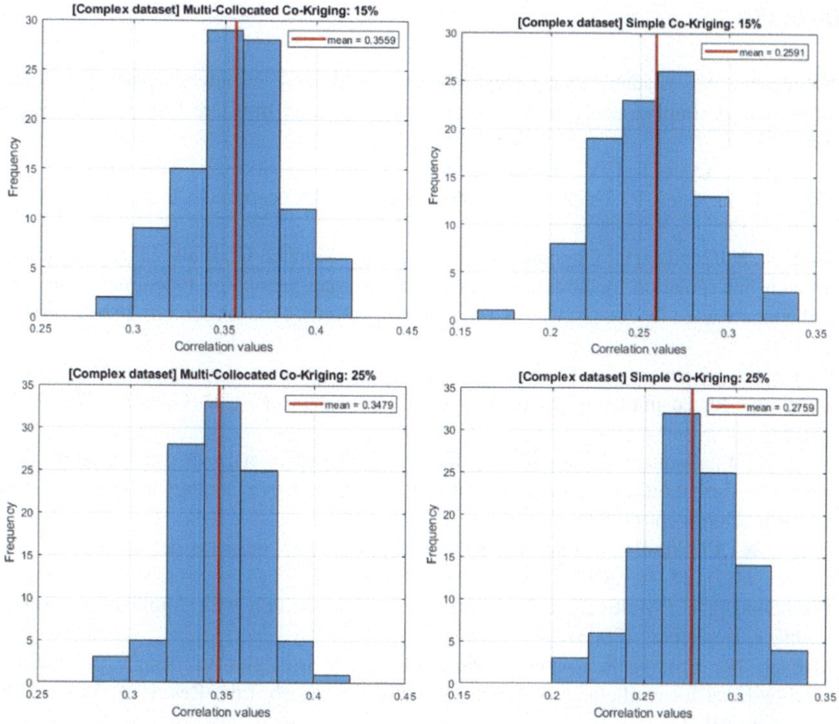

Fig. 4. SGCOSIM Simple Co-Kriging and SGCOSIM Multi-Collocated Co-Kriging imputation results for Complex dataset, red line represents the mean value, for brevity, only the cases with 15% and 25% missing values are shown.

6 Conclusions

Overall, proposed stochastic approach, namely Sequential Gaussian Co-Simulation, for resource modeling offers flexible platform to carry out data simulation or imputation. The selected methodology, Simple Co-Kriging and multi-collocated Co-Kriging showed reasonable results in data imputation through the proposed homotopic datasets from limestone and gold-copper deposit. The results of imputed values at unsampled points were analyzed and correlation coefficients regarding the true values was calculated. In gold-copper deposit, although both options of SGCOSIM have presented acceptable results, the Multi-collocated Co-Kriging outperformed the Simple Co-Kriging technique. However, for second complex dataset the results are not as satisfying as the simple case study, because of complicated spatial distribution and low correlation values between the variables. Therefore, there is room for future research works on novel imputation methods, or at least there should be more investigation on other kind of datasets with different intrinsic characteristics.

References

1. Battalgazy, N., Madani, N.: Categorization of mineral resources based on different geostatistical simulation algorithms: a case study from an Iron Ore Deposit. Nat. Resour. Res. (2019)
2. Rossi, M.E., Deutsch, C.V.: Mineral Resource Estimation. Springer, New York (2014)
3. Chiles, J., Delfiner, P.: Geostatistics: Modeling Spatial Uncertainty, 2nd edn. Wiley, New York (2012)
4. Metahni, S., Coudert, L., Gloaguen, E., Guemiza, K., Mercier, G., Blais, J.F.: Comparison of different interpolation methods and sequential Gaussian simulation to estimate volumes of soil contaminated by As, Cr, Cu, PCP and dioxins/furans. Environ. Pollut. **252**(A), 409–419 (2019)
5. Paravarzar, S., Emery, X., Madani, N.: Comparing sequential Gaussian and turning bands algorithms for cosimulating grades in multi-element deposits. C.R. Geosci. **347**(2), 84–93 (2015)
6. Verly, G.W.: Sequential gaussian cosimulation: a simulation method integrating several types of information. In: Geostatistics Tróia 1992: Quantitative Geology and Geostatistics 5. Springer, Dordrecht (1993)
7. Emery, X.: Co-simulating total and soluble copper grades in an oxide ore deposit. Math. Geosci. **44**(1), 27–46 (2012)
8. Wackernagel, H.: Multivariate Geostatistics: An Introduction with Applications, 3rd edn. Springer, Heidelberg (2003)
9. Madani, N., Emery, X.: A comparison of search strategies to design the cokriging neighborhood for predicting coregionalized variables. Stoch. Env. Res. Risk Assess. **33**(1), 183–199 (2019)
10. Babak, O., Deutsch, C.: Collocated cokriging based on merged secondary attributes. Math. Geosci. **41**, 921–926 (2009)
11. Gómez-Hernández, J.J., Cassiraga, E.F.: Theory and practice of sequential simulation. In: Armstrong, M., Dowd, P.A. (eds.) Geostatistical Simulations. Quantitative Geology and Geostatistics 7. Springer, Dordrecht (1994)
12. Bleines, C., Perseval, S., Rambert, F., Renard D., Touffait Y.: ISATIS. Isatis Software Manual. 5th edn. Geovariances & Ecole Des Mines De, Paris (2004)

Block Modelling Based on Grade Domaining: Is It Reliable?

Nursultan Iliyas(✉) and Nasser Madani

School of Mining and Geosciences, Nazarbayev University, Nur-Sultan 010000, Kazakhstan
{nursultan.iliyas,nasser.madani}@nu.edu.kz

Abstract. The mineral resource classification is one of the steps in feasibility assessment of a mining project. It involves a quantitative evaluation of hard data obtained from exploratory drilling and sampling procedures. Several approaches are aimed to accurately predict spatial variability and complex relationships between cross- correlated variables. However, addressing the grade constraints, such as mining grade domains arrangement and respective boundaries uncertainty, meaningfully is still doubtful. In this study, continuous-categorical variables relationships were analyzed with respect to a Shubarkol coal deposit located in Kazakhstan. One of the common methodologies in mining is grade domaining, for which the grade of interest should be truncated into sub-domains. Each sub-domain introduces a homogenous area that can be considered as a container for grade estimation. For this study, the main variable of interest is ash content, varying from 0 to 90%. A cut-off value of 45% was set to the ash variable, thus forming two domains: low ash domain with Coal variable, and high ash domain with Waste variable. In this paper, we propose an integrative algorithm as an alternative methodology for coherent mineral resource estimation, comprising of sequential indicator simulation for categorical variables and turning bands simulation for continuous variables modelling. In contrast, a global geostatistical analysis of the deposit without domaining is presented as well. The resulting estimates of this research showed satisfying reproduction of the deposit structure free of grade domaining, which can be adopted to precisely estimate the volume of a coal mine deposit.

Keywords: Grade domaining · Continuous-categorical variables · Coal deposit

1 Introduction

Modern mining practices involve a comprehensive feasibility assessment of a mineral deposit, including geostatistical analysis. As a relatively novel branch of mining engineering disciplines, geostatistics is aimed to provide a mathematically justified block model of the deposit. The estimation accuracy of the variable of interest in the block model, such as grade, plays a crucial role in mineral resource classification. A common practice in mining industry for mineral resource/reserve estimation, is grade domaining approach, where the deposit is truncated into a number of adjacent domains based on grade variability throughout deposit [1]. This approach has pitfalls, where the precision of estimation/simulation is reduced by a number of limitations, including an

E. Topal (Ed.): MPES 2019, SSGG, pp. 183–190, 2020.
https://doi.org/10.1007/978-3-030-33954-8_23

inability to preserve spatial relations among domains and to account for uncertainty in the spatial extent of the domains [2, 3].

This study focuses on examining the reproduction of global variability parameters, such as maximum and mean value, from four different case studies. The objective of this work is to assess the accuracy of grade domaining approach (cases II-IV) in comparison with reference model (case I), i.e. global geostatistical modelling of the deposit irrespective of grade domaining. On top of limitations associated with grade domaining itself, there exists a shortcoming affecting the kriging variance (1, 3), [4]. Simple Kriging is incorporated in each case, being used as conditioning technique in turning bands method and sequential indicator simulation. However, the variance of grade is lowered in Cases II, III, and IV, as a result of forming grade intervals, thus leading to distorted variogram and the kriging variance. The failure to incorporate the boundaries uncertainty into the estimation variance may be critical in reliable grade estimation.

The reliability of the research output is discussed and validated further, showing that unique domaining outperforms the conventional grade based domaining approaches. The suggested methodology is introduced and discussed in this work through a case study from a Kazakhstani coal deposit.

2 Methodology

The concept of grade domaining in mineral resource estimation is to truncate the orebody into a number of domains defined by grade ranges [2, 5, 6]. The domains then can be analyzed and undergone geostatistical modelling separately in each domain.

However, this approach for geological modelling has pitfalls, caused by uncertainty in-between the domains and ignoring spatial dependency [2, 7]. The actual domain boundaries are irregular in nature, and the bias interpretation in terms of smooth contouring, coupled with separate analysis of each domain, lead to inaccurate results. The following equation is widely practiced to estimate the final grade:

$$Z^{*}(X) = \sum\nolimits_{n=1}^{N} P_{n}(X) Z_{n}^{*}(X) \tag{1}$$

where $\{P_n, n = 1 \ldots N\}$ is the probability of each grade domain and $\left\{Z_n^*, n = 1 \ldots N\right\}$ is the grade at the respective locations.

Kriging interpolator is a geostatistical tool aimed to estimate the value of the variable of interest at unsampled locations throughout the deposit. The interpolator is based on the random function model with a weighted combination of the available data [8]. The weights are assigned according to the distance from the target location to surrounding data, as well as to redundancies within dataset.

Simple Kriging (SK) estimator incorporates the mean value into estimation, and is based on modelling the spatial continuity by variogram. The algorithm assumes stationarity of the continuous variable, with its variability fluctuating around a mean value [9–11]. Despite its straightforwardness, SK may yield negative weights leading to negative estimates and have a smoothing effect, where the tonnage estimation may be affected by a shrinked dispersion of the estimates, compared to dispersion of the true values. The estimator of the algorithm as a linear combination of the data is following:

$$Z^*(x_0) = a + \sum_{a=1}^{n} \lambda_a Z(x_a) \tag{2}$$

where Z(x) is the regionalized variable, $\{x_a,\ a = 1\ldots n\}$ is the data location, x_0 is the target location, $\{\lambda_a,\ a = 1\ldots n\}$ is assigned weight and a is an additive coefficient. Indicator Kriging (IK) uses the above formulae for modelling the categorical variables, creating a deterministic model of the geo-domains, where the indicator variables are independently estimated. The output of the algorithm needs to be processed so that the values are non-negative and sum up to 1.

Sequential Indicator Simulation is a probabilistic approach based on IK, yet additional- ally producing a number of realizations. The realizations are constructed from categorical variables, where each data point is codified into indicator. The categories are mutually exclusive, that's there is only one available category at a given location [9, 12]. Generally, these variables can be expressed as:

$$i(\mathbf{u}; k) = \begin{cases} \mathbf{1},\ \textit{if category } k \textit{ is present at position } \mathbf{u} \\ 0,\ \textit{if otherwise} \end{cases},\ k = 1, \ldots, k \tag{3}$$

Indicator Kriging is incorporated to construct conditional distributions from indicator variograms. [12].

Turning bands method is an algorithm that simulates the variable of interest in random lines in 1D, after that combining them into 3D. Technically, simulation takes place on the lines that are spread throughout the space in different directions. It is based on non-conditional Multi-Gaussian model, further being conditioned by Kriging [13–15].

The computation of random numbers in 3D is carried out by taking the projection of the values simulated in 1D lines:

$$Y(x) = Y^{(1)}(<x|u>) \tag{4}$$

where $Y^{(1)}$ is a random function in one dimension, u is a vector in the Rd, and $<|>$ is a projection of location x in the line oriented by u.

3 Case Study

3.1 Presentation of the Dataset

The Shubarkol coal deposit is an important resource of coal in the Kazakhstan located in the Nurinsky district of the Karaganda region of the Republic of Kazakhstan.

The dataset is composed of 89,942 samples belonging to this deposit, where each sample has assay of Ash. Originally, ash content is the only variable of interest within the deposit, varying from 0% to 90%. In this study, for the purpose of grade based domaining the ash variable is divided into Coal, Waste, Indicator Coal and Indicator

Waste variables depending on grade value based on cut-off 45% of ash. These variables constitute a basement for estimation techniques, such as Simple Kriging and Turning Bands Simulation. The data is totally heterotopic, where the first grade domain, Coal variable, ranges from 0 to 45% Ash content, inclusively, and Waste variable includes Ash from 45 to 90% content. Therefore, there are 78,379 samples with ash content less than 45% belonging to Coal, and 11,562 samples with ash content higher than 45% belonging to Waste. Indicator Coal and Waste variables are used to implement simulation techniques, including Indicator Kriging and Sequential Indicator Simulation. The variable present for the given samples is assigned value of 1 if pertaining to coal and 0 if pertaining to waste.

The area of sampling covers approximately 6300 × 14000 × 169 m, with a typical elevation adjustment of 1 cm between the samples. Table 1 shows the global statistical parameters of all the variables within the entire deposit. Hereby, it is seen that the variance of original data (Ash) is very high due to wide spread of data points from the mean and from one another. The shape and spread of the sample data show a distribution with a positive skewness.

Table 1. Statistical parameters of variables under study (%)

Statistical Parameter	Ash	Coal	Waste	Indicator Coal	Indicator Waste
Minimum	0.05	0.05	45.02	0	0
Maximum	89.84	45	89.84	1	1
Mean	14.04	6.74	63.56	0.87	0.13
Variance	434.31	61.93	145.04	0.11	0.11

3.2 Geostatistical Modelling

The geostatistical modelling of the available data is carried out by means of conventional Grade Domain and Unique Domain approaches. In Case I, the Unique Domain is meant to represent the whole deposit without being splitted into domains, where the model is built by Turning Bands method. Here, we examine the closeness of the derived statistical parameters to original ones. On the other hand, in Cases II, III and IV grade domaining approach is examined separately through a combination of estimation and simulation techniques, such that we can compare the statistics with Case I and original dataset. Case II incorporates Simple Kriging and Indicator Kriging, Case III-Simple Kriging and Sequential Indicator Simulation, and Case IV- Turning Bands Simulation and Sequential Indicator Simulation. Modelling outputs of Simple Kriging, Indicator Kriging and Turning Bands Simulation are derived by utilizing the ISATIS software. For the purpose of conducting Sequential Indicator Simulation, SGeMS software is used.

Case I: Unique Domain by Turning Bands Simulation

Case I of the current study considers the Ash variable only. Due to irregular pattern distribution of data samples across the deposit, the variable should be declustered. This technique places the domain into a grid with a cell size of 1000 × 1000 × 10 m and assigns each sample data a weight depending on its proximity to surrounding neighbors within a common cell. The weight is calculated by taking inverse of multiplication of the number of data samples within one cell and overall number of non-empty cells. The assignment of weights makes sure that the spatial representivity of samples is taken into account. Following the declustering of Ash dataset, we should infer normal distribution model of frequency curve to run Turning Bands simulation from the original declustered Ash data. It is based on the normal score Gaussian distribution, with mean 0 and variance 1 at a given set of spatial locations. Next step towards the simulation is to describe quantitatively and model the spatial continuity of the regionalized variable, Ash. The spatial continuity is modelled firstly by calculating an experimental variogram on the basis of the avail- able data, and then by fitting a theoretical variogram:

Variogram: 0.83Sph(250 m, 250 m) + 0.14Sph(4000 m, 4000 m)

For the sake of simplicity, only approximately the half of the deposit area is taken for simulation grid, with a cell size of 50 × 50 × 5 m and centered target point. A moving neighborhood of 10 × 10 × 10 km size with 100 samples per sector was established. Finally, optimum number of turning bands equal to 1,000 was used in analysis. Figure 1(a) below depicts an average E type map out of 100 realizations, where it can be seen that there are soft boundaries respected in this case, with an approximate layer- ing of coal illustrated in black lines.

Case II: Grade domains by Simple Kriging and Indicator Kriging

This part of the study gives a start to a grade domaining based mineral resources estimation. Here, we analyze the combination of Simple Kriging as a deterministic methodology for totally heterotopic Coal and Waste variables, and Indicator Kriging to stochastically model the probability of Coal or Waste presence (Indicator Coal and Waste variables). An experimental variogram model for each variable is calculated by 100 lags with a value of 75 m and a tolerance of 0.5 m. Fitted theoretical model is then integrated into grade estimation by Kriging, with a moving neighborhood of 10 × 10 × 10 km size and 100 samples per sector. Equation 1 is used to calculate the final grade prediction in each block. The following direct variograms are inferred, where the variogram is nested and consists of two structures following spherical model. For example, 1st structure of the variogram for direct variogram of coal variable has a sill of 48,96 m and range of 320 m, whereas the 2nd structure has a sill of 16,96 m and 5650 m range:

Coal variogram: 48.96Sph(320 m, 320 m) + 16.96Sph(5650 m, 5650 m)
Waste variogram: 136Sph(120 m, 120 m) + 9Sph(1600 m, 1600 m)
Ind. Coal variogram: 0.102Sph(150 m, 150 m) + 0.012Sph(4200 m, 4200 m)
Ind. Waste variogram: 0.103Sph(250 m, 250 m) + 0.012Sph(5000 m, 5000 m)

As can be acquired from Fig. 1(b), Kriging yields visible contours of coal layer, but still the abrupt transition between domains remain visible.

Case III: Grade domains by Simple Kriging and SISim

The third case examines the combination of Simple Kriging, obtained in Case II, and Sequential Indicator Simulation. SISim algorithm is implemented in SGeMS software, with the same input parameters for direct variograms, as in ISATIS for Indicator Kriging.

The output matrix from SGeMS is then additionally transformed in MATLAB to reflect the probability of variable occurrence from 100 realizations. Figure 1(c) below is an estimation map based on average of 100 realizations, with less observable coal layer, but with a soft boundary, i.e. gradual transition between coal and waste zones.

Case IV: Grade domains by TBSim and SISim

Case IV of the research conjoins the two simulation algorithms, TBSim and SI- Sim, to estimate the Ash content throughout the deposit. Turning Bands simulation is taken into account to model the Coal and Waste variables, and Sequential Indicator

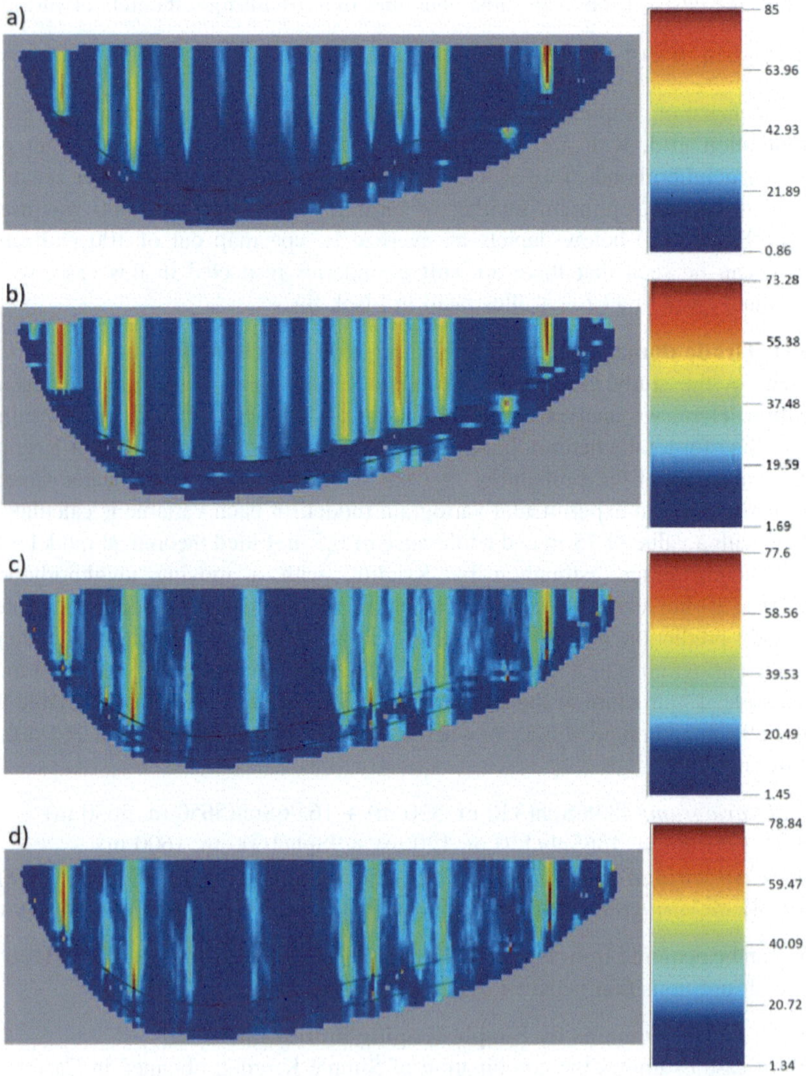

Fig. 1. (a) Case I (b) Case II (c) Case III (d) Case IV

simulation for probability description of indicator coal and waste variables. While SISim algorithm is already developed in Case III, the TBSim is performed newly. Normal distribution models of empirical density curve are inferred, based on Gaussian distribution (mean of 0 and variance of 1). Direct variograms of Coal and Waste variables are produced independently and respective fitted models are used further in algorithm. The same grid model is used to carry on the simulation with a cell size of 50 × 50 × 5 m and centered target point. 1000 turning bands are assigned to derive a reasonable simulation output. Figure 1(d) illustrates an average E type map out of 100 realizations with a much less delineation of coal layer and soft boundaries.

4 Validation of Algorithms

In this part, in order to make a statistical comparison between the presented cases, Table 2 is given. Based on the table, one can compare the closeness of a global variability indicators from the case studies to the original dataset of Ash. As it can be acquired, Case I yields the closest mean values of estimated ash to the declustered original ones. On the other hand, in terms of maximum and mean value, Case II produces the worst reproduction of borehole data. However, the minimum value of 0 indicates that Simple Kriging implies smoothing effect on grade distribution.

Table 2. Reproduced statistical parameters based on different case studies

Study	Minimum	Maximum	Mean	Variance
Original data	0.05	89.84	14.04	434.31
Case I	**0.86**	**85.00**	**17.29**	**32.61**
Case II	0.00	74.34	19.09	44.85
Case III	1.45	77.60	18.59	31.75
Case IV	1.34	78.84	17.40	24.88

5 Conclusion

A precise estimation of mineral resources and its classification is of paramount importance in a modern mining practice. A reliable interpretation of borehole samples forms an essential condition for trustworthy mineral resource estimation. Currently, global mining industry prefers estimating grades through the deposit using Grade Domaining approach. In this study, the reliability of this approach was examined in comparison to Unique Domain approach, where the area of interest is analyzed as one domain. A dataset taken from Shubarkol mine deposit was used to examine four case studies, including three examples on Grade Domaining approach and one study on Unique Domain approach. This research work shows that the global statistics of original grade distribution of ash is best reproduced in case of a single unified domain, Case I. It is shown in this study that discretizing the area into domains yield less reliable estimation of grade distribution, as shown in Cases II, III and IV and increases the mean value erroneously that may lead to biased tonnage calculations.

Acknowledgement. The authors are grateful to Nazarbayev University for funding this work via "Faculty Development Competitive Research Grants for 2018–2020 under Contract No. 090118FD5336." The second author acknowledges the Social Policy Grant (SPG) supported by Nazarbayev University. The authors also thank the ERG Company in Kazakhstan for providing the dataset.

References

1. Emery, X., Ortiz, J.M.: Estimation of mineral resources using grade domains: critical analysis and a suggested methodology. J. South Afr. Inst. Min. Metall. **105**(4), 247–255 (2005)
2. Emery, X., Ortiz, J.M.: Defining geological units by grade domaining. Technical report, Universidad de Chile (2004)
3. Ortiz, J.M., Emery, X.: Geostatistical estimation of mineral resources with soft geo- logical boundaries: a comparative study. J. South Afr. Inst. Min. Metall. **106**(8), 577–584 (2006)
4. Blackwell, G.: Relative kriging errors; a basis for mineral resource classification. Explor. Min. Geol. **7**(1–2), 99–105 (1998)
5. Romary, T., Rivoirard, J., Deraisme, J., Quinones, C., Freulon, X.: Domaining by clustering multivariate geostatistical data. In: Abrahamsen, P., Hauge, R., Kolbjørnsen, O. (eds.) Geostatistics Oslo, pp. 455–466. Springer, Dordrecht (2012)
6. Yunsel, T.Y., Ersoy, A.: Geological modeling of gold deposit based on grade domaining using plurigaussian simulation technique. Nat. Resour. Res. **20**(4), 231–249 (2011)
7. Madani, N., Ortiz, J.: Geostatistical simulation of cross-correlated variables: a case study through Cerro Matoso Nickel-Laterite deposit. In: The 26th International Symposium on Mine Planning and Equipment Selection, Luleå, Sweden (2017)
8. Cressie, N.: The origins of kriging. Math. Geol. **22**(3), 239–252 (1990)
9. Deutsch, C.V.: A sequential indicator simulation program for categorical variables with point and block data: BlockSIS. Comput. Geosci. **32**(10), 1669–1681 (2006)
10. Emery, X., González, K.E.: Incorporating the uncertainty in geological boundaries into mineral resources evaluation. J. Geol. Soc. India **69**, 29–38 (2007)
11. Emery, X., González, K.E.: Probabilistic modelling of lithological domains and it application to resource evaluation. J. South Afr. Inst. Min. Metall. **107**(12), 803–809 (2007)
12. Emery, X.: Properties and limitations of sequential indicator simulation. Stoch. Environ. Res. Risk Assess. **18**(6), 414–424 (2004)
13. Emery, X., Lantuéjoul, C.: TBSIM: a computer program for conditional simulation of three-dimensional gaussian random fields via the turning bands method. Comput. Geosci. **32**(10), 1615–1628 (2006)
14. Brooker, P.I.: Two-dimensional simulation by turning bands. J. Int. Assoc. Math. Geol. **17** (1), 81–90 (1985)
15. Mantoglou, A.: Digital simulation of multivariate two-and three-dimensional stochastic processes with a spectral turning bands method. Math. Geol. **19**(2), 129–149 (1987)

Multicriteria Analysis as a Fundamental Tool for Mineral Research. Case Study: Quadrilátero Ferrífero/Brazil

Pedro Benedito Casagrande(✉), Viviane da Silva Borges Barbosa, Pedro Henrique Alves Campos, Luciano Fernandes Magalhães, and Gilberto Rodrigues da Silva

Mining Department, Universidade Federal de Minas Gerais, Antônio Carlos Av. 6627, Belo Horizonte 31270-901, Brazil
pedrocampos@demin.ufmg.br

Abstract. The "Quadrilátero Ferrífero" (Brazil) had a geological influence for the first explorers who entered Brazil, to explore and to search for precious metals. The geological features and their geomorphological consequences led and influenced the beginning of the occupation of the country area in Brazil. Thus, the presence of mineral deposits conditioned the colonization of the region by the Portuguese in the late seventeenth century. Therefore, in this work, the data, treated by Geographic Information System (GIS), were analyzed by multicriteria, in order to obtain the Synthesis Model of the area. Multicriteria Analysis consists of raster-type structure operations in a GIS environment, which produce a new information. It is understood that the Multicriteria Analysis procedure is widely used in the creation of new information. This analysis is done by assigning weights and values, indicated by a specialist in the phenomenon under study, as well as a specialist in the variable in question. The results of this analysis demonstrate the region is suitable for mining and corroborates a better understanding of the territory use for this activity with the synthesis map for a study area based on variables used. For this work, the variables of existing mine pit, lithology and areas with mining requirements, collected directly from the geological survey of Brazil were used as initial data for the study.

Keywords: Geoprocessing · Big data · Quadrilátero Ferrífero · Iron Ore · Multicriteria Analysis

1 Introduction

Investigations of the potential of the territory for the management of areas of geological interest - whether of production, recognition of values or protection - was chosen as the case study of the Quadrilátero Ferrífero area, in Minas Gerais, Brazil, as a function of its emblematic role for the sector. The Quadrilátero Ferrífero is the heart of exploration and mineral production in Minas Gerais, Brazil, in which significant conflicts of interest: great economic value related to mining and significant urban expansion, an area of important natural and environmental resources, landscape values of the

E. Topal (Ed.): MPES 2019, SSGG, pp. 191–199, 2020.
https://doi.org/10.1007/978-3-030-33954-8_24

mountains of Minas Gerais. In this sense, the case study, due to its complexity, favors a wide discussion about the potential of planning the mining landscape.

With this fact, the region became a mining province, due to the exuberant amount of mineral commodities present in the area. The geological formations led and influenced the beginning of the atropisation of the interior of Brazil. This relationship can be seen in the reflection of [1], "a knowledge about the formative processes of our planet and its evolution in time," which has served as a guide for man since the beginning of its history.

The Quadrilátero Ferrífero is located in the Center-South of the State of Minas Gerais, Brazil, covering an area of approximately 7,000 km^2, located in a highland region, where it occupies a region surrounded by orthogonally arranged mountain ranges (Fig. 1). And the taxonomic origin of the region was named by Gonzaga de Campos [2, 3], due to the iron ore deposits found there.

According to [4], the Quadrilátero Ferrífero has its origin of structure associated with differential erosion, in which the quartzites and itabirites were located in areas of high altimetry. The phyllites are part of the rocky substrates of the median third of the slopes and, in the lower third, at the base of the geological substrate, the composition is rocks from the Belo Horizonte Complex, which are granite-gneiss rocks. Thus, in an interpretation of the space, the high altimetry is related to a set of ridges and places with erosive surfaces that have elevations composing as a whole a quadrangular form, associated with the fact that itabirite is a ferruginous rock of high content, name of the "Quadrilátero Ferrífero" region.

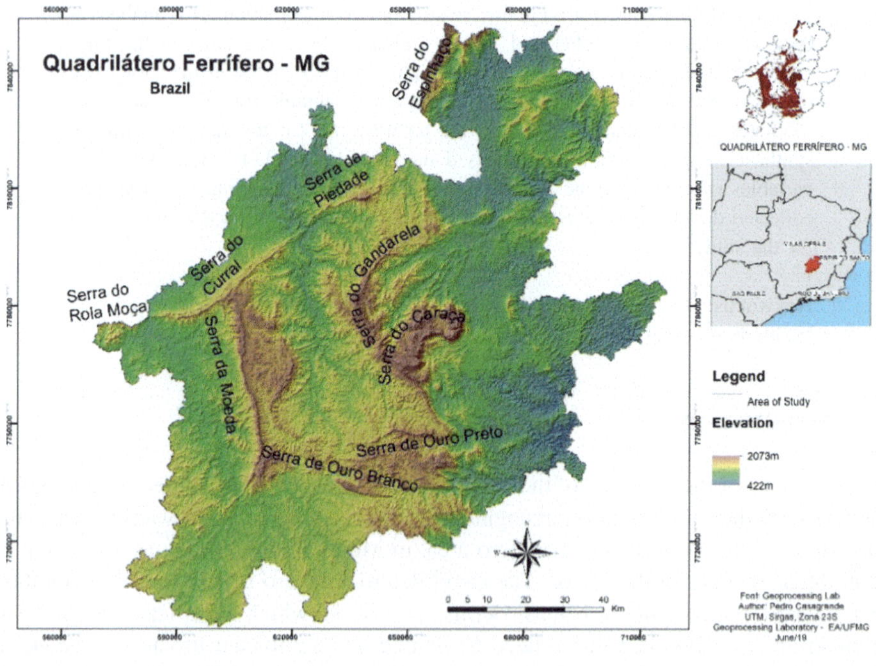

Fig. 1. Location of the Quadrilátero Ferríro and the elevation of the mountains in the area

The objective of this analysis is to identify mineral deposit prospecting targets from geoprocessing analyzes, since the structures of the physical environment correspond to each other when performed some spatial analysis procedure. Therefore, it is possible to identify targets for mineral research from this study procedure.

2 Methodology

The main goal of this analysis is to test the capacity of an innovative approach in the treatment of the (complex) context of mining research, with geology and its lithological variables related to the approach. The reasons for this study is to open up a new possibility for mineral research planning issues.

The work is based on the combination of variables by map algebra, with a view to creating an evaluation that results in a numerical and quantitative ranking, which indicates, for a given territorial cut, the potential of occurrence of a particular commodity, in the case the iron ore, in ascending order. On a relative scale, the territorial portions most favorable to the occurrence of this commodity are identified and more suitable for mineral research.

The mapping of targets for mineral exploration of iron results from the combination of the following variables: Geological Layers, Mining areas and Iron Ore Body.

The data, treated by GIS tools, underwent multicriteria analyzes, in order to obtain the Final Model. Multicriteria Analysis consists of raster-type structure operations in a GIS environment, which in turn produce new information [5].

The Multicriteria Analysis is a method based on Map Algebra, a term proposed by [6], in which the association of numerical (quantitative) values with spatial information is performed, providing mathematical operations. The application of this method has as objective the identification of the potential of the landscape through its uses and values from the thematic maps linked to the physical environment, being the whole procedure performed by geoprocessing techniques [7–9].

Multicriteria synthesis is characterized by the composition of an index resulting from the weighted sum of principal components that account for a research reason. The degree of importance of each variable, understood as the "weight", must be decided by a defensible criterion, either according to a bibliographic reference, or by consulting a group of experts, or even by measuring trends recognized in the territory. The processes of obtaining weights are classified by [10] into two groups: knowledge-driven evaluation (when consulting experts who give their opinions) or data-driven evaluation (when working with data resulting from the measurement of observed trends).

The methodological process can be summarized according to the following stages of work:

(a) Selection and mapping of the main components variables, with the definition of the values ("grade") of their subcomponent components from 0 to 10, according to the degree of pertinence according to experts opinion;

(b) In addition, we used the Delphi method, which is based on consensus maximization [5, 11], adding 100% for the combination in the Multicriteria Analysis by Weights of Evidence.
(c) Map composition is a synthesis by Map Algebra according to the weighted average, which was then symbolized according to the level of suitability for the mineral research use.

3 Results and Discussions of the Variables

The variables included in the index were Geological Layers, Mining areas and Iron Ore Body.

The maps were generated from open data created by the Brazilian Geological Service and prior knowledge of the study area, for the classification of geological characteristics, which evaluates the set of lithotypes according to their mineralogy and, consequently, according to the degree of possibility of iron deposition.

The assignment of grades from 0 to 10 to the geological components followed the indications of the legend of the geological map elaborated by the Brazilian geological service [12], noting that there is no perfect lithotype and totally homogeneous, which would result in note 10, or one so poor as to be note 0, since they are natural elements that can be altered by the geological history (Table 1).

Table 1. Grades for lithotypes.

Lithotype	Grades
Laterite	3
Dolomite	4
Phyllite	5
Schist	5,5
Sandstone	5
Itabirite	9
Granite/Gneisses	2

In general, itabirite occupies the top of the region, due to its resistance to erosion, which places it as of very high importance in the composition of the local landscape. This rock can be massive or powdery (friable), and when they are massive they present high mechanical resistance, [13].

Also according to [13], the mountain tops of the study area contain lithotypes of sandstone and other sedimentary rocks. These lithotypes are considered good aquifers and require the recommendation of preservation of these areas and the prevention of their waterproofing [12]. In such environments, caution should be exercised when cutting, embankment and deforestation, the consequences of which are exposure to rainwater, especially in hilly relief areas with concave and well drained surfaces, since these rocks are susceptible to erosion. In this sense, these aspects were expressed as shown in Fig. 2. This variable received a weight of 20% for the analysis.

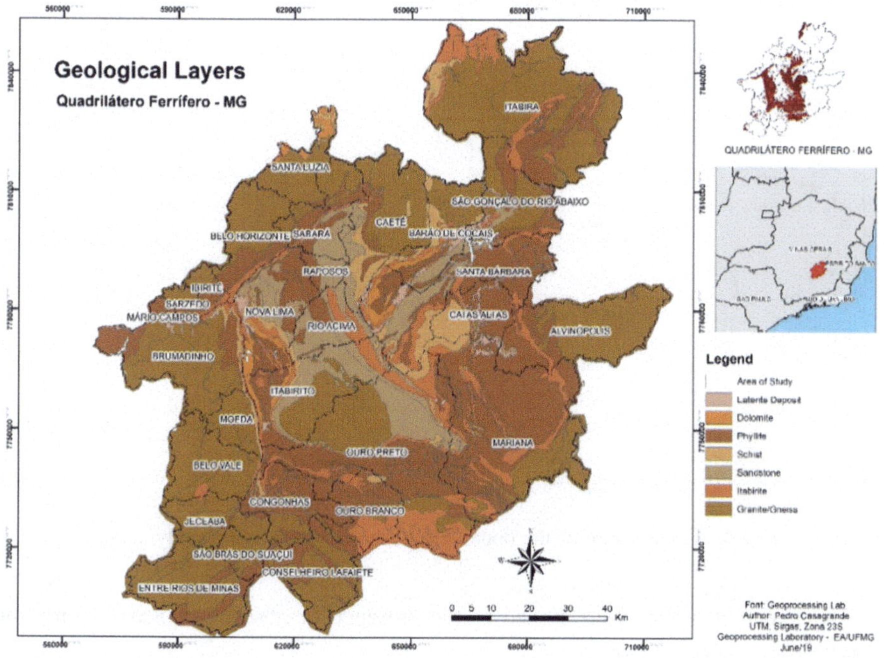

Fig. 2. Geological Layers of the study area.

The Mining areas map was elaborated considering the high concentration of mineral activity in some sectors of the region (Fig. 3), where the most mines are located on the surfaces dominated by hills whose altitudes are approximately 1000 m away. Their identification occurred by supervised classification of satellite with the use of geoprocessing software. The open-pit under study are currently active and this variable received weight of 40%. Once the pits were mapped, notes referring to the legend components of the variable were applied.

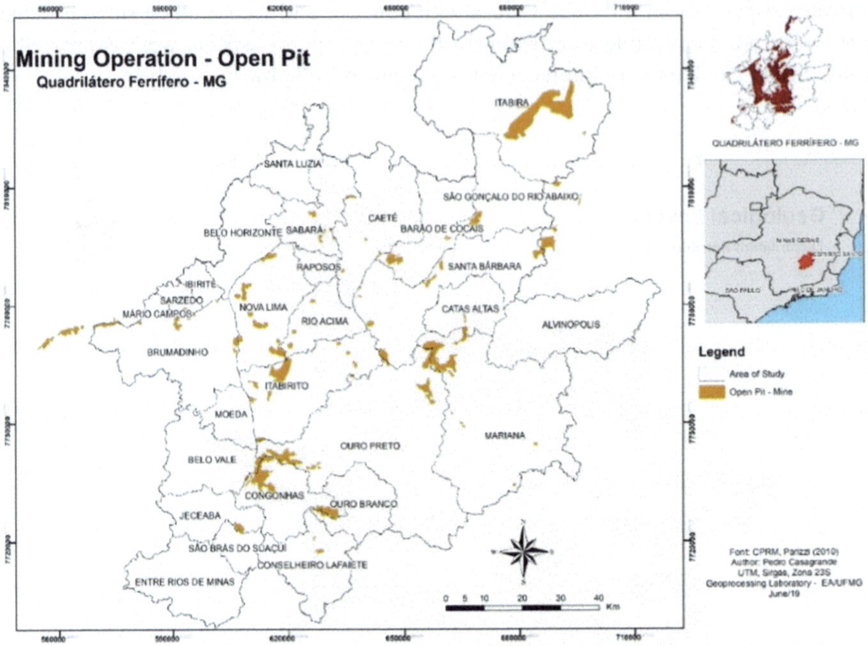

Fig. 3. Localization of the operation mine (open-pit) at the study area.

The assigned grades were linked to the presence or absence of active iron mine (ope-pit), as shown in Table 2.

Table 2. Grades for open-pit.

Lithotype	Grades
Occurrence of open-pit	7
Non-occurrence of open-pit	1

The location map of the Iron Ore Body (Fig. 4) was elaborated considering lithology data in relation to its position (geographic and geological). In general, occurring in the ridges of the region, and the variable received weight of 40%. The weights are in Table 3.

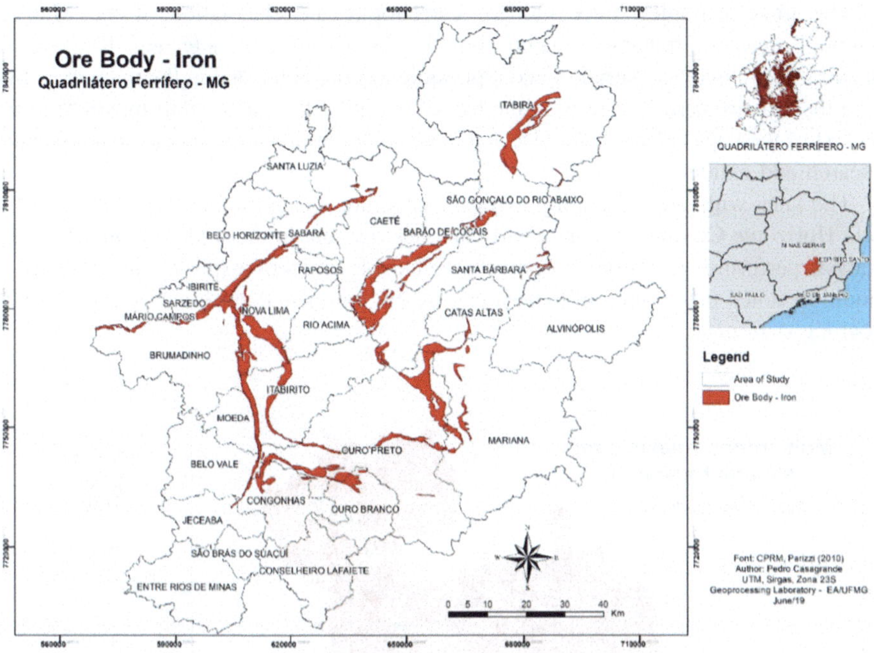

Fig. 4. Location of the body of iron ore in the study area

Table 3. Grades for ore-body.

Lithotype	Grades
Occurrence of ore-body	9
Non-occurrence of ore-body	1

Once all the partial maps were composed, with their caption components classified according to the degree of pertinence for mineral research, and the relative importance of each variable decided, the final composition process was promoted according to the general table of weight values (Table 4).

Table 4. Grades for lithotypes.

Variable	Weight (%)
Geological Layers	20
Mine (Open-Pit)	40
Ore-Body (Iron)	40

The three variables, once unified, corroborate the elaboration of the Map of Potential areas for mineral research (Fig. 5). As a result, the site with the greatest potential for mineral exploration and a possible mining is located in the main axes that form the Quadrilátero Ferrífero, at the top of the hills. It is also worth mentioning the context of the edges of the hills, places that occur, even in a specific way, of targets for research and extraction of iron.

The sites with low possibility of occurrence are located under the context of the Belo Horizonte Complex (Granite and Gneisses) present in the study region. Thus, the analysis performed is likely to condition the axes for mineral exploration. However, it should be noted that several of these areas are environmental preservation according to local legislation.

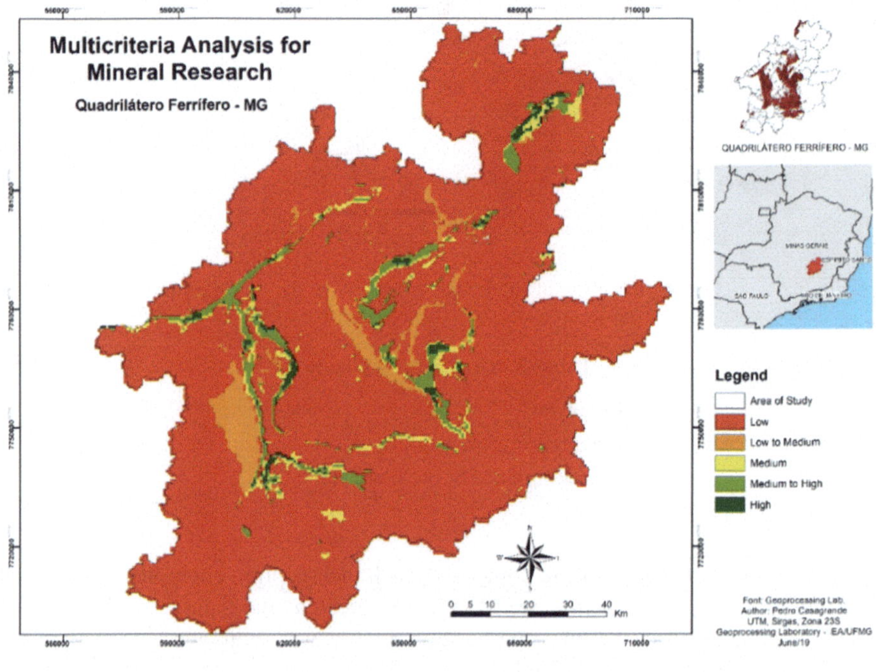

Fig. 5. Synthesis map for iron-ore mineral research targets.

4 Conclusion

Geology is the basis of mineral research, which in turn is an element of financial risk because the entire procedure is high financial cost. Thus, if it is possible to induce targets for mineral exploration from previously known and public data, the likelihood of success and, therefore, savings in exploration costs can be of significant value. Although, the studies of public agencies of geological services of each country can help companies to identify targets for mineral prospecting and to be more assertive in prospecting new mining ventures.

This methodological process can be uses as tool to mineral research, but it has some limitation such as a lack of data source or incompatible between scales of the data source.

References

1. Paraizo, P.L.B.: A construção do conhecimento nas ciências geológicas: contribuições do pensamento de Gaston Bachelard. Rio de Janeiro (2004)
2. de Azevedo, U.R.: Patrimônio Geológico e Geoconservação no Quadrilátero Ferrífero, Minas Gerais: Potêncial para a Criação de um Geoparque da UNESCO. Belo Horizonte (2007)
3. Scliar, C.: Geologia da Serra da Piedade. In: Horta, R.D. (Org.) Serra da Piedade. CEMIG, Belo Horizonte (1992)
4. Harder, E.C., Chamberlin, R.T.: The geology of Central Minas Gerais. J. Geol. **23**, 341–424 (1915)
5. Moura, A.C.M.: Reflexões metodológicas como subsídio para estudos ambientais baseados em Análises de Multicritérios (2007)
6. Tomlin, C.D.: Geographic Information Systems and Cartographic Modeling. Englewood Cliffs, Prentice-Hall, New Jersey (1990)
7. Moura, A.C.M.: Geoprocessamento na gestão e planejamento urbano. Belo Horizonte (2005)
8. Magalhães, D.M.: Análise dos espaços verdes remanescentes na mancha urbana conurbada de Belo Horizonte - MG apoiada por métricas de paisagem (2013)
9. Rocha, A.R., Casagrande, B.P., Moura, A.C.M.: Análise Combinatória e Pesos de Evidência na Produção de Análise de Multicritérios em Modelos de Avaliação. Revista Geografía y Sistemas de Información Geográfica. Argentina (2018)
10. Bonham-carter, G.F.: Geographic Information Systems for Geo-Scientists: Modelling with GIS. Pergamon/Elsevier, New York (1994). 398 p.
11. Vichas, R.P.: Complete Handbook of Profitable Marketing Research Techniques. Englewood Cliffs and Pretice-Hall, New Jersey (1982)
12. Silva, C.R.: Geodiversidade do Brasil: conhecer o passado para entender o presente e prever o future. CPRM, Rio de Janeiro (2008)
13. Parizzi, M.G.: Mapa de Unidades Geotécnicas da Região Metropolitana de Belo Horizonte. In: Programa Diretor de Desenvolvimento Integrado da RMBH: Relatório de Geoprocessamento. RMBH, Belo Horizonte (2010)

Artificial intelligence and Mine Automation

Mine Planning and Selection of Autonomous Trucks

Richard Price[1(✉)], Mitchell Cornelius[2], Lachlan Burnside[2], and Benjamin Miller[3]

[1] F.AusIMM, MTGA, 8 Cohn Street, Carlisle, Australia
richard@mtga.com.au
[2] MTGA, 8 Cohn Street, Carlisle, Australia
{mitchellc,lachlanb}@mtga.com.au
[3] M.AusIMM / SME, MTGA, 8 Cohn Street, Carlisle, Australia
ben@mtga.com.au

Abstract. Autonomous trucks achieve greater productivities (up to 30%), and do so with reduced fuel, maintenance and tyre wear. This is in addition to the lower costs for personnel, offset by costs of communication networks and control room operators.

This paper discusses the mine planning and selection of rendering trucks autonomous, including:

- What is an Autonomous Haulage System (AHS) and how they work;
- What changes in the mine design process;
- How does the mine planning and scheduling change;
- What are the benefits, both perceived and actual from implementing an AHS;
- What are the selection criteria, and how does a mine site make the decision to implement an AHS; and
- The decision-making process to complete the path to automation.

The paper will also summarise the current status and future state of general vehicle automation in the mining industry.

1 Introduction

Automation is an emerging key value driver within the mining industry. Automation was used extensively within mineral processing and metallurgical functions prior to even the use of solid-state electronics. However, Autonomous Haulage Systems (AHSs) only began field testing in open pit mines in the early 1990s. Commercial adoption happened nearly 20 years later in 2008. Technologies for vehicle autonomy advanced rapidly in recent years, which has moved the automation of mining fleets within reach of many mining operations.

On-board systems of an Autonomous Haulage Truck (AHT) are similar to other autonomous vehicles including a Google Car or Tesla's system (Fig. 1). An autonomous vehicle system requires a vehicle control unit that takes in sensory perception,

E. Topal (Ed.): MPES 2019, SSGG, pp. 203–212, 2020.
https://doi.org/10.1007/978-3-030-33954-8_26

communication, and outputs the appropriate vehicle controls to the actuators within the vehicle system.

Typical sensors utilized in autonomous vehicles include:

- Radar
- LiDAR
- High Precision GPS
- Cameras
- Odometers
- Inertial Measurement Units (IMUs)

Typically, actuators utilized in autonomous vehicles are already onboard a manned vehicle. They are accessed via a Controller Area Network (CAN) bus, which is an internationally accepted standard designed to allow microcontrollers and devices to communicate with each other in applications without a host computer. Most modern on and off-road vehicles are primarily drive-by-wire devices. They take inputs from a sensor that the operator/driver controls and then electronically control the vehicle functions such as steering, throttle, and braking.

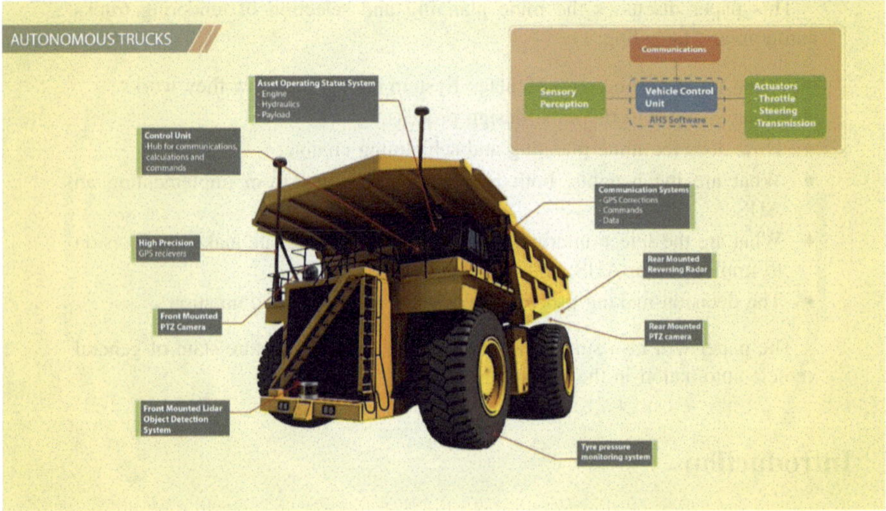

Fig. 1. Generic AHT onboard AHS equipment (Miller 2016).

AHSs are more than trucks and onboard sensors. AHSs consists of:

- Mining trucks equipped with commercial and proprietary electronic devices.
- Software to command, control and track vehicle movements and interactions.
- A communications network with complete coverage to all active mining areas.

- High precision Global Positioning System (GPS) transponders.
- A team of control room operators and site support staff to manage vehicles, devices, and the network.

Integration of AHSs at a mine is a complex task that includes many of the safety hazards of a conventional surface mine as well as the specific safety risks associated with AHSs. Prior to implementation of an AHS, risk analysis and risk mitigation must be undertaken during development of the system and must continually occur during the commercial operation and improvement of AHSs.

AHSs are operating safely and effectively within the Pilbara region of Western Australia for over a decade. Safety incident and accident data show very few events have occurred and are primary limited to human errors such as failing to yield to traffic or follow operational procedures.

2 The Economic Case for AHS

AHS conversion from a manned fleet is most easily described as a CAPEX (capital expenditure) to OPEX (operating expenditure) trade off. Current AHS sites see significant unit cost improvements over a manned site. The principal benefit of AHS is increased productivity in the AHTs due to increased utilization. Typically haul trucks in an open pit mine are scheduled for 5,000–6,000 h per year. AHS sites achieve significantly higher utilization rates that can result in annual scheduled operating hours of 7,000 h for AHTs. This is an increase of 16–20%.

> *"Rio Tinto's autonomous fleet accounted for about a quarter of the total material moved across our Pilbara mines. On average, each autonomous truck was estimated to have operated about 700 h more than conventional haul trucks during 2017 and around 15% lower load and haul unit costs."* Rio Tinto Media Release *(RTIO* 2019*).*

> *"AHS trucks have moved over half a billion tonnes of material and have achieved a greater than 30% increase in productivity levels."* FMG Media Release *(FMG* 2018*).*

Increases in utilization and operating hours are due to the reduction in variability that is achieved through the deployment of AHS. James Petty presents an effective utilization analysis in his 2017 AusIMM Bulletin article. His comparison shows how the removal of variability due to truck operators, removal of shift changes, removal of breaks, and the impacts of surging and queuing can vastly improve the consistency and efficiency of a haul truck fleet. Improved consistency through the introduction of machine control has the effect of transitioning the mining environment from managing discrete components, such as people, to managing a continuous process (Fig. 2).

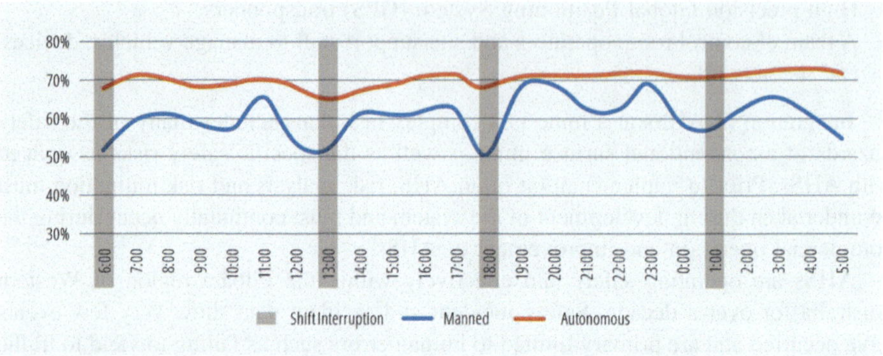

Fig. 2. Effective Utilization curve showing manned versus autonomous vehicles. Adopted from (Petty 2017)

Additionally, the cost to operate an AHS truck is reduced from a manned fleet and reflects a reduced operator count at the mine.

> *"If you think about a manually-operated truck, you need about four-and-a-half people. That's the metric that we've used to help articulate not only the safety component but also the economics behind labour. Within the last month we've dropped below 0.8 people per truck."* – Craig Watkins, Caterpillar commercial manager for a surface mining technology 2018 *(McCrae* 2018*).*

However, cost savings are not only limited to a reduction in operators. AHS fleets are experiencing reduced fuel consumption, reduced maintenance costs, and extended tyre lives.

> *"These autonomous trucks, by the way, have reduced fuel use by 13 percent and hence improved environmental performance by 13 percent."* Bill Walsh, former CEO of Rio Tinto Iron Ore, 2018 *(Barbaschow* 2018*).*

Mine sites implementing AHS typically achieve significantly longer tyre lives. In the manned Pilbara mines, typical tyre life budgets are 5,000 h at a manned operation while autonomous mining operations can potentially budget a 7,500 h tyre life. This extended tyre life benefit results from several factors at AHS sites:

- AHTs only operate on a programmed route.
- Vehicle collisions are greatly reduced.
- Impacts to rocks are reduced.
- Sidewall punctures due to rubbing tyres is greatly reduced.
- Wheel slip is nearly completely prevented due to AHS control.
- AHS sites have rigorous road maintenance programs to ensure AHS can operate unimpeded by hazards such as pot holes, water run-off pathways, and washboard road (Fig. 3).

Fig. 3. AHT tyre at retirement (Price 2017)

3 Mine Planning for AHS

Mine planning for an AHS mine is almost identical to a non-AHS mine. The single critical difference is that the haul roads are approximately 15% wider than a conventional mine, which results in a slight decrease in the overall slope angle of the mine wall. The batter angles, berms and catch bench width can remain the same.

A common truck utilised in many of the existing AHS installations is the Caterpillar 793F. A typical Caterpillar 793F is approximately 8.3 m wide. A typical haul road in a non-AHS mine is 3.5 Truck Widths (TW) wide, making a haul road for this truck 29 m wide. A haul road for a 793F on an AHS system is 32–34 m wide making it approximately 4 TW's wide. This increase in width of the road is due to the object detection system, which is less likely to produce a false positive from objects on the side of the road when they are wider (Fig. 4).

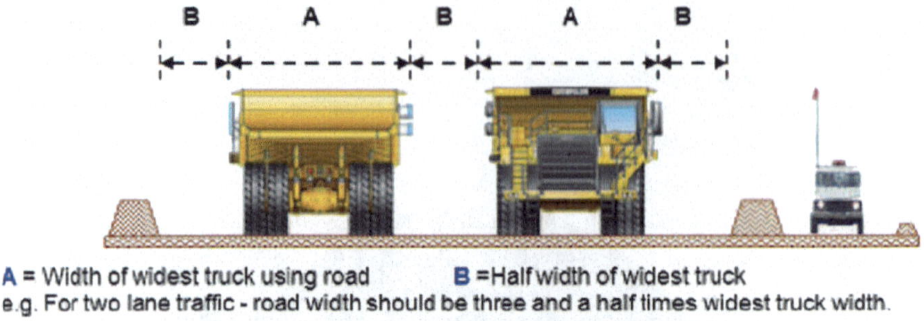

Fig. 4. Haul road truck widths (H&SA 2019)

Another minor difference to mine planning on an AHS site is that the windrow is pulled back from T-Intersections by 10–20 m. The purpose of this is so that the sensors do not think the windrow is an object when the truck is turning or driving past the windrow, and therefore halt the truck (via the Object Detection System). Dump designs for an AHS system are practically identical to that of a normal mine. The trucks prefer a higher overall (waste) dump; this is because their on-board communications systems benefit from the increased height.

4 Perceived and Actual Benefits of Implementing AHS

The clear perceived incentive in making any system or process autonomous is the associated increase in safety and efficiency, increasing the profit margin through lower cost and greater productivity. Implementing AHS is no different. While having a high initial cost (as discussed below), implementing AHS has a definite long term gain with productivity improvements of around 15-20%, along with maintenance savings, tyre life and equipment improvements and significant safety benefits (Bird *et al.* 2019).

The predictable nature of autonomous systems allows for the lifetime of a machine to be planned from implementation to the end of life of the machine, including downtime due to maintenance. This allows a cost analysis to be accurately implemented to compare the difference between autonomous and non-autonomous systems. While the cost benefit varies for each mine site, productivity due to machine utilisation has been increased by about 21% through a study on actual autonomous truck data (Brundrett 2014). The obvious advantage is the decrease in downtime due to human activity, such as breaks, shift change and accurate dispatching. With the need for an operator alleviated, the cost in staff is reduced per truck with each truck optimised for operation, with one or two supervisors capable of managing an entire fleet. Typical shift personnel are reduced by a headcount of 4, with an increase in technical and operating staff of about 1 (Bird *et al.* 2019). In addition, the wear on the machine is also predictable allowing maintenance to be scheduled accurately for machine wear and tear due to optimal driving and performance of the machine.

As mentioned above, one of the main advantages of autonomous systems is an increase of safety within the autonomous zone. There are several factors to why autonomy increases the safety of a mine site, such as;

- Removal of personnel (and therefore human errors and associated hazards) from autonomous zone
- Increased and shared awareness of obstacles
- Predictable nature and path of machinery

The obvious advantage in making a system autonomous is the removal of any staff in the autonomous zone, creating an environment where if any incidents were to occur, injures would be avoided. Most accidents are caused due to fatigue, stress and lack of competency of an individual. AHS remove human error and judgment, relying upon the overall network and individual sensors mounted on each truck to guide the vehicle and avoid collisions. The ability to rely upon multiple sources greatly increases the accuracy and precision of the vehicle, allowing navigating and avoiding accidents before line of sight confirmation is achieved. Caterpillar, one of the leading manufacturers of autonomous mining equipment, boasts zero lost-time injuries from 27.3 million kilometres driven across 5 years of their autonomous trucks (Harry 2018).

With the addition of 'smart' systems, trucks can be updated with improved predicted pathways capable of navigating around obstacles in real time. This creates a stable, overall greater net income with a higher yield per hour than previously manned systems, with 7 autonomous trucks hauling the same quantity of ore as 9 manned trucks (Brundrett 2014). In addition, the expected behaviour of autonomous trucks and how they respond in certain situations allows for pre-determined responses, potentially before an incident occurs.

5 Negatives Associated with AHS

Just as there are many perceived benefits of implementing AHS onto a mine site, there are also many perceived negatives associated with both the idea of autonomous technology and the implementation of autonomous systems. As with most technological upgrades, there are a multitude of initial costs that must be accounted for with the implementation of a new system. AHS is no different.

AHS key components:

- Mining trucks fitted with both commercial and proprietary electronic devices.
- Software that commands, tracks and controls vehicle interactions and movements.
- Communications network (both hardware and software) with network coverage to all areas.
- Control room operators and support staff managing and maintaining the vehicles and associated systems.

Once AHS has been implemented onto a mine site, higher costs are associated to the increased use of graders and dozers that now have a higher run time. General maintenance costs also increase with faster wear and tear of the vehicles due to the increased operating run time. A flow on effect is made from this, with roads also having to be maintained at a greater frequency. In addition, the continual upkeep of the network and software license fees also add an increased ongoing cost.

With mining making a move to increased implementation of autonomous systems, the role of operators is made redundant, however the experience required for assisting in programming appropriate responses to the environment is of high value. While there is an added benefit of employment in support staff, training must be undertaken for support staff and any staff that may be required to access autonomous areas, with the associated training costs included.

A mine conversion to an AHS system is a slow process and the ramp up time of the system is generally around 12 months in which the utilisation of the trucks will slowly increase over this time until it reaches the highest possible utilisation. The main reason for this is the site-specific programming of the AHS system which includes haul routes, load locations and dump locations. The site must learn the most efficient way of programming an AHS system that utilizes the site resources optimally. The next reason is the training of the mine staff, where the AHS operators, pit crew ("pit patrol"), AHS Technicians, loading unit operators and maintenance staff require extra training in order to be able to work effectively around an AHS installation. The final reason for a slow ramp up is the required cultural change of the site, once the site is used to AHS, the system will run smoother.

Social impact of autonomous systems is significant, with potential adverse community and government mindset constraining the uptake of the technology and increase implementation costs (Moffat & Zhang). One perceived negative regarding AHS is the safety of personnel and equipment within autonomous zones. This perception, while mostly just perception, creates its own issues as the world and workforce adjusts to the ideas associated with autonomous mobile equipment. Autonomous equipment allows for human error to be decreased, however this error cannot be fully negated. Horberry and Lynas (2012) state that while operator acceptance, up-skilling, user-centred design and integration are of key importance, trust is also required for the technology to be effective. One only has to look at previous technology that has been implemented but phased out due to adverse public opinion. The super-sonic Concorde for example, met with negative public opinion after a safety incident occurred, resulting in the aircraft ceasing production and implementation. A similar incident may result in the same outcome for autonomous systems. Therefore, positive press, and public workforce opinion is of high importance for this industry to succeed.

While the safety concept of an autonomous system is to negate the need for direct human interaction, due to unforeseeable circumstances or general maintenance on site, some human interaction is required to maintain the system. This is the greatest concern for safety as the autonomous system needs to be navigated safely in order to accomplish manned tasks, creating a risk for autonomous and manned interaction to take place. The development of 'Site Awareness Systems' or 'GPS-Equipped Manned Vehicles' and 'Autonomous Emergency Stops' aim to provide safe passage into the autonomous zone through creating a safe 'bubble' around the manned vehicle, which interfaces into the autonomous network. This alerts autonomous trucks to the presence of a manned vehicle and provides the safety criteria that the truck must meet. As the name suggests, the Autonomous Emergency Stop shuts down any local autonomous machines in the case of an emergency. Both of these are in addition to the peripheral sensors mounted on autonomous trucks.

Unfortunately, a potential negative of making one part of a system more efficient is the flow on effect causing the transfer of stress onto other areas. What was once the choke point within the mine system has now been transferred over to other sections, such as the processing plant. While the increase of efficiency is not necessarily a negative aspect, it does create stress on other areas that may be restricted either by human or machine operating factors or both. This may require future upgrades or be the limiting factor in the productivity of mine sites as they strive for the most economical and efficient system.

6 Current and Future Status of Mining Automation

Currently the state of AHS in the mining industry is in a transformation stage where Original Equipment Manufacturers (OEM) and automation companies are reporting that AHS is going from a push to a pull.

> *"One of the key changes that's happened over the last year or so is that we have gone from a push scenario where we've been trying to educate the industry of the value and the benefit of automated hauling solutions" – Craig Watkins, Caterpillar commercial manager for a surface mining technology 2018 (McCrae* 2018*)*

This means that AHS manufacturers aren't having to push their product on customers anymore, but the clients are coming to them for their product. This signals a major change in the mining industries outlook on AHS, that the proven improvements and results from the pioneers of AHS have made other companies see the benefits.

Due to the price of automation dropping, the price of labour increasing and the proven increase in safety (Amin 2018) the amount of companies interested in AHS and the number of manufacturers producing AHS systems has increased in recent years with smaller companies announcing their plans to invest in AHS. Soon AHS will be manufactured by more companies, bought by more customers and be implemented in more countries than ever before. Moving it from a niche product that only the big mining companies can afford to an essential commodity and in the future, the standard for mining.

Mining companies are already planning their mines around AHS, which will soon be the standard for mining. This means mine planning, infrastructure, vehicles and AHS specific operating areas will be designed into a mine from the beginning rather than it being a retrofit on an existing mine.

Automation of other mining equipment is the next step in automation in the mining industry with the next biggest hurdle being the automation of diggers (Stevens 2019). The challenges here will be ensuring the autonomous diggers don't overload or underload the trucks and that they do not slip off benches being mined.

Acknowledgements. Acknowledgements to Aidan Ayres of AMC Consultants for assistance with this paper. We further acknowledge our clients and their dedication to technology investments.

References

Amin, C.: Mining industry automation: let the robots take our dangerous and dirty jobs. Retrieved from Create Digital, 10 December 2018. https://www.createdigital.org.au/mining-industry-automation-robots/

Barbaschow, A.: Rio Tinto preparing for the mine of the future with automation. Retrieved from ZDNet.com, 26 February 2018. https://www.zdnet.com/article/rio-tinto-preparing-for-the-mine-of-the-future-with-automation/

FMG. Fortescue's autonomous truck fleet surpasses one billion tonne milestone, 13 September 2019. https://www.fmgl.com.au/in-the-news/media-releases/2019/09/13/fortescue%E2%80%99s-autonomous-truck-fleet-surpasses-one-billion-tonne-milestone

H&SA. Construction of Haul Roads. Retrieved from Health and Safety Authority of Ireland: Health and Safety Authority of Ireland (2019)

McCrae, M.A.: Selling autonomous trucks moves from push to pull: Caterpillar. Retrieved from Mining.com, 29 July 2018. http://www.mining.com/autonomous-trucks-sales-move-push-pull-caterpilllar/

Miller. Autonomous Haul Systems Savings: It's not about less truckies but better truckies! AusIMM, Perth, WA (2016)

Petty, J.: Getting the best out of autonomous mining fleets. Retrieved from AusIMM Bulletin, December 2017. https://www.ausimmbulletin.com/feature/getting-best-autonomous-mining-fleets/

Price. Autonomous haualge systems - the business case. AusIMM, Perth, WA (2017)

RTIO. Rio Tinto's autonomous haul trucks achieve one billion tonne milestone. Retrieved from Rio Tinto, 30 January 2018. https://www.riotinto.com/media/media-releases-237_23991.aspx

Stevens, M.: Rio's smart trucks book their own sickies. Retrieved from AFR, 8 May 2019. https://www.afr.com/business/mining/rio-s-smart-trucks-book-their-own-sickies-20190507-p51kya

Bird, D., Beal, C., Thomson, A., Vinson, C.: Report 2 - Autonomous Mining Equipment (2019). http://www.rfcambrian.com/wp-content/uploads/2019/04/RFCA-NTI-Report-2-Autonomous-Mining-Equipment-May-2019.pdf

Brundrett, S.: Industry Analysis of Autonomous Mine Haul Truck Commercalization. (Master of Business Administration), University of Birtish Columbia (2014). summit.sfu.ca/system/files/iritems1/14425/SBrundrett_Capstone.pdf

Harry, R.: Exclusive: caterpillar reveals the impact of autonomy on the industrial sector. Industrial Vehicle Technology International, June 2018

Horberry, T., Lynas, D.: Human interaction with automated mining equipment: the development of an emerging technologies database. Ergon. Aust. **8**(1) (2012)

Moffat, K., Zhang, A.: Resources Policy **39**, 61–70 (2014). https://doi.org/10.1016/j.resourpol.2013.11.003

Intelligent Enterprise with Industry 4.0 for Mining Industry

Narendra K. Nanda(✉)

President Indian Institute of Mineral Engineers, Hyderabad, India

Abstract. The mining industry is under transformation as digital and automated technologies transform the traditional process of extracting ore from rock. With high global demand of raw materials, reduced ore grades, stringent environmental legislation and less profit margin, the mining industry has to improve its productivity through smart mining to survive in the competitive market.

In smart mining, it is imperative to have a real-time flow of information between enterprise level systems and shop floor systems through application such as ERP. A mining company needs instantaneous visibility on production, quality, cycle times, machine status, and other variables in order to achieve optimum operations which necessitates intelligent enterprise with industry 4.0.

Industry 4.0 technologies provide deeper understanding of resource base, optimization of material and equipment flow, improved anticipation of failure, increased automation such as automated surveying using drones and 3D laser technology, automated drills, automated mobile fleets using fleet management system, automated plants. It also helps in monitoring real time performance vs plan and safety of men and machineries. By using Internet of Things (IoT), Big data, Machine to Machine (M2M), data analytics, intelligent sensors, robotics, drone the smart mining will be up graded with industry 4.0 in mining activities such as surveying, drilling, blasting, excavation. Haulage, processing and transportation.

The orientation of the paper revolves around various action plan for intelligent enterprise with industry 4.0 for improving their productivity to become more competitive globally.

Keywords: Internet of Things · Big Data · Mining · Operation research

1 Introduction

In today's competitive world the mining industry has to shift focus from production to productivity to survive. The boom and bust cycle of mining has been the same for decades, productivity has not improved rather declined over the decades. In this context, need of the hour is intelligent enterprise with industry 4.0 which have the power to unlock new ways of improving the productivity through the large-scale adoption of modern technologies. Intelligent enterprises have broad horizons with long-term perspectives. The enterprise can be intelligent in two ways. It can behave intelligently, or it can utilize "intelligence," to achieve its goals. The embedding of sensors in machinery

E. Topal (Ed.): MPES 2019, SSGG, pp. 213–218, 2020.
https://doi.org/10.1007/978-3-030-33954-8_27

to collect data and enable communications between machines is increasingly affordable and accessible.

1.1 Present Mining Technology

Fully mechanized mining operations have increased production and productivity in comparison to semi-mechanized and manual mining but to remain competitive productivity and standard of safety is to be increased many fold.

1.2 Challenges with Present Mining Technology

The present mining technology talks about physical entity, repetitive job, less productivity, less safety of men and machineries, less automation, less data analysis etc.

2 Need of the Hour- Intelligent Enterprise with Industry 4.0

Mining companies generate vast amounts of data, and extracting relevant data about machinery, processes and ore bodies is now more important than ever before. Complex mining tasks such as geo-modeling, day-to-day scheduling and predictive maintenance are increasingly handled by smart analytics software packages, while smart phones and other handheld devices have transformed the way that workers interact not only with each other but with machines. Work clothing can also incorporate sensors that transmit employee locations and trigger warnings about hazardous situations, improving safety standards.

Advances in robotics and sensor technology are also now making guided equipment much more affordable and effective. The use of tele-remote, assisted control and fully autonomous equipment is becoming increasingly widespread in the mining industry.

These technologies will enable a fundamental shift in the way we mine. There will be reduced variability in decision-making and more centralised automated operations that reduce variability in execution. Through a better understanding of the resource base, knowing exactly what is in the ground and integrating geological information into one universal database enables operators to optimize drilling and blasting operation, creates better mine plans and helps avoid resource quality issues. Deploying remote-controlled machinery such as Unmanned Ariel Vehicles (UAV) or underground vehicles with laser scanning technologies can cause a change in productivity, and the 3D modelling data provided by these machines can inform engineers in remote locations and avoid sending manpower to unsecured mine faces.

Real-time data and better analysis tools also make possible improved scheduling and processing decisions, while fitting smart sensors to equipment facilitates the prediction of failure of components.

2.1 Industry 4.0

The first industrial revolution used water and steam to mechanize production, the second used electric energy to create mass production and the third used electronics and information technology to automate production. Today a fourth industrial revolution builds upon the third revolution and the digital revolution that has been taking place since the middle of the last century. This fourth revolution is characterized by merging technology that synthesizes the lines between the physical, digital and biological spheres to completely uproot industries all over the world. The extent and depth of these changes are a sign of transformations to entire production, management and governance systems.

The fourth industrial revolution "Industry 4.0" is with big data, machine to machine connectivity and automation. The current level of innovation and strategic partnerships within the mining technology space demonstrates the agility of businesses for the future of the industry.

The genesis of industry 4.0 rests with the emergence of the **Internet** which is the mother of digitalization. This digitalization enables us to build a new **virtual world** from which we can steer the physical world.

The industry of today and tomorrow aim to connect all production means to enable their interaction in real time. **Factories 4.0** make communication among the different players and connected objects in a production line possible due to technology such as **Cloud**, Big Data Analytics and the Internet of Things.

The applications for the industrial sector are already enormous: predictive maintenance, improved decision-making in real time, anticipating inventory based on production, improved coordination among jobs, etc. Day after day, all these improvements are gradually optimizing production tools and revealing endless possibilities for the future of industry 4.0 at crossroads for an interconnected global system.

Industry 4.0 could be the first to deviate from the energy grid trend in terms of non-renewable resources because it has been integrating more and more possibilities to power production processes with alternative resources.

2.2 Big Data

In each organization, a vast amount of data is generated that contains various information from internal and external sources such as data transactions, corporate documents, social media, sensors and other devices. Companies can take advantage of analyzing their data to satisfy customer needs, optimize their operations or obtain new sources of revenue. Big Data encourages the analysts to take the complexity and diversity of the world into consideration when they examine the data sets instead of endeavoring to reach punctual and perfectly accurate results from analyses made in an artificial and controlled environment.

Four dimensions need to be considered to get value out of big data: volume, velocity, validity and value. Volume is the amount of data. To gain insight from this mass of data, special tools are required. Velocity means that data does not have a stable state, but it is always changing, and new data is generated and transferred in a matter of milliseconds. Validity means that whether data is correct and accurate for the intended use. Value refers to the outcome, the value that can be extracted from the data sources with big data analysis.

Machine data is generated in every industry from healthcare equipment through handheld devices to industrial machines, and they can be used to find patterns and clusters or to predict trends. It has several business use cases, such as performance analysis, root-cause analysis, predictions etc.

2.3 Data Analytics

Big Data Analytics is now a big blip on the radar of the Mining industry. Big Data Analytics would spur the next wave of efficiency gains in ore extraction, analysis, transportation and processing by enabling faster and better informed decisions at all levels.

In a competitive market, every effort to improve margins using operational intelligence is necessary. That is why analytics is expected to play a major role in driving better asset utilization, boost productivity, and address material flow delays. Helping achieve this goal are sensors embedded across mining operations. These sensors are generating vast amounts of geo scientific, asset condition and operational data in real time. Improvements in internet speeds are enabling real-time collection of data from the extraction point right up to the final transportation of ore to plants. This data can be analyzed using massively parallel processing and faster distribution of intelligence. It is possible to do this because modern Big Data platforms can assimilate vast amounts of heterogeneous, real-time inputs from multiple sources. These, in turn, extract real-time predictive and prescriptive analytics to drive operational excellence.

Intervention Across Mining Processes: Material process flow plays a big role in the mining value chain. This includes analyzing impact of unscheduled events owing to mechanical breakdowns Dumpers, shovels and critical transportation medium, queuing time, and such overheads. There are a number of other causal variables that can be analyzed for impact on production throughput on a daily/monthly basis using techniques such as Machine Learning, Continuous Pattern Matching and Statistical Predictive Model.

Big Data Analytics Platform, equipped with these models, can leverage the value, volume, velocity and variability of data, delivering several benefits across extraction, intermediate transportation and final transport to plants.

2.4 Application of Big Data Analytics in Mining Process Through Operation Research:

Different operation research techniques are available to achieve this objective [1].

- Queuing model for transport system
- Markov Model for maintenance system.
- Reliability model for production system and equipment performance

The Mining industry can derive several critical business benefits from Big Data Analytics. These include:

- Ensuring continuous flow of material from ore extraction point to the processing plant
- Maximizing ores hauled by optimizing bottlenecks in production
- Reducing non-productive time between unit operations such as unscheduled maintenance, delays, wastage and waiting time
- Helping management make informed decisions on the "as-is" production process, covering the value chain from extraction to delivery at plants and beyond
- Providing on-line assay results and interpretation analysis to field geologists to take informed decisions

2.5 Internet of Things (IoT)

The Internet of Things (IoT) is the connection of objects such as computing machines, embedded devices, equipment, appliances, and sensors to the Internet. This emerging network technology can potentially transform the mining industry by creating new ways of maintaining mine safety and improving productivity. The technology involves connecting equipment, fleets, and people based on radio frequency identification device (RFID) and sensor technologies [2].

How can IoT help the mining industry:

1. **Automate maintenance and operations of machines** – Leads to creation of newer collaboration models with OEMs for monitoring via cloud connectivity and networks.
2. **Standardize processes** – Helps build newer business models and highly agile processes at the operations level.
3. **Improve traceability and visibility** – Lets users automatically transfer and receive data over a network without requiring human intervention. Moreover, remote monitoring of operations ensures maximum efficiency, improved safety, decreased variability, and better identification of performance issues.
4. **Ensure safety of people and equipment** – Integrates mine automation system with automated physical elements to create a real-time, multi-dimensional model from a variety of data sources including the sensors on equipment as well as geological and other data. The system can then be used to optimize and coordinate the mine's layout, operation, and vehicle paths to ensure high efficiency and safety.
5. **Move from preventive to predictive maintenance** – Prevents equipment failure using machine to machine sensors that can detect the status of the equipment (like temperature, pressure, vibration, speed), collect maintenance history, and determine external weather conditions. Provides analytics to predict failures before they occur, giving the ability to react at the right time. Spare parts can also be ordered well in advance and procurement of machines can be based on Life Cycle Cost analysis.
6. **Get real-time data and analytics** – Uses tools to provide 3D displays of the mine and other related data for use by pit controllers, geologists, drilling/blasting teams, mine planners, and supervisors. Mining vehicles have built-in sensors to measure oil temperature, contamination, tire pressure, bearing rotation, vibration, frame rack, bias and pitch, engine speed, and brake pressure. The data is transmitted remotely and used to recommend maintenance schedules and alert teams about potential trouble before it takes place.

Current IoT Usage

The opportunity to visit operation center of Rio Tinto at Perth, Australia where mining majors like Rio Tinto and BHP Billiton have set up their integrated remote operations centers for monitoring operations in the iron ore mines of Pilbara, about 1,500 km away. Rio Tinto also opened a processing excellence center in Brisbane to monitor and analyze the processing data in real time from seven of its operations in Mongolia, the United States, and Australia with the help of huge, interactive screens. From the data collected, a team of experts in mineral processing suggests solutions for optimizing mineral processing at these seven sites.

Leveraging IoT to Drive Bottom-Line Growth

The IoT-led journey towards business transformation in mining has just begun. By making mining safer, more efficient, and more automated, IoT is making mining jobs more high tech and allowing people to work remotely, with fewer workers in hazardous roles at the mine site.

3 Summary

Slow to change and reluctant to address the economic, environmental, and safety impacts, the mining industry has been struggling to keep up with the ever-progressing technological world. If it does not embrace the fast-advancing technical developments seen today, it will quickly lose its standing in this world of regulations and competition. To survive in this competitive environment, every effort to improve margins using operational intelligence is imperative. That is why Intelligent Enterprise with Industry 4.0 is expected to play a major role in driving better asset utilization, boost productivity, address material flow delays and helps mining industry to survive in the comimg era.

This is the beginning of new era in mining industry; we have to go a long way ahead.

Acknowledgement. The author is thankful to all colleagues who interacted for formulating this paper Thanks are also due to organisers of the conference for giving the opportunity to share the views.

References

1. Nanda, N.K.: Optimisation of mine production system through operation research techniques. In: Application of Technology for Sustainable Mining, Topic 9, pp. 115–130 (2017)
2. Singh, D., Tripathi, G., Jara, A.J.: A survey of Internet-of-Things: future vision, architecture, challenges and services. In: 2014 IEEE World Forum on Internet of Things (WF-IoT), pp. 287–292 (2014)

The Optimization of Cemented Hydraulic Backfill Mixture Design Parameters for Different Strength Conditions Using Artificial Intelligence Algorithms

Ehsan Sadrossadat[1], Hakan Basarir[1(✉)], Ali Karrech[1], Richard Durham[1], Andy Fourie[1], and Han Bin[2]

[1] The University of Western Australia, Perth, Australia
hakan.basarir@uwa.edu.au
[2] University of Science and Technology Beijing, Beijing, China

Abstract. Cemented hydraulic backfill (CHB) is widely used in underground mine backfilling, especially when regional stability is required. One of the critical backfill properties is uniaxial compressive strength (UCS), the maximum axial compressive stress that the sample can withstand before failing. CHB is expected to have a certain UCS in a specific time so that the adjacent stope can be mined out according to the production plan. To reach the desired strength in a specific time, there exit parameters which are must be well adjusted. To increase the strength of CHB, either the cement dosage could be increased or a longer curing time could be allowed using less cement and more tailings. However, increasing cement content significantly increases the operational mining cost. If there is enough curing time for the planned production then less cement and/or more tailings can be added to get the desired strength at a reduced cost. This paper investigates the applicability of artificial intelligence (AI) algorithms to optimise key parameters of CHB design so that the desired strength would be reached in a specific time. Genetic programming (GP) is used to generate models relating the UCS factor to CHB's key parameters using an experimental database. The generated GP models are then used by a particle swarm optimisation (PSO) algorithm in order to determine the amounts of CHB's parameters which can satisfy the specified UCS conditions in a planned time. Some examples are presented to emphasize the benefits of the optimization of CHB mixture design. Using the presented approach, it is possible to optimize CHB design parameters by considering mine production plan, requiring certain UCS at specific ages, and to reduce the cost.

Keywords: Cemented hydraulic backfill · Uniaxial compressive strength · Constrained optimization · Genetic programming · Particle swarm optimization

1 Introduction

Tailings and waste rock are the main byproducts of mining operations generated during the extraction of ores and mineral processing. These wastes must be treated according to environmental and safety concerns. A cleaner waste management solution is to send

E. Topal (Ed.): MPES 2019, SSGG, pp. 219–227, 2020.
https://doi.org/10.1007/978-3-030-33954-8_28

these materials back into the mine in order to fill the mined out stopes [1, 2]. These wastes are mixed with a binder, usually cement, in order to provide local support and stability allowing for a safer and more efficient mining operation. In this regard, cemented hydraulic backfill (CHB) is widely used, which is composed of tailings, crushed waste rock, water and a small dosage of cement to satisfy the strength conditions required in the project [1, 3].

In underground mining operations CHB must remain stable during the production of any adjacent stope. If it does not have enough strength, adjacent faces may relax and fall into the open stope. To measure the strength, the uniaxial compressive strength (UCS) test is commonly used. In designing CHB, both short and long term strengths are important from the mine planning perspective. If CHB does not gain enough strength when the adjacent stope/s are mined out, the possible consequences are mine dilution, collapse of a stope, and even large scale hanging wall or footwall failure threatening the global stability of the mine. The solid content (SC), cement dosage (CD), aggregate to tailing ratio (ATR) and time are the key parameters controlling the UCS of CHB. Cement is the most expensive component of CHB and it can be considered as a notable percentage of the total cost of mining operation [1, 3]. To decrease cement cost, it is desirable to use more tailings as replacement material.

From the mine production perspective, to mine out the adjacent stope without jeopardising the overall stability and production plan, there might be two or more minimum UCS required, at different timeframes. For example, a UCS of over Y_1 (kPa) might be required in short term (T_1 days) and the long term (T_2-day) UCS should be more than Y_2 (kPa). In order to have a specific UCS, an appropriate mixture design is required. Conventionally, the amounts of CHB's ingredients for such problems are obtained by doing tests based on a process of trial and error. Given that mine planning is a dynamic process depending on the mine production plan, a large number of trial tests should be carried out until the best batch is formulated for each case or stope. Thus, the recipe of backfill mixture design may be changed or optimized frequently during a mining project depending on the production plan. This is a time consuming and costly operation. However, without such optimisation more cement than necessary can be consumed leading to serious economic loss. Theoretical and mathematical methods can be utilized to find near-optimal solutions, to reduce the number of trial mixtures and to have a more economical and engineered mixture design during a project.

The indicated problem is an optimization problem which is the process of finding the best solutions when one or more objectives are achieved. There exist several steps for formulating and solving such an optimization problem. Firstly, a proper problem statement must be developed. Then, the problem should be formulated in which the verbal or qualitative statement of the problem is converted into a mathematical or quantitative form. The next step is to collect data and information relevant to the problem. Then, the designer should utilize appropriate methodologies to solve the problem which may involve different numerical modelling and optimization methods. This paper aims to demonstrate the applicability of artificial intelligence (AI) algorithms to find feasible solutions to such problems. For this aim, an example problem is formulated to be resolved by AI algorithms. The problem assumed here is to determine possible CHB batches yielding a desired UCS at a specific time with reduced cost for different conditions. Genetic programming (GP) is used to find the models existing between different variables in the database. The generated GP models are then used as objective functions in an optimization process using a particle swarm optimisation (PSO) algorithm, in order to make a back analysis and find near-optimal quantities of CHB's ingredients.

2 Basic Concepts of Constrained Optimization

Mathematically, a single-objective optimization is the selection of the best-input values from within a range when an objective is being maximized or minimized. In the process of mathematical optimization, objectives and constraints are the mathematical models which might be linear or nonlinear in engineering problems. By means of the constraints, the space of solutions is narrowed down, thereby the algorithm is directed to the fittest solution. Constraints can be functions or relationships between two or more design variables considered in a problem. A typical engineering constrained optimization problem can be defined as follows [4, 5]:

Design variables: $X == [x_1, x_2, x_3, \ldots, x_i, \ldots, x_n]^T$
Objective: minimize or maximize f(X)
Subject to: $a_i \leq x_i \leq b_i$
$g_k(X) = 0 \; (k = 1, 2, \ldots, l)$
$h_j(X) \leq 0 \; (i = 1, 2, \ldots, p)$

where a_i and b_i are lower and upper bound of x_i, n is the number of design variables, l is the number of equality constraints which can be functions of X represented here by g_k and p is the number of inequality constraints which can be functions of X represented here by h_j.

3 Genetic Programming

Here, genetic programming (GP) is utilized to develop models which are then used as objectives in the optimization process. GP uses the concepts of the theory of evolution and genetic inheritance to recognize and produce the fittest linear or nonlinear models existing between input and output variables in a database [2, 6, 7]. Technically, these models which are computer programs (so-called tree-expressions (ETs)), have tree-shaped structures and can be developed in computer using programming languages e.g. LISP [6, 7]. GP models can be simply converted to mathematical expressions, which can be used for further calculation aims. This is a useful superiority of GP over other AI algorithms such as support vector-machines (SVMs) or artificial neural networks (ANN). Figure 1 illustrates an ET and the converted mathematical equation.

As is shown in Fig. 1, an ET is composed of different nodes. Operation nodes may do unary operations such as abs, sin, cos, tan, sinh, cosh, exp and log, or binary operations such as add (+), multiply (×), and divide (/). Nodes without children are called leaf nodes or terminals that represent input values or evolved constant values within the system [2, 6, 7].

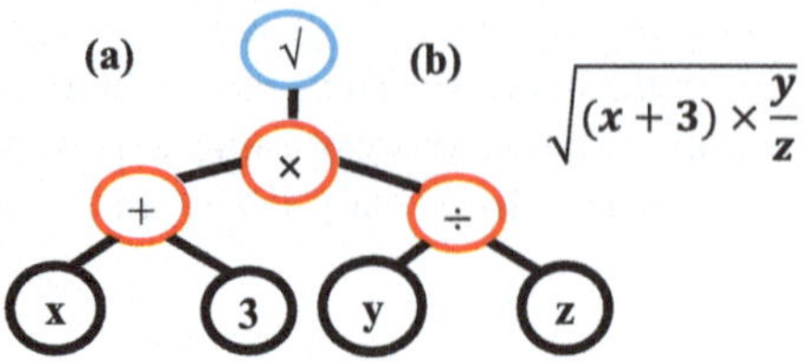

Fig. 1. A GP program in the form of (a) an ET (b) a mathematical equation.

Generally, GP starts with the initialization of a random number of models. After evaluating the fitness of programs and ranking them, those with better fitness are reproduced or selected as a parent for the next generation. The programs are randomly changed into new program mutations, recombination or crossovers. The crossover operation changes random parts of selected pairs (parents) to generate new and different programs (offspring). Mutation involves changing some random parts in only one program. Finally, the fitness and the accuracy of programs are assessed and the process stops once the desired fitness is reached or repeats until finding the fittest program or model [2, 6, 7].

4 Single-Objective Particle Swarm Optimization Algorithm

In the present study, a single-objective particle swarm optimization (PSO) is used to find optimal solutions to the problem. PSO is a population-based search algorithm, which is inspired from the social behaviour of birds or fish seeking food [5]. Each particle (member) of a swarm (population size) updates its search pattern in line with its own experience and the success of the others in its group or in the neighbourhood. A particle, p_i, in PSO has a position (x_i) and velocity (v_i). Assume that $\mathbf{x}_{i(t)}$ represents the position of particle p_i, at iteration t, with a velocity of v_i. x_i represents a solution obtained by the particle and v_i shows the rate of changes of the current particle [5]. x_i and v_i are randomly initialised. Then, the position of p_i is updated by adding the velocity $\mathbf{v}_i(t)$ value to previous the p_i position as follows:

$$x_{i(t)} = x_{i(t-1)} + v_i(t) \tag{1}$$

The velocity vector reflects the exchanged information and is defined as the following equation:

$$v_i(t) = Wv_i(t-1) + C_1r_1\left(x_{pbest} - x_{i(t)}\right) + C_2r_2\left(x_{gbest} - x_{i(t)}\right) \tag{2}$$

where x_{pbest} is the best position found so far by the particle and x_{gbest} is the best position found by the neighbouring particles. C_1 and C_2 are learning factors defined as constants, r_1 and $r_2 \in [0, 1]$ are randomly generated values, and W is the inertial weight defined within the algorithm. These definitions and indicators have also been extended for the vectors of X_i and V_i in a N-dimensional space as is represented by [4, 5].

A single objective PSO is basically programmed to find solutions for merely one function. The accuracy of each solution is examined by a fitness function. In the process of searching the solution space, if the fitness of a solution is better than x_{pbest} or x_{gbest} or even both, their value will be replaced. The process of updating is done after each iteration until the termination or if the stopping criterion is reached which might be the predefined value for the maximum iteration is achieved or target solution is attained. Eventually, the optimal feasible solution is found by PSO [4, 5].

5 Experimental Database and Parameters

In this paper a database presented in recently published paper by the authors [1] was used. The database consists of UCS test results conducted on CHB cylindrical specimens after 7, 14 and 28 days curing times, here presented by UCS_7, UCS_{14} and UCS_{28}, respectively. The percentage of solid content in the mixture (SC), aggregate to tailing ratio (ATR) and cement dosage (CD) are the dominant parameters impacting the UCS of CHB at different curing times and they can be used to have an optimal mixture design [1]. For each curing time 547 tests have been conducted where the amount of three parameters were changed, i.e. SC, ATR and CD. The waste rock and tailings used to prepare the specimens were collected from Jinfeng underground gold mine, which is located in Southwest Guizhou province of China. The details of the properties of materials and testing procedure are presented in detail by [1]. A summary of descriptive statistics of the database is given in Table 1 for further consideration.

Table 1. Statistical description of variables in the database.

Parameter		Mean	St. Dev.	Range	Min.	Max.
Input variables	SC (%)	75.42	1.25	9	67	76
	ATR	4.88	1.09	9.99	1.46	11.45
	CD (%)	11.66	3.99	28.06	1.92	29.98
Output variables	UCS_7 (kPa)	551.2	309.19	2438.3	26.67	2465
	UCS_{14} (kPa)	850.5	508.65	3969.7	49.67	4019
	UCS_{28} (kPa)	1260	724.44	6268.3	67.67	6336

6 Problem Formulation, Analyses and Results

6.1 Problem Formulation

The scenario of the problem considered here is to find feasible and near-optimum CHB mixture batches, i.e. amounts of design variables which are SC, ATR and CD, for a UCS_7 of about Y_1 (kPa), when the UCS_{14} should not be less than Y_2 (kPa) and/or the UCS_{28} should be more than Y_3 (kPa). Considering the main objective of this paper, each UCS is a model, which may be used either as an objective function or as a constraint. Actually, Y_j is considered as a limit of UCS here. The lower and upper bound of design variables are their minimum and maximum values in the database, which is given in Table 1. Note that several different other scenarios may be constructed. An example of the considered problem can be presented as follows:

Objective: To acquire ($UCS_7 \approx Y_1$)
Subject to: $67 \leq SC \leq 76$, $1.46 \leq ATR \leq 11.45$ and $1.92 \leq CD \leq 29.98$
$UCS_{14} \geq Y_2$ and $UCS_{14} \geq Y_3$

6.2 Determination of UCS Models by GP

Three different GP models are developed using the key design parameters for predicting the UCS of CHB specimens with 7, 14 and 28 day curing time, respectively, UCS_7, UCS_{14} and UCS_{28}. These models will be used as objective functions or constraints in the optimization process. The tree-shaped GP models are converted to mathematical equations to be used for hand-calculation and further aims. GP models presented here have been selected after conducting several runs. In each run, the model was developed using a set of data and then it was tested on some unseen data to examine the accuracy of the model. Finally, the optimal models are presented here. The GP-based models are as follows:

$$UCS_7 = CD\sqrt{1.15SC \times CD\sqrt{ATR}} \tag{3}$$

$$UCS_{14} = CD\sqrt{ATR + 5.43(SC + 2ATR)CD} \tag{4}$$

$$UCS_{28} = CD(0.45ATR \times CD + CD + SC) \tag{5}$$

In order to assess the estimation performance of the models, the error existing between the predicted and experimental output values are examined. To do so, some well-known performance indicators, i.e. the correlation coefficient (R), root mean square error (RMSE) and mean absolute error (MAE), are calculated to evaluate the model [1, 7]. To examine the capability of GP, the performance of GP and linear multiple regression (LMR) models are compared. The LMR models are as follows:

$$UCS_7 = -3.74SC + 13.64ATR + 65.84CD \tag{6}$$

$$UCS_{14} = -7.24SC + 27.33ATR + 107.77CD \tag{7}$$

$$UCS_{28} = -10.92SC + 59.18ATR + 153.44CD \tag{8}$$

According to Figs. 2 and 3, it can be said that GP models are able to estimate UCS factor better than LMR models. It should also be mentioned that the LMR models give some negative results. GP models are also able to consider the nonlinear relationship between variables. Furthermore, GP algorithm can generate models which may be converted to equations to be used for further aims whereas those models obtained by other AI algorithms such as ANFIS, SVMs and ANNs are usually black-box or too complex to be converted to mathematical equations [6, 7].

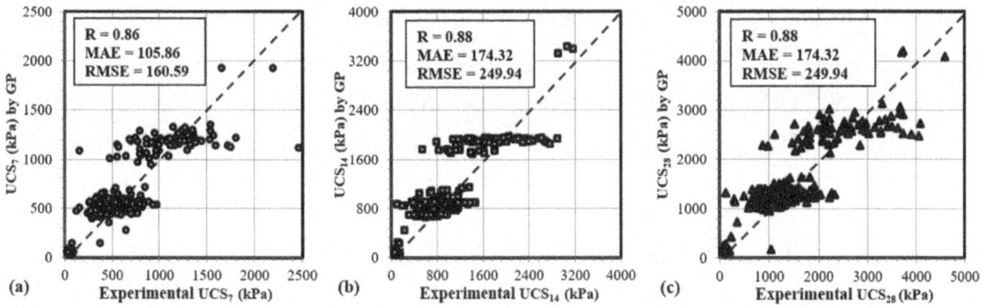

Fig. 2. GP versus experimental values of UCS for (a) 7, (b) 14 and (c) 28 day curing time

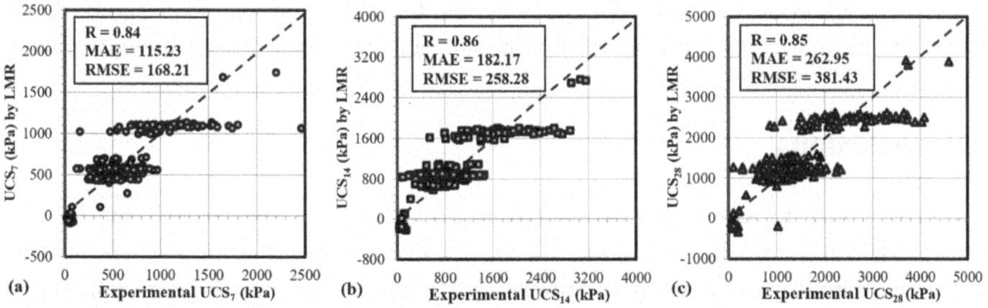

Fig. 3. LMR versus experimental values of UCS for (a) 7, (b) 14 and (c) 28 day curing time

6.3 Finding CHB Mixture Design Parameters for a Specific UCS by PSO

After formulating the optimization problem and finding the objectives, the next step is to perform the process of optimization with a PSO algorithm. In each run, the PSO algorithm provides one feasible solution, i.e. the amounts of the design parameters, which satisfy the objective and constraints. Since different solutions might be produced in each run, one way is to consider the average of their values as the final solution. Here, a number of runs were performed to obtain feasible CHB mixture design parameters for different problem formulations subject to different UCS limits Y_j. For example, in the third case it is assumed that a UCS_7 around 300 kPa is required subject to that UCS_{14} should not be less than 300 kPa and the UCS_{28} should be greater than 1000 kPa. The average value of design variables after 10 runs is considered as the final PSO solution and some results are represented in Table 2.

As can be seen from the presented cases #1–3 the amount of cement and tailings change significantly depending on the required strength and time. As shown by the last two extreme cases, when very high strength is needed in a week (#4) quite high cement (26%) and less tailings (ATR = 8.7) should be used, whereas to get the same strength in 28 days (#5) significantly less cement and more tailings can be used. When the proposed batches are compared with the experimental data they are quite close. It should be noted that in optimization problems, the accuracy of solutions is related to several factors such as the accuracy of the objective functions and the type of

Table 2. Obtained CHB mixture parameters for the considered UCS conditions by PSO.

Item	UCS limit			Proposed batch by PSO			Predicted UCS_i by GP		
	UCS_7, Y_1	UCS_{14}, Y_2	UCS_{28}, Y_3	SC (%)	ATR	CD (%)	UCS_7	UCS_{14}	UCS_{28}
1	200	300	–	71.78	3.62	6.34	200.06	330.89	560.76
2	300	–	700	73.69	3.02	8.49	300	490.1	776.95
3	500	700	1000	74.58	4.23	11.23	500	798.2	1204.1
4	0	0	2000	74.82	2.52	17.75	873	1555	2000
5	2000	0	0	70.16	8.73	25.54	2000	2824	5020

constraints and these can be finally solved by considering a good scenario for solving the problem. Of course, it must always be remembered that the engineer and decision-makers are ultimately responsible for finding and selecting the proper plan and strategy.

7 Conclusions

This paper aimed to investigate the capability of AI-based prediction and optimization to find CHB mixture batches yielding a certain UCS at a specified time as are required by a mine planning engineer. It is demonstrated that the optimization problem should be properly formulated so as to find appropriate solutions. GP is used to produce nonlinear models using an extensive database containing SC, ATR, CD, time and corresponding UCS. The developed GP models are used as objectives and/or constraints to find optimum CHB mixture variables for considered UCS conditions by a PSO algorithm. Several examples, showing the practical application of the presented approach, are provided. The presented procedure can provide the engineer with a better idea of an appropriate mixture design for a desired UCS in order to avoid several trial tests. The optimization procedure presented will result in reducing the costs of CHB, by using optimum quantities of costly materials during the project.

References

1. Basarir, H., Bin, H., Fourie, A., Karrech, A., Elchalakani, M.: An adaptive neuro fuzzy inference system to model the uniaxial compressive strength of cemented hydraulic backfill. Min. Miner. Deposits **12**(2), 1–12 (2018)
2. Qi, C., Tang, X., Dong, X., Chen, Q., Fourie, A., Liu, E.: Towards Intelligent Mining for Backfill: a genetic programming-based method for strength forecasting of cemented paste backfill. Miner. Eng. **133**, 69–79 (2019)
3. Sivakugan, N., Veenstra, R., Naguleswaran, N.: Underground mine backfilling in Australia using paste fills and hydraulic fills. Int. J. Geosynthetics Ground Eng. **1**(2), 18 (2015)
4. Aziz, N.A.A., Alias, M.Y., Mohemmed, A.W., Aziz, K.A.: Particle swarm optimization for constrained and multi objective problems: a brief review. In: International Conference on Management and Artificial Intelligence, IPEDR, vol. 6, pp. 146–150 (2011)

5. Wang, D., Tan, D., Liu, L.: Particle swarm optimization algorithm: an overview. Soft. Comput. **22**(2), 387–408 (2018)
6. Sadrossadat, E., Ghorbani, B., Oskooei, R., Kaboutari, M.: Use of adaptive neuro-fuzzy inference system and gene expression programming methods for estimation of the bearing capacity of rock foundations. Eng. Comput. **35**(5), 2078–2106 (2018)
7. Tajeri, S., Sadrossadat, E., Bazaz, J.B.: Indirect estimation of the ultimate bearing capacity of shallow foundations resting on rock masses. Int. J. Rock Mech. Min. Sci. **80**, 107–117 (2015)

Application of Deep Learning Approaches in Igneous Rock Hyperspectral Imaging

Brian Bino Sinaice[1(✉)], Youhei Kawamura[1], Jaewon Kim[1], Natsuo Okada[2], Itaru Kitahara[3], and Hyongdoo Jang[4]

[1] Graduate School of International Resource Sciences, Akita University, 1-1 Tegatagakuenmachi, Akita, Akita Prefecture 010-8502, Japan
bsinaice@rocketmail.com

[2] Faculty of International Resource Sciences, Akita University, 1-1 Tegatagakuenmachi, Akita, Akita Prefecture 010-8502, Japan

[3] Center for Computational Sciences, University of Tsukuba, 1-1-1 Tennodai, Tsukuba, Ibaraki 305-8577, Japan

[4] Western Australian School of Mines Minerals, Energy and Chemical Engineering, Curtin University, Kalgoorlie, WA 6430, Australia

Abstract. Hyperspectral imaging has been applied in remote sensing amongst other disciplines, success in these has triggered its extensive use. Hence, it comes as no surprise that we took advantage of this technology by conducting a study aimed at the spectral analysis of several igneous rocks, and to deduce the spectral signatures of each rock unit using neural networks. Through visual observations and comparisons of these spectral signatures, parameters such as band curvature(shape), tilt(position) and strength were used for lithological discrimination. Even with this said, there often exists similarities in rocks, which are rather difficult to differentiate by means of visual or graphical analysis. However, with numerous technologies making new waves in today's era and artificial intelligence (AI) being at the forefront of these developments, it was best fitting to employ deep learning, often referred to as a subset of AI; to train/learn from these hyperspectral signatures with a goal aimed at classifying these rocks. Deep learning has networks such as the convolution neural network (CNN), which has algorithms that excel in feature representation from visual imagery; taking into account that the more data is fed into the training process and later used as a database for further training, the higher the future prediction accuracy. Gathered outcomes from the CNN show exceptionally high prediction accuracy capabilities of 96%; suggesting viable field and laboratory usage of these systems as a unit for mining and rock engineering applications.

Keywords: Deep learning · Hyperspectral imaging · Convolution neural network

1 Introduction

Light as we see it via our human eyes does not let us detect material information at wavelengths beyond the visible light range (Tompkins and Pieters 1999), it is for this reason that this paper employs the use of wavelength ranges from 380–1000 nm,

E. Topal (Ed.): MPES 2019, SSGG, pp. 228–235, 2020.
https://doi.org/10.1007/978-3-030-33954-8_29

usually referred to by the science community as the 'Visible-Near-Infrared (VNIR)' range, as it gives out more information due to its spectral range (Pieters et al. 2009). Another aspect that comes into action is spectral resolution (Kruse 2012), this is the number of spectral bands (125 for this study) present within a given spectral range. Simply put, hyperspectral imaging is the collection of image spectra from 100s of spectral bands.

Dealing with this much data requires artificial intelligence (AI), under AI, machine learning comes into play, a subset of this machine learning is deep learning, this has networks such as the convolution neural network(CNN) employed for this study; which has proven able to learn from multitudes of imagery data (Xing et al. 2017), keep it, and reemploy whatever leant as a database for times in which the CNN comes across similar or unrelated data types (Poria et al. 2016), in this case, hyperspectral signatures of igneous rocks.

For industrial viability and economically sound mineral production, quick and accurate rock classification is significantly important; to achieve this, hyperspectral imaging of rocks can be used to differentiate rock units. To study these technologies and how effective they are, it requires the use of closely related rocks. A total of 8 igneous rock units were used, 4 plutonic rocks and 4 of their equivalent volcanic counterparts. Capturing and visualization of their hyperspectral images, classification through deep learning techniques and the subsequent prediction accuracy score (Jia et al. 2016) of these combined technologies are suggested as a vital tool to be employed in current and future mining activities seeing as that the world is moving into a technology and highly computer oriented phase, such as shown in Fig. 1.

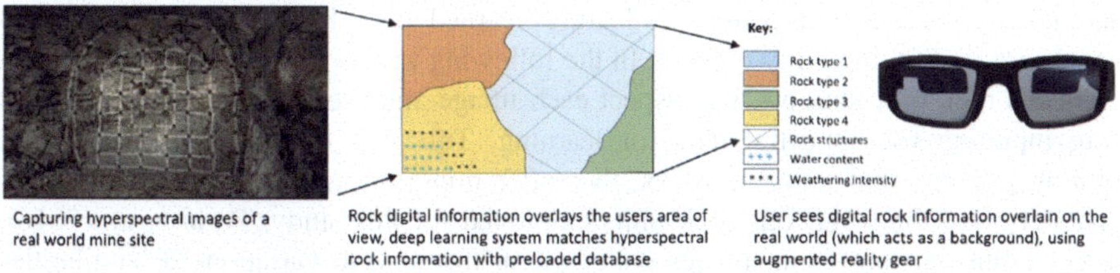

Fig. 1. How hyperspectral imaging and deep learning could be employed in future mining; combined with augmented reality, this system could revolutionize the mining industry.

2 Background

Arial compositional mapping of locations based on water content, lithology, soil and vegetation varieties (Meer 2006) are very important aspects of what helps scientists understand and at times predict what has or will take place from location to location, hence find ways to best engineer such localities. This requires deep understanding, detection and correct classification of rocks.

2.1 Hyperspectral Imaging

It is known that the regions beyond the visible light portions provide the most diagnostic absorption features (Kruse 2012), combined with a high spectral resolution, the collective spectral range (380–1000 nm) forms a robust, highly distinguishing method to identify and illustration spectral signature differences between rocks. The huge advantage of employing high resolution spectral methods is in its ability to draw data from an image input on a pixel-by-pixel basis from hundreds of narrow, adjacent spectral bands, thereby allowing thorough extraction of data (Pieters et al. 2009).

Spectral reflectance, which is the backbone of spectral signatures, varies from material to material due to the degree of energy differences along a given wavelength range (Meer 2006). Also referred to as the ratio of reflected energy to incident energy as a function of wavelength (Kruse 2012). Spectral bands where the signatures show a prominent downward deflection are said to be absorption bands of that particular material, thereby making it easy for material discrimination.

2.2 Deep Learning

There are a lot of proposed methods out there aimed at image data recognition, amongst these, deep learning has been at the forefront of current developments due to its reliability and near perfect recognition capabilities relative to the rest. A great deal of this capability comes from its 'CNN'.

Xing et al. (2017), point out that different granularities can be learnt and discriminated by CNN models directly from data, this gives them an advantage when dealing with multitudes of problems at the same time. Jia et al. (2016), also point out that higher-level structural patterns in features learned at the last layers are what CNNs models are said to implicitly capture. In the following text, we focus our discussion on 8 igneous rock unit datasets, we inspect each image and preform data pre-processing afore inputting rock spectral data for learning. This was done in two ways, first, splitting plutonic and volcanic rocks, thereafter importing and classifying each rock separately into the 8 igneous rock units employed for this study. Poria et al. (2016), further point out that from images collected, at times wide variations or similarities which may not be visually obvious to the human eye are effortlessly pinpointed by CNNs. As pointed out by authors such as Simonyan et al. (2013), the model architecture is also very important for the deep rock classification problem, hence models are continuously developed for various studies in which they are intended for application.

3 Methodology

Spectral techniques have become famous in the compositional differentiation of rocks, like any other science discipline where diagrams are used to represent a phenomenon, hyperspectral signature shapes depend of three main variables. Tompkins and Pieters (1999); Pieters et al. (2009) mention that the first variable; band tilt, shows the band absorption sensitivity. The second variable; band strength, shows the intensity of light being reflected. The third variable; band curvature, is a representation of how light is affected by a few adjacent and connected bands such that a wavy amplitude or trough across that section becomes evident (Kruse 2012). All these are based upon a 'white reference' which creates a straight line anomaly or standard from which all subsequent data emanates.

3.1 Capturing the Hyperspectral Image

8 igneous rock units from Akita mining museum were employed for this experiment, known beforehand and named accordingly before data collection.

To carry out pixel by pixel hyperspectral data collection from the 8 igneous rock units in question, an experimental setup was put in place, this consisted of a hyperspectral camera (380–1000 nm, 125 bands), a studio-like stage and a computer to operate the whole undertaking as shown in Fig. 2. Capturing the hyperspectral images entails placing a white reference board on the illuminated stage and capturing its hyperspectral image, hence creating the above mentioned background reference (Meer 2006). With that done, the board is replaced with a rock sample of the rock in question and its hyperspectral image captured.

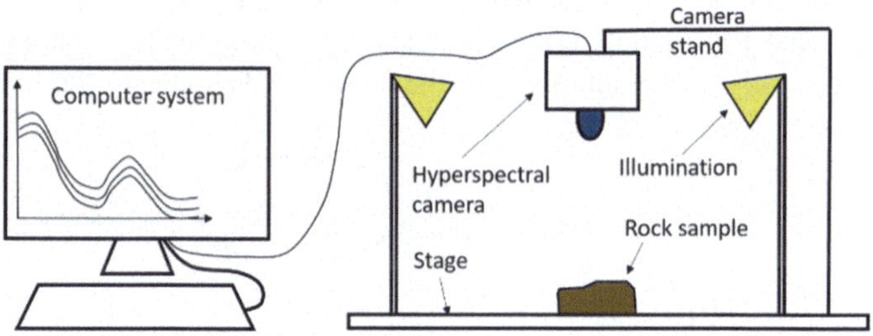

Fig. 2. Main components of a hyperspectral imaging experimental setup.

3.2 Pixel Data Collection

The analogue data captured by the camera requires some conversions before the digital analysis can begin, this entails the use of a hyperspectral analyzing software. Figure 3 shows the manner in which this is done by allowing the user to select how much area of the captured image they would like to employ for analysis, by selecting multiple smaller sized image subsets of the whole image so as to gather spectral information in fine detail. This creates hundreds of anomalies from smaller-sized subsets of the entire image, hence each anomaly representing each image partition's interactions with light across the spectral range. From this data, we can re-manipulate it to get hyperspectral signatures of each rock.

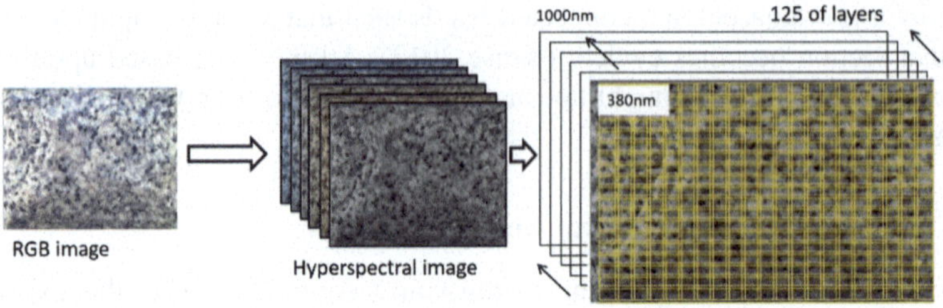

Fig. 3. Image splitting into smaller portions in order get representative data of each small area from the whole hyperspectral image. The RGB image has less spectral resolution (1 layer) compared to the hyperspectral image with 125 spectral channels.

3.3 Data Pre-processing

Spectral signatures from the hundreds of individual subsets of each rock image provide enough data for machine learning to take place. However, Simonyan et al. (2013), state that data pre-processing provides the best solution by adding or subtracting data through over and under-sampling to balance the underlying data, hence this process (Fig. 4). Numerous other data scientists such as Jia et al. (2016), vouch 80% learning, 10% validation and 10% testing as the ultimate ratio to yield the best output. The CNN is set to automatically and randomly split the complete dataset into these ratios.

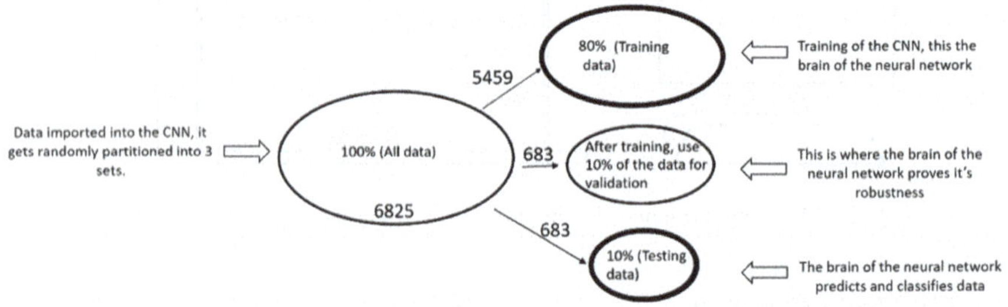

Fig. 4. Data pre-processing at rations of 80% learning, 10% validation and 10% testing from the 100% full dataset.

4 Results and Discussion

As previously hypothesized, hyperspectral images do indeed amplify rock differences by means of spectral signatures as shown by Fig. 5. This spectral technology combined with CNNs form a great tool from which future mining activities could be based due to their robustness.

4.1 Hyperspectral Signatures

From the 8 igneous rocks, the volcanic rocks (Fig. 5, lower row), 'rhyolite, andesite, basalt and dacite', which are rather difficult to distinguish in the field due to their fast cooling history, are easily distinguishable by visual observation of their hyperspectral

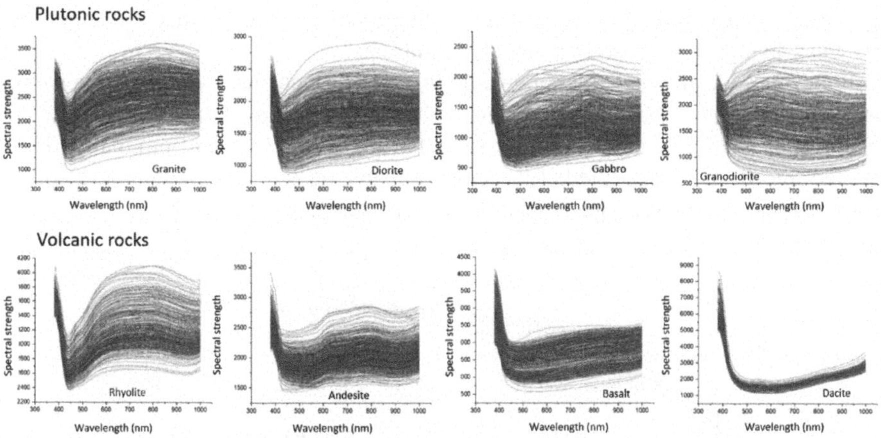

Fig. 5. Hyperspectral signatures of 8 plutonic and volcanic rock types, band shapes differ from rock to rock due to differences in interaction with light energy at given wavelengths.

signature band tilt, strength and curvature. However, their plutonic equivalents (Fig. 5, upper row), 'granite, diorite, gabbro and granodiorite', have different but slightly similar spectral signatures, hence the need to employ computer vision to thoroughly distinguish rocks with the option of database expansion.

4.2 Deep Learning Analysis

Our input, holds raw pixel values of each spectral image with a depth representing the entire VNIR. Simply, a CNN such as one in Fig. 6, consists of an input layer where data is fed, multiple hidden layers where the brain of the network is located, and an output layer where results are delivered. The network does this classification across all 125 layered bands since our input data contains 125 layers from the VNIR.

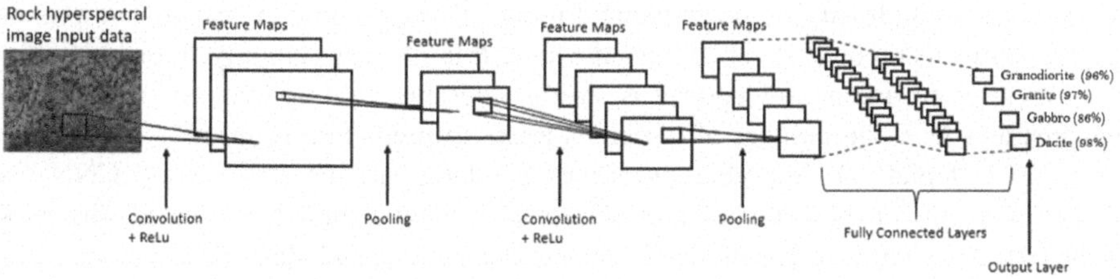

Fig. 6. Simplified architecture of a CNN with hyperspectral images as input, and rock classification as output.

The network presents 2 vital results after network training has been completed. The first, validation accuracy; communicates the recognition capabilities of the network at making predictions, the higher the validation output value, the more accurate the network is (Poria et al. 2016) at making predictions. The second vital variable is the test output result, this shows how capable the network is at running data which is unlabeled or whose origin is unknown against its database and predicting what that input data may most likely to be, based on already learnt data.

From Fig. 7A, the network shows a high validation accuracy of 96%, this simply means the system is able to recognize each rock's spectral data imported into the system, suggesting that indeed deep learning has the ability to learn from multitudes of data fed into it, thereafter, predict the most probable answer centered upon the previously mentioned database, whilst simultaneously expanding its database.

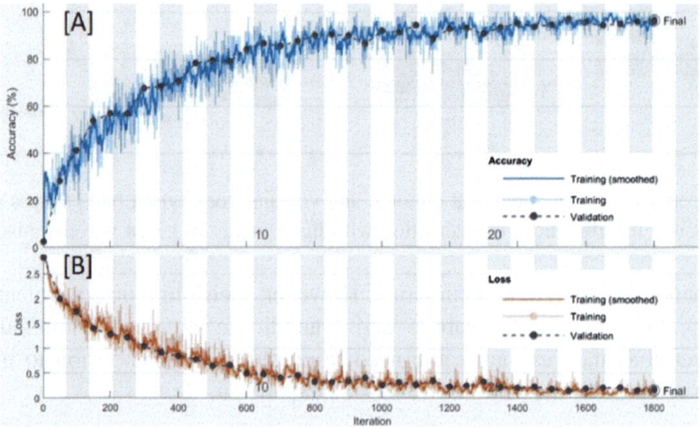

Fig. 7. CNN results. The network is highly capable of rock predictions as shown by the high validation percentage (96%) during network training, and maintenance of a low prediction loss.

During training, the CNN is also able to show the relationship between the number of training steps/iterations, with training batches, also known as the prediction loss. The lower this value, the more accurate the CNN's prediction stability, also, the longer it preserves this low value, the more stable and reliable the network is (Xing et al. 2017) at predicting and classifying phenomena. Figure 7B shows how the network improves and maintains its prediction capabilities the more training steps employed; this variable, at time referred to as the epoch, is easily alterable to improve the network, it demonstrates how the number of iterations relate to prediction robustness accuracy.

Table 1 depicts all important information relating to the state of the CNN, its prediction precision percentages and correctly recalled values from testing the rock characterization capabilities of the CNN. These results and their database can furthermore be applied to any rock engineering works which may need rock classification, database expansion is also one of the great advantages of deep learning as more data can be fed at any particular time to make predictions even more accurate and precise.

Table 1. CNN output results showing the test output predictions of each of the 8 igneous rock types.

Plutonic rock CNN results				
Rock type	Granite	Diorite	Gabbro	Granodiorite
Recalls/wrong scores	99/3	84/18	88/14	98/4
CNN precision percentage (%)	97	82	86	96
Volcanic rock CNN results				
Rock type	Rhyolite	Andesite	Basalt	Dacite
Recalls/wrong scores	102/0	149/5	100/2	100/2
CNN precision percentages (%)	100	97	98	98

5 Conclusion

From the presented proposal, it has been observed that hyperspectral imaging of rocks is highly capable of distinguishing rocks due to its high spectral resolution. Differences amongst rocks which the human eye is unable to observe are easily amplified by the use and application of hyperspectral imaging, making it easier for rock classification to take place in mining environments and beyond. This technology coupled with the deep learning CNN, has succeeded in creating a database from which later predictions can be based. CNN results show a rock classification capability prediction accuracy of 96%, hence this system is viable and has met above mentioned aims and objectives of this study by flawlessly classifying all 8 igneous rocks, hence worthy for consideration for mine planning and equipment selection.

References

Jia, F., Lein, Y., Lin, J., Zhou, X., Lu, N.: Deep neural networks: a promising tool for fault characteristic mining and intelligent diagnosis of rotating machinery with massive data. Mech. Syst. Signal Process. **72–73**, 303–315 (2016)

Xing, J., Li, K., Hu, W., Yuan, C., Ling, H.: Diagnosing deep learning models for high accuracy age estimation from a single image. Pattern Recogn. **66**, 106–116 (2017)

Simonyan, K., Vedaldi, A., Zisserman, A.: Deep inside convolutional networks: visualising image classification models and saliency maps (2013)

Kruse, F.A.: Mapping surface mineralogy using imaging spectrometry. Geomorphology **137**, 41–56 (2012)

Meer, F.: The effectiveness of spectral similarity measures for the analysis of hyperspectral imagery. Int. J. Appl. Earth Obs. Geoinf. **8**, 3–17 (2006)

Pieters, C.M., et al.: The Moon Mineralogy Mapper (M3) on Chandrayaan-1. Curr. Sci. **96**(4), 500–505 (2009)

Poria, S., Cambria, E., Gelbukh, A.: Aspect extraction for opinion mining with a deep convolutional neural network. Knowl.-Based Syst. **108**, 42–49 (2016)

Tompkins, S., Pieters, C.M.: Mineralogy of the lunar crust: results from Clementine. Meteor. Planet. Sci. **34**, 25–41 (1999)

Development of an Underground In-Situ Stress Monitoring System for Mining Safety Using Multi Sensor Cell and Wi-Fi Direct Technology

Hajime Ikeda[1(✉)], Youhei Kawamura[1], Hyongdoo Jang[2], Nur Ellisha Binti Mokhtar[3], Jun Yokokura[3], and Zedrick Paul L. Tungol[1]

[1] Graduate School of International Resource Sciences, Akita University, Akita 010-8502, Japan
ha2ikeda@gmail.com

[2] Western Australian School of Mines, Curtin University, Kalgoorlie, WA 6430, Australia

[3] Faculty of International Resource Sciences, Akita University, Akita 010-8502, Japan

Abstract. The increasing global demand for minerals contributes to the necessity of mineral extraction at greater depths. However, the increase of rock in-situ stress with depth leads to higher risk and increasingly dangerous working conditions faced by mining workers. The presence of shafts, tunnels and other excavations necessary in mine expansions further increase the complexity of underground mines. This complexity of underground stress conditions increases the importance of monitoring and analysis of underground strata conditions, as early detection is crucial in the prevention of rock failure and the occurrence of fatal accidents. A better comprehension of the underground stress conditions in a mine is vital in considering mine design and supports that need to be installed. The development of an efficient monitoring system that can obtain and transmit data is necessary. This paper suggests the utilisation of a multi sensor cell that combines the functions of an accelerometer, gyroscope and a magnetometer, as well as strain gauge displacements to continuously measure the stress conditions of bedrock. The obtained data is then conveyed to the surface using a Wi-Fi Direct communication system and analysed to comprehend the changes in the underground stress conditions. The latter part of this paper also describes the experiments conducted to verify the ability of the proposed monitoring system.

Keywords: Monitoring system · In-situ stress · Wi-Fi Direct

1 Introduction

In underground conditions, bedrock is subjected to stresses resulting from the load of the overburden and tectonic movements. Understanding the stress and strain (in-situ stress and displacement) conditions of the rock is important in the design and excavation of underground mines. Although, recent years have seen great progress in numerical computational methods such as FEM which allows more accurate numerical

E. Topal (Ed.): MPES 2019, SSGG, pp. 236–244, 2020.
https://doi.org/10.1007/978-3-030-33954-8_30

calculations, there are many factors such as support conditions which are difficult to describe quantitatively. The increase in the depth of underground mines result in a higher demand for safety, and consequently, in-situ stress and strain monitoring. Although, in-situ stress and strain measurement has an advantage over numerical modelling methods as it can evaluate the actual stress and strain conditions of a mine, most methods can only measure the stress and strain conditions before excavation, and not throughout the whole mine development. Long-term stress monitoring is necessary to continue observing the stress and strain conditions around the excavation.

This research on the implementation of a monitoring system to detect the changes in the underground in-situ stress conditions was separated and carried out in two main components. One component is the sensor that detects the changes in the underground stress and strain conditions. The second component is the utilisation of the Ad Hoc communication (Wi-Fi Ad Hoc and Wi-Fi Direct) system to transmit the data from the sensor unit to mobile devices that then transmit data to above ground. Experiments were conducted to evaluate the functionality of the proposed technology.

2 Proposed Underground Rock Stress Monitoring System

The usage of wireless sensor networks (WSNs) for underground mine monitoring systems solves problems that would usually occur when using wired communication systems, for example in the occasion of a collapse or machinery malfunction [1]. Mining companies use WSNs for locating mining workers and environmental monitoring (smoke and dust detection, the concentration of toxic gases, measurement of temperature and humidity). Figure 1 shows a collapsed section of an underground mine. As the system tries to reach equilibrium after excavation, the opening deforms and eventually collapses. A monitoring system could be employed to track these deformations that lead up to the collapse. Some examples of WSNs currently used are Wi-Fi, Bluetooth, and ZigBee [1–5]. The authors have tested the reliability of the ZigBee communication system in a Western Australian underground mine [4, 5].

Fig. 1. Collapse of underground mine.

The underground rock stress monitoring system proposed in this research can be divided into two parts. The first part is sensing, which is the differential measurement of in-situ stress and strain. The second part of the system is the transmission of data utilising Ad Hoc communication (Wi-Fi Direct). As shown in Fig. 2, the monitoring system proposed in this research utilises Ad Hoc communication and a sensor unit that is installed in a borehole which is afterwards filled. Once blank holes are drilled, stress release occurs. Embedding the sensor units in interspersed locations will enable the detection of changes in stress applied from the rock based on the value of the strain gauge. It is also necessary to know the orientation of the sensor unit after it is embedded to detect the direction from which stress is being applied.

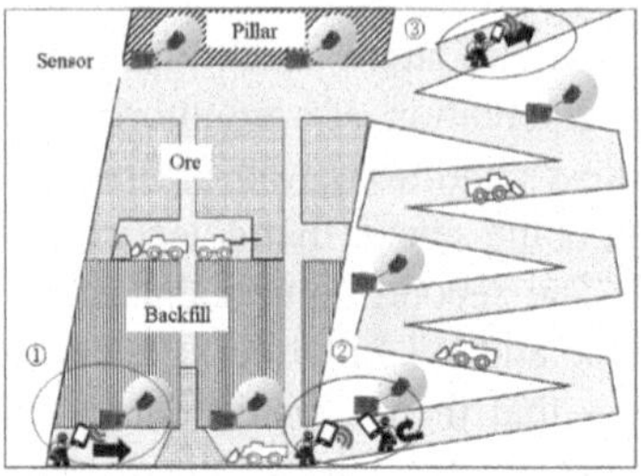

Fig. 2. System design of rock stress monitoring system.

Figure 2 shows mining workers obtaining measurement data from the sensor unit at ①. ② shows the passing of measurement data from other sensor units and mining workers within the communication range. Measurement data is passed along from mining workers to the ground office when a worker carrying the data exits the underground mine at ③. The development of the monitoring system was regarded separately. The performance of the communication system was investigated in 3, and the sensor unit was investigated in 4.

3 Ad Hoc Wireless Communication System

Currently, communication in underground tunnels is carried out using Walkie-talkies for verbal communication, and transmission of monitoring system data is still carried out using wired communication systems (Ethernet cables, etc.). It goes without saying that, in a constantly changing environment such as an underground mine, a wireless communication system would be more suitable as there would be no need for re-routing wires. By using WSNs for computerisations in underground mines, it is possible to detect and prevent underground collapses, machinery malfunctions, aiding in the automation of mining equipment and the aggregation of information on various measurement instruments to create big data. WSNs can be configured as autonomous networks just by setting up a large number of nodes that can connect automatically via wireless connections. WSNs can support reliable monitoring and communication systems that can include many nodes, compared to wire communications that are very likely to fail due to cable damage.

Some examples of WSNs currently being studied include ZigBee and Wi-Fi Ad Hoc. However, although ZigBee allows many connectable terminals and multi-hop, its transmission speed is slow [3–5]. On the other hand, in addition to the advantages of WSNs, Wi-Fi Ad Hoc allows the P2P (Peer to Peer) connection between devices to be established without needing an access point, enabling the transmission of information via movement of users carrying these devices.

Communication tests on the Wi-Fi Ad Hoc system were conducted in the Osarizawa Mine in Akita Prefecture to measure the received signal strength indicator (RSSI) and transmission speed of the system [3]. The RSSI is a measurement of the power present in a received radio signal [6]. The verification of RSSI allows to test the reliability of wireless communications in a real setting. The signal reception quality is deteriorated due to wave interference. To calculate the power of the receiving antenna and consider the free-space basic propagation loss formula, Eqs. (1) and (2) are generally used:

$$RSSI = 20\log\frac{PTGtGr\lambda}{4\pi d} \quad (1)$$

$$G_n = 4\pi(A_e)/\lambda^2 \quad (2)$$

Where, *RSSI* is the received signal strength indicator in dBm. *PT* is the transmission powers in milli-watts, respectively, G_t and G_r (G_n) are the absolute gain of the transmitting and receiving antennas in decibels relative to an isotropic radiator. λ and d are the wavelength and distance in meters. The maximum *PT* of the base unit was about 1.5 W. A_e is the effective aperture in square meters, it is related to the physical size of the antenna. RSSI value of −80 dBm was defined as a reference of the minimum value for a stable connection.

Figure 3(A) shows RSSI according to distance. Figure 3(B) Throughput according to distance. It was concluded that communication is possible at a communication speed of 2.3 MB/s at a distance of up to about 110 m in a straight line of arched tunnels of length 2 m [3]. The Ad Hoc communication feature enables the proposed monitoring system to expand communication range by using the movement of passing miners. This research proposes a communication system in which mobile receivers and sensor units can communicate with each other through a relay system and transmit data from the underground to the surface (or between different points on the surface)

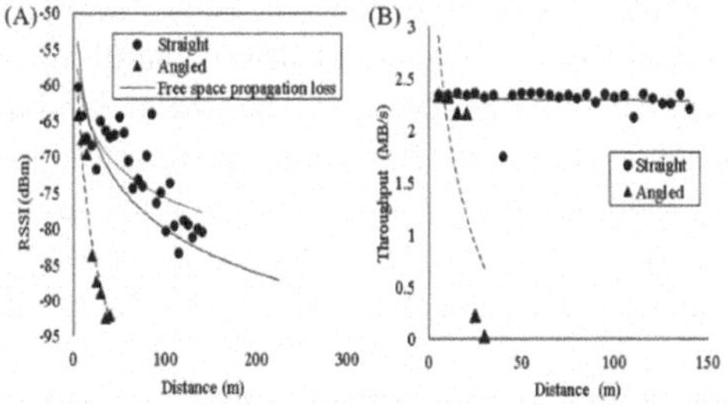

Fig. 3. Communication test results. (A) RSSI according to distance, (B) Throughput according to distance.

Throughout this research, Wi-Fi Direct, a type of Ad Hoc communication methods was utilised. Wi-Fi Direct not only has the same advantages as Wi-Fi Ad Hoc but also simplifies the complexity of the settings of an Ad Hoc system. Wi-Fi Direct estimates that similar communication performance can be obtained from using the same Wi-Fi standard, output, and 2.4 GHz band radio waves as Wi-Fi ad hoc.

4 Differential Measurement of In-Situ Stress and Strain Conditions

The following sections demonstrate techniques employed and designed for the sensing of in-situ stress; other aspects to be elaborated include stress measurement methods, the performance of the sensor unit, determination of sensor unit orientation and finally the determination of stress exerted the on the sensor unit.

4.1 Stress Measurement Methods

A conventional in-situ stress measurement method is the hydraulic fracturing method in which the magnitude of stress is determined by injecting high-pressure fluids into boreholes that cause tensile failure; hence determining the principal stresses of the surrounding rock [7]. The in-situ stress measurement method requires a moderate cost efficiency as increase with depth drastically increases difficulty in implementation [7]. Another method is the over-coring method, where the physical strength of the rock can be determined by measuring the differential strain experienced by the surrounding rock when a pilot hole is over-cored and stress release occurs. However, the hydraulic fracturing method is considered to be undesirable in terms of long-term safety and stability around the tunnel after excavation because it may damage the bedrock at the measurement points.

Disadvantages of the over-coring method include the difficulty of fixing a strain gauge to the measurement point. The over-coring method also requires the elastic modulus of the surrounding rock, which at most times is oversimplified and assumed to be homogenous. Also, both the hydraulic fracturing method and the over-coring method can be used to determine the in-situ stress/strain conditions during the time of measurement, however, they cannot be used to measure the subsequent changes in the stress/strain conditions. Because the stress and strain conditions of underground mines are constantly changing as mining operations progress, in-situ stress measurement methods that can only measure in-situ stress and strain at only one point of time are insufficient. Therefore, methods of measuring differential stress after the release of stress (during excavation) are preferred. In other words, a long-term, non-destructive measurement method that can measure the difference in the in-situ stress conditions is required.

There are two methods in measuring in-situ stress, non-contact methods where the displacement of the object to be measured is measured relative to a predetermined reference space, or contact methods where sensors are directly attached to the surface of the object which its stress is to be determined. Except for situations where the vibrations of the object to be measured can be affected by the mass of the sensor, or cases when the sensor cannot be directly attached to the surface due to high temperatures, contact sensors are generally easy to use. Contact methods are also preferred because, although non-contact methods can measure displacement, they cannot measure stress conditions directly. Therefore, to determine the changes in in-situ stress that occur throughout the progress of the mine, research was focused on the development of a sensor unit that can be installed in a drill hole drilled with a support drill bit (φ60 mm).

The sensor unit developed was a solid rectangular box made of a stainless-steel plate. In addition to the strain gauges attached to the inside of the sensor on the XYZ plane, a 9-axis sensor (LSM9DS1) that included an accelerometer, gyroscope and a magnetometer to determine the orientation of the sensor after its installation was also fixed in the sensor unit. The obtained data is transmitted to the data logger through a built-in wire cable.

4.2 Determining Sensor Unit Orientation

Figure 4(A) shows a sensor unit affixed with a portable Ad Hoc communication terminal. The purpose of the 9-axis sensor (LSM9DS1) is to determine the orientation of the sensor after it has been installed, by calculating the acceleration (accelerometer), change in angular velocity (gyroscope) and the direction of true north (magnetometer). The sensor fusion of these three sensors compensate for the disadvantages of the respective sensors and improve the accuracy when determining the orientation of the sensor after installation.

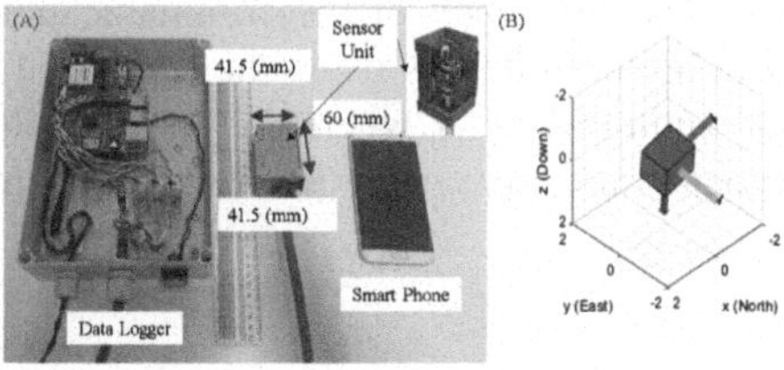

Fig. 4. Sensing equipment and orientation calculation results. (A) Sensor unit and mobile device, (B) Orientation of sensor unit.

In this research, a software was used to determine the orientation of the sensor unit. Figure 4(B) illustrates the actual orientation of the sensor, confirming that the real-time orientation of the sensor can be determined accurately. Data from the readings of the stress measurements by the strain gauges attached to the inside of the sensor surfaces are combined with the sensor orientation to determine both the direction and magnitude of stress applied on the sensor. Figure 5(A–C) shows X and Y, Z axis reliability with linearity error of less than ±0.1%. Figure 5(D) is a disassembly view of the unit housing.

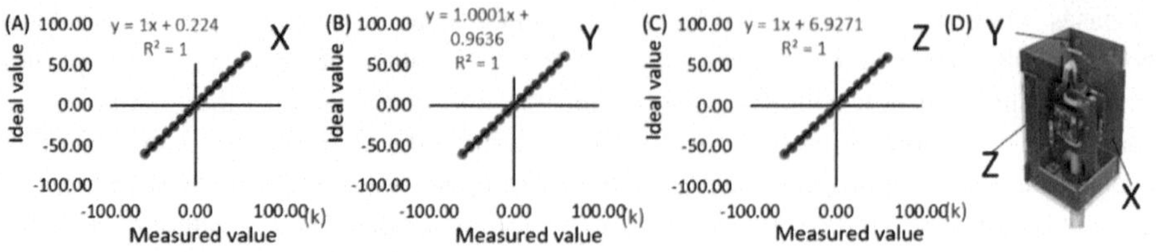

Fig. 5. Sensing unit reliability. (A) X axis reliability, (B) Y axis reliability, (C) Z axis reliability, (D) Sensor unit disassembly view

4.3 Determination of Stress Exerted on Sensor Unit

A uniaxial compression test was conducted on the sensor unit, and the stress (in-situ stress) applied to the sensor unit is determined from the load and displacement curve obtained. The relationship between stress and strain was determined from the value of the strain gauge. The in-situ stress conditions of a mine-site can be determined from the strain of the sensor unit embedded in the borehole.

During this research, indoor experiments were conducted as basic experiments. A manual uniaxial compression tester (with a capacity of 200 N) was used to conduct a load response test. The load exerted on the sensor unit was increased with increments of 0.5 N, and the readings from the strain gauge were used to form a load-displacement curve (which shows elastic deformation) and data such as the acceleration, angular velocity and the direction of true north (based on geomagnetism) were obtained from the 9-axial sensor. The data acquired at a sampling frequency of 1 Hz was transmitted from the sensor unit data logger to a mobile device using the Ad Hoc communication (Wi-Fi Direct) mentioned above. The frequency of the communication between the mobile device and data logger can be controlled by using the developed application. The load-displacement curve is shown in Fig. 6(A), where the line is plotted linearly. The Fig. 6 (B) shows the displacement-strain curve, where the X's line is plotted linearly.

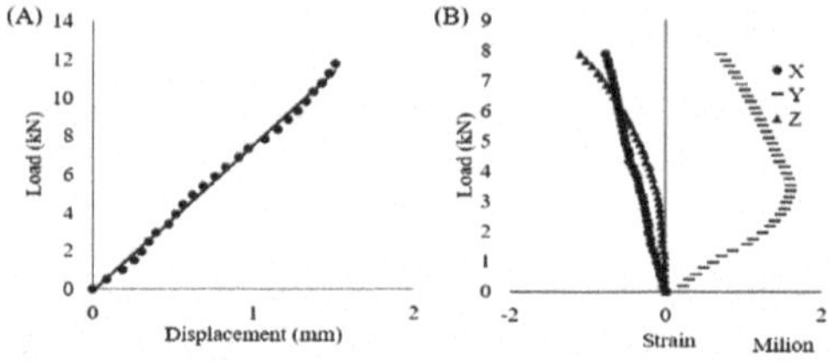

Fig. 6. Uniaxial compression test results. (A) Load-displacement curve, (B) Displacement-strain curve.

Up to 5 MPa of load was applied to the sensor unit during the load response test. The data from the 9-axis sensor and the 3-axis strain gauges change with time. Based on the values from the strain gauge, it is possible to measure the in-situ stresses in underground mines. In order to determine the in-situ stress and displacement that occurs in underground mines, it is important to not only be able to detect the changes in the magnitude of in-situ stress but to also know the orientation of the sensor to determine the direction that the force is being applied from.

The total amount of data transmitted from the sensor is about 13 MB/day, which is fairly small compared to the amount of data that can be transmitted by using the proposed communication system. The results of these tests verify the feasibility of the proposed monitoring system.

5 Conclusion

The implementation of better communication systems will improve the productivity and safety of underground mines. This research proposed a monitoring system that was separated into two components, sensing and data transmission. Among the several WSNs available, this research focused on Wi-Fi Direct is considered to be suitable for informatisation in underground mines. Communication tests were carried out to investigate the performance of the proposed communication system. By measuring the RSSI and the transmission speed, the distance and transmission speed that data can be stably transmitted were determined. As a result of the conducted communication tests, it was determined that during stable wireless communication (at an RSSI of more than −80 dBm), it is possible to communicate at a transmission speed of 2.3 MB/s at a point 95 m away from the data logger in a straight line.

The efficient measurement and understanding of stress distribution underground are important in the safe design of underground mines. Although, it is possible to determine the in-situ stress/strain measurement beforehand, it is important to be able to monitor the long-term changes in the stress and strain conditions. This research proposed sensing the in-situ stress conditions based on the rock stress changes resulting from mine development. Sensor fusion of a 9-axial sensor installed inside the sensor unit is used to determine the orientation of the sensor unit post-installation, while the stress exerted onto the sensor unit can be measured using the strain gauges attached on the inside of the sensor unit. By combining these results, the magnitude and direction of the in-situ stress can be determined. The amount of data regarding in-situ stress totals to about 13 MB/day, which is relatively small compared to the amount of data that can be transmitted by the proposed communication system. Based on the results of the tests carried out, the functionality and feasibility of the proposed monitoring system have been verified.

References

1. Moridi, M.A., Kawamura, Y., Sharifzadeh, M., Chanda, E.K., Jang, H.: An investigation of underground monitoring and communication system based on radio waves attenuation using ZigBee. Tunn. Undergr. Space Technol. **43**, 362–369 (2014)
2. Kawamura, Y., Dewan, A.M., Veenendaal, B., Hayashi, M., Shibuya, T., Kitahara, I., Nobuhara, H., Ishii, K.: Using GIS to develop a mobile communications network for disaster-damaged areas. Int. J. Digit. Earth **7**, 279–293 (2014)

3. Ikeda, H., Kawamura, Y., Tungol, Z.P.L., Ito, Y., Jang, H.: Development of underground space communication systems using Wi-Fi Ad Hoc for smart mining. Jpn. Soc. Geoinf. **29**, 3–11 (2018)
4. Moridi, M.A., Kawamura, Y., Sharifzadeh, M., Chanda, E.K., Wanger, M., Okawa, H.: Performance analysis of ZigBee network topologies for underground space monitoring and communication systems. Tunn. Undergr. Space Technol. **71**, 201–209 (2018)
5. Moridi, M.A., Sharifzadeh, M., Kawamura, Y., Jang, H.: Development of wireless sensor networks for underground communication and monitoring systems (the cases of underground mine environments). Tunn. Undergr. Space Technol. **73**, 127–138 (2018)
6. Sun, H.Y., Bi, L.J., Lu, X., Guo, Y.J., Xiong, N.X.: Wi-Fi network-based fingerprinting algorithm for localization in coal mine tunnel. J. Internet Technol. **99**, 1–10 (2015)
7. Funato, A., Ito, T.: A new method of diametrical core deformation analysis for in-situ stress measurements. Int. J. Rock Mech. Min. Sci. **91**, 112–118 (2017)

Spatio-Temporal Change Detection of North Antelope Rochelle and Black Thunder Coal Fields of US Using Multi-temporal Remote Sensing Satellite Data

Muhammad Ahsan Mahboob[1(✉)], Bekir Genc[1], and Iqra Atif[2]

[1] School of Mining Engineering, University of the Witwatersrand, Johannesburg, South Africa
868654@students.wits.ac.za, ahsan.igis@gmail.com
[2] School of Advanced Geomechanical Engineering, National University of Sciences and Technology (NUST), Islamabad, Pakistan

Abstract. Opencast mining activities including mineral exploration, extraction, waste dumping and stockpiling affect the nearby environment and land cover. To minimize the negative impacts of mining on the environment, it is highly recommended to adopt sustainable mining practices along with proper and regular monitoring of mining activities. In this research study, 32 years of Landsat satellite data from 1984 to 2016 was used to map the spatiotemporal changes in coal mining activities of the World's largest North Antelope Rochelle and Black Thunder coal fields of the United States of America. Digital image processing of satellite images was done and the Maximum Likelihood Classification algorithm was applied to quantify the land cover changes in the region. The most appropriate band selection was done for the identification of coal seams through satellite images. The results showed that the Shortwave Infrared-2, Near Infrared and Green bands of Landsat satellite, highlighted as Red, Green, and Blue respectively are proven to be very effective in the identification of areas of coal mining. The classification results showed that since 1984, coal mining has been expanded by 74,000 ha with an increase of more than 87% in thirty-two years. Whereas almost 94% decrease in the vegetation cover of the region was also observed during the same period of time. The methodology adopted in this research study is efficient, time-saving and inexpensive to monitor the detailed surface mining activities over a large period of time, and is applicable to other mining regions, as well.

Keywords: Remote sensing · Sustainable mining · Coal mining · Change detection

1 Introduction

Opencast mining activities including mineral explorations extractions, waste dumping, and stockpiling affect the environment and land cover [1]. On the other hand, the mining of precious metals plays a significant role not only in the global economy but also in the economic situation of any country. The negative consequences caused by

E. Topal (Ed.): MPES 2019, SSGG, pp. 245–254, 2020.
https://doi.org/10.1007/978-3-030-33954-8_31

the mining activities are usually irreversible and devastating, which damage several other natural resources of any country [2]. To minimize the negative impacts of mining on the environment it is highly recommended to adopt sustainable mining practices and to improve the existing practices. In order to achieve this, proper monitoring of mining areas on a regular basis should be done to get comprehensive information about ongoing mining activities and changes.

Traditional field based mining survey and monitoring techniques are not only time consuming and laborious but also very expensive. Whereas due to recent advancement in the field of remote sensing technologies, it is quite easy to get not only the past information of the mining areas but also the current up-to-date data through commercial and open source satellites [3]. Several researchers have already shown the effectiveness of satellite remote sensing data for sustainable mining in the modern era. For example, Schueler, et al. [4] used remote sensing satellite data to assess the impacts of gold mining on land use in Ghana. The study highlights that from 1986 to 2002, mining caused 58% of deforestation and 45% reduction of agricultural farmlands. Another study conducted by Schmidt and Glaesser [5] successfully used Landsat satellite data for quantitative analysis and change assessment of several different surface mine features. Some research studies such as done by Petropoulos, et al. [6] combine remote sensing and other advanced data science technologies like machine learning for change detection due to surface mining in Greece. Coal mining is a very big industry in the United States of America (USA), where the mining of coal is dated back to the 1300s [7]. Although recently there is a decline in the overall coal mining industry such as in 2017 where there was 33% less coal production than in 2006. The main consumption of coal is in the energy sector for electricity production, as coal mining contributes 30% in electricity generation of the country in 2016 [8]. Besides all this declination of the coal mining industry, the USA owns some of the world's largest surface coal mines, which may cause negative impacts on the environment.

The main objective of this study is to identify, quantify and analyze the spatial change detection of two of the world's largest surface coal mines located in the USA from 1984 to 2016 using open source multi-temporal remote sensing satellite data, in order to assess the impact of surface coal mining on the surrounding environment.

2 Materials and Methods

2.1 Study Area

In this research, two of the world's largest coal fields known as North Antelope Rochelle and Black Thunder [9] situated in the United States of America (USA) were selected as a study area (Fig. 1). North Antelope Rochelle was opened in 1983 and the Black Thunder in 1977. Both are located in Powder River Basin, which has coal-rich geological structure and the coal is believed to have formed more than 60 million years ago [10]. There are a total of 2.25 and 1.4 billion tons of Proven Coal Reserves in the North Antelope Rochelle and Black Thunder coal mines respectively. In 2003, the coal produced from North Antelope Rochelle and Black Thunder coal fields were 110.9 and 100.7 million tons respectively.

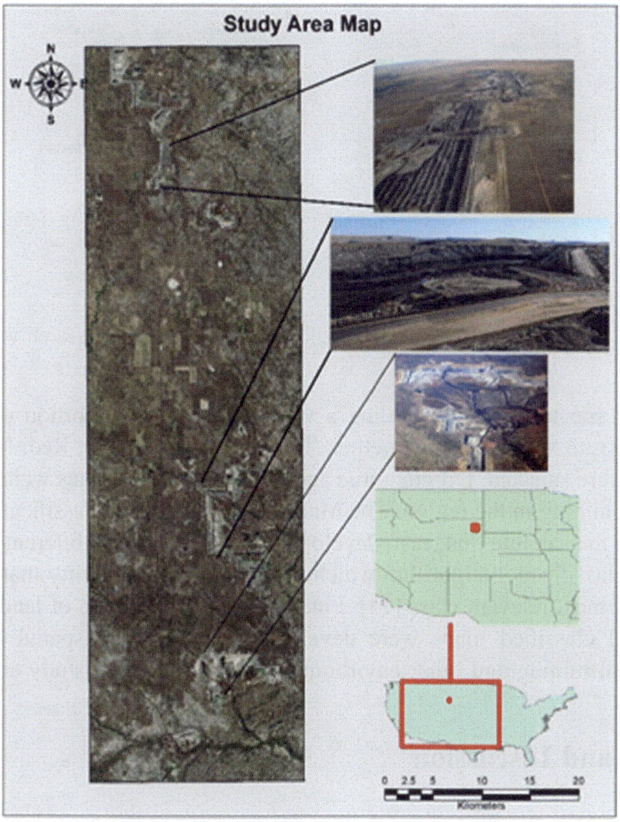

Fig. 1. The location map of the study area.

2.2 Satellite Data

Long term Landsat satellite data with good spatial, spectral and temporal resolution was utilized in this research. A total of eight satellites have been launched under the Landsat satellite mission since 1972 and the latest one, Landsat-8, was launched in 2013 [11]. In this research, seven satellite images covering 32 years of the temporal history was acquired and processed.

2.3 Research Framework

The detail research framework of the current study is given in Fig. 2. The satellite images of Landsat satellites from 1–8 were acquired and preprocessed.

Usually, the Landsat satellite images have several distortions due to sensor operations, atmosphere, topographic and sometimes solar activities. By preprocessing satellite data, the effects of these distortions can be minimized, and satellite images can be used for further analysis [12]. In this research, the dark object subtraction (DOS) algorithm was applied for the atmospheric correction of Landsat images [13].

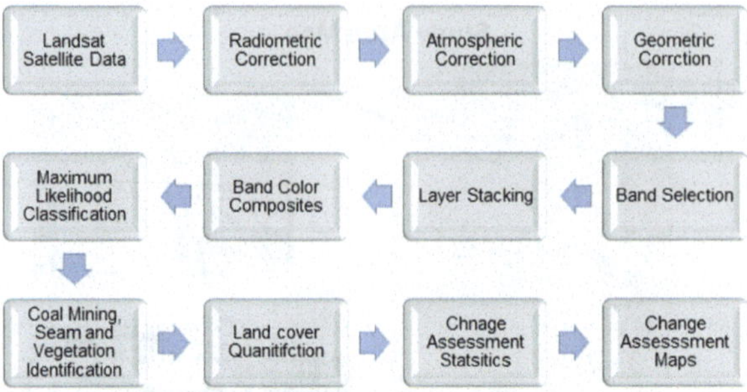

Fig. 2. Comprehensive methodological framework of the research work.

A total of six spectral bands including a visible and infrared portion of the electromagnetic spectrum were stacked together. The band Blue, Green, Red, NIR, SWIR-1, and SWIR-2 were selected. Different true and false color composites were developed to highlight coal mining in the region. The Maximum Likelihood Classification algorithm was applied to the satellite images to develop land cover maps of different years. This is a parametric classification algorithm which works on the probability that each pixel is assigned to its most relevant class [14]. Finally, the quantification of landcover classes was done and classified maps were developed to assess the spatial and temporal changes in coal mining and other environmental features of the study area.

3 Results and Discussion

Based on the DOS algorithm, the satellite images were preprocessed for atmospheric corrections. The coal mining areas and other land cover features were more prominent after the correction was done. Barakat et al. [15] also applied the dark object subtraction algorithm for atmospheric correction of Landsat satellite data and to map land cover changes.

Several band combinations as given in Table 1 were tested and applied in order to highlight the coal mining in the region. The only true color composite developed was with the first three visible bands of Landsat satellite i.e. Red, Green and Blue spectral bands highlighted as Red, Green, and Blue [16]. Except for this combination, all other band combinations were false color composites [17]. All the five false color composites highlight some features, but composite No. 5 highlights the coal mining areas effectively.

In composite No. 5, the SWIR-2, NIR and Green bands of Landsat satellite were highlighted as Red, Green, and Blue. The light magenta color represents the coal mining regions whereas the dark magenta color highlights the coal seams. The vegetation is highlighted as green and water in black color. The red color represents the mining tracks possibly used by the shovels, dumpers and other mining vehicles to access the mine. Further, the advanced digital image processing techniques were applied on composite No. 5 to extract the coal seams for each year. The results showed that the histogram adjustment [18] made the coal seams quite clear as shown in Fig. 3.

Table 1. True and false color composites of the Landsat satellite images.

Sr. No	Band Composite as (Red-Green-Blue)	Composite Type	Effect on Satellite Image
1	Red-Blue-Green	True Color	
2	Blue-Green-Red	False Color	
3	NIR-SWIR1-Red	False Color	
4	SWIR2-Blue-Green	False Color	
5	SWIR2-NIR-Green	False Color	
6	SWIR2-SWIR1-Green	False Color	

The (A) is composite 5 (false color composite) of band combinations whereas (B) represents the coal seams in green color after histogram adjustment.

The results of the maximum likelihood classification were also promising as shown for different years. The classified images of 1984 and 1988 are shown in Fig. 4, the area of coal mining, vegetation and water were 85,000, 77700 and 1000 ha respectively. The coal mining in the region started in 1977 as the Thunder coal mine started operation in that year. The classified map of the year 1988 (Fig. 4) showed that the area

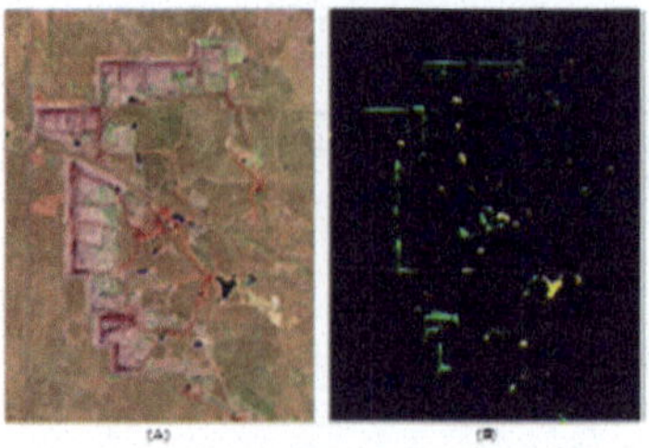

Fig. 3. Coal seams highlighted by the histogram adjustment image processing technique. (A) is the false color composite, (B) is the coal seams in green color.

of total coal mined, vegetation and water were 102,900, 60,000 and 800 ha, respectively. The total area of coal mining and vegetation increased and decreased by about 21% and 22%, respectively, since 1984.

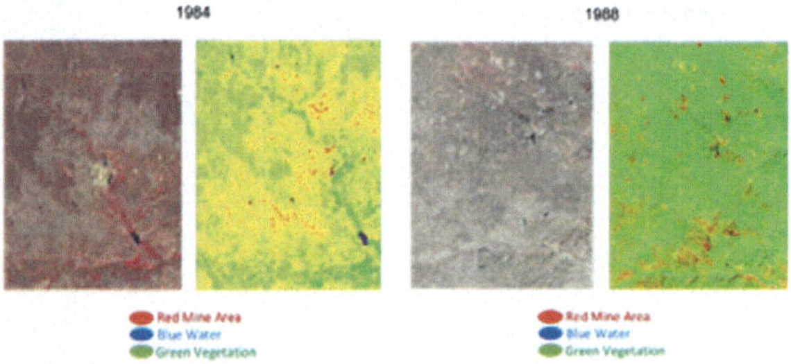

Fig. 4. Landsat satellite image of 1984 and 1988 along with the classified map of similar year, representing the spatial location of features.

The classified map of the year 1995 showed that the area of total coal mined, vegetation and water were 119,000, 42,700 and 2000 ha, respectively. The total area of coal mining and vegetation increased and decreased by about 15.6% and 28.8% respectively, since 1988. Similarly, the classified map of the year 2000 (Fig. 5) showed that the area of total coal mined, vegetation and water were 131,000, 31,500 and 1200 ha respectively. The total area of coal mining and vegetation increased and decreased by about 10% and 26%, respectively, since 1995.

The classified map of the year 2005 (Fig. 5) showed that the area of total coal mined, vegetation and water were 139,000, 23,800 and 900 ha, respectively. The total area of coal mining and vegetation increased and decreased by about 6% and 24%, respectively, since 2000.

The classified map of the year 2010 (Fig. 6) showed that the area of total coal mined, vegetation and water were 147,000, 15,400 and 1,300 ha, respectively. The total area of coal mining and vegetation increased and decreased by about 5.7% and 35.2%, respectively, since 2005.

The classified map of the year 2016 (Fig. 6) showed that the area of total coal mined, vegetation and water were 159,000, 4,300 and 400 ha, respectively. The total

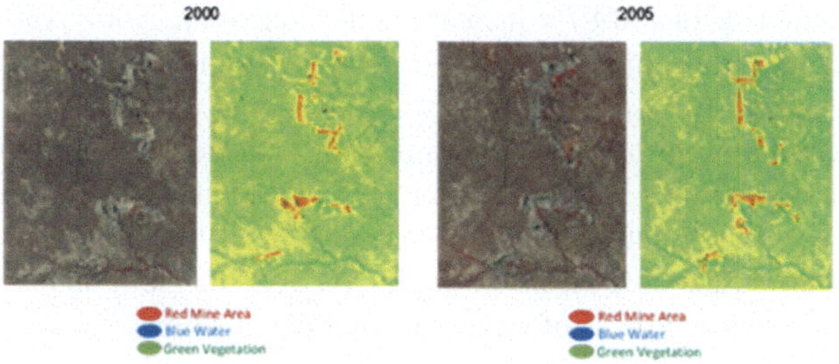

Fig. 5. Landsat satellite image of 2000 and 2005 along with the classified map of similar year, representing the spatial and temporal variations of features.

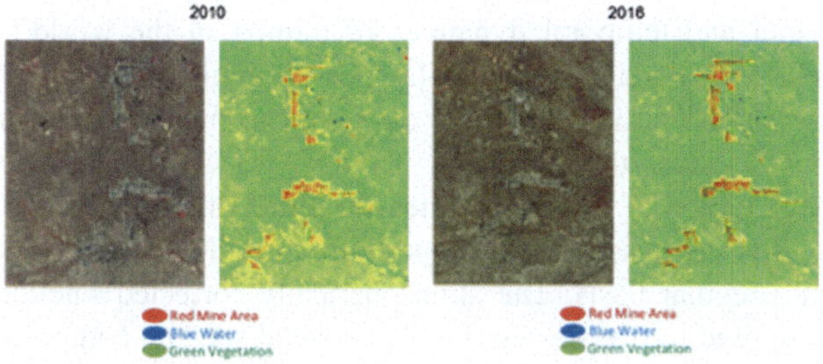

Fig. 6. Landsat satellite image of 2010 and 2016 along with the classified map of similar year, representing the spatial and temporal variations of features.

area of coal mining and vegetation increased and decreased by about 8.1% and 72%, respectively, since 2010.

Since the year 1984, the vegetation cover is continuously decreasing and coal mining is increasing over time. As shown in Fig. 7, after 1988, there was a sharp decrease and increase in vegetation and coal mining, respectively. In 32 years, from

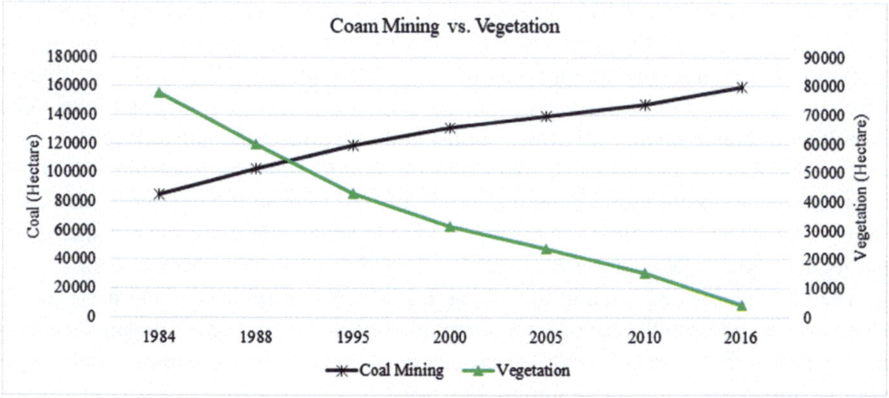

Fig. 7. 32 years temporal changes in the coal mining and the vegetation cover area from 1984 to 2016.

1984 to 2016 there is a total of 87% increase in the area of the North Antelope Rochelle and Black Thunder coal mining and 94.4% decrease in the vegetation cover over the same period of time.

Although there are no major agricultural lands situated in the region, and no significant commercial farming is going on, but the vegetation cover including the natural vegetation is very important for the environment. It is time that mining companies should practice the sustainable means of mining all over the globe so that the world can get their proper benefits instead of environmental degradations.

4 Conclusion

This research study utilized the long term multi-temporal Landsat satellite data to quantify the spatial and temporal dynamics of mining in the world's largest North Antelope Rochelle and Black Thunder coal. Open-pit mining, particularly coal mining, has some negative impacts on the nearby environment including the dust, air, water, and soil pollution. Besides this pollution, another very significant impact of surface mining is the degradation of the environment in terms of land consumption. For sustainable mining, it is very essential to properly monitor and record the changes in the mining areas on a regular basis. The atmospherically corrected satellite images from 1984 to 2016 were used for the development of several spectral composites to highlight the coal mining and coal seams in the region. The Shortwave Infrared-2 (2.107–2.294 μm), Near Infrared (0.851–0.879 μm) and Green (0.533–0.590 μm) spectral bands of Landsat satellite were proved to be very effective for identification of coal mining. Through this false-color composite, the mining area is usually highlighted as the tone of magenta color. Advanced digital image processing algorithm in which the histogram of satellite data was adjusted in order to highlight the coal seams, and using the same false color composite, the coal seams were successfully identified and highlighted in the tones of green color.

The maximum likelihood classification scheme was applied on multi-temporal remote sensing data and classified maps were generated for each year. By the year 2016, there was a total of 87% increase and a 94% decrease in coal mining and the vegetation cover respectively of the region since 1984. The significant change in the vegetation cover and coal mining was observed after the year 1988. The significant increase in coal mining of the region can cause several other environmental issues like air pollution, contamination of water bodies both at surface and subsurface, the temperature increase in the area, etc. Although the North Antelope Rochelle and Black Thunder coal are providing a large proportion of coal to meet the energy needs of the country. However, coal mining in the region should be done in a sustainable manner so that its negative impact can be controlled and minimized in the coming years.

The methodology adopted in this research is efficient, time-saving and inexpensive which can be successfully used for long term monitoring of land cover dynamics due to mining activities. The results of this study can be used to reassess the mining policies in order to implement sustainable mining practices.

Acknowledgment. The work presented here is a Ph.D. research study in the School of Mining Engineering at the University of the Witwatersrand, Johannesburg, South Africa.

References

1. Demirel, N., Emil, M.K., Duzgun, H.S.: Surface coal mine area monitoring using multi-temporal high-resolution satellite imagery. Int. J. Coal Geol. **86**, 3–11 (2011)
2. Shrestha, R.K., Lal, R.: Changes in physical and chemical properties of soil after surface mining and reclamation. Geoderma **161**, 168–176 (2011)
3. Prakash, A., Gupta, R.: Land-use mapping and change detection in a coal mining area-a case study in the Jharia coalfield, India. Int. J. Remote Sens. **19**, 391–410 (1998)
4. Schueler, V., Kuemmerle, T., Schröder, H.: Impacts of surface gold mining on land use systems in Western Ghana. Ambio **40**, 528–539 (2011)
5. Schmidt, H., Glaesser, C.: Multitemporal analysis of satellite data and their use in the monitoring of the environmental impacts of open cast lignite mining areas in Eastern Germany. Int. J. Remote Sens. **19**, 2245–2260 (1998)
6. Petropoulos, G.P., Partsinevelos, P., Mitraka, Z.: Change detection of surface mining activity and reclamation based on a machine learning approach of multi-temporal Landsat TM imagery. Geocarto Int. **28**, 323–342 (2013)
7. Arcadia Publishing: Evolution of the Coal Industry in America, 28 July 2017. https://www.arcadiapublishing.com/Navigation/Community/Arcadia-and-THP-Blog/October-2017/%E2%80%8BEvolution-of-the-Coal-Industry-in-America
8. Douglas, S., Walker, A.: Coal mining and the resource curse in the eastern United States. J. Reg. Sci. **57**, 568–590 (2017)
9. Godby, R., Coupal, R., Taylor, D., Considine, T.: Potential impacts on Wyoming coal production of EPA's greenhouse gas proposals. Electr. J. **28**, 68–79 (2015)
10. Heffern, E., Coates, D.: Geologic history of natural coal-bed fires, Powder River basin, USA. Int. J. Coal Geol. **59**, 25–47 (2004)
11. Schleeweis, K., Goward, S.N., Huang, C., Dwyer, J.L., Dungan, J.L., Lindsey, M.A., et al.: Selection and quality assessment of Landsat data for the North American forest dynamics forest history maps of the US. Int. J. Digit. Earth **9**, 963–980 (2016)
12. Mahboob, M.A., Genc, B., Celik, T., Ali, S., Atif, I.: Mapping hydrothermal minerals using remotely sensed reflectance spectroscopy data from Landsat. J. South Afr. Inst. Min. Metall. **119**, 279–289 (2019)
13. Gilmore, S., Saleem, A., Dewan, A.: Effectiveness of DOS (Dark-Object Subtraction) method and water index techniques to map wetlands in a rapidly urbanising megacity with Landsat 8 data. In: Research@ Locate 2015, pp. 100–108 (2015)
14. Cabral, A.I., Silva, S., Silva, P.C., Vanneschi, L., Vasconcelos, M.J.: Burned area estimations derived from Landsat ETM+ and OLI data: comparing genetic programming with maximum likelihood and classification and regression trees. ISPRS J. Photogramm. Remote Sens. **142**, 94–105 (2018)
15. Barakat, A., Ouargaf, Z., Khellouk, R., El Jazouli, A., Touhami, F.: Land use/land cover change and environmental impact assessment in Béni-Mellal District (Morocco) using remote sensing and GIS. Earth Syst. Environ. **3**(1), 113–125 (2019)
16. Amara, B.N., Aissa, D.E., Maouche, S., Braham, M., Machane, D., Guessoum, N.: Hydrothermal alteration mapping and structural features in the Guelma basin (Northeastern Algeria): contribution of Landsat-8 data. Arab. J. Geosci. **12**, 94 (2019)

17. Singh Bramhe, V., Kumar Ghosh, S., Kumar Garg, P.: Extraction of built-up areas from Landsat-8 OLI data based on spectral-textural information and feature selection using Support Vector Machine method. Geocarto Int. 1–18 (2019)
18. Yücer, E., Erener, A.: GIS based urban area spatiotemporal change evaluation using landsat and night time temporal satellite data. J. Indian Soc. Remote Sens. **46**, 263–273 (2018)

An Automated Underground Space Monitoring and Communication System Based on Wireless Sensor Networks

Mohammad Ali Moridi[1](✉), Mostafa Sharifzadeh[1], Hyongdoo Jang[1], and Youhei Kawamura[2]

[1] Western Australian School of Mines: Minerals, Energy and Chemical Engineering, Curtin University, Kalgoorlie, WA, Australia
mohammad.moridi@curtin.edu.au
[2] Graduate School of International Resource Sciences, Akita University, Akita, Japan

Abstract. In the challenging environment and Cutting-edge technology in mining industry, reliable and effective communication is a high-stake issue, along with the objectives of safe and efficient underground mining operations. Automation through remote and automatic systems has delivered improvements in workplace health and safety for employees, operational management, energy and cost-effectiveness, and real-time response to events. In this context, Wireless Sensor Networks (WSNs) have been widely employed in underground monitoring and communication systems for the purpose of environmental monitoring, the positioning of workers and equipment, operational monitoring and communication system. Considering the capabilities of WSNs, a ZigBee network is adopted in this study.

The aim of this study is to propose a reliable and effective monitoring and communication system in underground environments, using WSN nodes were developed to sense environmental attributes and texting emergency messages. A trigger action plan for monitored attributes above normal and threshold value limits is programmed in the surface GIS management server. The system will provide multi-users surface operation and 3D visualization for realistic understanding of underground environment and miners' conditions.

Keywords: Mine automation · Safety · Wireless Sensor Networks · ZigBee · GIS

1 Introduction

Underground excavation safety and health remain challenging issues in the mining industry. Death toll statistics in China's coal mines have gradually reduced from 5798 to 2631 between 2000 and 2009 [1] but fatality still occurs. The number of occupational mining fatalities in the United States' underground metal mines has fluctuated from 40 to 46 during the years 2001 to 2010. Most importantly, 33.8% of the deaths have resulted from ignitions and explosions of gas or dust [2] in underground mining. In April 2014, two men were killed when a wall collapsed in an underground coal mine

E. Topal (Ed.): MPES 2019, SSGG, pp. 255–261, 2020.
https://doi.org/10.1007/978-3-030-33954-8_32

in New South Wales, Australia. Figure 1 illustrates the death toll of underground mining in some countries such as USA, India and China between 2000 to 2013. It shows that the underground mining occupations remains one of the most dangerous activities. Human errors were concluded from reports as the most significant reasons for mining fatalities. Thus, safety is always a significant concern in mining and tunnels operation. Some studies have recently focused on improving the health for underground miners.

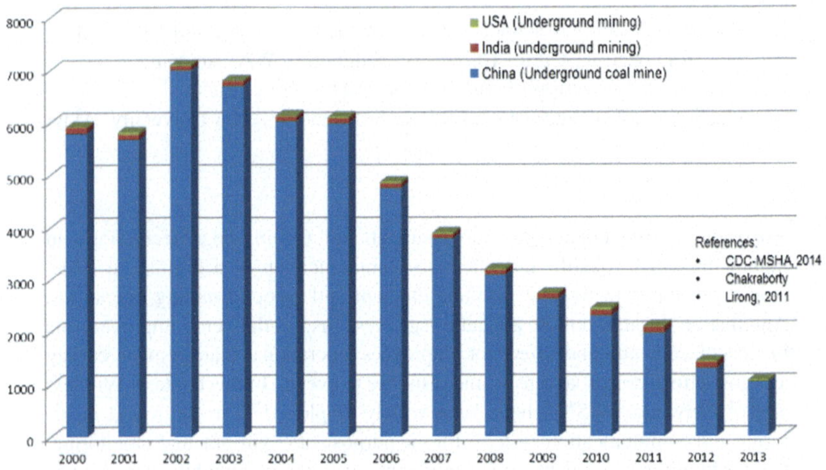

Fig. 1. Death toll of underground mining in some countries between 2000 to 2013

In this study underground safety and health concerns are significantly mitigated based on the system integration which incorporates ventilation management and emergency message texting. The applications of the ZigBee network for underground communication and monitoring is illustrated in Fig. 2. These applications considering sensor nodes' abilities are classified as follows:

(a) **Safety and health approaches**

- Air quality and quantity measurements
- The determination of workers' location
- Emergency and safety communications
- Gas detector and fire alarm
- Geotechnical monitoring

(b) **Operations management and control**

- Real-time monitoring underground mine operations from surface control centre
- Improving the underground operation cycles (scheduling)
- Traffic control (Signals) and Ventilation on Demand (VOD)

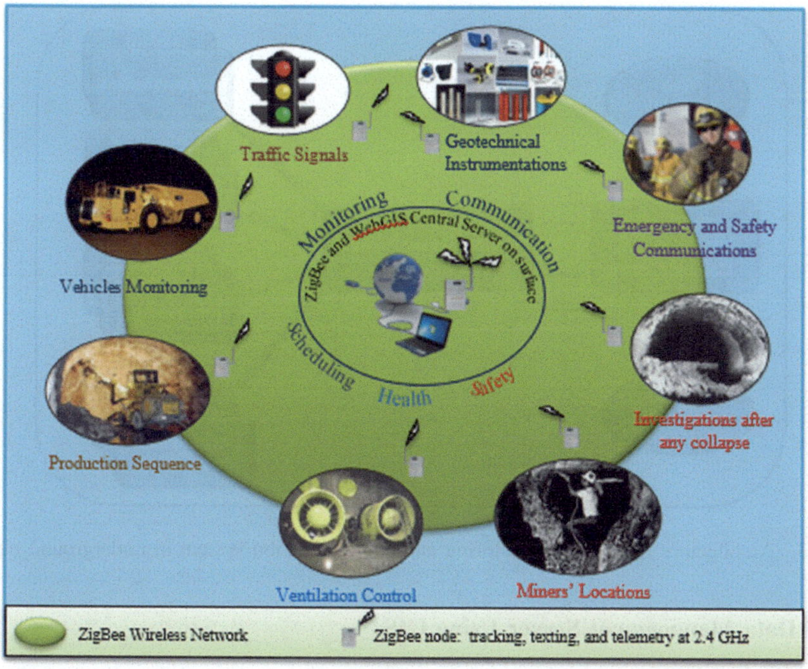

Fig. 2. Applications of WSNs for underground communication and monitoring

2 Underground WSN Data Management

In response to these challenges, mine automation by new technologies such as wireless sensor network (WSN) assisted with geographic information system (GIS) has been widely utilised in underground environments to enhance safety and health, productivity and reduce operational costs. GIS is new technology used for spatial data analysis in order to capture, store, analyze, manage, and present data that is linked to locations [3]. GIS allows users to view, understand, question, interpret, and visualize data in many ways that reveal relationships, patterns, and trends in the form of maps, globes, reports, and charts. Web-GIS is an inevitable trend which helps solve the problems of spatial information integration and sharing in technical aspect of web media [4]. The architecture of underground monitoring and communication for the system integration is illustrated in Fig. 3. Temporal ZigBee data including messages and environmental attribute readings such as temperature, humidity and gases concentration are transferred to GIS management server in the surface control centre. The transmitted data is received and stored by ZigBee program then provided for manipulation in the control centre. Risk situations are immediately identified and responded through a logical process of data analysis in the GIS management server before reaching dangerous (unsafe) levels and accidents occurring. The ventilation system management is also used for the workplace health and safety compliance and the optimisation of mine site power usage.

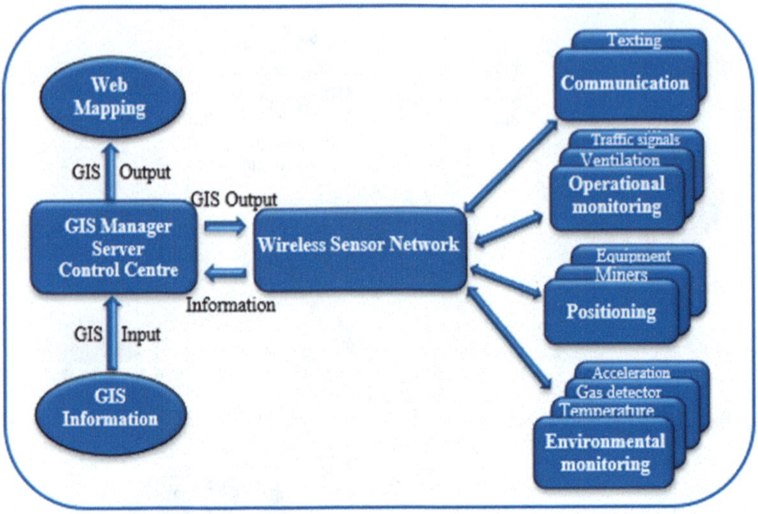

Fig. 3. Architecture of a WSN monitoring and communication system in underground mines

2.1 Data Management Server Using GIS

In the challenging environment and changing topology of a mine, reliable and simplified communication is a high-stake issue with the objectives of safe and efficient mining operations. In response to these challenges, integration of technologies has a significant role in underground mining automation. According to WSNs' specific features of high reliability and multi-hop networking, ZigBee can create an integrated wireless network between nodes in the underground mine tunnels and the surface gateway. In this study, ZigBee's capability of monitoring underground environmental attributes is combined with geographic information to provide potential applications in communication, operational and environmental monitoring systems of underground mining.

In order to achieve such smart underground mine system, integrating maps information and spatio-temporal data from ZigBee nodes into a database at a control centre is required [5]. Figure 4 illustrates flow chart of data processing in the integrated system.

The network demanded in an underground mine must be capable of providing bilateral communications between the surface control centre and all underground wireless nodes interactively. According to the threshold limit values for the different variable parameters (V1, V2, …, Vn) of underground mine environment, the conditions of safe, transient and unsafe were set. Thus, the remote or automatic countermeasures in a GIS management server were arranged in order to control ventilation fans and send alert or alarm messages to relevant authorities. Additionally, immediate texting messages are bilaterally communicated between underground personnel and the surface operator in emergency conditions [6].

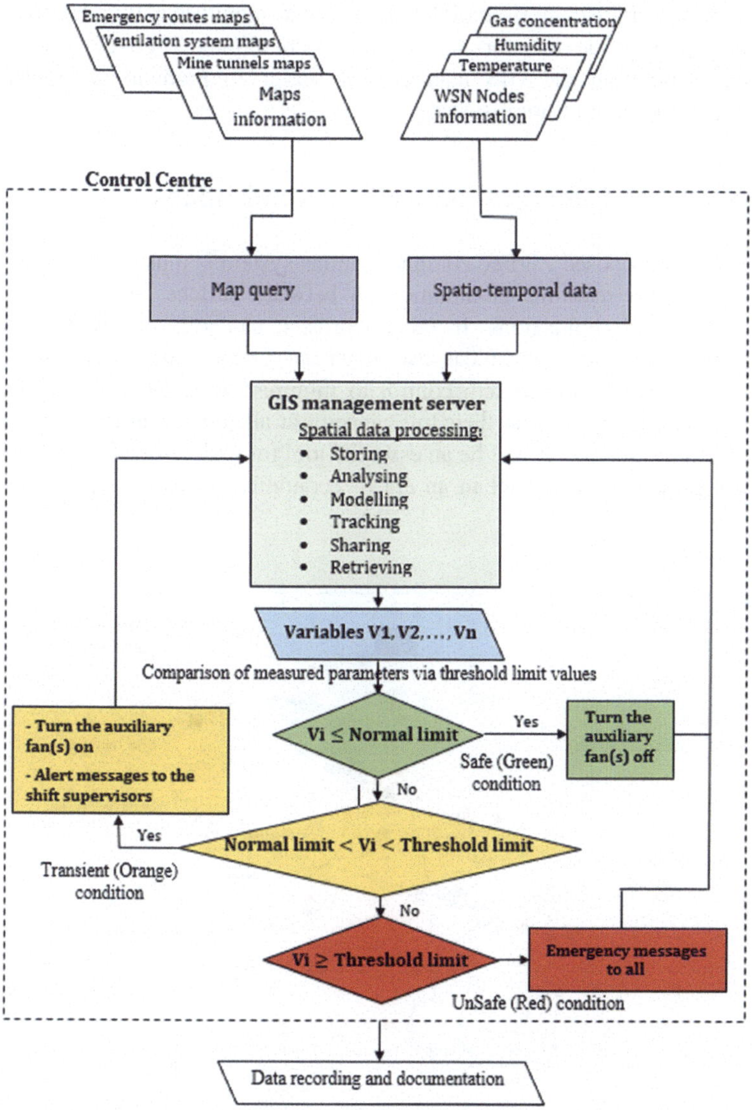

Fig. 4. Flow chart of data processing in the integrated system

Based on this system, near real-time monitoring data, remote and automatic controls, and communication by texting messages have achieved the required safety and health outcomes and improving underground mining operations.

2.2 System Outputs

Mine safety and health were improved by intelligent maps supporting spatio–temporal data and coordinate of ZigBee nodes in this experiment. The final outputs of GIS management server are comprised of 3D visualization monitoring of underground mine tunnels and messages texting for alert and alarm conditions. The web-GIS is another application supporting the GIS management server to promote the underground monitoring and communication system.

3 Prospects of Underground ZigBee Applications

The Prospect of proposed ZigBee communications system is illustrated in Fig. 5. This system enables for real-time communication between surface operators and underground miners and applications. In other words, it can join all sensors' data with different duties such as geotechnical instrument data, miners, trucks and other equipment locations, ventilation system control to minimise wastage of electrical energy caused by running vent fans at their full capacity at all times, and traffic signal management from the surface. It will be an essential tool for underground mine automation to improve project management in an era of economic, operation, health and safety optimization.

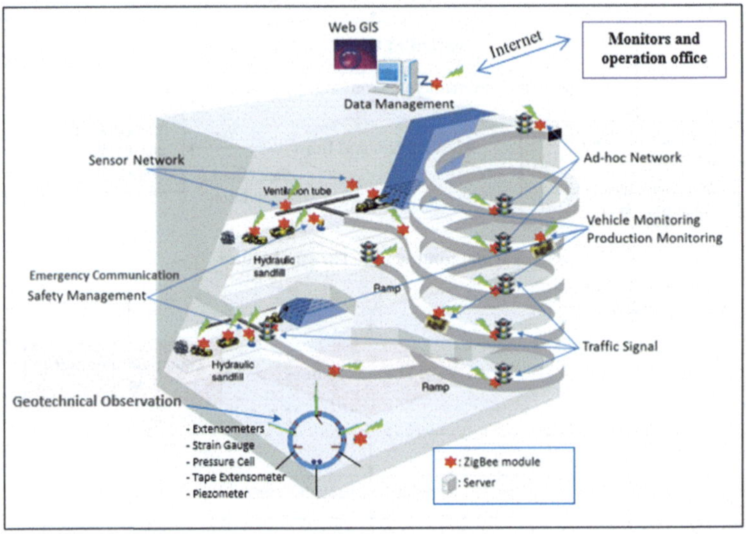

Fig. 5. Prospect of proposed automated underground mine monitoring and communication system based on WSNs

4 Conclusion

In this research it was shown that underground wireless networks could significantly improve the efficiency of environmental monitoring, workers and equipment locations, operational readings and communication system. In this research was shown that an integrated system based on the WSNs and GIS was introduced to automate underground mine monitoring and communication. The system is equipped with an automatic or remote trigger action plan for monitored attributes above normal and threshold value limits programmed in the surface GIS management server. The system also provides multi-users surface operation and 3D visualization for realistic understanding of underground environment and miners' conditions, and it could be a useful approach for high-tech underground mining.

References

1. Wu, L., Jiang, Z., Cheng, W., Zuo, X., Lv, D., Yao, Y.: Major accident analysis and prevention of coal mines in China from the year of 1949 to 2009. Int. J. Min. Sci. Technol. **21** (5), 693–699 (2011)
2. CDC: Centres for Disease Control and Prevention Mining Fatalities, Underground, USA (2010)
3. Moridi, M.A., Kawamura, Y., Sharifzadeh, M., Chanda, E.K., Jang, H.: An investigation of underground monitoring and communication system based on radio waves attenuation using ZigBee. Tunn. Undergr. Space Technol. **43**(0), 362–369 (2014)
4. ESRI: ArcGIS® for Emergency Management. USA (2012)
5. Şalap, S., Karslıoğlu, M.O., Demirel, N.: Development of a GIS-based monitoring and management system for underground coal mining safety. Int. J. Coal Geol. **80**(2), 105–112 (2009)
6. Moridi, M.A., Kawamura, Y., Sharifzadeh, M., Chanda, E.K., Wagner, M., Jang, H., Okawa, H.: Development of underground mine monitoring and communication system integrated ZigBee and GIS. Int. J. Min. Sci. Technol. **25**, 811–818 (2015)

Analysis of Mobile Communication Coverage and Capacity for Automation in Open-Pit Mines

Viviane da Silva Borges Barbosa(✉), Gilberto Rodrigues da Silva, Pedro Henrique Alves Campos, Pedro Benedito Casagrande, and Luciano Fernandes Magalhães

Mining Department, Universidade Federal de Minas Gerais, Antônio Carlos Av. 6627, Belo Horizonte 31270-901, Brazil
grsilva@demin.ufmg.br

Abstract. The search for greater safety and operational performance has pushed the mineral industry towards the automation of large scale operations, an important feature in intelligent mines. Wireless connection plays an essential role in the automation of open-pit mines, although representing a technical challenge. The growing number of features with high demands from networks and the transmission of real time videos increased data traffic from kilobits/second to megabits/second, requiring wireless networks of greater capacities. Additionally, as working benches distance from the communication infrastructures in open-pit mines, increased coverage is required throughout the life of a mine. This work investigates the coverage and capacity of a wireless communication network in an open-pit mine for a 7-year period, addressing the influences of topography, technology and fleet size. The results indicate the integration of mining and network planning is a natural requirement for mining automation, necessary to guide the positioning and scaling of communication structures in all phases of a mining project.

Keywords: Open-pit mines · Wireless communication · Intelligent mines

1 Introduction

Communication is always present in mining environments and can take place as voice transmissions and data transmission (digital applications). In the former, messages can travel between people using radio devices, *e.g.*, walkie-talkies, as for the latter, messages travel between equipment or software.

Intelligent mines integrate technologies which support remote and automated loading and hauling operations, aiming at increasing both the fleet and the operators' productivity as well as meeting safety standards [1]. Data transmissions are necessary for applications in fleet management (dispatch), telemetry and automation, and take place without the use of cables, *i.e.*, they are modulated in radiofrequency (RF) signals which travel between transmitters and receptors. As decisions are taken by Control Center algorithms, which are capable of compiling large amounts of data from all mining operations, intelligent mines need their own robust communication

E. Topal (Ed.): MPES 2019, SSGG, pp. 262–270, 2020.
https://doi.org/10.1007/978-3-030-33954-8_33

infrastructure [2, 3]. Brucutu (Vale, Brazil) [4], Chuquicamata (Codelco, Chile) [5], KazAtomProm (Kazakhstan) [6] and Jimblebar (BHP Billiton, Australia) [7] are examples of intelligent open-pit mines which aggregate technologies to operations depending on data transmission between equipment in the mining site and the Control Center.

The Vale S.A. Brucutu iron ore mine is located in the municipality of São Gonçalo do Rio Abaixo, Brazil. It is classed as a large size mine having produced 28.7 million metric tons of concentrate in 2013, with a 0.9 stripping ratio. In order to achieve this production rate, the mine had a fleet which could be divided into large size equipment (16 trucks with a transport capacity of 235 tons, 6 loaders and 1 electric excavator) and small size equipment (16 trucks with a transport capacity of 35.5 tons and 3 excavators) [8]. As of 2018, the Brucutu mine kept its iron ore concentrate production rate by making some changes in its fleet, with the addition of seven 235 tons autonomous trucks, besides auxiliary equipment, *e.g.*, autonomous drill and semi-autonomous bulldozers [4].

The implementation of equipment with automated skills require continuous monitoring and stricter systems for data transmission, which demands greater investments in the mine's wireless infrastructure. Planning the communication network is an essential phase for the mine planning of intelligent mines, although this task is non-trivial. The mine site is a hostile environment, exposed to sunlight, rain, dust, mist and formed by an irregular topography (benches and faces), making it difficult to ensure the expected network performance. The position of the fixed infrastructures (*e.g.*, the macro-cell) and the mobile infrastructures (*e.g.*, the access-points), are key elements to ensure a reliable communication signal and, more importantly, a proper operation of fleet management, telemetry and automation. Additionally, the wireless network infrastructure should match the continuous distancing of the working benches regarding the macro-cell and the growing number of machines.

This work indicated that the integration of network planning and mine planning processes can provide the knowledge necessary to design an adequate communication infrastructure, ensuring a high performance and personalized service for mining applications which rely on wireless networks. Although the mine site is a scenery of intense modifications, there is predictability for the evolution of each alteration in a yearly, monthly or even daily basis, which can be used to prepare for the forthcoming wireless network working conditions and requirements.

2 Methodology

2.1 Macro-cell and Access-Points

This work presents data from Brucutu Mine and premises of the Brazilian iron ore industry [4, 8] to show how mine planning and network planning are a natural unfolding of automation in the mining industry, necessary to guide the location and scaling up of communication structures (*e.g.*, macro-cell and access-points). Brucutu mine was chosen for case of study because of data availability and the ongoing implementation of autonomous fleet. This expansion required the implementation of a communication infrastructure to support the different technologies present in the mine site, allowing any expansion needed. This infrastructure consists basically of a macro-cell and its access-points, as shown in Fig. 1. The Control Centers, usually connected to the master antenna by cables, are fixed administrative centers which concentrate a large amount of information and support computers running high computational cost software. Although part of this network is cabled, the interaction between the equipment in the pit and the Control Center takes place necessarily through the wireless network, via radiofrequency waves which travel directly to the macro-cell or to the intermediate mobile access-points, as shown in Fig. 2 [9].

A

B

Fig. 1. Examples of typical macro-cell (A) and access-points (B).

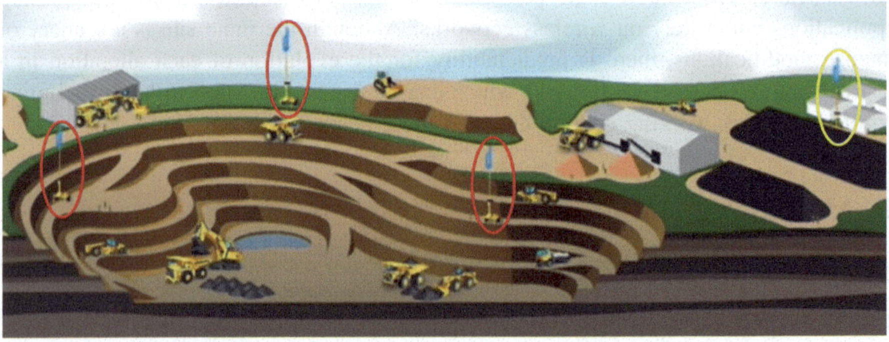

Fig. 2. Communication infrastructure: mobile access-points (in red) and macro-cell (in yellow).

2.2 Simulated Scenarios

This work was based on a surface digital model and in an estimated number of clients (equipment) of Brucutu mine in a 7-year period. It was considered that the equipment required have to send and receive data according to its telemetry, automation and fleet management applications. This inputs (*e.g.*, the mine topography, number of clients and data traffic estimation) was used to build the scenarios investigated in the Atoll® software, for radiofrequency predictions.

The software outputs indicate when the clients are able (or not) to download data from the macro-cell according to radiofrequency predictions. For example, when a semi-autonomous dozer is able to receive the data packet which describes the actions it should perform. The ability to download data is a function of the distance between customers and the antennas, so this capacity is reduced as customers move away from access-points or macro-cell.

The final goal is the estimation of Received Signal Strength Power (RSSP) that can indicate areas in the mine which have strong or weak signal. Figure 3 presents the flowchart adopted for this study showing the inputs and outputs expected for the simulation.

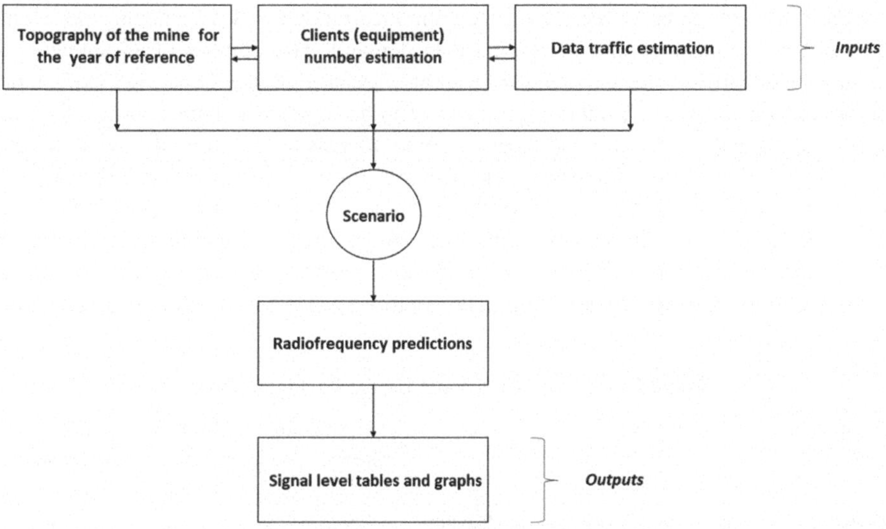

Fig. 3. Methodology flowchart.

In order to analyze the possible changes which may take place in a 7-year period due to new technology, changes in the mine morphology and fleet size, two scenarios were considered: (A) Conventional Mine scenario and (B) Intelligent Mine scenario. Their characteristics are summarized in Table 1.

The network planning software (Atoll®) employs models for wave propagation which relate the frequency, the terrain morphology and the loss during propagation to estimate the RSSP in each of the surface model pixels which represent the topography of the mine. The power level received refers to the throughput available to the

Table 1. The main characteristics of scenarios A and B.

Scenario	Year/Mine's area	Clients
A – Conventional Mine	Year 1 (2007)/2.12 km^2	3 dozers; 2 drills; 26 trucks; 6 loaders
B – Intelligent Mine	Year 7 (2014)/4.02 km^2	5 dozers; 4 drills; 32 trucks; 10 loaders

application according to its position under the signal coverage, *i.e.*, it refers to the maximum data the application will be able to receive from the macro-cell considering its position in the mine.

Scenario A. Mining operations only make use of telemetry and fleet management to conduct manual truck allocation, presenting an average throughput of 32 kb/s per equipment. If the most demanding scenario is considered, in which all the machines host applications download data simultaneously, the communication infrastructure must be able to properly meet a demand of 1,184 kb/s.

Scenario B. After 7 years, due to the topographic changes and the addition of autonomous equipment which presented greater network demands, it was investigated if the changes in the initial infrastructure provided an adequate communication service. In year 7, the area to be covered by the communication network was larger, as well as the number of machines and the global data demand. The number of equipment for loading and hauling was expected to increase because of the larger stripping ratio considered for the year 7, although the run-of-mine target was kept constant for this work. Additionally, the mining faces become increasingly more distant from the original dumping points, increasing the average transport distance and reducing the productivity (tons/h) of the trucks. All the machines used in the Intelligent Mine were considered to be in the automation state-of-the-art, leading to a global data demand 44 times greater due to the introduction of new application in the equipment hosts which, in this scenario, must run remotely guided operations (*e.g.*, dozers) and operations with

Table 2. Global data demand for the Intelligent Mine.

Equipment	Applications	Throughput kb/s	Number	Global demand kb/s
Dozer	Video, audio, commands, dispatch, telemetry, high precision GPS	3,500	5	17,500
Drill	Video, commands, telemetry, high precision GPS	3,600	4	14,400
Truck	Telemetry, dispatch, commands, high precision GPS	500	32	16,000
Loader	Video, telemetry, high precision GPS	500	10	5,000
	TOTAL			52,900

higher level of automation (*e.g.*, drills and trucks). Table 2 summarizes these considerations.

3 Results and Discussions

The first simulation considered the implementation of a Long Term Evolution 4G (LTE 4G) wireless network infrastructure able to properly meet the demands of all the clients in scenario A in year 1. This requires considering strategic choices, *i.e.*, the ideal or viable location for the macro-cell. As building its tower is a costly process, the antenna location should be carefully planned to avoid moving for the longer period possible. The location choice for the macro-cell in the simulation of scenario A assumed its real position at Brucutu mine from 2007 up to 2016.

Figure 4 shows the average Received Signal Strength Power (RSSP) levels for the morphology of Brucutu mine in year 1. It was observed, based on Monte Carlo simulations using the software Atoll®, that the macro-cell was able to meet the demands of all the clients in the Conventional Mine.

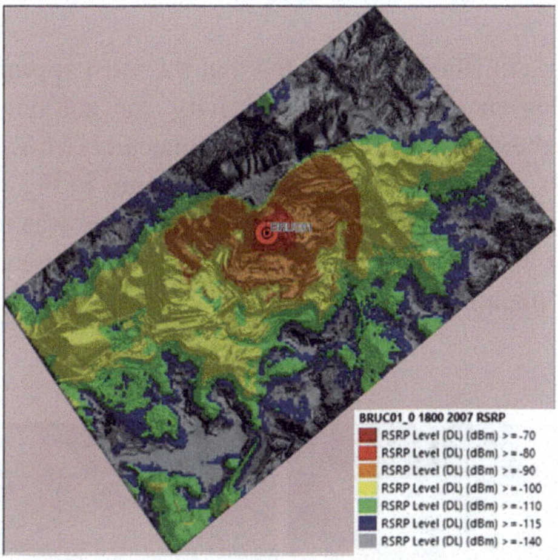

Fig. 4. RSSP in Brucutu mine for year 1 (2007).

Aiming at raising the capacity of the communication system in Scenario B, 4 access-points were included in the surroundings of the mining faces and the macro-cell location was altered. These changes increased the communication reliability since the master antenna coverage could then superimpose the coverage of the small cells, working as an alternative for connection in case the access-points failed or even as the main link in the areas not covered by them, as shown in Fig. 5.

The analyses with Atoll® indicated the proposed infrastructure was able to cover all the mine, providing adequate connectivity for at least 98.3% of the clients in the Intelligent Mine. The simulations also indicated that using the infrastructure of year 1

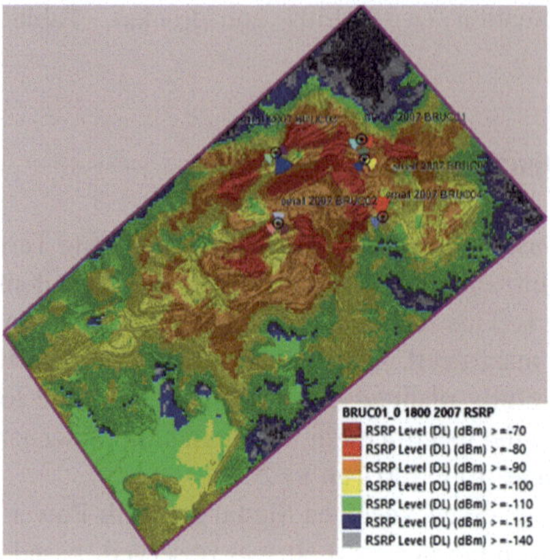

Fig. 5. RSSP in Brucutu mine for year 7 (2014).

in the scenario of year 7 would provide adequate connectivity to only 37.8% of the 51 equipment.

When this work was finished, it suggested that the most appropriate location for the master antenna, aiming at providing connectivity for autonomous applications in Brucutu mine, was at the eastern hill, at the coordinates UTM 670.450 E, 7.804.066 S (23K, Datum SAD 69). The Google Earth satellite images in Fig. 6 show the change in the antenna location by the time the autonomous truck fleet was implemented in 2018, which was consistent with the proposed alteration. The research integrating mine planning and network planning also resulted in patent applications in Brazil [10] and in Europe [11].

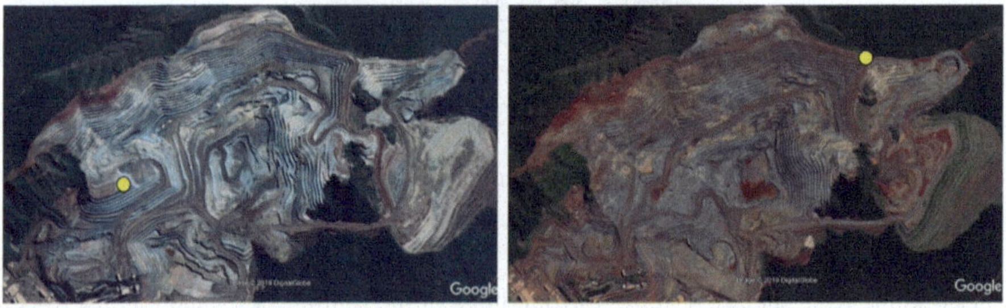

2016 2018

Fig. 6. Macro-cell location (yellow circle) in 2016 and 2018.

4 Conclusions

The characteristics of an open-pit mine which affect the performance of a wireless network infrastructure are not familiar to network engineers while the peculiarities of the required infrastructure are commonly not well understood by mining engineers. This gap led to the supply of non-optimum solutions in the market, resulting in networks designed to match immediate demands and expectations for scale-up which are greater than what the systems can offer. The solutions offered for wireless dead zones or low capacity are based on oversizing the number of access-points or trial and error methods which aim at positioning the cell at a location with acceptable coverage and capacity for the clients in the operational area.

The study indicated the addition of equipment with a higher level of automation, throughout the life of the mine, required reorganizing the network infrastructure. Four access-points were then included next to the working benches and the macro-cell was relocated, resulting in high connectivity for 98.3% of the clients. It is important to mention, however, that this performance is still not adequate for a full automation scenario, in which high connectivity should be offered to 99.999% of the clients to allow relying on autonomous equipment safely. It is also necessary to point that, even if the solution provides high connectivity to 100% of the clients, this would not be permanent as the topography and the lithology of the mine constantly changes, requiring continuous network planning to avoid lack of coverage and capacity for mining operations.

References

1. Mchattie, L.: Advances in mine engineering to enable information mobility for 'intelligent mining'. In: 23rd Mining Congress, Toronto (2013)
2. Barbosa, V.S.B., Garcia, L.U., Caldwell, G., Almeida, E.P.L., Rodriguez, I., Sørensen, T.B., Mogensen, P., Lima, H.: The challenge of wireless connectivity to support intelligent mines. In: 24th Mining Congress, Brazil (2016)
3. Garcia, L.G.U., Almeida, E.P.L., Barbosa, V.B., Caldwell, G., Rodriguez, I., Lima, H., Sørensen, T.B., Mogensen, P.: Mission-critical mobile broadband communications in open-pit mines. IEEE Commun. Mag. **54**(4), 66–69 (2016)
4. Vale Homepage (Vale will have the first mine operating only with autonomous trucks in Brazil). http://www.vale.com/brasil/EN/aboutvale/news/Pages/vale-tera-a-primeira-mina-operando-somente-com-caminhoes-autonomos-no-brasil.aspx. Accessed 01 June 2019
5. National Instruments Homepage (Cadetech improves monitoring of electromechanical shovels at world's largest copper mines). http://sine.ni.com/cs/app/doc/p/id/cs-11154. Accessed 01 June 2019
6. BNEWS Homepage (Kazatomprom launched a Digital Mine). http://bnews.kz/en/news/ekonomika/promishlennost/kazatomprom_launched_a_digital_mine-2016_05_30-1273836. Accessed 05 Oct 2016
7. BHP Billiton Homepage (BHP Billiton opens Jimblebar iron ore mine). http://www.bhpbilliton.com/media-and-insights/news-releases/2014/04/bhp-billiton-opens-jimblebar-iron-ore-mine. Accessed 01 June 2019

8. Martins, A.G.: Simulation of Brucutu's mining operations using a Linear Programming Algorithm to manage loading machines (Translated). Master thesis (Mining Engineering Department): Federal University of Ouro Preto, Brazil (2013)
9. Barbosa, V.S.B.: Analysis of wireless communication coverage and capacity for automation in open pit mines (Translated). Master thesis (Mining Engineering Department): Federal University of Ouro Preto, Brazil (2016)
10. Instituto Nacional de Propriedade Intelectual Homepage (Code: BR 10 2016 005371 4). https://gru.inpi.gov.br/pePI/jsp/patentes/PatenteSearchBasico.jsp. Accessed 01 June 2019
11. European Patent Office Homepage (Integrated open-pit or underground mines and wireless transmission networks). https://worldwide.espacenet.com/. Accessed 01 June 2019

Mine Equipment Selection, Utilisation and Maintenance

A Novel Approach to Safely Increase Dump Truck Pay-Load Capacity to Optimize Material Haulage

Khairulla Aben[1(✉)], Yerkin Orazaliyev[1], and Fidelis T. Suorineni[2]

[1] KAZ Minerals, Almaty, Kazakhstan
khairullenok@gmail.com
[2] Nazarbayev University,
53 Kabanbay Batyr Avenue, Nur-Sultan 010000, Kazakhstan

Abstract. Equipment selection is an important component of mine planning. The number and types of mining equipment must match expected productivity and be compatible. For loading and hauling, the capacity of the equipment selected will also dictate the number needed. This paper presents a study in which haulage trucks manufactured to specifications by companies had their dump bodies changed from the originals to light-weight tray (LWT) dump bodies to increase their volumetric capacity and payload. The paper outlines the procedure adopted to safely change the original dump truck trays to light weight trays. The study resulted in fewer number of haulage trucks required in the fleet and reduced haulage cost by about 12%. An additional benefit of the study is that the exercise resulted in longer life and increased time to body replacement for the LWTs compared with the trucks with the original trays. The economic impact on the operations was significant, based on both OPEX and CAPEX.

Keywords: Dump truck · Payload · Haulage · Light-weight body

1 Introduction

For the last several years, mineral reserves with high grades and favourable mining conditions have been constantly depleting. Therefore, modern mining companies actively seek and implement innovative technologies and engineering solutions to increase efficiency, productivity and safety of their operations and to decrease costs.

KAZ Minerals is a fast growing and aggressively expanding copper producer based primarily in Kazakhstan. The company currently operates three open pit mines (Bozshakol and Aktogay in Kazakhstan and Bozymchak in Kyrgyzstan) and three underground mines in the Eastern region of Kazakhstan (Fig. 1). The company also has large properties of copper deposits at the feasibility study stages in Kazakhstan and Russia.

Bozshakol and Aktogay are two KAZ Minerals state-of-art mines. The company has significantly increased overall production rates at these two mines, with the aim of becoming one of the top copper producers in the World.

Bozshakol produces 25 million tonnes of sulphide and 5 million tonnes of kaolinized ore per year, and Aktogay produces 25 million tonnes of sulphide and 15

E. Topal (Ed.): MPES 2019, SSGG, pp. 273–280, 2020.
https://doi.org/10.1007/978-3-030-33954-8_34

Fig. 1. Map of KAZ Minerals operating mines [1].

million tonnes of oxide ore per year. Aktogay is also going through expansion with a plan to double the sulphide ore mining and processing. Future of the company includes starting a similar size Koksay mine in Kazakhstan, at one of the largest deposits in the World, the Baimskaya deposit in Russia that was acquired recently.

The mining equipment fleet at the two mines were selected during the feasibility studies to match the productivities. Ore and waste loading at both Aktogay and Bozshakol are done using 21 m^3 bucket shovels, and hauling is by using 136-tonne trucks. With increasing depths at the mines, the plan was to increase the fleet. Because both mines are trucks-constrained in terms of productivity, priority was placed on hauling. However, before spending a significant capital cost to simply increase the number of trucks, which would lead to further increase in operating costs through the increase in the number of operators, maintenance works and other associated logistics, the engineering team of the KAZ Minerals decided to look at other options of improving haulage efficiency.

Invariably, hauling is often a significant portion of the cost per tonne of rock mined, with fuel cost being the major component, forming about 50% of the hauling cost (Fig. 2). Therefore, optimizing hauling could lead to significant operating cost savings. It is hypothesized that hauling can be optimized by replacing existing dump bodies with light-weight dump bodies with the ability to increase truck volumetric capacity. Such a strategic step also has benefits for environmental protection by hauling more tonnage with less fuel consumption. Additionally, the ability to safely increase truck haulage tonnage could result in use of fewer trucks for the same output, and hence, less CAPEX.

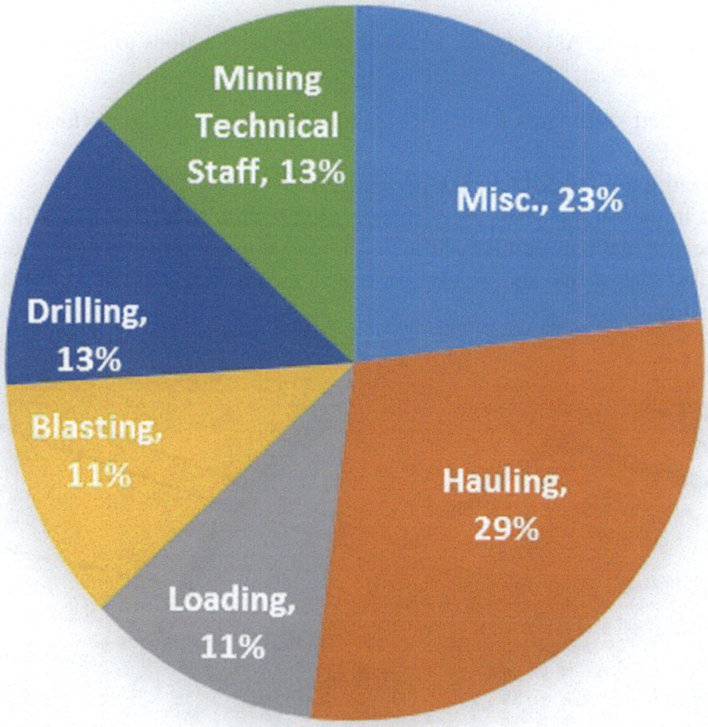

Fig. 2. Typical breakdown of mining cost estimate [2]

2 Truck Productivity Improvement Opportunity

2.1 The Need for Truck Payload Improvement

Even though a Gross Machine Operating Weight (GMOW) constraint of 249.5 tonnes allows payload of up to 136 tonnes, considering the truck volumetric capacity (78 m^3) of the existing dump bodies, this maximum payload becomes constrained by the volume of the dump body, and could be lower than specified even when 100% fill factor is applied.

For the given truck specifications, in order to achieve the allowable payload of 136 tonnes or more, requires replacing the original dump bodies with light-weight dump bodies. Light-weight dump bodies are offered by manufacturers such as Duratray [3], Westech [4], and Ground Force Worldwide (RWB) [5], CAT (MSD II) [6] or DT Australia [7] which provide an opportunity to increase the overall truck productivity by increasing the payload up to between 140 to 147 tonnes. The exact achievable payload for a given light-weight body used depends on the specific body type.

2.2 Initial Measures for Productivity Increase

First, KAZ Minerals Aktogay team has taken following actions to improve the overall efficiency of the fleet:

- implemented sideboards ("hungry" boards) that increased the payload of the truck by 10% (Fig. 3);
- optimized the number of trucks per shovel;
- started implementing "hot-seat" change (quick shift change program) to decrease the "dead" time between the shifts and improve the equipment utilization;
- reprofiled the roads and safely increased the average speed of the trucks by approximately 10%.

Fig. 3. Truck body with sideboards.

The above measures improved the overall productivity of the mine. However, further measurements and actuals showed that there was more potential to improve truck payloads, if light-weight trays (LWT) were implemented.

For Aktogay Mine, eliminating volumetric constraint was the key factor that contributed to the potential payload increase, rather than the difference in the weight of the dump bodies.

2.3 Further Productivity Enhancement Strategy

The actual payloads of the trucks due to the material densities were in the range of 117 to 123 tonnes for truck volumetric capacities of 78 m^3 (Eq. (1)) [6]. Thus, the actual payload is below the nominal value of 136 tonnes, which could be potentially reached within Gross Vehicle Weight (GVW) limits. This was due to the volumetric constraint of the existing dump bodies considering 2.6 m^3/t rock density and average 1.5 swell factor.

$$\text{Payload} = \text{Body Volume} * \text{Rock Loose Density} * \text{Fill Factor} \tag{1}$$

3 Identification of the Appropriate Light-Weight Body as Replacement

3.1 Approach

Extensive research was conducted on the different light-weight tray (LWT) products compatible with the existing trucks to understand the advantages and disadvantages on the operating and capital costs, and maintenance. Also, feedbacks on the use of different LWT's were requested from several open pits, including a visit to Kumtor mine in Kyrgyzstan.

Based on the reviews of the light-weight tray products and information obtained from interviews of operations already using these products, it was concluded that:

1. for all the LWT's, increase in payload is possible due to the:
   - difference in weight of the original dump body and light-weight dump body;
   - Increase in the effective volumetric capacity of the dump body by replacing it with a lightweight body.
2. Although light-weight dump bodies existing in the market differ in terms of design and material, the key benefits claimed by the manufacturers are similar:
   - payload increase
   - reduced carryback
   - reduced wear
   - less spillage.

In general, 15–20% larger bodies with corresponding higher payload of around 147 tonnes would give approximately 15–20% increase in payload (Eq. (1)), which in turn would significantly contribute to reducing Operating Costs and enable reducing the number of required trucks thus saving Capital Expenditures.

3.2 Testing of Light-Weight Dump Bodies

Considering all the potential advantages of the LWT's and having done research on the potential suppliers, four Hercules light-weight bodies from DTHiload were purchased by the mine for test trials.

First the bodies were installed on the trucks, and several test runs were performed. For several months, the bodies were tested with both ore and waste materials with constant measurements and scaling. This was important as different materials have different densities, hardness and clay content, which in combination give different wear, carryback and payloads. Overall, the bodies have shown stable payload increase with either type of material. The wear after several months of use was minimal, which proved potentially longer life and increased time to truck replacement.

3.3 Results

In this study various LWTs were identified from different suppliers and evaluated as replacements for exiting dump trucks trays. The methods of evaluation consisted of interviews of companies with experience in the use the technology, field trials with sample trucks using different muck with different physical characteristics such as density and particle size distributions. The study results have indicated the following:

- payload has increased by 15–17% to around 145 tonnes;
- fuel consumption has incrementally increased per truck by 1–2% - meaning that specific consumption per tonne has decreased;
- maintenance did not observe any serious issues;
- total number of trucks required can be decreased almost proportionally in accordance with the required truck hours (Fig. 4).

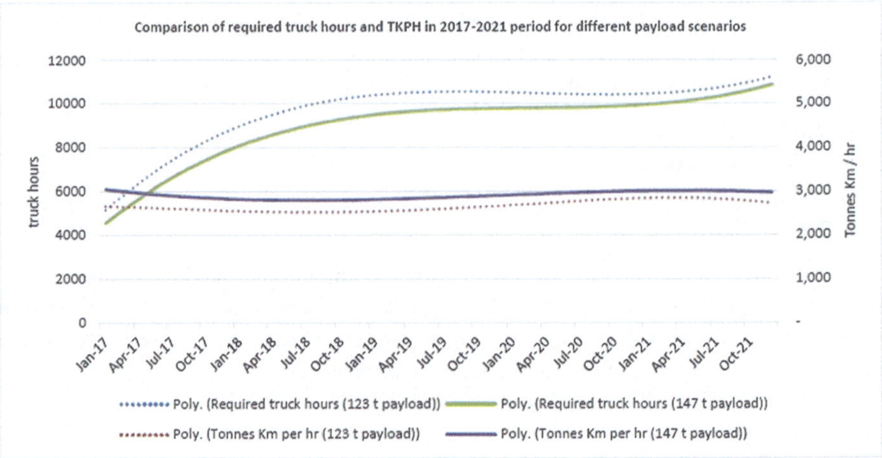

Fig. 4. Comparison of required truck hours with a regular body and LWT.

4 Economic Impact

Replacing the existing dump bodies with light-weight dump bodies positively impacted hauling costs by increasing the payload. Overall, the following savings are being observed:

- reduction in hauling operating costs by approximately 12%;
- reduction in total number of required trucks and thus saving on CAPEX.

In general, the economic justification of the LWT's was calculated using Eq. (2) below:

$$\text{Value} = -\ (\text{CAPEX for LWT's}) - (\text{Maintenance}) + (\text{OPEX savings}) + (\text{CAPEX savings}) \tag{2}$$

After the trials, a regular NPV approach was adopted to get more accurate results.

5 Conclusions

Haul truck productivities at KAZ Minerals Aktogay, in terms of tonnes per hour were constantly increasing due to improving utilization, haul roads redesign and optimization of the amount of trucks per shovel. Further installation of sideboards increased payload (volumetric capacity) of the trucks by approximately 10–12%. However, because the volumetric capacity of the dump bodies (rather than the Gross Vehicle Operating Weight) was still the constraint for further increase in payload there was a research conducted and the results showed that there is a possibility to implement light-weight trays.

Light-weight dump bodies with larger volumetric capacity could increase the truck payload up to 147 tonnes, thus giving extra 0.5–0.6 Mt per year per truck. The improved payload estimation is based on the tray weight reduction yielding increased load tonnage, along with the larger contributor of volume increase from the current trays to the specific design trays.

Considering key factors such as safety, payload advantage and maintenance, a specific body was selected as the best option among 6 different light-weight bodies that were evaluated as part of the study.

Replacement of the existing dump bodies by light-weight dump bodies added economic value to the company by:

- Increasing truck productivity thus reducing hauling cost by about 12%;
- Decreasing specific operating costs;
- Reducing the total number of required trucks, thus saving several million USD in capital expenditures.

References

1. KAZ Minerals Homepage. https://www.kazminerals.com/. Accessed 10 Feb 2019
2. US Securities Homepage. http://www.sec.gov/archives/. Accessed 10 Feb 2019
3. Duratray Homepage. https://www.duratray.com/. Accessed 10 Feb 2019
4. Westech Homepage. http://www.westech-inc.com/. Accessed 10 Feb 2019
5. Ground Force Worldwide Homepage. https://gfworldwide.com/. Accessed 10 Feb 2019
6. CAT (MSD II) Homepage. https://www.cat.com/. Accessed 10 Feb 2019
7. DT Australia Homepage. https://www.dthiload.com/. Accessed 10 Feb 2019

Prediction of Mining Railcar Remaining Useful Life

Mohammad Javad Rahimdel[1(✉)], Behzad Ghodrati[2], and Amir Taghizadeh Vahed[2]

[1] Department of Mining Engineering, Faculty of Engineering, University of Birjand, Birjand, Iran
rahimdel@birjand.ac.ir
[2] Division of Operation and Maintenance Engineering, Lulea University of Technology, 97187 Lulea, Sweden

Abstract. Railcar or Rolling stock that is referring to all vehicles moving on railway, is one of the most important component of the rail transport system. In-operation failures of the railcar results delays in transportation and therefore predicting the remaining useful life (RUL) and accordingly considering the preventive maintenance activities are essential. The RUL estimation is a key factor in predictive maintenance that has an influence on the maintenance scheduling, spare parts prediction and also the operational performance of the assets. This paper aims to model and analyze the RUL of railcars at LKAB, a Swedish Mining Company. To achieve this goal, first the failure behavior of the critical subsystems was analyzed and discussed. Then, with considering the effective operational factors (covariates), proportional hazard model (PHM) is applied to calculate the reliability functions. In this concept, RUL of the railcars can be obtained and discussed at the various initial survival times.

Keywords: Rolling stock · Failure behavior · PHM · RUL

1 Introduction

Railway transportation is one of the most usual passenger and freight transferring system in many countries. Rolling stock, refer to all vehicles that move on a railway, is one of the most important component of the rail transport system, which affect the efficiency and long operation life of railway transportation. In-operation failures of the rolling stocks may result in transportation delays and therefore, being aware of them and accordingly considering the preventive maintenance activities are essential. Monitoring the health condition of rolling stocks is important in terms of reliability, availability, safety and also punctuality and efficiency. Therefore, various inspections and maintenance methodologies need to be performed on the rolling stock equipment to meet the mentioned performance measures.

Remaining Useful Life (RUL) estimation is one of the key factors in predictive maintenance (PdM), prognostics and health management that affect the maintenance planning, spare parts provision, operational performance and the profitability of the asset owner. Failure behavior and reliability analysis of the rolling stocks have already

E. Topal (Ed.): MPES 2019, SSGG, pp. 281–288, 2020.
https://doi.org/10.1007/978-3-030-33954-8_35

been evaluated in different studies. Conradie and Treurnicht [1] studied in-service failures of rolling stock in South African Passenger Rail Company to determine the contribution of different failure modes. Results of this study showed that, about 25% of the travel cancellations caused by the rolling stock department. 85% of delays caused due to failures in electrical control equipment, high voltage, switch equipment, traction/axillary machines and controls. Fourie et al. [2] created the reliability block diagram for three motor coach sets of the passenger rail service of South Africa in the form of four main subsystems including power generation, compressed air, vacuum and propulsion systems. As the results of this study, the axillary equipment had the lowest reliability, while the compressor and vacuum exhauster were the most reliable components. In a recent study, Ghodrati et al. [3] studied and prioritized the failure modes of a mining rolling stock. In this study, first, the analytical hierarchy process method in the fuzzy environment was used to determine the importance degree of the failure modes and then the risk priority number was calculated to prioritize the various failure modes. Results of this study show that the crack and flat spots of the wheels and axles, profile erosion of wheel and wheel tire flat are the most hazardous failure modes which need to be under special consideration during the operation and maintenance activities.

Reviewing the past studies show that few attempts have been made by researchers to evaluate the failure behavior and remaining life of the rolling stock components. There are several tools and approaches that are currently used to evaluate the reliability behavior of engineering systems. Among these methods, the Homogenous Poisson Process and the Renewal Process are the most commonly used models where the times between failures are independent and identically distributed [4]. These methods are not suitable when the influencing factors have significant effects on reliability. Parametric and non-parametric regression models such as Proportional Hazard Model (PHM) and accelerated failure time models can be used to incorporate the effect of those influencing factors that are referred as covariates. This paper aims to estimate the RUL of rolling stock of the LKAB Company in Sweden using the PHM method. To achieve this aim, the Time between Failures (TBFs) are extracted and analyzed. Then, with considering the effective operational covariates, which have a significant effect on the reliability behavior, the PHM method is used to calculate the reliability functions and consequently the remaining useful life at different initial survival time.

The rest of the paper is organized as follows. In Sect. 2, the methodology of the remaining useful life estimation based on the PHM approach is presented. Then in Sect. 3, the RUL of rolling stock subsystems are calculated and discussed.

2 Methodology

The PHM was introduced by Cox [5] for estimation of the hazard (failure) rate of a system under the effects of different influencing covariates. In proportional hazard model (PHM), the hazard rate of a component is calculated using a baseline hazard rate and a functional term $\psi(z\beta)$ which describes how the hazard rate changes as a function of operating environmental factors. In $\psi(z\beta)$, z is a row vector consisting of the covariates and β is a column vector consisting of the regression parameters. The baseline hazard rate represents the hazards rate when all influencing factors are equal to zero. In PHM, the hazard rate ($h(t, z)$) is described as: [6, 7].

$$h(t,z) = h(t)\exp(z\beta) = h_0(t)\exp\left(\sum\nolimits_{j=1}^{q} \beta_j z_j\right), \tag{1}$$

where, $h_0(t)$ is the baseline hazard rate, $z_j; j = 1, 2, \ldots, q$ are the covariates associated with the system and $\beta_j; j = 1, 2, \ldots, q$ are the unknown parameters of the model, defining the effects of each q covariates. With considering $R_0(t)$ as the baseline reliability function, PHM in the form of reliability can be written as:

$$R(t,z) = [R_0(t)]^{\exp\left(\sum\nolimits_{j=1}^{q} \beta_j z_j\right)} \tag{2}$$

In remaining useful life estimation, $R(t) = P(T > t)$, where T is failure time. If it has been already known that the equipment has survived until t_0, then, the conditional reliability can be calculates as:

$$R(t|t_0) = P(T > t|T > t_0) = R(t + t_0)/R(t_0), \tag{3}$$

If the conditional reliability at different initial survival time (distance in our case) is considered, the mean residual life/remaining useful life can be calculated form the conditional exception value as [8–10]:

$$m(t_0) = \mathrm{E}(T - t_0|T > t_0), \tag{4}$$

where, $m(t_0)$ is the mean remaining lifetime at age t_0, T is the life time and t_0 is the initial survival time. With considering the current working age (t_0) and the state of covariates ($z_i(t)$), the RUL can be calculated as:

$$m(t,i) = \mathrm{E}(T - t_0|T > t_0, z_i(t_0)) \tag{5}$$

3 Remaining Useful Life Estimation of Rolling Stock

Rolling stock and infrastructure are two main sub-systems of each railway. Signal and power supply and rail tracks are the main parts of the infrastructure, while all vehicles that move on a rail track (e.g. coaches, wagons and locomotives) refer to the rolling stock. Rolling stock can be considered as the most important system of railways. The primary failure data of rolling stocks of LKAB Malmtrafik Company [11] shows that wheelset is the most critical subsystem of rolling stocks due to the high number of failure and also its maintenance cost. Wheel sets, wheel tires or steel hoops, disk brake, roller bearing, axle guide, and oil pipelines are the main parts of the wheelset subsystem in the bogie frame [12].

In this section, the failure behavior of wheelset subsystems is analyzed and discussed. To achieve this goal, at the first step, with considering the effective operational covariates which might have a significant effect on the reliability behavior, the PHM method is used to calculate the reliability functions. Then, the remaining useful life of system at different initial survival time is obtained and discussed. In this study, the operational environment including temperature (Z_1), snowfall (Z_2), operator skill (Z_3), and maintenance quality (Z_4) are considered as the influencing covariates. In order to formulate the covariates, the traveled kilometers of the wheelset is considered as the TBFs.

The example of the TBF data, from 21 total number of failures, and the corresponding covariates collected from the daily operation and periodical maintenance reports are given in Table 1. In this Table, codes 1 and −1 for covariate Z_3, respectively means good and poor operator skill. Similarly, codes 1 and −1 for covariate Z_4 denote the good and poor maintenance qualities, respectively.

Table 1. Example of the TBF for the wheelset subsystem

TBF (km)	Status	Temperature (^{0}C), Z_1	Snowfall (cm), Z_2	Operator skill, Z_3	Maintenance quality, Z_4
109140	1	−4	5.1	1	1
68480	1	−5	5	1	1
72974	1	−3	4	1	1
118598	1	−4	4.2	1	1
93518	1	−8	3.8	1	−1
63986	1	−11	0	−1	1

The analysis of data was done using the SPSS Software with backward step Wald method [13]. The regression coefficient (β) was estimated and the significance of each β was tested by calculating the Wald statistics and its p-value. In this study, the p-value of 5% is considered as the upper limit to check the significance of the covariates. The results of the analysis show that only the operator skill (covariate Z_3) has a significant effect on the hazard rate and accordingly the reliability performance of the wheelset subsystem (with 95% confidence interval). The time dependency of covariate was also represented graphically by plotting the log-minus-log (LML) or hazard function versus time for different strata. The plotted curve for covariate Z_3, shown in Fig. 1, indicates that two LML curves for both qualities of operator skill are parallel. This means the covariate Z_3 is time-independent and the assumption of PHM is valid.

Regarding obtained results, the regression coefficient (β) of the covariate Z_3 is estimated 0.954 and therefore with using Eq. (1), the hazard rate of wheelset can be written as:

$$h(t,z) = h_0(t,z)\exp(-0.954Z_3) \tag{6}$$

To find the baseline hazard rate, the best-fitted distribution is obtained according to the Kolmogorov-Smirnov test. Results show that the baseline hazard rate follows Normal distribution function with the parameters $\sigma = 40166$ and $\mu = 68507$. Therefore, the reliability function is calculated using Eq. (2):

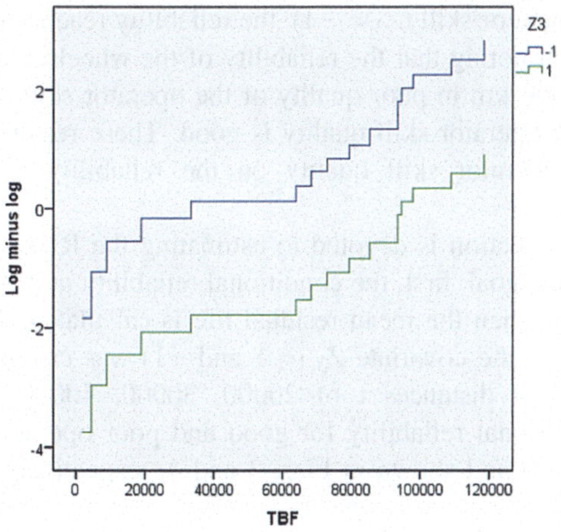

Fig. 1. The LML of the wheelset for both conditions of covariate Z_3

$$R(t, Z_3) = \left(1 - \Gamma\left(\frac{t - 68507}{40166}\right)\right)^{\exp(-0.954 Z_3)}, \tag{7}$$

where Γ is the Laplace integral (Gama function).

Reliability plots of the wheelset subsystem for two states of covariate Z_3, (= 1 and −1) is shown in Fig. 2.

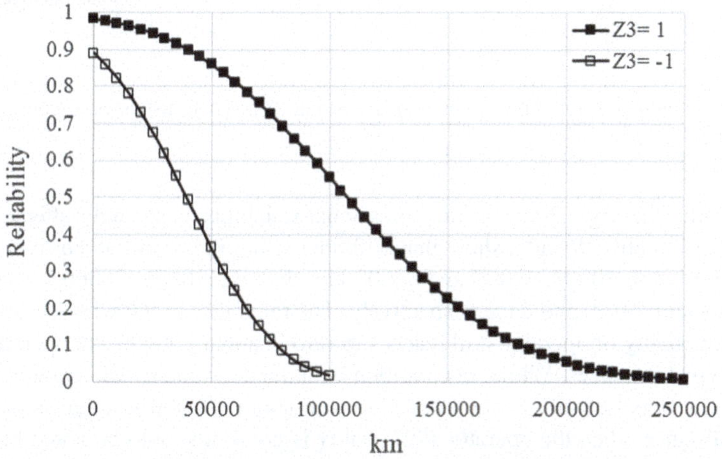

Fig. 2. The reliability plots for two states of the covariate Z_3

Regarding Fig. 2, when the operator skill is good ($Z_3 = 1$), after 50000 km the reliability of the wheelset subsystem decreases to 86.07%; While in the conditions of poor quality of the operator skill ($Z_3 = -1$), the reliability reaches to 82.22% only after 10,000 km. It is worth noting that the reliability of the wheelset subsystem reaches to zero after about 100000 km in poor quality of the operator skill while, it increases to 250000 km when the operator skill quality is good. These results indicate the significant effect of the operator skill quality on the reliability behavior of wheelset subsystem.

Remaining of this section is devoted to estimating the RUL of the wheelset subsystem. To achieve this goal, first, the conditional reliability at different initial survival distance is considered, then the mean residual life is calculated. The conditional reliability at both states of the covariate Z_3 (= 1 and −1) was calculated by considering different initial survival distances t of 20000, 30000, 40000, 50000, 60000 and 70000 km. The conditional reliability for good and poor operator skill qualities was calculated using Eq. (3) and shown in Figs. 3 and 4, respectively.

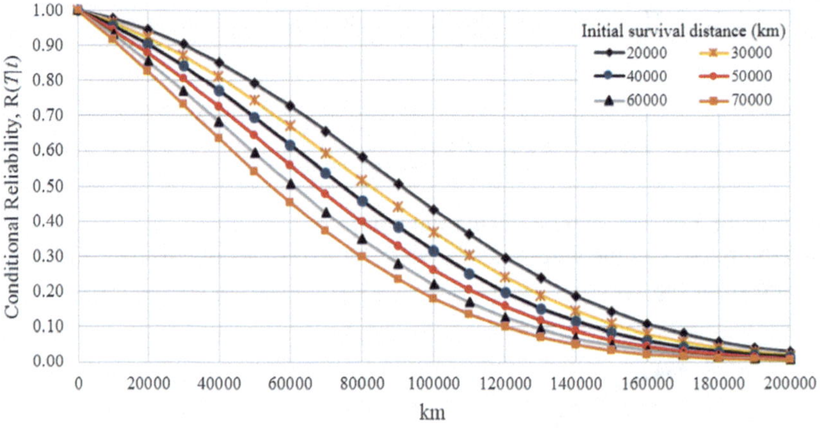

Fig. 3. The conditional reliability plots at different initial survival distances in good quality of operator skill

Regarding the Figs. 3 and 4, the conditional reliability is strongly affected by the operator skill quality. Results show that with increasing in the initial survival distance from 20000 km to 30000, 40000 and 50000 km, the reliability at 30000 km operation decreases about 40, 30 and 24%, respectively. The reliability of the wheelset subsystem in the good quality of operator skill reaches to zero in about 200000 km operation at all initial survival distances. While, in condition with the poor quality of operator skill, the reliability of wheelset subsystem reaches zero in about 100000 km that is half of the traveled distance when the operator skill quality is good. It should be noted that, when the operator skill quality is poor, the initial survival distance has more impact on the wheelset reliability than the good ones.

The remaining useful life in each state of the operator's skill quality was calculated by considering different initial survival distance (t) of 5000, 10000, 15000, 20000 and 25000 km. The results are given in Table 2.

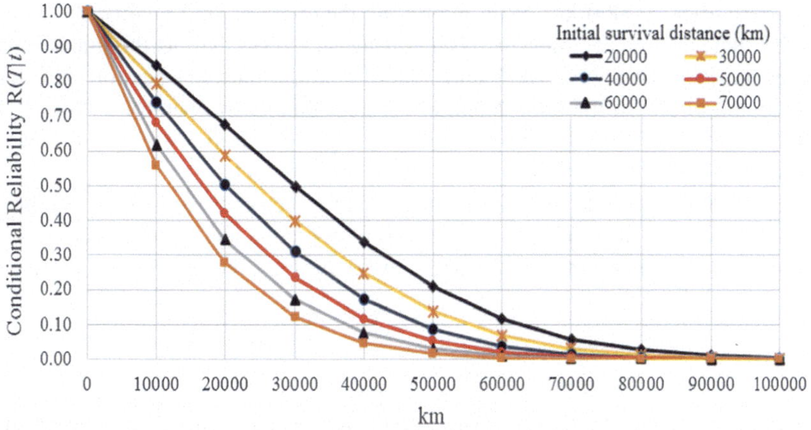

Fig. 4. The conditional reliability plots at different initial survival distances in poor quality of operator skill

Table 2. RUL of wheelset subsystem at different initial survival distances (km)

Initial survival distance (km)		5000	10000	15000	20000	25000
Quality of operator skill	Good	221006.6	211331.5	199686.96	191058.8	180481.71
	Poor	40106.14	37306.35	32376.24	30322.93	26078.5

Regarding the results, with improving the skill quality of operators, RUL of wheelset is increased by 83.53%, on average. With increasing in the initial traveled distance from 5000 to 10000, 15000 and 20000 km, remaining useful life in conditions with good quality of operator's skill is decreased by 4.58, 10.68 and 17.67%, respectively, while in poor skill quality of operators, it is decreased by 7.51, 23.87 and 32.26%. These results indicate the significant effect of operator skill quality on the RUL of wheelset subsystems.

4 Conclusions

In this paper, the remaining useful life (RUL) of rolling stock at the LKAB, a Swedish mining company, was analyzed and discussed. In the first step, it was found that the wheelset is a vital subsystem of the rolling stock because of their high failure frequency. To study the failure behavior, the operational and environmental factors which might have a significant effect on the failure behavior of rolling stocks were considered and analyzed. Then, the RUL of wheelset subsystem were obtained and discussed using the PHM method. The results of this study shows that the reliability behavior of the wheelset subsystem was strongly affected by the operator skill levels. In conditions in which the quality of operator skill was poor, the reliability of the wheelset subsystem reached zero after 100000 km. With improving the operator skill quality, the wheelset reliability increased to 250000 km. With considering 10000 km as the initial survival distance, the RUL was 211331.50 and 37306.35 km when the operator skill quality was good and poor. On the other hand, in the poor quality of operator's skill, RUL of wheelset was in the worst situation and decreasing in the initial survival distance had no considerable effect on the remaining life of rolling stock.

These results, persuade the mine manager and directors to consider a comprehensive training plan for improving the skill levels of operators and maintenance crews and also, are helpful to make the careful decision for running the project and production.

References

1. Conradie, P., Treurnicht, N.: Exploring critical failure modes in the rail environment and the consequential costs of unplanned maintenance. In: Computers and Industrial Engineering, vol. 42 (2012)
2. Fourie, C.J., Vlok, P.J., Treurnicht, N.F.: Quantifying system reliability in rail transportation in an ageing fleet environment. S. Afr. J. Ind. Eng. **26**(2), 128–142 (2015)
3. Ghodrati, B., Rahimdel, M.J., Vahed, A.T.: Fuzzy risk prioritization of the failure modes in rolling stocks. In: 2018 IEEE International Conference on Industrial Engineering and Engineering Management (IEEM), pp. 108–112. IEEE, December 2018
4. Rahimdel, M.J., Ataei, M., Khalokakaei, R., Hoseinie, S.H.: Maintenance Plan for a Fleet of Rotary Drill Rigs/Harmonogram Utrzymania I Konserwacji Floty Obrotowych Urządzeń Wiertniczych. Arch. Min. Sci. **59**(2), 441–453 (2014)
5. Cox, D.R.: A note on the graphical analysis of survival data. Biometrika **66**, 188–190 (1979)
6. Furuly, S., Barabadi, A., Barabady, J.: Reliability analysis of mining equipment considering operational environments-A case study. Int. J. Perform. Eng. **9**(3), 287–294 (2013)
7. Ghodrati, B.: Weibull and exponential renewal models in spare parts estimation: a comparison. Int. J. Perform. Eng. **2**(1), 135–147 (2006)
8. Ghodrati, B., Ahmadzadeh, F., Kumar, U.: Mean residual life estimation considering operating environment. In: International Conference on Quality, Reliability, Infocom Technology and Industrial Technology Management, Delhi (2012)
9. Finkelstein, M.: Failure Rate Modelling for Reliability and Risk. Springer, London (2008)
10. Wang, Y.F., Bao, Y.K., Zhang, H., Qiu, Q.W., Yan, Z.P., Guo, C.X.: Evaluating equipment reliability function and mean residual life based on proportional hazard model and semi-Markov process. In: 2014 International Conference on Power System Technology (POWERCON), 20 October 2014, pp. 1293–1299. IEEE (2014)
11. LKAB Malmtrafik. The Swedish railway company (2018). https://www.lkab.com/en/
12. The Railway Technical Website, adapted in (2018). http://www.railway-technical.com/trains/rolling-stock-index-l/bogies.html
13. SPSS. Statistical package for the social sciences, User's Guide, Version17 (2010). http://support.spss.com/ProductsExt/SPSS/Documentation/SPSSforWindows/index.htmlS

Determination of Transition Time from Truck-Shovel to an IPCC System Considering Economic Viewpoint by System Dynamics Modelling

Hossein Abbaspour(✉) and Carsten Drebenstedt

Institute of Mining and Special Civil Engineering, Freiberg University of Mining and Technology, 09599 Freiberg, Germany
hossein.abbaspour@student.tu-freiberg.de,
carsten.drebenstedt@mabb.tu-freiberg.de

Abstract. The conventional transportation system of mining projects is one of the significant parts of mining designs, which is considered as the most costly part of the mining operation with approximately 30%–50% of the total operating costs. Accordingly, it is worth to rethink about other transportation system alternatives that have not been properly investigated. Fixed (FIPCC), Semi-Fixed (SFIPCC), Semi-Mobile (SMIPCC) and Fully Mobile In-Pit Crushing and Conveying (FMIPCC) systems, as the crushing and transportation methods in mines, can be occasionally an alternative for Truck-Shovel system. Despite the general thinking about the high CAPEX of these systems, they impose much lower OPEX to the project. This finding is investigated in the literature but in a static way, which the time is not considered through evaluation. This study presents an economic comparison of Truck-Shovel and IPCC systems in a dynamic condition by system dynamics modeling and defining an economic index (EcI) that allows the designers to have better understanding about the whole of the project. As a result, it was shown whereas Truck-Shovel system is preferable at the first five years of the project, FMIPCC system shows a higher economic index in the rest of the mine's life.

Keywords: Truck-Shovel system · In-Pit Crushing and Conveying (IPCC) system · Economic index (EcI) · System dynamics modelling

1 Introduction

Economic viewpoint of the IPCC systems is one of the most significant concerns that persuade many researchers to focus on it. In the literature, reducing the operating costs by using IPCC systems is mentioned as one of the key features of these systems. As an example, utilizing the IPCC system in Bingham Canyon Mine showed the reduction of cost in this project [1]. Radlowski compared the all truck haulage and IPCC systems in case of capital and operating costs for an open pit mine [2]. He showed operating an IPCC system instead of all truck would result in 30% lower total costs [2]. He used simulation for building his model to evaluate the truck haulage in terms of the

E. Topal (Ed.): MPES 2019, SSGG, pp. 289–295, 2020.
https://doi.org/10.1007/978-3-030-33954-8_36

operating time and fuel consumption [2]. Some other researchers worked likewise on this issue and tried to investigate on the cost saving potential of the IPCC systems, which could be in a rage of 26%–50% [3, 4] (as cited in [2]). In another study, the use of high-speed conveyor belts, especially the side-slewable belt conveyors, with a combination of IPCC system was investigate [5]. It was shown that it could change the profitability of the project through reducing the operating costs resulting from the lower power consumption. Lieberwirth compared using conveyor belts in the IPCC systems instead of trucks for transporting ore and highlighted the economic advantages of using them [6].

As it can be seen in the literature, most of the studies merely focused on one part of the economic viewpoint of the IPCC systems e.g. operating costs, capital costs, cost saving potentials etc. However, an integrated system that can consider all these items still need to be investigated. This study presents the economic evaluation of the transportation alternatives including Truck-Shovel and IPCC systems through system dynamics modelling. In spite of a few works on the system dynamics modelling in mining [7, 8], especially in Truck-Shovel and IPCC systems [9–11], there is still the lack of a work regarding the economic evaluation of these system. As the final step of this work, the transition time from a conventional Truck-Shovel to an IPCC system will be presented by defining an economic index for transportation alternatives.

2 Economic Index Modelling in System Dynamics

In this section, the process of constructing economic index will be explained. Primarily, the most relevant, important and measurable items for defining economic index of different transportation systems must be determined. Accordingly, three items including income, total capital costs and total operating costs for each transportation system were constructed.

2.1 Income

Income is based on the production per year of each transportation system regarding its production plan. Accordingly, it could vary from one transportation system to another. It is constituted by multiplying the final product per year and ore price. In this model, the future trend of the ore price can be estimated by the linear or exponential trend.

2.2 Total Capital Costs

Total capital cost is the summation of the capital costs rate (trucks and conveyor belt) and the investment cost of IPCC procurement. In this study, it is assumed that buying the IPCC happens at the start of the project and its cost is dependent on the type of IPCC. The total capital cost of trucks is calculated as a function of purchasing and replacing trucks. It means that the capital costs would be incurred when new trucks should be added to the project or the old ones should be replaced by new ones. Similarly, this paradigm is consistent for the total capital costs of the conveyor belts.

2.3 Total Operating Costs

Total operating costs are resulted from three sub-categories of conveyor belt operating cost, trucks operating costs and IPCC relocation cost. Each of conveyor belts and trucks operating costs are calculated based on the following equations:

$$\begin{aligned}\text{Truck operating cost} &= \text{tire cost and repair} + \text{truck repair cost}\\ &\quad + \text{fuel cost} + \text{truck oil and lubricant cost} + \text{truck labor cost}\end{aligned} \tag{2}$$

$$\begin{aligned}\text{Conveyor belt operating cost} &= \text{conveyor maintenance and repair cost}\\ &\quad + \text{power(electricity)cost} + \text{conveyor oil and lubricant cost}\\ &\quad + \text{conveyor labor cost}\end{aligned} \tag{3}$$

2.4 Economic Index Equation

Since each transportation system with the higher income and the lower costs is preferable rather than others, accordingly economic index (EcI) is defined based on the division of income by the total costs:

$$\text{EcI} = \frac{\text{Income}}{\text{Total capital costs} + \text{Total operating costs}} \tag{4}$$

3 Case Study

A hypothetical copper mine with the ore reserve of 700 million tonnes was simulated during its life. Based on Taylor's method [12], the mine's life and the annual production were 32.53 years and 21.52 million tonnes respectively. The start of the ore extraction was considered 2016 (base year). The model was run for the transportation system alternatives including Truck-Shovel, FIPCC, SFIPCC, SMIPCC and FMIPCC. The economic and the technical parameters considered in the simulations are represented in Tables 1 and 2 respectively.

Table 1. Economic specifications of the hypothetical copper mine

Parameter	Quantity	Unit
Truck specification		
Truck price (base year)	550,000	$
Truck lifespan	25	years
Fuel price (base year)	1	$/L
Number of tires per truck	6	
Hourly wage of truck driver (base year)	20	$/h

(*continued*)

Table 1. (*continued*)

Parameter	Quantity	Unit
Conveyor belt specification		
Conveyor belt price (base year)	100	$/m
Conveyor belt lifespan	5	years
Conveyor set price (base year)	10,000	$
Conveyor set lifespan	15	years
Electricity price (base year)	0.07	$/kWh
Workforce per 1000 m conveyor belt	2	
Hourly wage of conveyor labor (base year)	20	$/h
IPCC specification		
Crusher station purchase price (base year)	60	M$
FIPCC purchase price (base year)	80	M$
SFIPCC purchase price (base year)	90	M$
SMIPCC purchase price (base year)	110	M$
FMIPCC purchase price (base year)	150	M$
Relocation cost in depth (base year)	1000	$/m
Rates, factors and copper price		
Inflation rate	2	%
Wage annual increment rate	1.5	%
Repair factor	0.5	
Depreciation rate	13	%
Copper price trend (since 1996)	$1513e^{0.096x}$	$/t

Table 2. Technical parameters in dynamic simulation of the model

Parameter	Quantity	Unit
Working days in a year	350	
Working hours per day	22	hours
FIPCC relocation's year	2030	
Depth of relocation of FIPCC	150	M
First year of SFIPCC relocation	2018	
Intervals of relocations of SFIPCC	4	years
Depth of relocation of SFIPCC	100	m
Last year of SFIPCC relocation	2040	
First year of SMIPCC relocation	2018	
Intervals of relocations of SMIPCC	2	years
Depth of relocation of SMIPCC	70	m
Last year of SMIPCC relocation	2040	
Average FMIPCC advancement rate in depth	40	m/year
Average faces advancement rate in depth	40	m/year

3.1 Total Capital Cost

As it was expected and shown in Fig. 1, the highest total capital costs belongs to FMIPCC system. This is mainly due to the high investment cost of purchasing FMIPCC systems. In the next ranks of the total capital costs are placed SMIPCC, SFIPCC, FIPCC and Truck-Shovel systems respectively.

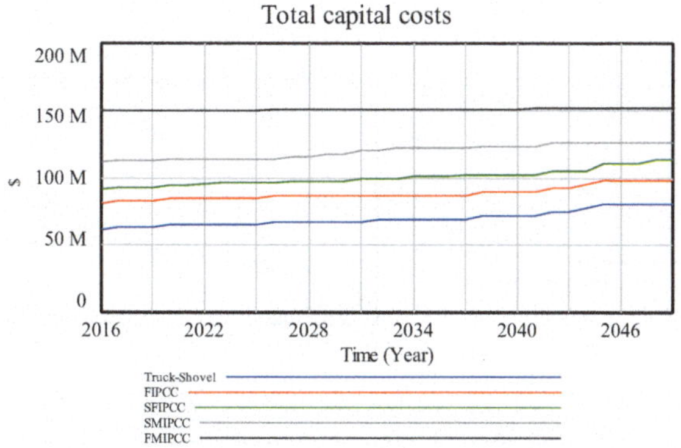

Fig. 1. Total capital costs for different transportation systems

3.2 Total Operating Costs

Total operating costs, which is the sum of the conveyor belt operating costs, truck operating costs and IPCC relocation costs, are represented in Fig. 2. As it is clearly visible, FMIPCC system shows the lowest total operating cost through the mine's life. Until 2030, Truck-Shovel and FIPCC systems present the highest operating costs. However, after 2030 and by relocating FIPCC and decreasing the number of trucks, its operating costs will be decreased.

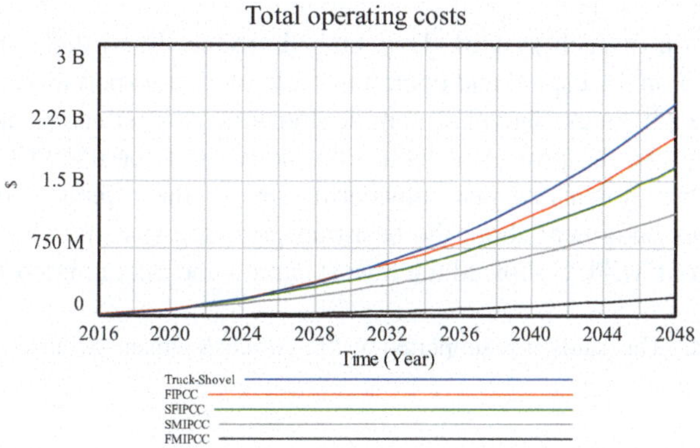

Fig. 2. Total operating costs for different transportation systems

3.3 Economic Index

Unlike the higher economic index in Truck-Shovel system from 2016 to 2021, FMIPCC represents the highest economic index from 2021 to 2048 (Fig. 3). This can be interpreted that capital cost of FMIPCC system is the highest at the start of the project; however, its operating costs is the lowest. Additionally, the rate of increment in capital and operating costs for this system is much lower than the others.

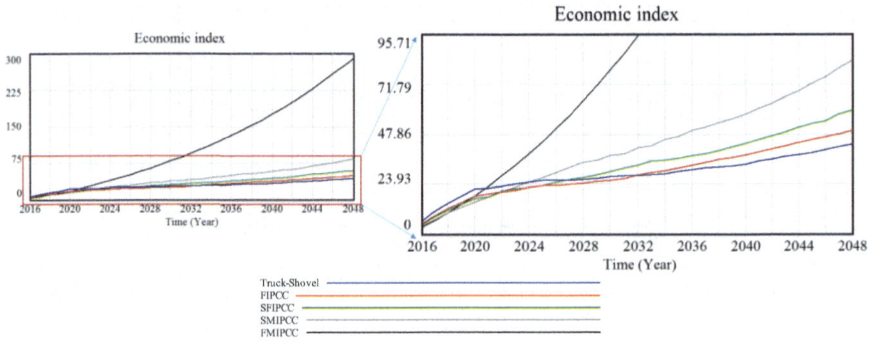

Fig. 3. Economic index for different transportation systems

4 Conclusion

Selecting the transportation system in a mining project is one of the most important processes in mine design, which can highly affect the profitability of the project. Truck-Shovel system as the conventional transportation system is mostly presume as the first option in mining. However, it imposes around half of the operating cost into the project. Accordingly, considering other transportation alternatives to resolve this issue is highly desirable. IPCC systems as a substitution for Truck-Shovel system are able to reduce costs by decreasing the operating cost in contrast with the more capital costs. Therefore, considering both capital and operating costs can get a better view of the economic situation of each system. This study introduced and defined an Economic Index (EcI) based on the capital and operating costs of all transportation systems. It was shown that although Truck-Shovel system demonstrates higher economic index within the first five years of the project, it would be the highest for FMIPCC system afterwards. It was also shown that just considering one of the capital or operating costs could lead to a misjudgment about the best transportation system, which in this study Truck-Shovel and FMIPCC showed the lowest capital and operating costs respectively.

Acknowledgment. This study was supported by “Friedrich-Naumann-Stiftung für die Freiheit”.

References

1. Kammerer, B.A.: In-pit crushing and conveying system at Bingham Canyon Mine. Int. J. Surf. Min. Reclam. Environ. **2**(2), 143–147 (1988)
2. Radlowski, J.K.: In-pit crushing and conveying as an alternative to an all truck system in open pit mines. The University of British Columbia, Cracow, Poland (1988)
3. Martin, T.W.: Large rock conveying systems and their application in open pit mines. In: Mini Sysmposium: Open Pit Haulage Systems in the 80's, SME AIME Anuual Meeting, Chicago (1981)
4. Utley, R.W.: Component analysis for movable in-pit crushers. In: AIME 44th Annual Mining Symposium, 56th Annual Meeting of the Minnesota Section, Dulluth, Minn (1983)
5. Terezopoulos, N.: Continuous haulage and in-pit crushing in surface mining. Min. Sci. Technol. **7**(3), 253–263 (1988)
6. Lieberwirth, H.: Economic advantages of belt conveying in open-pit mining. In: Mining Latin America, pp. 279–295. Springer, Dordrecht (1994)
7. Inthavongsa, I., Drebenstedt, C., Bongaerts, J., Sontamino, P.: Real options decision framework: strategic operating policies for open pit mine planning. Resour. Policy **47**, 142–153 (2016)
8. Abbaspour, H., Drebenstedt, C., Badroddin, M., Maghaminik, A.: Optimized design of drilling and blasting operations in open pit mines under technical and economic uncertainties by system dynamic modelling. Int. J. Min. Sci. Technol. **28**, 839–848 (2018)
9. Abbaspour, H., Drebenstedt, C., Dindarloo, S.R.: Evaluation of safety and social indexes in the selection of transportation system alternatives (Truck-Shovel and IPCCs) in open pit mines. Saf. Sci. **108**, 1–12 (2018)
10. Abbaspour, H., Drebenstedt, C.: Environmental comparison of different transportation systems—Truck-Shovel and IPCCs—in open-pit mines by system dynamic modeling. In: Widzyk-Capehar, E., Hekmat, A., Singhal, R. (eds.) Proceedings of the 27th International Symposium on Mine Planning and Equipment Selection - MPES 2018, pp. 287–305. Springer Nature, Switzerland (2019)
11. Abbaspour, H., Drebenstedt, C.: IPCC systems as a bulk material handling method in mines: a review regarding the technical, economic, environmental, safety and social factors. In: Sladkowski, A. (ed.) Proceedings of the 8th International Symposium of Young Researchers Transport Problems 2019, pp. 785–796. The Silesian University of Technology, Katowice, Poland (2019)
12. Hustrulid, W., Kuchta, M., Martin, R.: Open Pit Mine Planning and Design. Taylor & Francis Group, Boca Raton (2013)

Discrete-Event Simulation of a Maintenance Policy with Multiple Scenarios

Merve Olmez Turan[1] and Onur Golbasi[2(✉)]

[1] Colorado School of Mines, Golden, CO 80401, USA
molmez@mymail.mines.edu
[2] Middle East Technical University, 06800 Ankara, Turkey
golbasi@metu.edu.tr

Abstract. The current study offers a discrete-event simulation algorithm developed to reveal the effects of different maintenance work packages on the failure and the cost profiles of the mining systems. The algorithm is stochastic with two possible optimization criteria as maximizing system availability and minimizing maintenance cost for a given period. A numeric example is also provided to highlight how the effectiveness of a maintenance policy may differ with its content. In the example, 42 alternative maintenance policies were applied to an earthmover where corrective maintenance, preventive maintenance in regular inspections and opportunistic maintenance were located combinatorially for different inspection intervals. The total maintenance cost was dropped to $913,481 with an achievable production of 7,304 h for the optimal policy that included only corrective and opportunistic maintenance in the system-level, and excluded regular inspections.

Keywords: Maintenance policy · Mining machinery · Discrete-event simulation

1 Introduction

Increasing rate of production and the technology adapted to systems have evolved maintenance policy requirements in the recent decades. Corrective maintenance, which stands for run-to-failure approach and repair after component failure, had a high priority in maintenance policies a couple of decades ago. Corrective repairing or replacement of components may be performed immediately after the failure or with a delay when the defect is not major. However, global industrial competition motivated especially after the world wars has necessitated preventive actions in maintenance policies in such a way that high production rates and system health can be satisfied simultaneously. In addition, improvements in sensor technology have enabled the utilization of remote monitoring systems that can warn operators and managers about the approaching failures. These systems continuously check some indicators such as vibration, heat, pressure, tension, and revolution in the related components so that predetermined thresholds for each indicator may give a notice about the minor defects that can turn to failures. These systems are called condition-monitoring systems, and categorized under preventive maintenance. Since condition-monitoring is not always

E. Topal (Ed.): MPES 2019, SSGG, pp. 296–303, 2020.
https://doi.org/10.1007/978-3-030-33954-8_37

applicable, its financial and operational benefits need to be inquired. One another branch of preventive maintenance is predetermined maintenance where calendar-clock or engine hour counter is taken as a basis in defining which components need to be replaced or repaired with which intervals. The last type of preventive maintenance is opportunistic maintenance. In case that any mandatory or pre-scheduled maintenance activity for any component in the system creates any opportunity time, and the required resources, i.e. spare parts and crew capacity, are good enough, then, an opportunistic maintenance may be performed for the operable but deteriorated components in the meanwhile.

Maintenance work packages frequently applied for production systems can be viewed in Fig. 1. Single or multiple of these activities may be included in a maintenance policy if applicable. Although the implementation of preventive maintenance increases scheduled system halts, it decreases unplanned downtimes due to component failures and the resultant system deterioration. In this basis, properly located preventive measures in a maintenance policy may have a reduction in total maintenance cost by 8 to 12% [2].

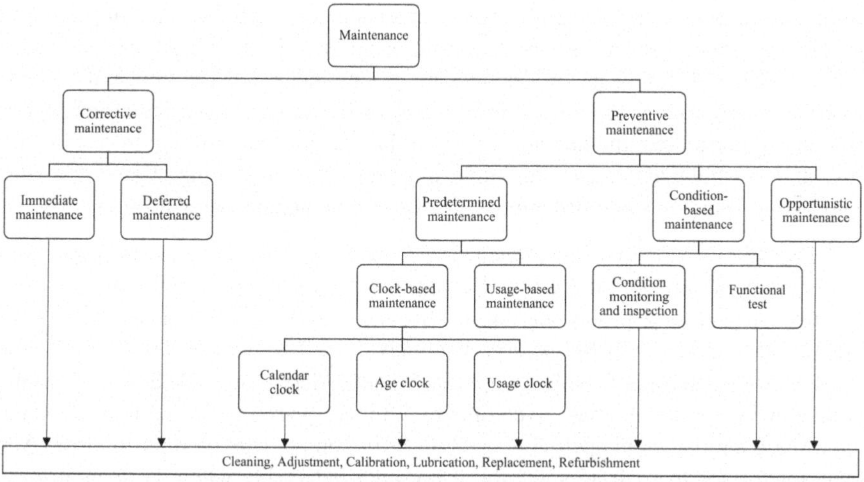

Fig. 1. Sub-branches of a maintenance policy for production systems [1]

A vast majority of the literature studies concentrates mainly on predetermined maintenance while corrective maintenance has a remarkable weight in industrial applications (Fig. 2). In this sense, future researches need to evaluate more than one maintenance type to have more applicable and comparable results with the industry. In addition, there is still a lack of enough literature about which maintenance work packages need to be selected specific to mining machineries, and what the overall benefit or loss for different policies is. The current study offers a discrete-event simulation algorithm that is capable of calculating annual availabilities and overall maintenance costs of mining machines sensitive to the scopes of maintenance policies.

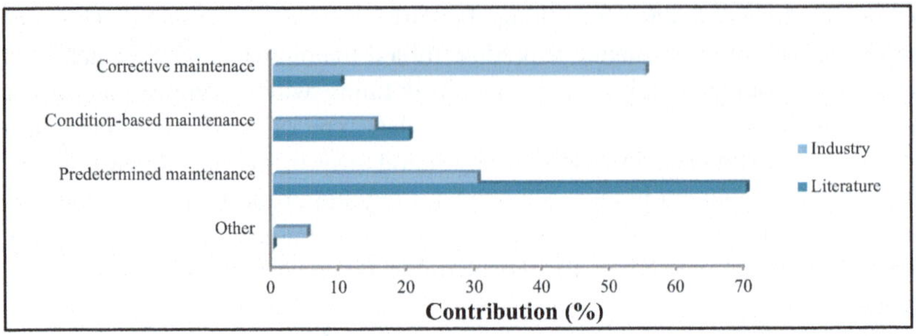

Fig. 2. Weight of maintenance activities in the US industry [2] and the literature researches [3]

2 Multi-scenario Maintenance Model

This study intends to develop a dynamic maintenance model to reveal the sensitivity of system outputs to the scope of applied maintenance policies. System outputs will be system availability, production rates, failure profiles of components or subsystems and total maintenance cost including physical maintenance cost and the value of production loss. The developed model regards stochastic behaviors of operating times and failure downtimes when deciding on the changes in system status that can either of (i) under corrective maintenance, (ii) under preventive maintenance in a regular inspection, (iii) under opportunistic maintenance, (iv) production interruption due to a scheduled break, and (v) no maintenance required and production resuming. Using the study methodology (Fig. 3), three different scenarios can be generated as follows:

- 1st Scenario - Corrective maintenance and preventive maintenance in inspections: This policy assumes that the failed components will be recovered with corrective maintenance during production shifts. Second, any deterioration detected during regular inspections will be fixed preventively. The deterioration levels of all components are determined with an approach called delay-time. Delay-time assumes that some components may give some preliminary warnings for various anomalies and these anomalies happen after a while from components' certain lifetime. After the appearance of such an anomaly, it becomes detectable when the crew makes a visual check on that component during any system halt.
- 2nd Scenario - Corrective maintenance, preventive maintenance in inspections, and opportunistic maintenance in system level: Corrective and preventive maintenance parts of this scenario are same as the first scenario. Additionally, opportunistic maintenance will be applied for the deteriorated components if corrective maintenance of any other failure gives enough time for the opportunistic maintenance of any non-failed but deteriorated component in the system. For instance, let's suppose that a component called Comp01 fails and the algorithm assigns a random corrective repair time of 1.5 h. Here, the algorithm also performs a simultaneous check for the deterioration levels of other components. If a component called Comp04 is detected to be deteriorated and its required maintenance time is less than 1.5 h, this component is also maintained opportunistically.

- 3rd Scenario - Corrective maintenance, preventive maintenance in inspections, and opportunistic maintenance in subsystem level: All the policy items work in the same way with the second scenario. Only difference here is that opportunistic maintenance is applied in sub-system level. It means that the algorithm checks the probability of performing an opportunistic maintenance for the components of subsystem where a failure happens. The second and the third scenarios assume that there is not any restriction for crew capacity and competency, and inventory level.

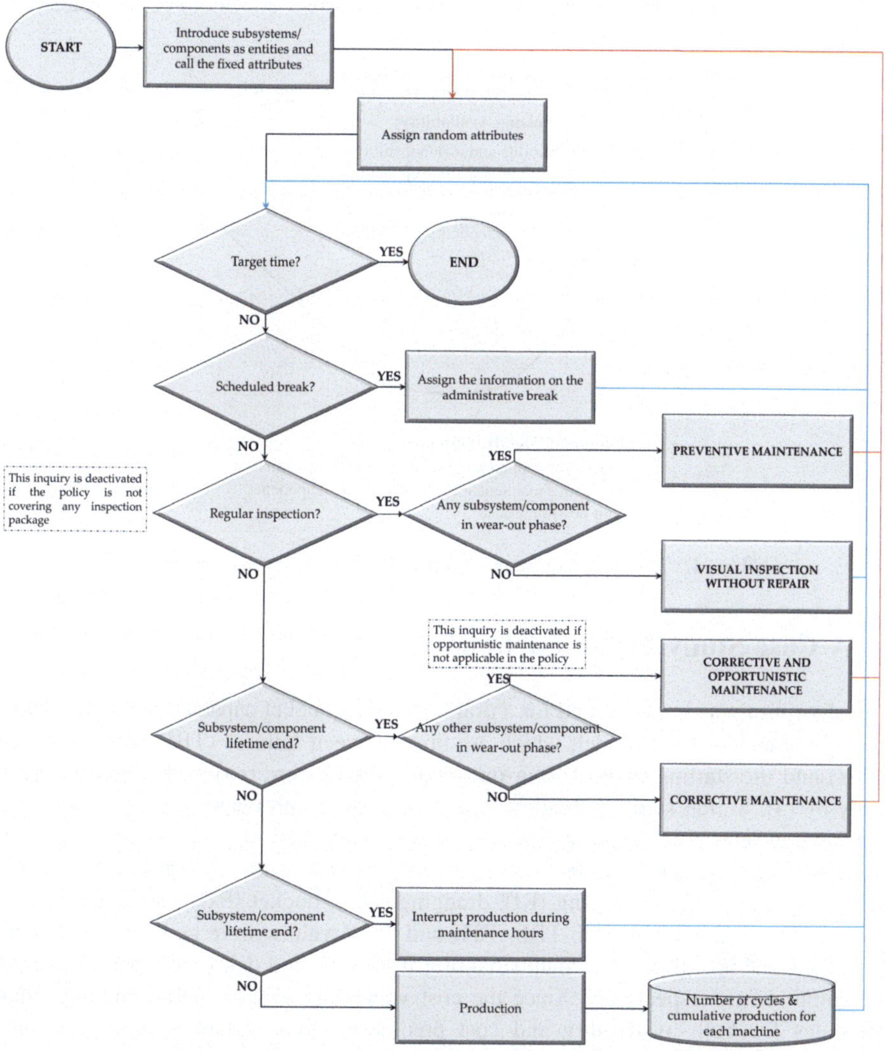

Fig. 3. The algorithm methodology

The algorithm briefly discussed in Fig. 3 is implemented in Arena® software, which is a discrete-event simulation software used in various industries to reveal the resource allocation and production flow in a cycle. In this basis, the current model requires an earlier acquisition and determination of some random and deterministic attributes. Since the model output can be either of maximizing availability or of minimizing cost, data requirement may show a change. As given in Fig. 4, an application just on system availability can be carried out if there is an available knowledge about component failure and repair times, scheduled breaks, times and durations of planned maintenance activities, and component deterioration rates. In case that cost values are also available, the algorithm can compute to see the best maintenance policy that minimizes the total maintenance cost including direct and indirect cost values.

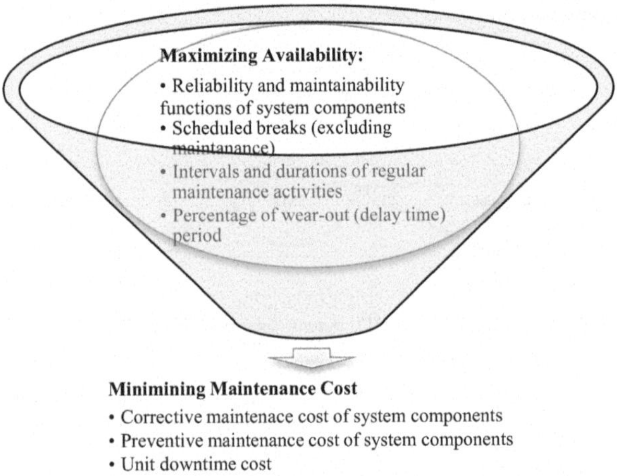

Fig. 4. The data requirement depending on the model objective

3 A Case Study for an Earthmover

The algorithm was implemented for a dragline with a bucket capacity of 30 yd^3. Three main parameters of the model, which are time between failures (TBF), time to repair (TTR) and the starting time of wear-out phase (TSW) were retrieved from a research study by [4]. In that study, a dragline was decomposed into twenty-seven components that have at least five failure records for an operating period of 11 years. The components were clustered under six main subsystems with a serial dependency: Machinery house (MH), rigging (RI), dragging (DR), bucket (BU), movement (MO), boom (BO). In the simulation, TBF, TTR and TSW values were generated randomly from the reliability functions, maintainability functions, and delay-time percentages of the components, respectively. Since the cost values are also available, the algorithm computes for both availability and cost predictions that variate depending on the maintenance policy type.

As discussed in the previous chapter, all three scenarios cover corrective maintenance and inspection that are the work packages observed in mining areas frequently. Here, the algorithm allows not just comparing three policies, also tries to find out the inspection interval that minimizes the total maintenance cost. In this section, all three policies are evaluated for an inspection interval changing between 16 h and 320 h with an increment of 24 h. Therefore, 14 different alternatives in each scenario, 42 alternatives in total, are evaluated and compared. According to the simulation results, the best alternative in each scenario is given in Table 1.

Table 1. The best alternatives for each scenario in terms of the total maintenance cost.

Scenario ID	TBI (h)	Direct cost ($)	Indirect cost ($)	Total cost ($)	Production time (h/year)
1	208	112,088	869,650	981,738	7,157
2	Very large	121,272	792,209	913,481	7,304
3	304	113,516	861,159	974,675	7,174

The results show that, in the first scenario where only corrective maintenance and regular inspections are applied, time between inspections (TBI) with a value of 208 h will drop the total maintenance cost to $981,738 where a total production time of 7,157 h per year will be achieved. In the second scenario where opportunistic maintenance in system level is applied additionally, the algorithm minimizes the total cost to $913,481 with a decision of no inspection. Here, the algorithm takes a comparatively very large value for TBI to eliminate the inspection package from the maintenance policy. Therefore, the second scenario is reduced to a policy including corrective maintenance and opportunistic maintenance alone. The third and the last scenario decides to apply an inspection interval of 304 h to minimize the total maintenance cost to $974,675. It should be remembered that the third policy is separated from the second one with its application area such that opportunistic maintenance is applied only in subsystem levels.

Among all three scenarios and 42 different alternatives, the optimal policy was chosen to have only corrective maintenance and opportunistic maintenance in system level under the second scenario. A representative example of how annual production time and total maintenance cost (includes physical costs and production losses) variate with different inspection intervals for the first scenario can be examined in Fig. 5. In Fig. 5(a), the surface is colored considering the variations in total maintenance cost where the same surface is colored with a legend regarding the changes in annual production hour for different TBI values in Fig. 5(b). The total maintenance cost minimizes and the system availability maximizes at a common point, TBI = 208, for the first scenario. For this TBI value, an annual production of 7,157 h can be obtained with a total maintenance cost of $981,738 per year as highlighted in Fig. 5(b) and (a), respectively.

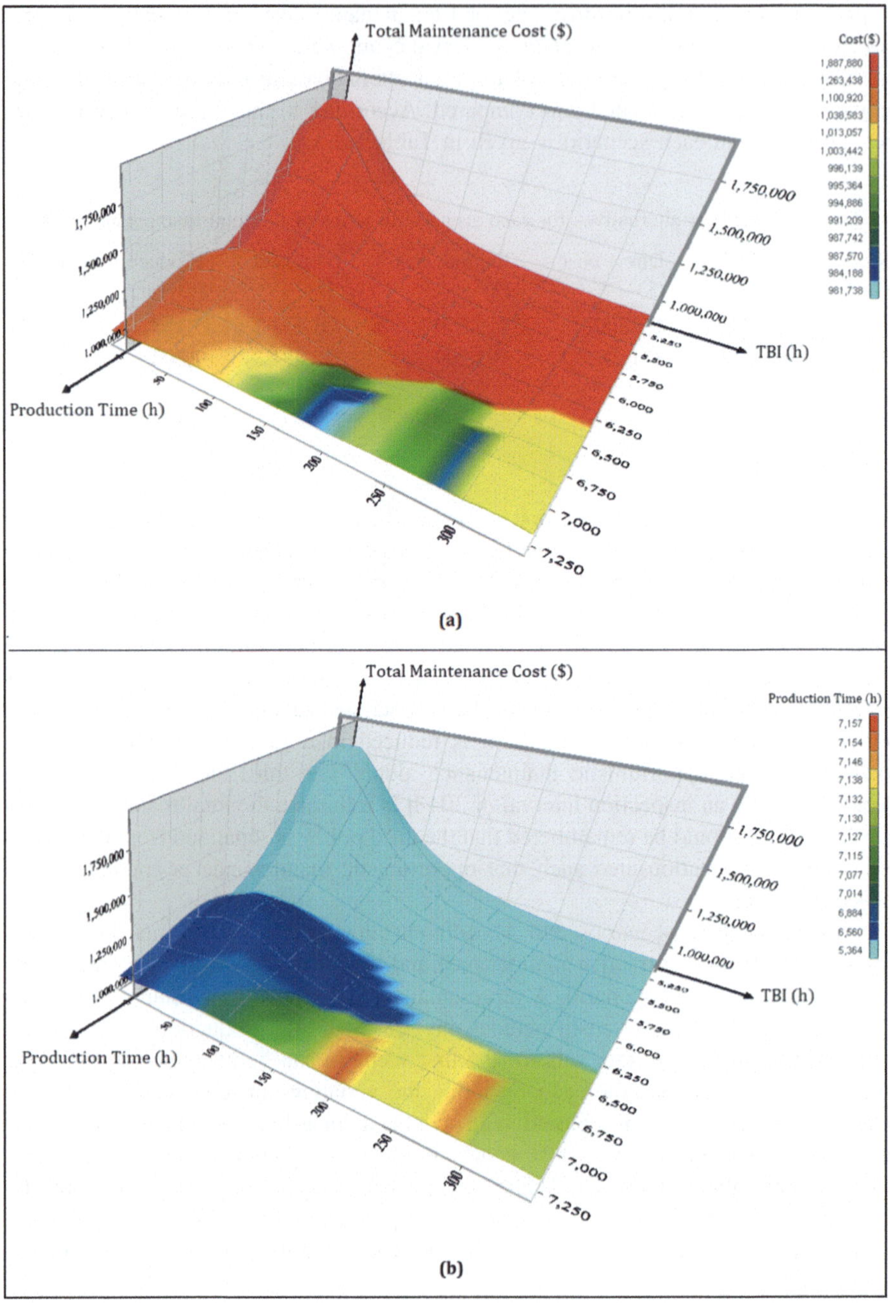

Fig. 5. Variation of annual production time and total maintenance cost with TBI values for Scenario-1: Thematic illustration considering the cost (a) and the production time (b)

4 Conclusions

This study allows a comparative evaluation of multiple scenarios in maintenance policies applied for production systems. Two model objectives, which are maximizing system availability and minimizing total maintenance cost, can be achieved with the developed simulation algorithm if the required datasets are available. Operating and downtime characterization of system components and their responses to maintenance work packages in terms of cost and availability can be observed as model outputs in system, subsystem and component levels. Corrective, preventive (regular inspections) and opportunistic maintenance applications, which are common work packages in maintenance activities, can be analyzed combinatorially with the algorithm for different inspection intervals. In the current paper, three different scenarios with 42 different alternatives were evaluated for an earthmover case study. The results revealed that application of corrective and opportunistic maintenance without inspection might minimizes the total maintenance cost and maximizes the availability in case that maintenance crew capacity and competency, and inventory level are good enough to perform opportunistic maintenance as a preventive action.

References

1. Ben-Daya, M., Kumar, U., Murthy, D.P.: Introduction to Maintenance Engineering: Modelling, Optimization and Management, 1st edn. Wiley, Chichester (2016)
2. Sullivan, G.P., Pugh, R., Melendez, A.P., Hunt, W.D.: A Guide to Achieving Operational Efficiency. https://www.energy.gov/sites/prod/files/2013/10/f3/omguide_complete.pdf. Accessed 29 May 2019
3. Alrabghi, A., Tiwari, A.: State of the art in simulation-based optimization for maintenance systems. Comput. Ind. Eng. **82**, 167–182 (2015)
4. Golbasi, O., Demirel, N.: A cost-effective simulation algorithm for inspection interval optimization: An application to mining equipment. Comput. Ind. Eng. **113**, 525–540 (2017)

Distribution of the Main Operational Costs Due to the Size of the Loading and Haulage Fleet: Brazilian Reality

Augusto Ribeiro Lages[1], Viviane da Silva Borges Barbosa[1], Pedro Henrique Alves Campos[1(✉)], Rodrigo Correia Barbosa[2], Gilberto Rodrigues da Silva[1], Pedro Benedito Casagrande[1], and Luciano Fernandes Magalhães[1]

[1] Federal University of Minas Gerais, 6 627 Antônio Carlos Avenue, Belo Horizonte, Brazil
pedrocampos@demin.ufmg.br

[2] ABM Technology, 144 Prof. Miguel de Souza Street, Belo Horizonte, Brazil

Abstract. Throughout the history of mining, the increase in the size of trucks and excavators allowed the mining of materials located in areas farther away from the dumping points at a reduced cost. Therefore, in pre-feasibility studies, it is possible to carry out several simulations, considering a production target to be reached and the use of trucks with capacities from 30 to 500 tonnes for the operation phase of a mine. In this study, a production target was set and for each of the fleet size considered, the main operational costs of loading and haulage materials were estimated, such as machine operators cost, diesel consumption and tire replacement. The results showed that the increase in the volumetric capacity of excavation or haulage fleet does not necessarily imply reduction of operational costs. Moreover, it can be concluded that labour costs are more significant for small fleets (up to 100 tonnes) than for large ones (more than 100 tonnes) in Brazil, showing that the automation of these small fleets may be an economical alternative in the future.

Keywords: Operational costs · Trucks · Excavators

1 Introduction

Most mining companies commercialize commodities, products with strict standards of physical-chemical homogeneity, whose prices are controlled by external agents, such as stock exchanges. Therefore, methods to add value to final products are very limited, as they require large investments in processing plants or metallurgy. Reducing operational costs is the most immediate way to increase profitability in mines. In this sense, loading and transport operations are strong candidates for minimizing operational costs, since they consist of 60% of mining costs [1, 2]. The greatness in the mining industry is that the use of large equipment results in lower cost per tonne transported.

For many decades, technologies of loading and transportation have been directed at increasing the size of the machines. Before the end of the 19th century, wheelbarrows, pickaxes, small rail-drawn or animal-drawn vehicles were the main means of excavation

E. Topal (Ed.): MPES 2019, SSGG, pp. 304–311, 2020.
https://doi.org/10.1007/978-3-030-33954-8_38

and transportation in mining. With the invention of steam powered engines, the first large excavators, built by Bucyrus-Erie Shovel Company (1882) and Marion Steam Shovel Company (1884) began to appear. These types of equipment were in high demand for many large projects, including road construction, railroads and also stood out in the construction of the canal of Panama (1908) and in the iron and cooper mining of the United States. At that time, the capacity of buckets varied between 2 and 6 m^3 [3].

Around the 1940s, the development of diesel and electric excavators with bucket capacities of up to 25 m^3 stand out, enhanced by the progress of the coal industry in the United States. In the late 1960s, the company Marion produced the largest excavator in history, the Marion 6360, with a capacity of 140 m^3 [3].

The trucks tried to replicate the magnitude of the excavators, but they were limited initially by tire technology. Over decades, this technology has advanced, enabling the payload capacity to be extended to more modern standards. Around the 1920s, trucks were built to transport between 1.5 and 7 tonnes. Currently, the world's largest truck, the Belorussian Belaz 75710, has a load capacity of approximately 500 short tonnes or 450 metric tonnes. A larger number of models for excavators can also be considered, since they have sizes proportional to the modules they are loading [4, 5].

However, does a larger proportion of equipment necessarily mean that the operation cost is lower? When considering geometries or track conditions, it is easy to realize that the above sentence is untrue. The choice of truck-excavator set which results in a lower cost of loading and transportation, local costs of labor, fuel and tires are preponderant in the decision process.

The objective of this article is to propose an applicable methodology for economic pre-feasibility projects to determine the proportions that result in the lowest operational costs. This may demonstrate that, when local economic factors are considered, the larger proportion is not always the most economical.

2 Methodology

Fundamentally, the bias of the economic analysis to be discussed in this article is comparative. The development of the study consisted of the accomplishment of a simplified dimensioning of trucks and hydraulic excavators of different capacities. In addition, an economic analysis at the level of pre-feasibility studies, aiming to raise relevant operational costs, practiced in Brazil, and to compare them to different sized fleets. The goal is to verify if the largest fleet of equipment represents the lowest operational costs of loading and transportation.

Therefore, the maximum excavation and transport capacity of the excavators and trucks was considered, to determine the minimum theoretical amount of equipment that would result in the desired annual movement. For this, global variables such as operational efficiency, average transport speed, average transport distance and hours effectively worked had to be established to make the sizing possible [6]. The Caterpillar Performance Manual [7] was also embraced to determine the different proportions of trucks and excavators, as well as other operational parameters.

The assumptions adopted in this article were based on the typical iron ore production of large companies in Brazil, especially those operating in the Iron Quadrangle region in the State of Minas Gerais, and are presented in Table 1. The companies considered as large are those that produce more than 1 million tonnes of ore per year [8].

Table 1. Global assumptions adopted for fleet sizing.

Parameters	Numbers
Annual movement (waste + ore)	42 millions of t
Operational efficiency (Physical Availability × Use)	0.87 × 0.89 = 0.77
Average transport speed (*Vm*)	22.6 km/h
Average transport distance (*DMT*)	1.55 km
Working hours per year	6,745 h

Assuming the parameters of Table 1 for fleet sizing and compatibility between the excavation modules and transportation, the minimum number of trucks and excavators required to meet the hourly moving target was determined (6,227 t/h).

2.1 Considered Trucks and Excavators

The loading and transport vehicles used for the sizing study were trucks with payloads ranging from 40 to 400 t and hydraulic excavators with bucket volume between 7 and 52 m^3.

The truck cycle time (Tc) is the amount of time required to complete 1 cycle. It considers the time that the module spends to transport the material from the extraction zone, the time it takes to return empty to the place of loading and fixed times (Tf) spent in queues, maneuvers, positioning, loading, and tipping. The Tf considered for all fleets was 12 min. The Eq. 1 was adopted for the average cycle time of the trucks.

$$Tc = \frac{DMT}{Vm} \times 2 + Tf \tag{1}$$

For the excavator, the cycle time consists of the sum of loading times, loaded hopper rotation, tipping time and hopper rotated time. The total cycle time of the excavator, also called the pass time, will depend on the size of the machine, the operating conditions and the operator's ability. For this work, it was considered the pass time for hydraulic excavators in moderate condition of excavation, as shown in Fig. 1.

The definition of possible combinations between the loading and transport vehicles was determined by the number of passes required to fill the dumpster of a truck. It was considered that an excavator would be compatible with a truck model if it could fill it with 3 to 5 passes, a quantity that also reflects the physical compatibility between the machines.

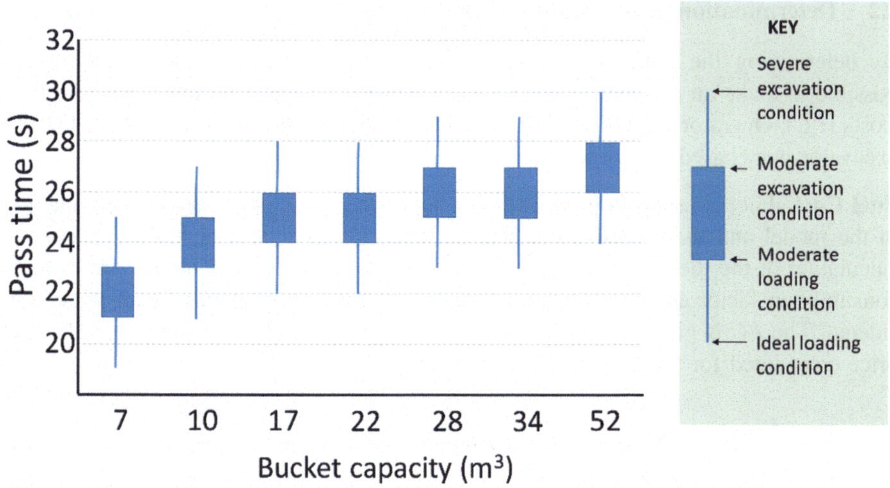

Fig. 1. Table for estimating the cycle time of hydraulic excavators [7].

Table 2 shows the number of passes required to fill the dumpster of the 15 pairs of equipment considered compatible for the actual work. The trucks are identified by their maximum carrying capacities and the excavators by a letter followed by the volume of the bucket. The last lines of Table 2 show the number of trucks and excavators required to meet the productivity of 6,227 t/h.

Table 2. Number of passes between excavators (column) and trucks (row) compatible, and the quantity of equipment to meet production.

Truck (t) Shovel (m^3)	40	50	70	100	150	200	250	345	400
A-7	4	4							
B-10		3	4						
C-17				4	5				
D-22					4	5			
E-28					3	4			
F-34						3	5		
G-52							3	4	5
Numbers of trucks	53	42	30	21	14	11	9	7	6
Numbers of shovels	A: 6	A: 5 B: 4	B: 4	C: 3	C: 3 D: 2 E: 2	D: 2 E: 2 F: 2	F: 2 G: 1	G: 1	G: 1

It is important to note that, as it was done for the trucks, the theoretical minimum number of excavators required to meet the production goal was calculated, i.e. to reach a productivity of 6 227 t/h. Due to the fact that excavators are generally the bottleneck of mine production, a design factor of 1.3 was considered for the quantity of this equipment over the calculated number.

2.2 Determination of the Main Operational Costs

By determining the number and sets of compatible trucks and excavators, it was possible to make an economic analysis involving the 3 largest mining costs: the Fuel Costs (*FC*), Operator Costs (*OC*) and Tire Costs (*TC*), according to the selected truck-excavator pair, based on economic values in Brazil.

Fuel Cost. Fuel consumption, in litres per hour (l/h), truck or excavator, is a function of the model and the consumption classification (low, medium or high) [7]. For the calculation of the diesel consumption of each truck and excavator model, the average consumption factor and the average of the values presented for this factor were considered. The *FC* is given by Eq. 2 for the fleet of trucks and excavators, in R$/h. The price considered for the litre of diesel was R$ 3.55.

$$FC = N^{\underline{o}}\ vehicles \times consumption \times diesel\ price\ \left(\frac{R\$}{l}\right) \tag{2}$$

Cost of Operators. In the determination of *OC*, for the salary (*S*) of an operator a monthly amount of R$ 3,000.00 was considered and a factor (f) on the salary of 2.5 to account for common indirect expenses in Brazil, such as food, insurance life, funeral plan, health plan, vacations, among others. It was also considered that, on average, each operator works 22 days (*D*) a month and, on each day of 8 h worked, 6.16 h (*h*) are effective for the operation. The determination of the cost of fleet operators per hour worked can then be estimated by Eq. 3. The unit of measurement of *OC* is R$/h.

$$OC = N^{\underline{o}}\ operators \times \frac{S \times f}{D \times h} \tag{3}$$

Cost of Tires. The *TC* of the trucks is another extremely important factor to be considered in transport costs in open pit mines. It was considered that each tire would have 3 life expectancies, with the first life expectancy being 3,000 h, the second of 2,000 h and the third of 1,745 h. In order to determine the cost of repair (*RC*) of the tires after each life expectancy, it was considered as the value of 30% on the cost of swap (*SC*) for new tires.

As a result of the calculation of the number of hours worked per year for each truck, it was estimated that the tires of each vehicle undergoes 3 annual repairs and are changed once at the end of the year. The unit of measure of *TC* is R $/h given by Eq. 4.

$$TC = N^{\underline{o}}\ Trucks \times \frac{RC + SC}{h} \tag{4}$$

3 Results and Discussion

From the results obtained, it was possible to create a series of graphs that compare the 3 operational costs (labor, fuel and tires) of the 15 pairs of excavator-trucks. The total cost per tonne moved (R$/t) is shown in Fig. 2. It demonstrates that the lowest cost is R $ 1.20/t (fleet of 200 t trucks with 22 m^3 excavators).

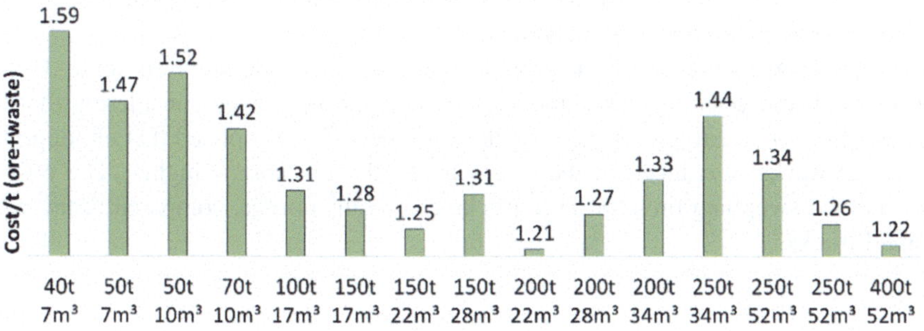

Fig. 2. Cost (R$/t) in function of the fleet.

Figure 2 also shows that there is a tendency to decrease the cost of the tonne moved with the increase in the proportion, however, the costs increase again when considering the application of larger excavators, such as 28 m^3. The fuel consumption of these big machines is mainly responsible for the cost increase: excavators of this magnitude consume more than 250 l/h.

In Brazil, the diesel cost is the most significant operational cost. In this economic analysis, it varied between 59 and 74% of the total costs, with a trend of growth from the lowest to the highest equipment, as can be seen in Fig. 3. In all cases, the consumption of the fleet of trucks is higher than the fleet of excavators. Figure 4 shows, in

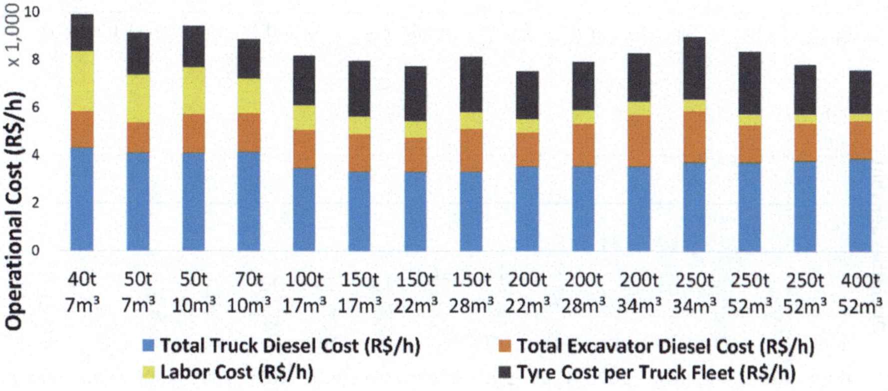

Fig. 3. Composition of total diesel costs per truck, total diesel by excavator, manpower and tires, according to the fleet.

detail, the consumption partition of the smallest fleet and the largest fleet in question: the percentage increase in fuel costs is noticed (from 60 to 72% of the total), while the labor costs decrease dramatically (from 25 to 4% of the total). In fact, the salaries of Brazilian operators are not influenced by the size of the machines, and it is reasonable to assume a constant value independent of the proportion.

The diesel costs are always significant, while the cost of labor is more considerable for small fleets. Figure 5 shows the trend of reducing labor costs with the growth of the fleet. While for the smaller fleets an operational team of almost 60 people is needed, in the larger fleets this number is theoretically less than 10.

The tire costs increase with the growth in the fleet size, and represent up to 31% of total costs. These costs increased moderately with the size, even considering that the number of tires decreases with the growth of the fleet. It was noticed that although the number of trucks is reduced in the larger fleets, the tire cost is higher. The cost of repairing and swapping tires for large trucks can be the double, compared to the costs of small trucks.

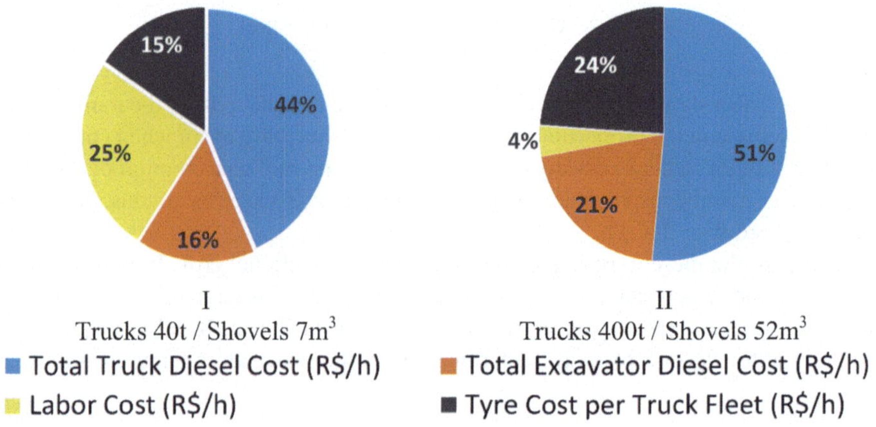

Fig. 4. Detail of the cost partition for the smallest fleet (I) and for the largest fleet (II).

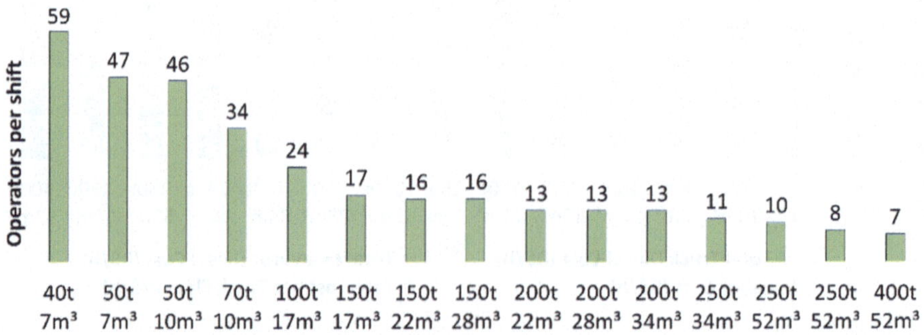

Fig. 5. Number of operators per shift according to equipment size.

4 Conclusion

There is a tendency of reducing costs, as the size increases; however, it is necessary to consider the compatibility between truck and excavator and, mainly, the estimated fuel consumptions that always represented more than 50% of the total cost. In the analysis performed by this article, the most economic model is the one represented by the 200 t truck with the 22 m^3 bucket excavator.

The graphs show that the diesel consumption of trucks represents, in general, the highest percentage of the total operating cost of the loading and transport operation, ranging from 41% to 51%. For the smaller truck fleets, the operator cost represented a large share of the total costs. For the combination of smaller sizes, this cost represented 25% of the total operating cost, while for the larger fleet; this cost represented only 4%.

An interesting point is observed when analyzing the tire cost among the studied fleets. It was noticed that although the number of trucks is reduced in the larger fleets, the tire cost is higher. This shows that despite the smaller number of tires to be restored and replaced, the cost of repairing and replacing large truck tires may be double compared to the cost of small trucks.

Smaller fleets, such as those with 40, 50 and 70 tonne trucks, had higher operational costs than larger fleets. However, it was found that these had an interesting economic range between 150 and 200 t, demonstrating lower operating costs than larger equipment fleets.

In order to apply the conclusions in this article and complete the feasibility studies, the authors suggest that the investment costs, equipment life expectancy, economic life of the equipment, resale value, capacity and deadlines to deliver the machines, spare parts availability and specialized manpower for the maintenance and operation of the equipment, as well as the technical characteristics of the mine may limit the applicability of any of these combinations.

References

1. Caterpillar: A reference guide to mining machine applications. Publisher, Location (2009)
2. Correa, M.M., et al.: The loss of productivity and costs associated with the premature scrapping of mining trucks tires. In: 24th Mining Congress, Brazil (2016)
3. Orleman, E.: Power Shovels: The World's Mightiest Mining and Construction Excavators. Motorbooks International, St. Paul (2003)
4. TUCSON Homepage (Mine Tales: Huge haul trucks changed the face of the mining industry). https://tucson.com/news/local/mine-tales-huge-haultrucks-changed-the-face-of-the/article_b7294736-5e4c-5d65-b095-cbe422c834ab.html. Accessed 02 June 2019
5. TUCSON Homepage, (Mine Tales: Transportation was a big issue in mining industry). https://tucson.com/news/local/mine-tales-transportation-wasa-big-issue-in-mining-industry/article_b2b2b4a4-7425-53fe-bb17-4cd94241592d.html. Accessed 02 June 2019
6. Lages, A. R.: Pre feasibility study of fleet size on operational costs of an open pit mine. Mining Engineering Department of Federal University of Minas Gerais, Brazil (2018)
7. Caterpillar Performance Handbook 47. Peoria, Illinois, USA, January 2017
8. Revista Minérios E Minerales Homepage (The 600 mines in Brazil 2018). http://revistaminerios.com.br/edicao-especial-200-minas-2018/. Accessed 05 June 2019

Mining Economics

Analysis of Capital Allocation by Mining Companies

Jacob van der Bijl[1(✉)] and Tinashe Tholana[2(✉)]

[1] Growth and Business Improvement, Large Diversified Mining Company, Johannesburg, South Africa
jacquesvdbijl@gmail.com

[2] School of Mining Engineering, University of the Witwatersrand, Johannesburg, South Africa
tinashe.tholana@wits.ac.za

Abstract. Correct timing of capital allocation in relation to commodity price is one of the key drivers of long term success for capital intensive companies such as mining companies. Nonetheless, historically mining companies have in general been poor at capital allocation decision making. This is mainly because they allocate most of their capital during the top half of a commodity price cycle which has often proved to be the incorrect time of the commodity pricing cycle to invest in growth projects. This in turn has led to poor returns on capital invested, impairments, underperformance in the share price of mining companies relative to other industries and ultimately destruction of value. Better capital allocation strategies are required for mining companies to unlock more value from invested capital. This paper identifies factors in the capital allocation decision-making process that can unlock value in mineral projects. Five international mining companies were analysed from 2001 to 2017, a period that covers a full commodity price cycle to determine the correlation between commodity price and capital allocation by mining companies. It was found that during periods of higher commodity prices, the analysed mining companies primarily focussed on volume growth. In a declining commodity price environment they focussed on rationalisation of mining projects and operations, and disposal of assets that do not meet minimum investment criteria. Mining companies should consistently invest capital throughout the commodity price cycle, and be cautious of over allocating capital towards growth projects during periods of high commodity prices.

Keywords: Capital allocation · Return on Capital Employed · Return on Invested Capital · Shareholder value · Mineral commodity price cycle

1 Introduction

Mineral commodity price is one of the key drivers of mineral companies' value [1] but, mining companies are price takers which means that they have no influence over the commodity prices they receive from the sale of their mineral products. Because of that reason [2] emphasized that mining companies should have differential value creation strategies beyond commodity price changes. In addition to commodity price amongst

E. Topal (Ed.): MPES 2019, SSGG, pp. 315–325, 2020.
https://doi.org/10.1007/978-3-030-33954-8_39

other value drivers, it has been theorized that capital allocation decisions is one of the key drivers of long term success for capital intensive companies such as mining companies [3]. Nonetheless, historically mining companies in general have been poor at making capital allocation decisions [4]. This has significantly contributed to poor returns on investment, and better capital allocation by a company can unlock value relative to other mining companies [4]. The consequences of poor capital allocation decisions include:

- Poor financial returns on capital invested;
- Significant capital write-downs;
- Underperformance in the share price of mining companies relative to companies in other industries; and
- Destruction of value.

It is therefore the aim of this paper to identify factors that mining companies should consider during the capital allocation decision-making process that can unlock value in mineral projects. The paper utilized a framework provided by [3] to determine the financial success of historical capital allocation of a few selected large diversified mining companies from 2001 to 2017. The period from 2001 to 2017 was selected as it is representative of one full commodity price cycle. Therefore, this period provides an opportunity to review mining companies' capital allocation decisions throughout the cycle and determines if there is a correlation between the commodity price cycle and the capital allocation decisions that were made during that period. Four large diversified mining companies and one single commodity focused growth mining company were selected for analysis in this study. The selection was based mainly on the following criteria:

- The companies' historical financial statements must have been available on the public domain for the full period of analysis;
- Companies need to have a diversified commodity production portfolio to be considered representative of the mining industry; and
- Companies must have announced substantial capital write-downs during the period under review. This provided the opportunity to evaluate the major factors that contributed to the write-downs.

Based on these criteria the four large diversified companies that were selected for analysis are; Anglo American, BHP, Rio Tinto, and Vale S.A. (Vale). In addition First Quantum was selected as an outliner to represent a growth company that has allocated capital towards growth projects throughout the commodity price cycle.

2 Value and Value Creation in Mining Through Capital Allocation

Economic value is defined as the monetary worth of something [5]. In mineral project valuation, a common measure of economic value is the net present value (NPV), which is the sum of the future cash flows in real terms discounted by a project specific risk adjusted discount rate [6]. There are many ways in which mining companies can create value for stakeholders. One of the ways is though efficient capital allocation, in which the return on capital employed is higher than the cost of capital plus an appropriate risk premium to compensate investors for the risk taken in investing in the project or operation [7]. There are a number of alternative options for the allocating capital that are listed by [3] and these are discussed in the next sub-sections.

2.1 Capital Allocation to Mergers and Acquisitions

The largest percentage of the overall capital allocated by mining companies is utilised for mergers and acquisitions (M&A) [3]. M&A activities usually increase when commodity prices are relatively high. This is reflective of the 2004 to 2007 commodity price boom when the value of M&A transactions increased by 227% between 2004 and 2007 [8]. M&A activities typically increase in value and frequency during times of higher share price valuations [3]. M&A typically happen in waves within an industry [9]. Early movers in the merger and acquisition wave tend to conclude deals when commodity prices are in the bottom half of the cycle, whilst late movers conclude deals at the top of the cycle. There are potentially significant advantages gained by being an early mover in the merger and acquisition wave [9].

2.2 Capital Allocation to Expand and Sustain Operations

Expansionary capital and sustaining capital attracts the second highest amount of capital allocation [3]. Mining companies are by nature capital intensive, with a significant amount of capital required per annum to sustain production output.

2.3 Capital Allocation Through Divestment

Value can also be created by disposing company assets if the assets are deemed to be more valuable in the hands of the buyer. Divestment or sale of assets are frequently used by mining companies to balance their portfolio of mines [10]. Historically mining companies have used divestment as a tool to diversify their commodity production and thereby reduce risk [10]. However, the most recent divestments concluded by mining companies between 2013 and 2016 have been due to a requirement to rebalance their balance sheets and reduce debt [10]. Mining companies could potentially unlock more value through adopting a more long-term strategic view of the preferred list of commodities produced, and use divestment as one of the tools to achieve the strategic objective [10]. A recent example of the disinvestment strategy is Anglo American who over the period from 2013 to 2016 divested 26 operations, which reduced the number of mines from 68 to 42 [11].

2.4 Capital Allocation to Shareholders

Dividends are another form of capital allocation where portions of profits are paid to shareholders. Dividends and share buy backs are the two main capital allocation methods that companies use to return cash to shareholders. [3].

3 Measures of Value Realised from Capital Invested

The best accounting metric to measure capital efficiency is Return on Invested Capital (ROIC) [12] which is a measure of how efficiently a company employs its overall capital. ROIC is calculated using Eq. 1:

$$\mathrm{ROIC} = \frac{NOPAT}{Sales} \times \frac{Sales}{Invested\ Capital} \tag{1}$$

where NOPAT is Net Operating Profit after Tax which is calculated using Eq. 2 [12]:

$$\mathrm{NOPAT} = \mathrm{EBITDA} - \mathrm{Cash\ taxes} \tag{2}$$

where EBITDA is Earnings Before Interest, Tax, Depreciation and Amortisation. Invested capital is calculated using Eq. 3 [12]:

$$\begin{aligned} \text{Invested capital} = {} & \text{Current assets} \\ & - \text{non-interest-bearing current liabilities} \\ & + \text{Net property, plant and equipment} \\ & + \text{Goodwill} \\ & + \text{Other operating assets} \end{aligned} \tag{3}$$

In addition to ROIC [13] mentioned that Return on Capital Employed (ROCE) is also used to calculate return on capital invested or employed and is calculated using Eq. 4:

$$\mathrm{ROCE} = \frac{Earnings\ before\ interest\ (EBIT)}{Capital\ employed} \tag{4}$$

Where capital employed is calculated using Eq. 5 [13]:

$$Capital\ employed = Total\ assets - current\ liabilites \tag{5}$$

Equations 1 and 4 show that ROIC calculates return on capital employed after tax, whilst ROCE calculation is before tax. In this paper both ROIC and ROCE were used to measure the success of capital allocation decisions. ROIC was used because of the argument provided by [12] that ROIC is the best accounting metric to measure capital efficiency. ROCE was used because by excluding the impact of tax on the returns on capital achieved, it provides a more direct comparison on the operational performance between mining companies across different tax jurisdictions.

4 Methodology

The framework provided by Mauboussin and Callahan (2014) was used as a guideline to determine the capital allocation success of the five selected mining companies from 2001 to 2017. The framework follows the following five steps:

1. Reviewing of capital allocation;
2. Reviewing of revenue, basket commodity price and production volume growth;
3. Reviewing of capital commitments;
4. Calculating the capital invested as a function of the prevailing commodity prices;
5. Calculating the return on capital employed and return on invested capital.

5 Results from Analysis of Capital Allocation Decisions

5.1 Review of Capital Allocation for the Five Companies

Figure 1 shows the results of the review of capital allocation for the five companies. In Fig. 1 it can be seen that all five mining companies allocated significant portions of capital to the purchase of assets, represented by the "Buy" red lines in the Figure,

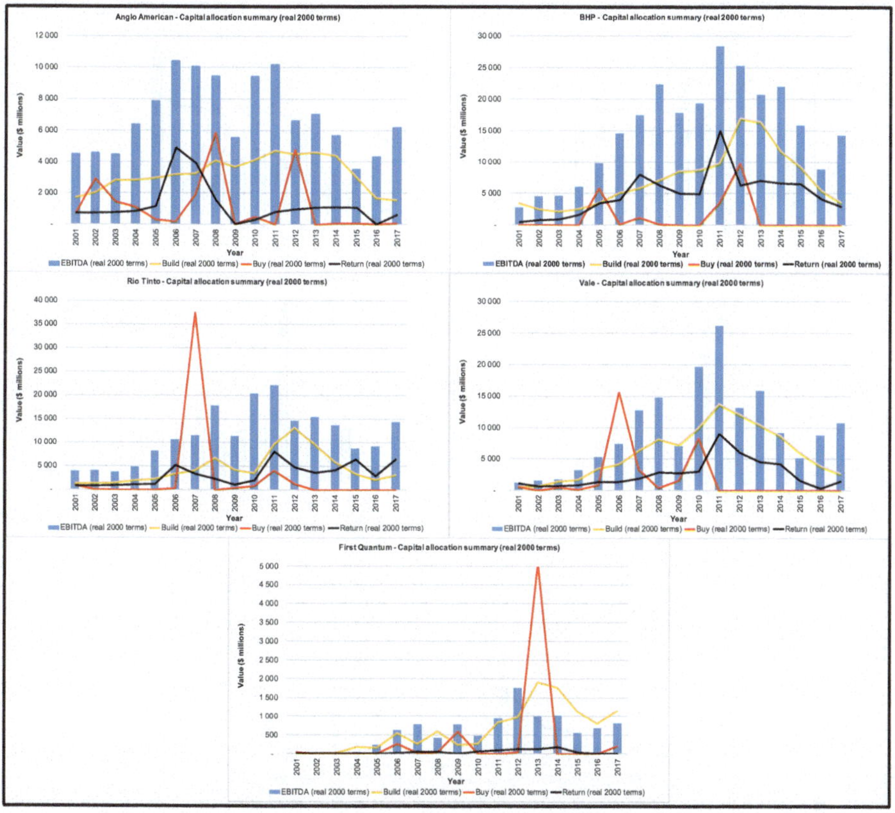

Fig. 1. Capital allocation of five selected mining companies [11, 14–17]

during the two periods of high commodity prices in 2006 to 2008 and 2010 to 2012. The two stand-out acquisitions indicated in Fig. 1 are Rio Tinto's US$38.7 billion acquisition of Alcan Inc. in 2007 and First Quantum's US$5.1 billion acquisition of Inmet Mining Corporation in 2013. During the subsequent period of lower commodity prices from 2013 to 2017 very little to no capital was allocated to the purchase of new assets. All five mining companies allocated the highest portion of capital towards building of new mining operations represented by the yellow line in Fig. 1. The second highest allocation of capital was to returning money to shareholders through either dividends or share buy-backs.

5.2 Review of Production Volume Growth for the Five Companies

Figure 2 shows the production volume growth achieved by the five mining companies over the period from 2001 to 2017. It can be seen that First Quantum achieved the highest production volume growth percentage, followed by Vale, BHP, Rio Tinto and Anglo American in descending order. First Quantum achieved the highest growth by allocating a larger percentage of capital towards investment activities. Similarly, Anglo American achieved the lowest production volume growth by allocating the lowest percentage of capital towards investment activities as percentage of operating costs.

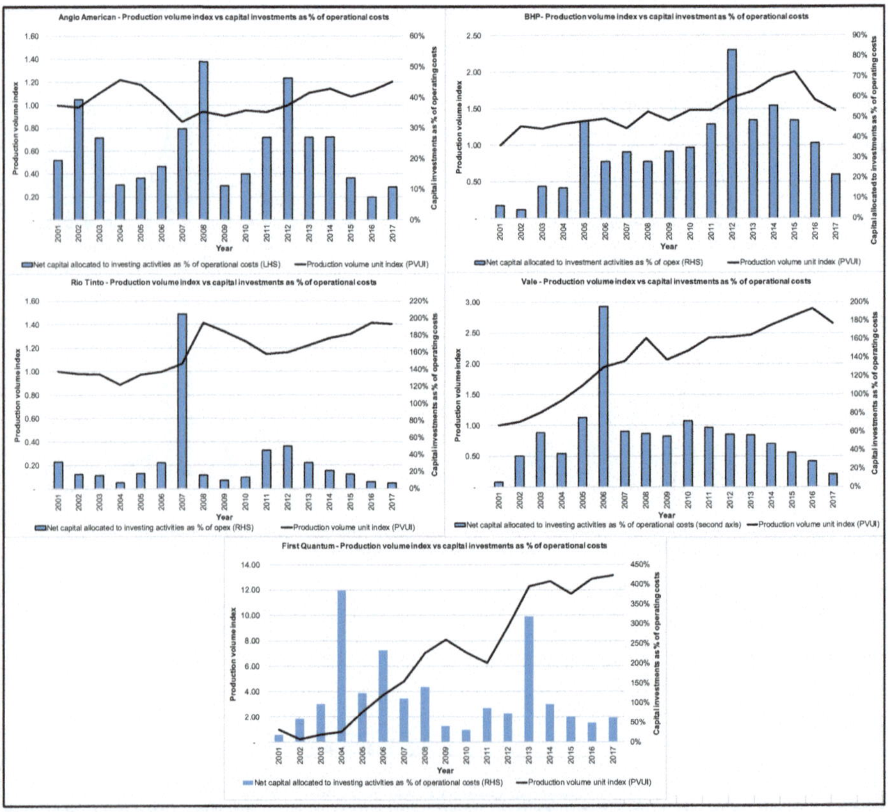

Fig. 2. Production volume growth index [11, 14–17]

5.3 Results from Review of Capital Allocation as Function of Commodity Price

Capital commitments towards stay-in-business (SIB) capital and expansionary capital in real 2000 terms, relative to the basket commodity price index for the five companies are indicated in Fig. 3. The basket commodity price index for each company was determined by analysing the sales contributions per commodity produced by each company then weighting each commodity's contribution to sales with the average real terms commodity price over the cycle. Figure 3 indicates a very strong correlation between capital commitments and the prevailing basket commodity price index in real 2000 money terms for four of the five mining companies. The Pearson correlation coefficient between capital commitments towards SIB capital and expansionary capital in real 2000 terms, and the prevailing basket commodity price index are 0.76, 0.82, 0.86, 0.94, and 0.49 for Anglo American, BHP, Vale, Rio Tinto and First Quantum respectively. With the exception of First Quantum there was a high to very high correlation. Anglo American, BHP, Vale and Rio Tinto allocated more capital towards SIB and expansionary capital during periods of high commodity prices. In contrast, First Quantum typically allocated capital towards growth initiatives throughout the commodity price cycle.

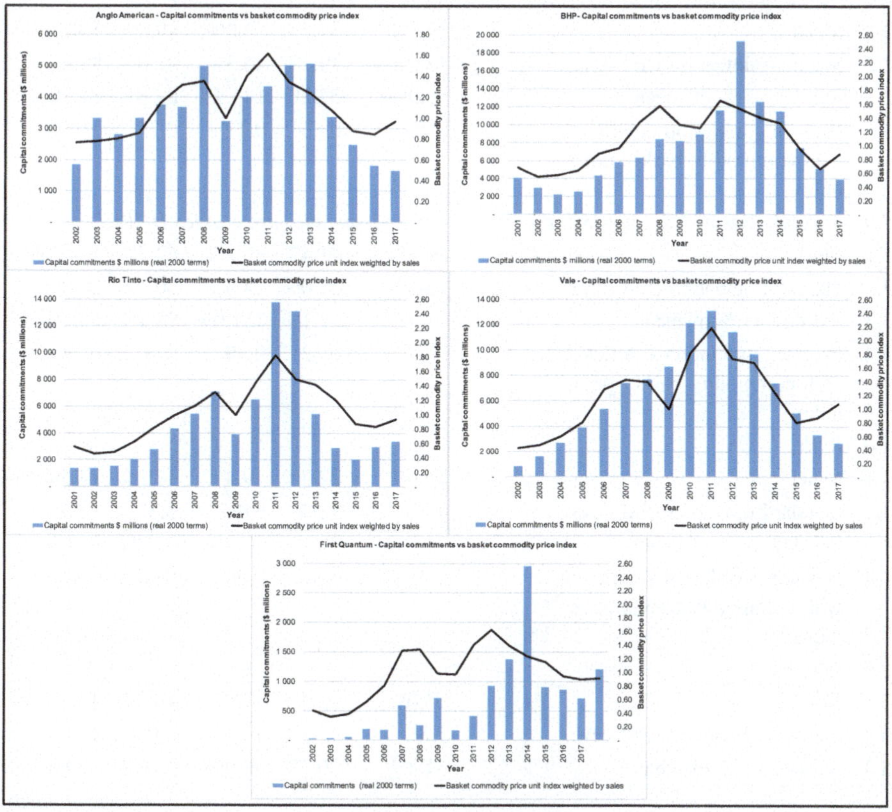

Fig. 3. Production volume growth index [11, 14–17]

5.4 Summary of ROCE and ROIC Results

Table 1 provides a summary of the results obtained from the capital allocation analysis performed on the five mining companies from 2001 to 2017. This summary provides the capital allocated to capital expenditure and investment activities as a percentage of total operational costs, and the ROCE and ROIC achieved. The graphical results of the analysis could not be presented because of space limitations.

Table 1. Summary of results obtained from the capital allocation study for the period 2001 to 2017 [11, 14–17]

No.	Description	Unit	Anglo American	BHP	Rio Tinto	Vale	First Quantum
1.0	*Capital expenditure as % of operational costs*						
1.1	Mean	%	21.3%	34.2%	23.3%	48.9%	82.1%
1.2	Standard deviation (Std dev)	%	6.4%	14.6%	9.0%	14.8%	80.0%
1.3	Std dev as % of mean	%	30.0%	42.7%	38.8%	30.3%	97.5%
1.4	Pearson correlation coefficient with commodity basket price index		0.78	0.58	0.54	0.64	−0.17
2.0	*Net capital allocated to investment activities as % of operational costs*						
2.1	Mean	%	23.2%	34.5%	32.0%	49.0%	107.7%
2.2	Standard deviation (Std dev)	%	13.0%	19.7%	46.3%	41.0%	88.6%
2.3	Std dev as % of mean	%	55.9%	57.2%	144.7%	83.6%	82.3%
2.4	Pearson correlation coefficient with commodity basket price index		0.42	0.66	0.21	0.25	−0.05
3.0	*Capital allocated to investment as % of opex * commodity price index (CIPCI)*						
3.1	Mean	%	26.1%	41.4%	36.3%	64.7%	104.2%
3.2	Standard deviation (Std dev)	%	18.6%	32.0%	54.8%	60.3%	80.0%
3.3	Std dev as % of mean	%	71.3%	77.1%	151.0%	93.1%	76.8%
3.4	Pearson correlation coefficient with commodity basket price index		0.69	0.82	0.39	0.60	0.44
4.0	*ROIC*						
4.1	Mean	%	15.0%	22.2%	19.4%	17.2%	13.5%
4.2	Standard deviation (Std dev)	%	4.8%	7.3%	5.4%	5.5%	12.5%
4.3	Std dev as % of mean	%	31.7%	33.1%	27.9%	31.7%	92.0%
4.4	Pearson correlation coefficient with commodity basket price index		0.42	0.62	0.06	0.09	0.42
5.0	*ROCE*						
5.1	Mean	%	15.0%	22.4%	17.9%	16.1%	14.2%
5.2	Standard deviation (Std dev)	%	7.8%	11.3%	7.2%	7.0%	17.0%
5.3	Std dev as % of mean	%	51.8%	50.7%	40.0%	43.6%	120.0%

(continued)

Table 1. (*continued*)

No.	Description	Unit	Anglo American	BHP	Rio Tinto	Vale	First Quantum
5.4	Pearson correlation coefficient with commodity basket price index		0.41	0.67	0.15	0.12	0.57
6.1	Increase in production volume index	%	21%	47%	40%	165%	1217%
6.2	CAGR - Production volume index	%	1.2%	2.5%	2.1%	6.3%	17.5%
7.1	Shareholders equity at end 2017	$ millions	22 972	57 258	44 711	44 772	8 960
7.2	Increase in real terms shareholders equity	%	29%	178%	350%	597%	14290%
7.3	CAGR - shareholders equity (real terms)	%	1.6%	6.6%	9.9%	12.9%	36.4%
8	Rank - shareholders equity growth		5	4	3	2	1

The success of the capital allocation is measured by the compounded annual growth rates (CAGR) achieved on production volume growth and shareholder equity, respectively. For each of the metrics the mean, standard deviation and Pearson correlation coefficient with the basket commodity price index are provided. In item number 1.0 on the Table only stay-in-business and expansion capital were included in the calculation. In item number 2.0 on Table 1, total capital expenditure and net cash flows allocated towards acquisitions and disposals were included in the calculation. The companies are ranked based on the CAGR achieved in shareholder equity growth.

Table 1 shows that First Quantum achieved the highest CAGR in shareholders' equity of 36.4%. However, this was achieved from a very low base in comparison with the other four mining companies. First Quantum also achieved the highest CAGR in the production volume index with 17.5%, which is substantially better than the next highest of 6.3% from Vale. This was achieved due to First Quantum allocating the highest average annual capital to investment activities equal to 107.7% of annual operational costs.

First Quantum also had the lowest Pearson correlation coefficient between net capital allocated to investments and prevailing commodity prices. This indicates that First Quantum allocated capital more consistently toward investment activities throughout the commodity price cycle. This indicates that on average more capital allocations were made at times when the prevailing basket commodity price was lower than the average basket commodity price over the period. The other four mining companies, respectively all have a higher mean CIPCI indices than the mean percentages of capital allocated to investment as percentage of operational costs. This indicates that Anglo American, BHP, Rio Tinto and Vale allocated more capital towards investments during periods when the prevailing commodity prices were higher than the average basket commodity price over the analysis period.

Anglo American achieved the lowest annual growth rate in the production unit index with a CAGR of 1.6%. This was achieved through a mean capital allocation of 23.2% of annual operational costs which was the lowest of all the five companies. From this result one can reason that circa 20% of annual operational costs are required to be allocated to stay-in-business and expansion capital in order to maintain steady production rates. This observation is supported by results from BHP and Rio Tinto where capital allocations of circa 30% of annual operational costs toward growth investments realised a marginally higher production volume index CAGR of approximately 2%.

Notably, whilst First Quantum achieved the highest shareholder equity CAGR, it also achieved the lowest mean ROCE and ROIC. This could potentially be due to the quality of the existing large long life operational assets already in operation in Anglo American, BHP, Rio Tinto and Vale. In order to determine a better understanding of the return on incremental capital invested one would have to extract the earnings contribution from the existing assets and only focus on the incremental operational returns from the assets in which the capital investments were made.

6 Conclusion

This paper found that during periods of higher commodity prices, the analysed mining companies primarily focussed on volume growth. For all five companies there is a very strong correlation between the total value of capital approvals and the average basket commodity price of the minerals produced. In a declining commodity price environment the mining companies focussed on rationalisation of mining projects and operations, and disposal of assets that do not meet minimum investment criteria. During these periods there were limited capital allocation towards growth projects, and most capital allocation decisions were directed towards reduction of net debt. The paper recommends that mining companies should have clear capital allocation frameworks which are guided by the respective companies' strategic objectives. Also, mining companies should clearly identify the minimum investment criteria to be met before capital is allocated to an investment. Lastly mining companies should aim to consistently invest capital throughout the commodity price cycle, and be cautious of over allocating capital towards growth projects during periods of high commodity prices.

References

1. MacDiarmid, J., Tholana, T., Musingwini, C.: Analysis of key value drivers for major mining companies for the period 2006–2015. Resourc. Policy **56**, 16–30 (2018)
2. Boston Consulting Group: Value Creating in Mining-More than Commodity Prices: The 2010 Value Creators Report (2011)
3. Mauboussin, M.J., Callahan, D.: Capital Allocation, Evidence, Analytical Methods, and Assessment Guidance. INTERNET (2014). www.valuewalk.com/wp-content/uploads/-2014/08/document-1036635381.pdf. Accessed 19 Nov 2016

4. Duval, D.: Management Issues, Poor Allocation of Capital, Confound Investors in Global Mining Sector. INTERNET (2014). http://www.kitco.com/ind/DDuval/2014-04-01-Management-Issues-Poor-Allocation-of-Capital-Confound-Investors-in-Global-Mining-Sector.html/. Accessed 19 Nov 2016
5. Merriam-Webster (2019). https://www.merriam-webster.com/dictionary/value
6. O'Regan, B., Moles, R.: Investment decisions of international mining firms: policy approaches. Simulation **78**(6), 363–379 (2002)
7. Pidun, U., Stange, S.: The art of capital allocation. INTERNET (2017). http://image-src.bcg.com/Images/BCG-The-Art-Capital-Allocation-Mar-2017_tcm9-149411.pdf. Accessed 7 July 2018
8. PwC: Mine 2007, Riding the wave. INTERNET (2007). https://www.pwc.co.za/en/assets/pdf/pwc-mining-review-07.pdf. Accessed 23 June 2018
9. McNamara, G., Haleblian, J., Dykes, B.J.: The performance implications of participating in an acquisition wave: early mover advantages, bandwagon effects, and the moderating influence of industry characteristics and acquirer tactics. Acad. Manag. J. **51**(1), 1–18 (2008)
10. Ernst & Young: Divestments: extracting hidden value. INTERNET (2016). https://www.ey.com/Publication/vwLUAssets/EY-divestments-extracting-hidden-value/$FILE/EY-divestments-extracting-hidden-value.pdf. Accessed 11 Oct 2018
11. Anglo American: Annual Report 2001 to 2017. INTERNET (2001–2017). https://www.angloamerican.com/investors/annual-reporting. Accessed 21 July 2018
12. Mauboussin, M.J., Callahan, D.: Calculating Return on Invested Capital. INTERNET (2014). https://doc.research-and-analytics.csfb.com. Accessed 19 Nov 2016
13. Faulkenberry, K.: Return on capital – calculations and ratios. INTERNET (n.d.). http://www.arborinvestmentplanner.com/return-on-capital-ratios-calculations/. Accessed 29 July 2018
14. BHP: Annual Report 2001 to 2017. INTERNET (2001–2017). https://www.bhp.com/media-and-insights/reports-and-presentations?q0_r=category%3dAnnual%2bReports. Accessed 21 July 2018
15. Rio Tinto: Annual Report 2001 to 2017. INTERNET (2001–2017). https://www.riotinto.com/investors/results-and-reports-2146.aspx. Accessed 23 July 2018
16. Vale: Annual Report 2001 to 2017. INTERNET (2001–2017). http://www.vale.com/EN/investors/information-market/annual-reports/Pages/default.aspx. Accessed 23 July 2018
17. First Quantum: Annual Report 2001 to 2017. INTERNET (2001–2017). https://www.first-quantum.com/Investors-Centre/Financial-Reports/default.aspx. Accessed 22 July 2018

Economic and Environmental Impacts of Utilising a Pre-concentration Process in Underground Metal Mining

Farzad Sotoudeh(✉), Micah Nehring, Mehmet Kizil, and Peter Knights

School of Mechanical and Mining Engineering, The University of Queensland, St Lucia, QLD 4072, Australia
{f.sotoudeh,m.nehring,m.kizil,p.knights}@uq.edu.au

Abstract. The mining industry has increasingly implemented new technologies to continue to achieve greater efficiencies and cost reduction. The integration of mining and pre-concentration technologies is considered to be one of the most logical approaches to improve the technical and economic efficiencies in mining. The subsequent flow-on effects have also generated positive environmental and social implications to further enhance the overall sustainability of the operation. There exist many efforts regarding the implementation of pre-concentration processes into open pit mines together with the respective economic impacts of this particularly in the reduction of grinding and processing costs. On the other hand, few attempts have been reported regarding the implementation of pre-concentration processes into underground operations. The benefits of the integration of a pre-concentration process into underground mining are not only reflected by significant decreases in surface processing and flotation costs, but also via the more effective management of waste through its handling and subsequent transportation costs, which is the main focus of this paper. This paper reviews the main impacts of utilising a pre-concentration process in underground mining and also presents a conceptual model for the investigation of waste management in sublevel stoping method (SLS) for backfilling purposes across various scenarios. The results are investigated and further developments are recommended.

Keywords: Underground mining · Pre-concentration · Sustainability · Backfilling · Sublevel stoping · Waste management

1 Introduction

In recent years, mining industry has faced with a considerable reduction in head grades for Copper, Gold, Zinc and other commodities. [1]. According to Walters [2], the feed grade of copper in Chilean mines has been decreased by 25 to 50% from 1991 to 2012 or another example is declination of head grade by 20% in Platinum grades [3]. The main consequence of declining head grades are an exponentially increase in energy usage and as a result overall production costs. This is mainly due to increased grinding time required for segregation of metal entire the extracted rocks [4, 5]. Therefore,

E. Topal (Ed.): MPES 2019, SSGG, pp. 326–332, 2020.
https://doi.org/10.1007/978-3-030-33954-8_40

In order to reduce the additional costs of grinding and metal energy intensity due to presence of more gangue, incorporation of new technologies is required to overcome this issue. One of the most productive and efficient approaches is pre-concentration of ore materials in the pit. Pre-concentration can be defined as a process of rejecting the uneconomic material prior to main processing step [6, 7]. Benefits of pre-concentration have been pointed out in a number of references [8–10]. The most important advantages of pre-concentration are:

- increased resource utilization,
- reduced of mining costs,
- improvement of grade control,
- decreased comminution energy,
- reduced environmental footprints and
- increased feed grade [11].

This Paper focuses on application of pre-concentration processes in underground mine operations. Hence, a conceptual model consist of 9 stopes is chosen to demonstrate implications and advantages of this integration approach.

2 Underground Pre-concentration (UPC)

Conventionally, all processing plants always have been located on surface regardless of the mining method employed in a surface or underground mine. With increasing of underground mines' depth, waste handling poses a challenging issue while the handling for the surface becomes more difficult and suffers a high transportation cost [10, 12]. In order to overcome this issue, (waste disposal) utilization of waste for filling of mined voids is an appropriate approach during the life cycle of the mine [13, 14]. One of the most common efforts for minimizing of waste disposal is establishment of a pre-concentration plant close to extracted area. Therefore, it gives a significant opportunity for usage of waste for backfilling targets rather than sending waste to the surface [14, 15]. In Fig. 1, a schematic view of an underground operation, consist of 15 stopes and a pre-concentrator in the middle of level 2 and 3 is illustrated. As seen in this figure, Stopes 2, 4, 6, 10, 13 and 15 are under extraction operation to satisfy plant capacity. The extracted ore from the lower levels of pre-concentration plant are handled by drifts and the above stopes can enjoy the gravity these stopes come to pre-concentration plant. Finally, the pre-concentration plant divide the total extracted ore into two separate materials including waste for backfilling purposes and concentrated ore to the surface.

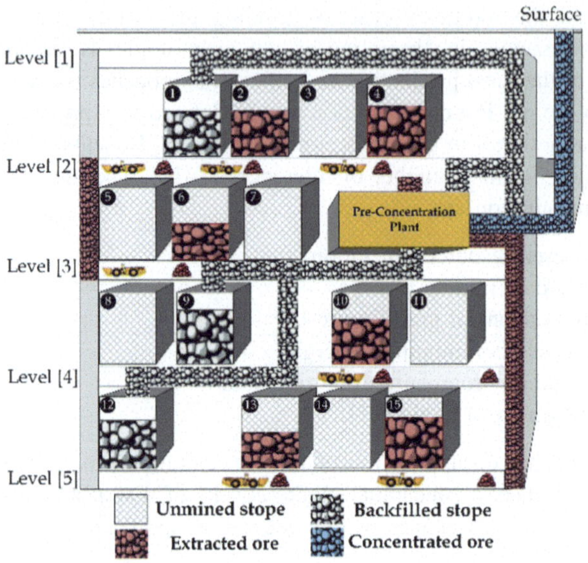

Fig. 1. Schematic view of an underground mine operation integrated with pre-concentration plant

3 Conceptual Model

With implementation of a pre-concentration process in underground mine, a specific proportion of feed, which contains low grade material, is rejected by pre-concentrator and the rest of the material is upgraded in grade for the main processing plant. In Fig. 2, an example of this metallurgical balance is illustrated. In order to demonstrate the benefits of implementing a pre-concentration process in an underground mine, a conceptual model consists of 9 stopes in a sublevel stoping operation was selected.

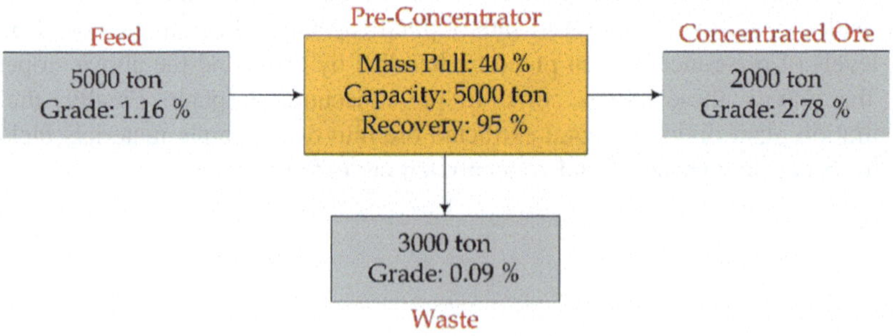

Fig. 2. Example of pre-concentration with variable plant capacity

The required data and sequence of underground stopes' extraction are based on Little, Nehring and Topal's research paper [16]. The size, shape and volume of these stopes are equal and each stope is assumed to be mined in 4 separate phases including: Preparation, Extraction, Void and Backfilling. For this example, it was assumed that 60% of feed materials is removed by pre-concentrator through 94% recovery and 40% mass pull (40% of Extracted ore will be transported to surface as concentrated ore). As a result, the average feed grade for the main processing plant is upgraded for each stope as shown in Fig. 3.

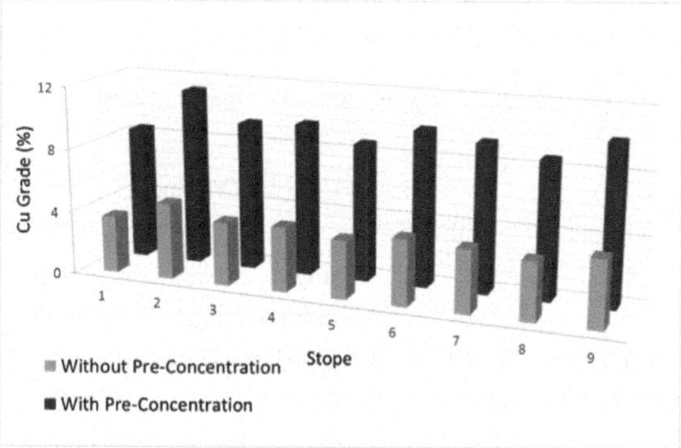

Fig. 3. Comparison of average grade for each stope with/without pre-concentration

Furthermore, new capital costs (CAPEX) and operating costs (OPEX) were calculated in presence of underground pre-concertation plant.

By incorporating a pre-concentration process, capital and operating costs will be reduced due to reduction of ore tonnage transferred to the surface and downstream facilities. Traditional cost estimation equations, shown in Eqs. 1 and 2 for capital and operating costs respectively, were used to measure the reduction of costs and then creating cash flow tables for two scenarios (with/ without pre-concentration).

$$C_2 = C_1(T_2/T_1)^{2/3} \tag{1}$$

$$C_2 = C_1(T_2/T_1) \tag{2}$$

Where C_2 is the revised cost, C_1 is the existing cost, T_2 is new throughput and T_1 is the old throughput. Since only 40% of materials will be transferred to the surface, these costs will be changed as shown in Table 1. As seen in this table, the capital savings can be observed in hauling, hoisting, milling and tailing. However, an additional cost related to the establishment of a pre-concentration plant has been accounted for, in the order of 2.6 million AUD. It should be noted that capital savings can be compensated if a low proportion of material is rejected by the pre-concentrator. On the other hand, there is only reduction of milling and tailing costs per tonne of concentrated ore.

The reason behind this is the fact that in pre-concentration step, coarse and hard waste materials are rejected and as a result fine materials with a higher grade will be transferred to milling process. Therefore, less energy is needed for processing of pre-concentrated materials [10–13].

Production scheduling of this conceptual model is determined by solving a mixed integer programming model as discussed by Little, Nehring and Topal [16]. The sequence of extraction of these stopes are scheduled in 18 time periods. Two scenarios are applied on this schedule including without pre-concentration and integration of pre-concentration and the final Net Present Value (NPV) of two scenarios are compared as illustrated in the cash flow, in Fig. 4.

Table 1. CAPEX and OPEX cost comparison for underground mine with/without pre-concentration

CAPEX (million AUD)			OPEX (AUD/tonne)		
	Without PC	With PC		Without PC	With PC
Mining	5.8	5.8	Mining	43.25	43.25
Hauling	2.3	1.24	Mucking/haulage	10.54	10.54
Hoisting	4.6	2.49	Crushing/conveying	5.46	5.46
Milling	8.2	4.43	Hoisting	2.78	2.78
Tailing	1.1	0.59	Milling	8.43	3.38
Pre-concentrator	–	2.6	Tailing	3.23	1.29
			Pre-concentration	–	7.43
			Backfilling (AUD/m^3)	24.38	11.7

4 Discussions

The production schedule without incorporation of a pre-concentration process produces a NPV of 46.38 AUD. However, the integration approach produces a NPV of 51.34 AUD. In comparison of these two scenarios, it can be observed that implementation of pre-concentration has improved the NPV by approximately 5 million AUD or 10.67%. In this case, the plant capacity was assumed as variable one. It means that pre-concentration approach can be used in an existing operation or a new operation which is still in the feasibility study stage. Figure 4 shows the cumulative discounted cash flows for these scheduled stopes with and without pre-concentration. As it is clear in this figure, the production schedule ends higher than the production schedule without pre-concentration approach in the 18th time period.

In terms of underground operating costs, a saving of up to 30% can also be achieved if a pre-concentration plant is installed. The most important cost saving can be observed in backfilling operation, which is reduced by 52%. This is due to elimination of a large amount of waste materials transferred from the surface to fill the void stopes. Figure 5 shows a comparison of total operating cost for each activity with/ without pre-concentration system. According to Norgate and Haque [4], the required energy for copper crushing and concentration is equal to 3277 Mj/tonne-ore and 140 Mj/tonne-ore respectively with a significant influence in global warming potential. The released CO_2 from the beforementioned processes is equal to 293 $KgCo_2$/tonneore and 19.2 $KgCo_2$/tonneore respectively. It is obvious that with implementation of pre-concentration technology a huge amount of CO_2 will be decreased.

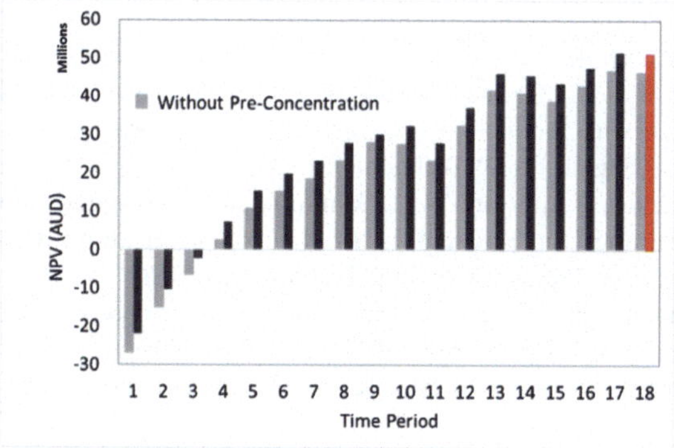

Fig. 4. Cumulative discounted cash flow comparison

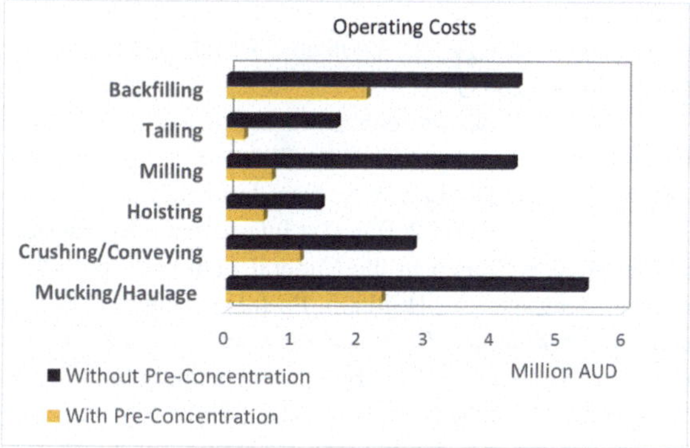

Fig. 5. Comparison of operating costs for underground mine with/without pre-concentration

5 Conclusions

In underground mining, as the operation goes deeper, cost savings become more critical due to an increase in energy, unit costs and material handling. Application of pre-concentration technology can help to overcome these kinds of difficulties by rejection of unwanted materials prior to main processing step on the surface. In this paper, a conceptual model was utilised to illustrate the influence of pre-concentration on NPV of an underground mine. It was observed that by rejection of 60% of extracted materials from stopes, NPV can be increased by 10.67% and reduction of total cost by 30%. It should be noted that, in this paper the plant capacity was assumed as a variable one. However, for existing operations, which are under the extraction, implementation of pre-concentration requires more extraction and production rate to satisfy the fixed plant capacity.

References

1. Batterham, R., Elvish, R.: Smarter mineral processing, or, what do mill operators think? In: Tenth Mill Operators Conference Proceedings, Adelaide, SA, pp. 12–14 (2009)
2. Walters, S.G.: Driving Productivity by Increasing Feed Quality Through Application of Innovative Grade Engineering® Technologies (2016)
3. Rule, C.M., Fouchee, R.J., Swart, W.C.: Run of mine ore upgrading–proof of concept plant for XRF ore sorting. In: Proceedings of the 6th International Conference on Semi-Autogenous and High Pressure Grinding Technology (2015)
4. Norgate, T., Haque, N.: Energy and greenhouse gas impacts of mining and mineral processing operations. J. Clean. Prod. **18**(3), 266–274 (2010)
5. Norgate, T., Jahanshahi, S.: Low grade ores–smelt, leach or concentrate? Miner. Eng. **23**(2), 65–73 (2010)
6. Tapia, C.E.C.: Development of geometallurgical tests to identify, rank, and predict preferential coarse size by size Au deportment to support feed preconcentration at Telfer Au-Cu mine, Newcrest Western Australia. MPill thesis, University of Queensland (2013)
7. Bowman, D.J., Bearman, R.A.: Coarse waste rejection through size based separation. Miner. Eng. **62**, 102–110 (2014)
8. Gray, A.H., Delemontex, G., Grigg, N., Yeomans, T.: InLine Pressure Jig Pre-concentration Plant at the Pirquitas Mine (2011)
9. Denysschen, D.F., Wagner, B.N.: Pre-concentration of low grade lateritic sulphide nickel ore. In: Base Metals Conference, pp. 291–306 (2009)
10. Klein, B., Hall, R., Scoble, M., Dunbar, W.S.: Simulation of integrated underground mining-processing. In: Proceedings APCOM 2003: 31st International Symposium on Application of Computers and Operations Research in the Mineral Industries, pp. 481–486 (2003)
11. Duffy, K., Valery, W., Jankovic, A., Holtham, P.: Integrating bulk ore sorting into a mining operation to maximise profitability. In: Proceedings of Metplant (2015)
12. Klein, B., Dunbar, W.S., Scoble, M.: Integrating mining and mineral processing for advanced mining systems. CIM Bull. **1057**, 63–67 (2002)
13. Bamber, A., Klein, B., Morin, M., Scoble, M.: Reducing selectivity in narrow-vein mining through the integration of underground pre-concentration. In: Proceedings of the Fourth CANMET International Symposium on Narrow Vein Mining Techniques (2004)
14. Scoble, M., Klein, B., Scott-Dunbar, W.: Mining waste: Transforming mining systems for waste management. Int. J. Surf. Min. Reclam. Environ. **17**(2), 123–135 (2003)
15. Bamber, A.S., Klein, B., Morin, M., Scoble, M.J.: Integration of pre-concentration underground: Reducing mining costs. In: Proceedings XVII International Conference on Mine Planning and Equipment Selection (2005)
16. Little, J., Nehring, M., Topal, E.: A new mixed-integer programming model for mine production scheduling optimisation in sublevel stope mining. In: Proceedings-Australian Mining Technology Conference, Twin Waters. The Australasian Institute of Mining and Metallurgy (2008)

Financial Performance of Johannesburg Securities Exchange Traded Gold Mining Companies: Du Pont and Economic Value Added Analyses

Sihesenkosi Nhleko(✉) and Paskalia Neingo

School of Mining Engineering, University of the Witwatersrand, Johannesburg, South Africa
Sihesenkosi.Nhleko@wits.ac.za

Abstract. Analysis of financial performance of mining companies offers management the opportunity to scrutinize company performance against its strategic goals. Furthermore, it assists investors in investment decision-making. Gold mining plays an important role in the South African economy through taxation, foreign investment income, employment and skill development. Hence, the financial performance of gold mining companies for the past 10 years was analysed based on their gold assets only. The analysis was conducted by using the commonly applied tools for analysis of financial performance viz Economic Value Add (EVA) and Du Pont Analysis. Data were sourced from individual company annual reports. Based on these measures, Sibanye-Stillwater was the better performing gold mining company, while AngloGold Ashanti came second in all fields of measurement except for profit, which averaged the highest over the period under study. Harmony Gold was the worst performing company of all the three. Company performance fluctuates on year-on-year basis, hence the importance of conducting financial analysis for asset optimisation, return maximisation and investment opportunities. While mining companies report a number of performance measures in silos, it is important that comparison per sector is done for investment purposes. This is supported by the fact companies compete for scarce capital investment.

Keywords: Gold · Mining · Finance · EVA · Du Pont · Investment · Evaluation

1 Introduction

Gold is important for various reasons but gold provides a good hedge against inflation [1]. South Africa has the largest known gold resource in the world, which is hosted in the Witwatersrand Basin [2]. Figure 1 shows the location of some of the gold mining companies operating in South Africa. However, South African gold mining contributes about 4.2% of the global gold production [2]. The low production account can be alluded to challenges presented by mining at great depth, low-grade deposits and reduced investment in the sector due to high sovereign risk.

E. Topal (Ed.): MPES 2019, SSGG, pp. 333–345, 2020.
https://doi.org/10.1007/978-3-030-33954-8_41

Sovereign wealth funds, pension funds and other investors use crediting rating to assess the credit worthiness of a country for investment opportunities. Therefore, it has an impact on the cost of borrowing for a country. Table 1 shows the crediting rating of South Africa over the period under study. Fitch as well as Standard & Poor's agencies downgraded South Africa to a junk status in 2017 while Moody's ratings were last reported at lower medium investment grade in 2018. Based on the credit rating agencies, South Africa is a high-risk investment destination. Trading Economics assign a score between 100 (riskless) and 0 (likely to default) for credit worthiness of countries. This score is forward looking taking into account factors such as economic indicators, and financial markets. South Africa scores a 50 making it a risky investment destination [3].

Table 1. Credit worthiness of South Africa (Adapted from 3)

Agency		2009	2010	2011	2012	2013	2014	2015	2016	2017	2018
Fitch	Rating	–	–	BBB+	BBB+	BBB	BBB	BBB−	BBB−	BB+	–
	Outlook	–	–	Stable	Negative	Stable	Negative	Stable	Negative	Stable	–
Moody's	Rating	A3	–	A3	Baa1	–	Baa2	–	Baa2	Baa3	Baa3
	Outlook	Stable	–	Negative	Negative	–	Negative	–	Negative	Under review	Stable
Standard & Poor's	Rating	–	–	BBB+	BBB+	–	BBB−	BBB−	–	BB	–
	Outlook	–	–	Stable	Negative	–	Stable	Negative	–	Stable	–
Credit rating scales											
Fitch, Standard & Poor's	BBB+ BBB BBB−	Lower medium grade									
	BB+ BB	Non-investment grade speculative ("Junk")									
Moody's	A3	Upper medium grade									
	Baa1 Baa2 Baa3	Lower medium grade									

South Africa introduced the Mining Charter in 2004 which was reviewed twice during the period of study in 2010 and 2018. The Mining Charter sets out what mining companies needs to do in order to achieve the objectives of the government such as targets and timeframe for effecting the full participation of historically disadvantaged South Africans (HDSA) into the minerals industry. All mining companies, irrespective of their profit margins are expected to fully comply with the requirements of the charter such as ownership through black economic empowerment, employee and host communities. Companies are usually expected to fund the ownership. For example, the Charter requires shareholders to give up 10% of the company to employees and host communities at no cost to the beneficiaries [4]. Thus, in a stagnant economy this requirement becomes difficult to achieve without affecting the value of the mining companies. Consequently, negatively impacting the return to shareholders.

Electricity is also a concern in South Africa as the parastatal; Eskom has failed to keep the light on for the whole country over the years and continuously seeking to increase electricity tariffs from National Energy Regulator South Africa (NERSA). Increasing tariffs results in increased cost of production and may make mining companies that cannot afford alternative energy sources to be unprofitable and consequently, close. Premature closure will result to loss of invested capital, thus investors will be wary to invest in an environment where access to affordable energy is not guaranteed.

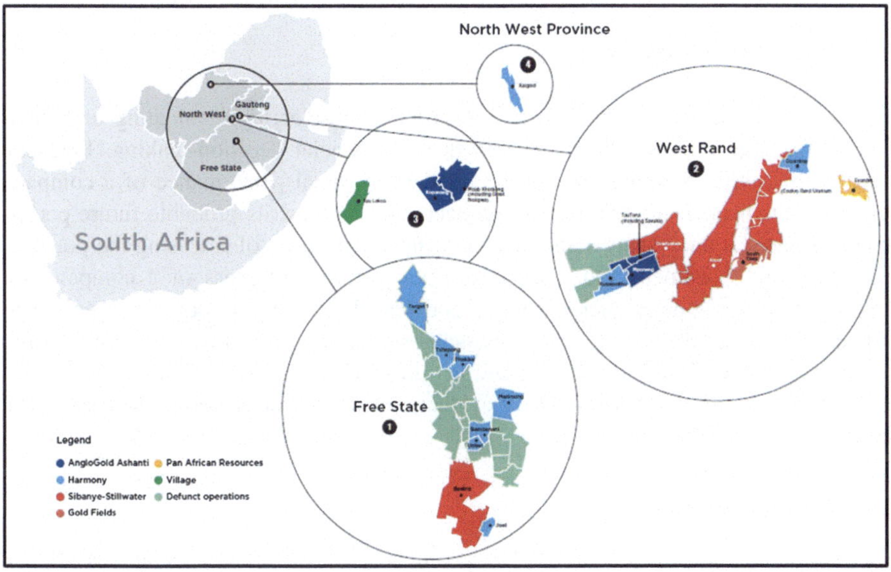

Fig. 1. Location of gold mining companies in South Africa [2]

Though the production contribution of South African gold mining has declined, gold mining still plays a critical role in the South African economic growth through taxes and employment. If the challenges faced by South African gold mines are resolved through adoption of new technologies and new and proposed policies, this will unlock the potential the sector has. Therefore, resulting in increased returns for the investors. This paper is aimed at analysing the financial performance of the Johannesburg Securities Exchange (JSE) traded gold mining companies for the period from 2009 to 2018. This period was selected because it covers the era post the Global Financial Crisis (GFC) and recession, which had adverse impact on the mining industry. The impacts of the two events are compounded by the sovereign risk introduced by changes in the Mining Charter and lack of access to efficient and low-cost energy.

The 2017 junk status from two agencies and the 2018 low medium investment grade make South Africa unappealing as an investment destination. This necessitates a need to frequently review financial performance of South African mining assets.

The companies that were selected are AngloGold Ashanti, Harmony Gold and Sibanye-Stillwater. Though all the companies considered have offshore assets and may have multi-commodity exposure, only South African gold assets were considered. These companies were selected based on their market capitalisation of R50.7 billion, R20.1 billion and R14 billion for AngloGold Ashanti, Sibanye-Stillwater and Harmony Gold respectively [5]. Gold Fields did not report company financial statements but group financial statement, hence it was not considered in the study. A comparative study of the financial performance of these companies was conducted using Du Pont analysis and Economic Value Added (EVA) metric.

2 Financial Analysis

Financial analysis can be defined as the process of selecting, evaluating and interpreting financial data to aid in investment and financial decision-making. Financial analysis is used as a tool to assess the trend in financial performance of a company relative to its historical performance, its peers and forecast its probable future performance. Financial analysis is commonly used for the purpose of assessing the potential for return on investment and associated risks, credit worthiness of a company for funding purposes, operational efficiency and due diligence for acquisition purposes. Numerous models are used to assess financial status of a company, the common ones being Discounted Cash Flow (DCF) and net present value (NPV), financial ratios, Du Pont analysis and EVA metrics. These evaluation tools are discussed in the subsequent sections.

2.1 Financial Ratios

Financial ratios model can be defined as a comparison between a financial measure and another. For example, current ratio where current assets are compared with current liabilities to measure. Financial ratios are categorized under profitability, liquidity, solvency, activity and market ratios. Usually financial ratios explain the performance of the company and do not provide explanations for the occurrences. Therefore, they should be used in combination with other measures that can consider the cause and effect. The measure limitations of financial ratios are that it [6]:

- Distorts the comparison findings if the companies considered use different accounting conventions;
- Does not provide accurate insight to the future performance of a company since it is based on ex post data; and
- Does not consider the effect of inflation on revenues and costs over time.

2.2 Du Pont Analysis

According to Soliman [7] and Doorasamy [8], Du Pont analysis decomposes return on net operating assets into two ratios viz profit margin, asset turnover and equity multiplier. These ratios measure different things and have different properties. There is a positive correlation between asset turnover and future earnings [9]. In a mining context. Profit margin measures the profitability where revenues exceeds the costs of mineral extraction and processing. Asset turnover measures the use of revenue generated from assets to create value yielding positive returns [10]. Du Pont Analysis is regarded as a robust tool for strategic decision making to increase return on equity (ROE). The information needed to generate the ratios for the Du Pont Analysis is shown in Fig. 2.

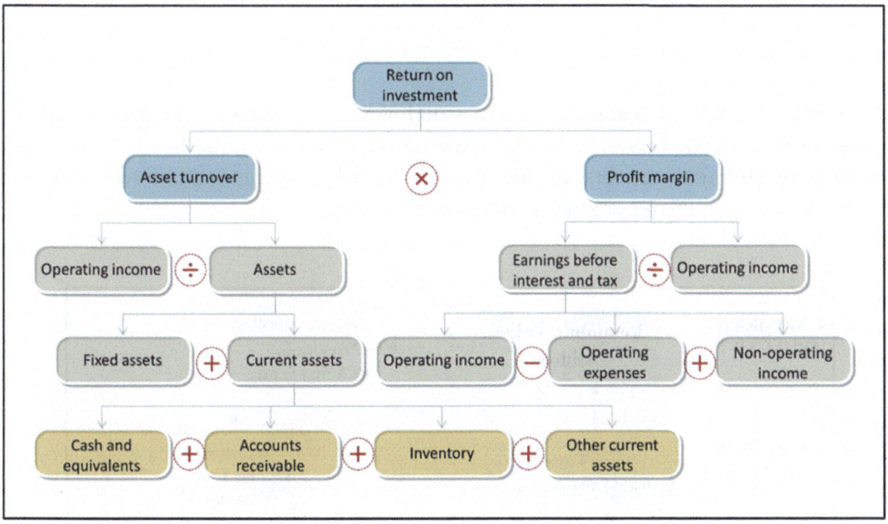

Fig. 2. Flowchart for the Du Pont Analysis [11]

2.3 Economic Value Add

Mining projects are characterised by capital intensive nature, volatile commodity prices (unless hedged) as well as volatile United States' Dollar South African Rand exchange rate. Additionally, South African gold sector is characterised by deep ore deposits requiring additional capital to mine innovatively and safely while creating value for all stakeholders in particular, shareholders. Miners need to shift the focus from mining for tonnages to mining for profits while carefully allocating capital when implementing long-term growth strategies [12]. While profit remain a good short-term measure of value, mining companies engage in cost cutting measures that may increase profit but do not create economic value add [13]. Profit is a commonly reported traditional measure of value that does not account for charge on invested capital. Charge on invested capital is the monetary value calculated by multiplying invested capital with the cost of capital as shown in Eq. 2. The cost of capital can be cost of equity, cost of debt or weighted average cost of capital (WACC) depending on the capital structure of the company. Thus, the concept of economic value add (EVA) which measures residual wealth is considered an appropriate measure of value in mineral resource management [14, 15].

The goal of EVA is to quantify the charge of investing into a project and assess whether the project generates enough value to be classified as good investment [16]. Bluszcz and Kijewska [17] citing Marshal (1920) stated that earning of management is what remains from the profit after subtracting charge on the invested capital. EVA is calculated as net operating profit after taxes less the capital charge as expressed in Eq. 1 [18, 19].

$$\text{EVA} = \text{NOPAT} - \text{Charge on invested capital} \tag{1}$$

$$\text{Charge on invested capital} = \text{Cost of Capital} * \text{Invested Capital} \tag{2}$$

In order to asses a company in terms of EVA, it is important to understand the fundamental value drivers of EVA. The value drivers of EVA result from management decision in different aspects of the project namely, operational, investment and financing structure of the project as illustrated in Fig. 3.

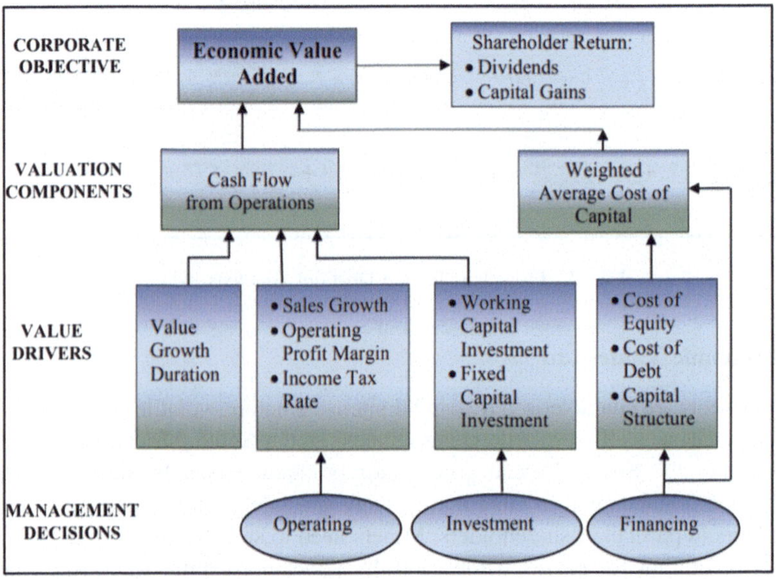

Fig. 3. Relationship between company goals and EVA [17]

From Fig. 2 and Eq. 1, EVA can increase if there is increase in sales offsetting capital investment, tax payments as well as cost of capital. Rappaport (1999) cited in [17] stated that there are seven value drivers including value growth duration, lower fixed and working capital. Rappaport has considered the effect of each of the seven drivers on assumption that the other drivers remain unchanged. However, multiple drivers can change at the same time due to the dynamic and complexity of the mining operating environment. Therefore, companies need to create EVA calculation models that enable iteration of different parameters to pro-actively evaluate management decisions.

3 Methodology

Data used for the analysis were sourced from annual reports of the individual companies over the 10-year period from 2009 to 2018 financial years. The annual reports comprise of income statement, balance sheet, statement of cash flows and justifications for company performance level. Sibanye-Stillwater was founded in 2012, thus its analysis begins in year 2012 for a period of six years. The financial performance models used in this study are DuPont analysis, economic value add and profit. From DuPont calculations returns on equity and return on assets were analysed. EVA was calculated using formula 1. Sibanye-Stillwater reported a 9.0% discount rate when they acquired Kroondal and other platinum assets; Harmony Gold reported a range between 8.35%–11.81% discount rates for South African operations [20, 21]. Other gold companies including AngloGold Ashanti, Pan African Resources and DRD Gold reported discount rates in the same range. On this basis, a discount rate of 10% was assumed in this paper. Invested capital is defined and calculated differently in different literature but in this paper invested capital was calculated as in Ernst & Young LLP (2017) cited in [13]:

$$\begin{aligned}\text{Invested capital} = {} & \text{Current assets} \\ & - \text{non-interest-bearing current liabilities} \\ & + \text{Net property, plant and equipment} \\ & + \text{Intangible assets} \\ & + \text{Goodwill} \\ & + \text{Other operating assets}\end{aligned}$$

4 Du Pont and Value Analysis

The graphs presented in this section were produced from data sourced from individual company income statements, balance sheets and discount rates reported in the annual reports. Some of the information regarding share price for individual companies was not reported in the annual reports thus Investing.com was used to supplement the missing information. AngloGold Ashanti share price information for the years 2013 to 2015 were sourced from Investing.com. Harmony Gold share price information for the years 2014 to 2016 was sourced from Investing.com.

4.1 AngloGold Ashanti

In Fig. 4, the overall trend for the return on equity and return on assets for AngloGold Ashanti shows a decline from 2010 financial year. The good performance observed in 2010 may be alluded to the gold price soaring from approximately US$1270/oz in January to US$1650/oz in December. The graph shows that AngloGold Ashanti is not providing expected returns to the investors, with its value eroding over the years.

However, the ROE for the company increased by 62.5% to 0.3% in 2018. Nonetheless, these returns are still not adequate to boost investor confidence in the company. The same trend is observed for ROA indicating that the company is not using its assets efficiently to generate value. The ROA improved by 66.7% in 2018 from −0.8% return on assets. The average ROE and ROA for the period under study are 15.1% and 11.8% respectively.

In years 2010, 2011, 2012, 2014 and 2016, the company generated profit however it failed to create economic value at the same level. The average EVA for these four years was R1.3 billion whereas the average profit was about R5 billion, a variance of 74%. The good performance over these four years can be attributed to higher net operating profit after tax as opposed to invested capital which on average it was higher than the other years where EVA was negative. For the other years, the company made a loss, with highest losses experienced in 2013. Although the company made a loss in 2013, Fig. 5 shows that the company paid dividends from 2009 to 2013. The dividends were paid at group level while the profit and EVA analysis are on South African assets. The share price of the company depreciated from 2011 until 2015 and thereafter it started to appreciate until 2018. Furthermore, no dividends were paid in 2014 and 2015 supporting that the economic value add was not good in the preceding years.

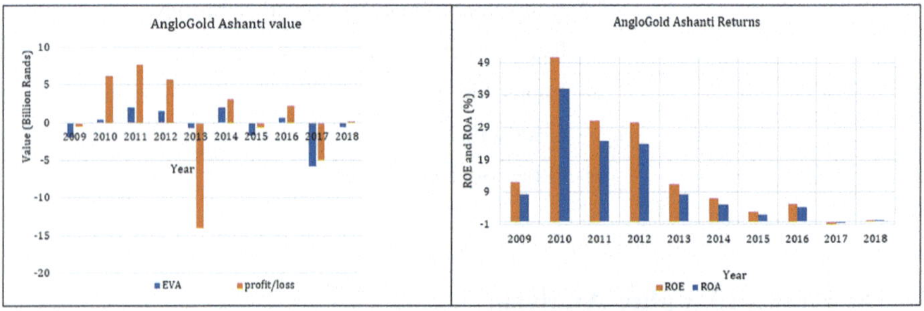

Fig. 4. AngloGold Ashanti financial performance

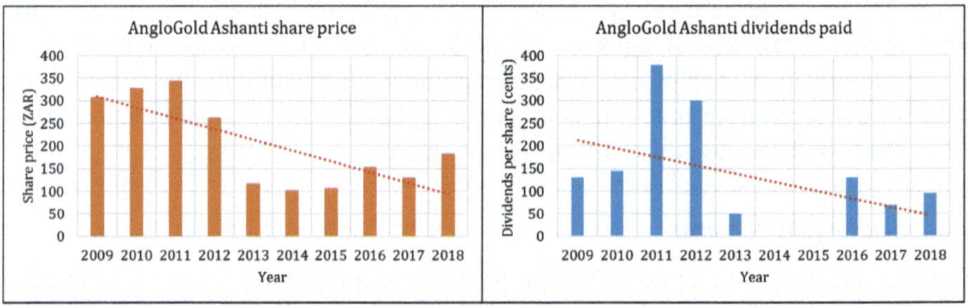

Fig. 5. AngloGold Ashanti share price movement and dividends payout

4.2 Harmony Gold

Figure 6 shows that Harmony Gold provided good returns for the investors from 2009 to 2013 with an arithmetic average of about 21% and 19% for ROE and ROA respectively. However, for the succeeding five years ending 2018 the ROE and ROA values have been negative thus eroding the value for investors. The failure of Harmony Gold to use its assets efficiently has reduced the ROE and ROA averages to 9.7% and 8.7% for the 10-year period. The poor performance may be alluded to low grade and unprofitable areas which resulted in scrapping of Harmony's carrying values [22]. In 2009, Harmony Gold made loss however, there was economic value added due to positive net operating profit after tax (NOPAT) paired with relatively low invested capital. The company only made profit in 2012 and 2013 but no economic value was added as indicated by EVA values. Although NOPAT was higher than that of 2009, the average invested capital in 2012 and 2013 was 480% more than the capital invested in 2009. Figure 7 shows that the share price of Harmony Gold has depreciated over the period under study and the company did not payout dividends in years 2014 to 2016 and 2018. Therefore, there were no returns to the shareholders in these four years. The depreciation in share price and lack of dividends payout support the overall lack of economic value add in the company.

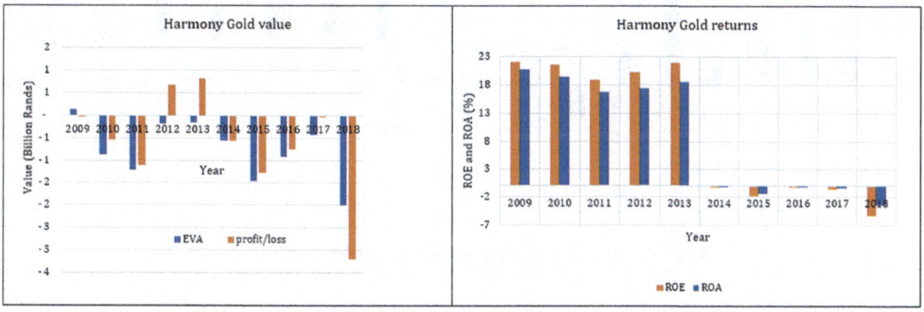

Fig. 6. Harmony Gold financial performance

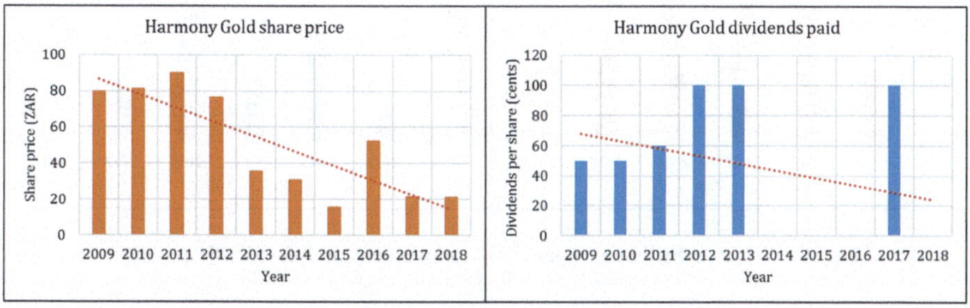

Fig. 7. Harmony Gold share price movement and dividends payout

4.3 Sibanye-Stillwater

Even though the ROE for 2012 was −33.8%, the performance of Sibanye-Stillwater was good as indicated by 17.4% return on assets (Fig. 8). This means that the company management was efficient in using the assets to generate earnings. This is further supported by the fact that from 2013 to 2017, the company generated positive return on investment. The company had negative ROE and ROA for year 2018. The average ROE and ROA for the 10-year period were 17.5% and 13.5% respectively. Generally, for the period under study, there has been reduction in the net property, plant and equipment from R16.3 billion in 2012 to R11.3 billion in 2018. Consequently, invested capital has also generally decreased over the period from R14.3 billion in 2013 to R9.2 billion in 2018. However, in 2018 the lower invested capital was paired with negative NOPAT resulting in economic value erosion. Figure 9 shows that the share price has remained relatively stable over the period. Sibanye-Stillwater has managed to create value for the shareholder through dividends payouts except in 2018, where no dividends were paid.

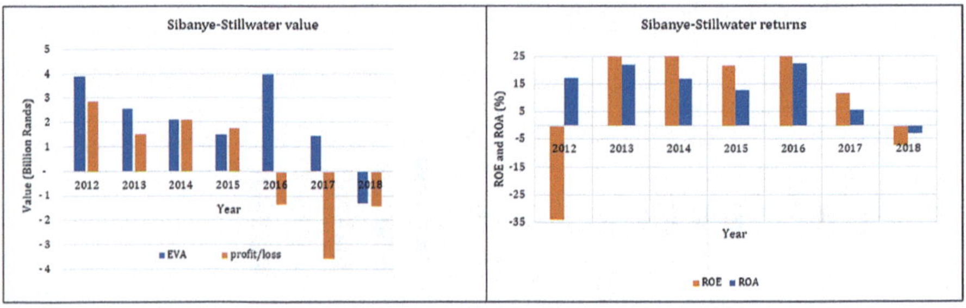

Fig. 8. Sibanye-Stillwater financial performance

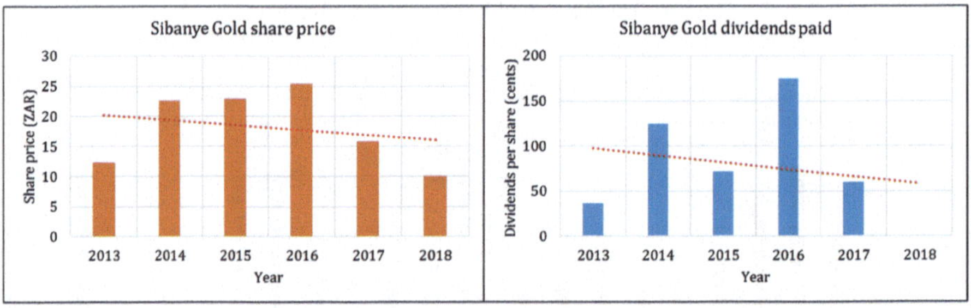

Fig. 9. Sibanye-Stillwater share price movement and dividends payout

5 Summary of the Findings

A colour code was used in order to compare the overall performance of the selected companies as shown in Table 2. Table 3 shows the average values of the measures of performance and their colour coding. Based on the colour coding Sibanye-Stillwater has performed better in terms of EVA, ROE and ROA. It is worth noting that Sibanye-Stillwater did not experience the negative impacts of the GFC and recession because it was founded in 2012. AngloGold Ashanti generated the highest average profit of R536 million. Per annum Year-on-year the performance fluctuated, thus it is important that financial performance is analysed on a yearly basis. In 2018, PriceWaterhouseCoopers highlighted the importance of capital allocation and transparency in mining project funding [12]. The share prices of the three companies fluctuated over the period but with a generally decreasing trend. Although share prices are at group level, this trend can be attributed to poor performance from the South African assets. The poor performance can further be attributed to low investor confidence as a result of sovereign risk evident from South Africa's credit worthiness from the credit rating agencies.

Table 2. Colour code for selected companies

Colour code	Interpretation
Green	Top performing company
Yellow	Middle performing company
Red	Lowest performing company

Table 3. Summary of the performance of the selected gold companies

Company	EVA (ZAR, million)	Profit/loss (ZAR, million)	ROE (%)	ROA (%)
AngloGold Ashanti	-440	536	15.1	11.8
Harmony Gold	-766	-597	9.7	8.7
Sibanye-Stillwater	2022	279	17.5	13.5

6 Conclusions

Most mining companies in South Africa are still dealing with the adverse impacts of the Global Financial Crisis of mid-2008 and recession as well as sovereign risk in South Africa. The impacts have affected the financial performance of the mining companies. This paper analysed the financial performance of three JSE traded gold mining companies over a 10-year period. The selected companies were AngloGold Ashanti, Harmony Gold and Sibanye-Stillwater. The analysis was conducted by using the commonly applied tools for analysis of financial performance viz. EVA and Du Pont Analysis. Data were sourced from individual company annual reports. The financial performance of all the companies fluctuated year-on-year. The measures used to assess the performance of the companies were EVA, profit/loss, ROE and ROA. Based on these measures, Sibanye-Stillwater was the better performing gold mining company. While AngloGold Ashanti came second in all fields of measurement except for profit, which averaged the highest over the period under study. Harmony Gold was the worst performing company of all the three.

Company performance fluctuates on year-on-year basis, hence the importance of conducting financial analysis for asset optimisation, return maximisation and investment opportunities. While mining companies report a number of performance measures in silos, it is important that comparison per sector is done for investment purposes. This is supported by the fact companies compete for the scarce capital investment.

References

1. Le Long, H., De Ceuster, M.J.K., Annaert, J., Amonhaemanon, D.: Gold hedge against inflation: the vietnamese case. In: International Conference on Applied Economics (2013). Procedia Economics and Finances 5(2013), 502–511
2. Minerals Council South Africa: Gold (2019). https://www.mineralscouncil.org.za/sa-mining/gold. Accessed 12 July 2019
3. Trading Economics: South Africa – credit rating (2019). https://tradingeconomics.com/south-africa/rating. Accessed 07 Sept. 2019
4. Government Gazette: Broad-based socio-economic empowerment charter for the mining and minerals industry, 2018 (2018)
5. SAShares: Top 100 JSE Listed Firms By Market Capitalisation (2019). https://www.sashares.co.za/top-100-jse-companies/. Accessed 21 July 2019
6. Makamba, T.: Evaluation of financial performance of South African gold, platinum, coal and iron ore mining companies for the period 2012–2016. Master of Science Research Report, University of the Witwatersrand (2018)
7. Soliman, M.T.: The use of DuPont analysis by market participants. Account. Rev. **83**(3), 823–853 (2008)
8. Doorasamy, M.: Using DuPont analysis to assess the financial performance of the top 3 JSE listed companies in the food industry. Invest. Manag. Financ. Innov. **13**(2), 29–44 (2019)
9. Herciu, M., Ogrean, C., Belascu, L.: A Du Pont analysis of the 20 most profitable companies in the world. In: 2010 International Conference on Business and Economics Research, vol. 1, no. 2011. IACSIT Press, Kuala Lumpur, Malaysia (2010)

10. Dodge, L.E.: A DuPont model approach to financial management: a case study of veterinary practices (2017). https://mountainscholar.org/bitstream/handle/10217/185727/Dodge_colostate_0053N_14547.pdf?sequence=1. Accessed 21 July 2019
11. YourFreeTemplates: DuPont Model ROE (2017). https://www.google.com/search?q=du+pont+analysis&rlz=1C1GCEU_enZA836ZA836&source=lnms&tbm=isch&sa=X&ved=0ahUKEwj1iprXjq_kAhVDL1AKHeNXBOMQ_AUIESgB&biw=1680&bih=939#imgrc=5jGicEeryPLKGM. Accessed 12 July 2019
12. Price Waterhouse Coopers: Mine 2018 Tempting times (2018). https://www.pwc.co.za/en/assets/pdf/mine-report-2018.pdf. Accessed 25 July 2019
13. Tholana, T., Neingo, P.N.: Economic value add analysis for mining companies. In: Proceedings of the 27th International Symposium on Mine Planning and Equipment Selection - MPES 2018. Springer (2018)
14. van der Poll, H.M., Booyse, N.J., Pienaar, A.J., Büchner, S., Foot, J.: An overview of the implementation of Economic Value Added (EVA™) performance measures in South Africa. South. Afr. Bus. Rev. **15**(3), 122–141 (2011)
15. MacFarlane, A.S.: Establishing a new metric for mineral resource management. J. South Afr. Inst. Min. Metall. **106**(3), 187–198 (2006)
16. Chen, J.: Economic Value Added (EVA) (2019). https://www.investopedia.com/terms/e/eva.asp. Accessed 20 July 2019
17. Bluszcz, A., Kijewska, A.: Factors creating economic value added of mining company. Arch. Min. Sci. **61**(1), 109–123 (2016)
18. Stewart, G.B.: The quest for value: A guide for senior managers. Harper Collins Publishers Inc., United States of America (1990)
19. Daraban, M.: Economic value added: a general review of the concept, "Ovidius" University Annals, Economic Sciences Series, volume XVII, Issue 1/2017 (2017)
20. Sibanye – Stillwater: Annual Financial Report 2017 (2017). http://reports.sibanyestillwater.com/2017/download/SBY-AFR2017.pdf. Accessed 20 July 2019
21. Harmony Gold: Harmony Gold Financial Statements FY2017 (2017). https://www.harmony.co.za/downloads/send/126-q4-fy2017/2529-financial-statements. Accessed 20 July 2019
22. Briggs, G., Abbott, F.: Quarter 2, FY2015 (2015). https://www.harmony.co.za/downloads/send/100-2015/1240-results-for-the-second-quarter-fy15-and-six-months-ended-31-december-2014. Accessed 13 July 2019

Ultimate Pit Limit Determination Considering Mining Royalty in Open-Pit Copper Mines

Hamid Mergani(✉), Morteza Osanloo, and Morteza Parichehp

Department of Mining and Metallurgical Engineering, Amirkabir University of Technology, Tehran, Iran
h.mergani@gmail.com, osanloo@aut.ac.ir, morteza_pariche@yahoo.com, mpariche@aut.ac.ir

Abstract. Mining royalty is a payment to the holder of mineral rights for utilization of the mineral resources. As well, the royalty is recognized as the amount of mining contribution in sustainable development. So far, there has not been a generalized world acceptable approach to examine the effects of royalty on mine design elements such as cut-off grade, strip ratio, and revenue as well as mineable reserve and its influence on Ultimate Pit Limit (UPL) determination. The present paper aims to investigate these effects. To this end, a data-driven simulation-based technique was developed to show how the mining royalty might be changed and how it might affect the open pit mine design. To this end, a regression model was developed to predict the royalty based on commodity price volatilities. A series of price scenario and the corresponding royalty values are created using the cumulative probability distribution function of the price data. These scenarios are not those fancied by the authors, indeed they are those happen in the real world. Thereafter, both price and royalty scenarios are used to the subsequent block economic value and UPL determinations. The technique was tested in Sungun copper mine of Iran. The results show that mining royalty often does not influence the mine design elements with a 72% probability of concurrence. Whereas, there is about 28% chance that the royalty reduces the mine life, the minable reserve, and the total revenue by 1.06% and 1.07% and 0.73% respectively, compared to the base case where the royalty is not considered. In addition, with the same chance, the stripping ratio and the average grade are to be relatively decreased by 0.23% and 0.34% compared to those by the base case.

Keywords: Mining royalty · Open pit mining · Mine design elements · Ultimate Pit Limit

1 Introduction

Natural resources are one main fundamental of the national economies around the globe. In most countries, including Iran, the value of mineral resources belongs to the nations; hence a portion of benefits resulting from mining the minerals venture must be paid to the government as royalty to create more national wealth.

In the last two decades, mining taxation has become a subject undergoing intense studies. Nowadays, with the experiencing of high metals/minerals prices, there is an

E. Topal (Ed.): MPES 2019, SSGG, pp. 346–358, 2020.
https://doi.org/10.1007/978-3-030-33954-8_42

increasing interest by host governments for participating via royalties as a special type of mining tax. The inappropriate tax laws imposed have far-reaching consequences and bring many disadvantages to both investors and civil societies. Specially, rising tax rates in order to make more money to the state undermine companies' incentive to invest in mining industries and more importantly have detrimental effects on the technical and economic factors in mine design that finally may result in the early closure of the mine. Therefore, in recent years, we have witnessed a limited number of researches on royalty considering how the economics of the ore-body is impacted by the additional costs associated with the mineral resource royalty around the world and therefore revisions in taxation legislation.

Otto (2002) studied the impacts of royalty taxes on the Peruvian gold mine [1]. He showed that the Internal Rate of Return (IRR) is decreased by raising the royalty rates on gross sales revenue. Thereafter, Cawood (2003) carried out a research report into the effects of mining royalties based on the investment rules of some countries, indicating that the best practice mineral royalties should range from 1 to 3% of net smelter returns because royalties more than 3% resulted in raising the cut-off grade in ore deposits and thereby led to the physical waste of the resources [2]. Also, in 2011, Cawood did his research for gold mines in South Africa and calculated the average royalty rates about 1.4% based on actual data history over the past 15 years, which confirms that the royalty more than average has the potential to significantly impact on government's gains from the mining industry [3].

Otto et al. (2006) provided a comprehensive analysis of royalty taxation concerning the pros and cons of various forms of royalties [4]. They presented how different types of royalties will have different impacts on project economics and government takes. They also showed that the calculated measures such as Net Present Value (NPV) and IRR are particularly useful for policymakers to understand how possible types and levels of royalty will affect company's profitability. In addition, it demonstrates the royalty impacts on paramount decisions, such as cutoff grade, mine life and mineral reserves by introducing several mine models. But considering the uncertain nature of the parameters involved, by the use of these quantitative models, we will not yield an optimally equitable royalty for all mines. In general, this analysis proves that for most cases, royalties are a relatively small proportion of the total fiscal flows from mining [4].

In addition, Grobler (2014) indicated that in spite of an increase in operating expenses, the overall impact of the mineral resource royalty on the economy is beneficial. This study also illustrates that imposing royalties has not affected the way the mining industry operates [5]. In recent years, Birch (2016) explained how the cost of the mineral resource royalty has increased cut-off grades in a series of mines and determines the impact on overall revenue generated for the state. He shows that each individual mine is affected by the additional cost associated with the mineral resource royalty differently [6]. In an effort to determine the optimal tax on mining, Tilton (2004) emphasized the difficulty of determining the tax rate and consequence of over- or under-estimation of that on both society and mining company [7]. It is worth mentioning that in these researches, the impacts of royalties on mine design elements are based on just some fixed price scenarios (that does not escalate over the project life) which must be obtained by sampling over the entire distribution of uncertain commodity price. Due to uncertainty in mineral/metal price, real options valuation might be considered in the assessment of royalty tax in order to measure mine design elements more realistically than before [8]. Samis et al. (2007) and MacKie-Mason (1990) showed that these considerations are vital to any quantitative assessment of the impacts of taxation [9, 10].

The main purpose of this study is to provide a simulation-based procedure for determining the impacts that the mining royalty as an additional operating cost can have on mine design elements such as Ultimate Pit Limit (UPL), cut-off grade, stripping ratio and revenue as well as the mineable reserve. This study differs from previous studies found during the literature review since it provides a comprehensive model to generate real royalty scenarios by Monte Carlo simulation and also presents a guideline to determine the proper level of royalty by comparing the royalty rate in copper mines around the world.

In the next section, we will present different systems of royalty measurements and the main contributing factors affecting the royalty rate. Then, in Sect. 3, our methodology will be explained in detail. The methodology will be applied in a copper mine in Iran and the results will be presented in Sect. 5. Finally, after a relatively short discussion, we will end this research paper with the main remarks in Sect. 7.

2 Royalty Measurement Systems

Generally, there are four different prevalent types of royalties: (1) Unit-based royalty which is the oldest form, (2) Value-based royalty which is the most common form, (3) Profit-based royalty which is considered the most favorable by investors and (4) Hybrid system which is a combination of profit-based and value-based royalties systems [4]. Through an analysis by the authors on the data gathered from copper mines of 36 different countries [4], it was identified that about 68% of these countries have applied value-based method. The profit-based method is the second common approach with a frequency of 19%. Such systems are more prevalent in those nations with many mines where a well-trained and well-equipped tax administration has been developed. The unit-based methods and hybrid systems are less common and are in the next ranks by 12% and 1%, respectively (see Fig. 1).

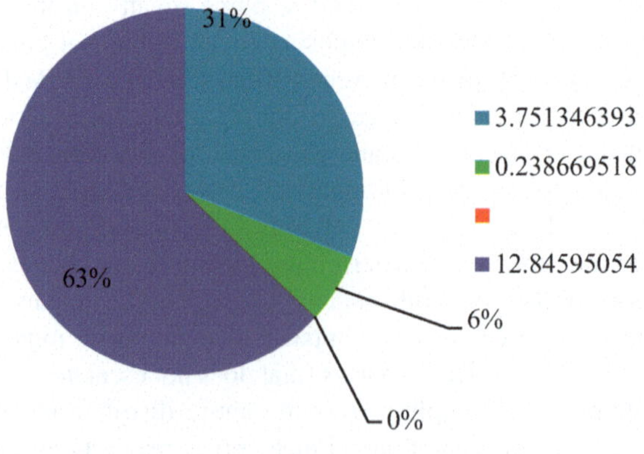

Fig. 1. Common royalty types

It is obvious that most of the policymakers are to prepare a mine specific system for the royalty. The advantage of using a unique assessment method for each mine is that it can be tailored to fit the unique characteristics of the deposit being exploited. These essential characteristics which are recognized the most important contributing factors to determining an appropriate royalty rate could be divided into four categories (see Table 1).

Table 1. Contributing factors for determining an appropriate royalty rate

Factor type	Factor		
Geological	– Mineral type – Ore body shape – Size – Thickness	– Depth – Average grade – Hydrologic condition – Geomechanic condition	– Structural geology condition – Topography
Economic	– *Commodity price* – Capital and operating costs – Cut off grade	– NPV – IRR – Environmental costs – *Mineable reserve*	
Technical	– Production rate – Technologies (equipment) – Mining method	– Stripping ratio – Recovery – Dilution – Life of mine	
Administrative	– Government perspective – Laws and regulations – Mine owner	– Environmental laws – Mine location – Accessibility	

Actually, these site-specific factors control the design, planning and the value of any mining project. The more valuable ore deposits necessitate more amount of royalty in term of monetary value. Generally, the total amount of royalty that the state might have in mind for a given deposit is a paramount thing which should be defined first. Thereafter, the yearly royalty that the mine must pay is determined based on the royalty type [4]. As shown in Fig. 2, the above said contributing factors have been intertwined in a complex form. Anyway, in order to model the royalty, identifying the independent factors is important. In such situations, cognitive maps can be helpful.

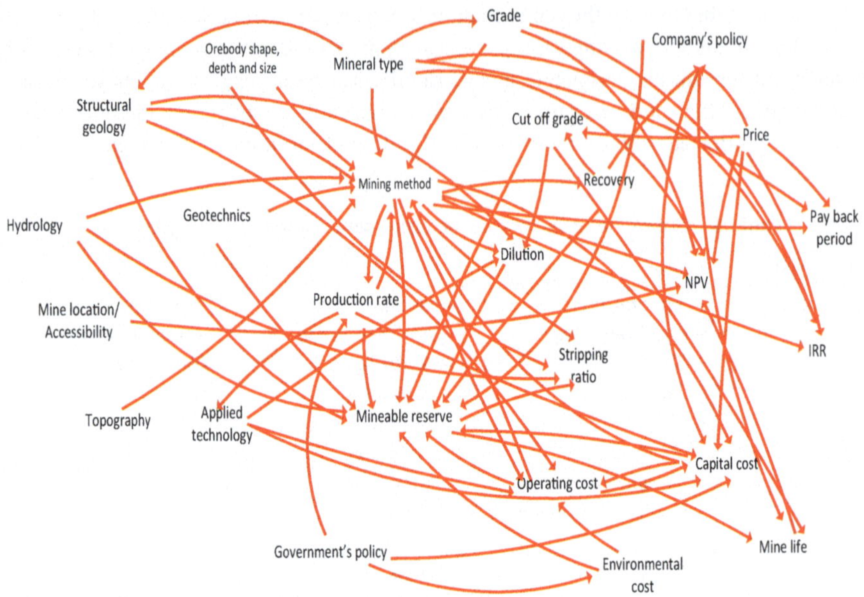

Fig. 2. Cognitive map of effective factors for determining an appropriate royalty rate

As shown in Fig. 2, the arrows come into dependent factors (like mineable reserve); and inversely, they flow out of independent factors (like price). By this way, seven factors of (1) mineral type, (2) grade, (3) physical properties of ore body like shape and depth, (4) price, (5) production rate, (6) mine location/accessibility and (7) government's policies are recognized as the main independent parameters.

3 Methodology

3.1 Royalty Prediction

The royalty is changed by time while some of these independent factors such as ore body characteristics and mine location are unique. The factor 'production rate' also can be assumed constant. Therefore, in order to develop a reliable method of royalty prediction, one must consider the volatile factors of price and government policy. As well, the other constant factors should be considered implicitly. To this end, we gathered historical data of royalties for different mines, mineral types and geographical locations in Iran. The Ministry of Industry, Mine and Trade which is an Iranian government body has the responsibility for making regulation and implementation of royalty policies applicable to all mines in the country. The most common way the ministry assesses the royalty is unit-based setting mine-by-mine. The data show that based on the above said independent factors; there is a unique system of royalty for each mine. However, for each unique mine the royalty changes by price volatilities on a yearly basis. Figure 3 shows the strong linear correlation found between the annual copper price and the annual royalty data from 2001 to 2017 for Sungun Copper Mine (SCM) in Iran.

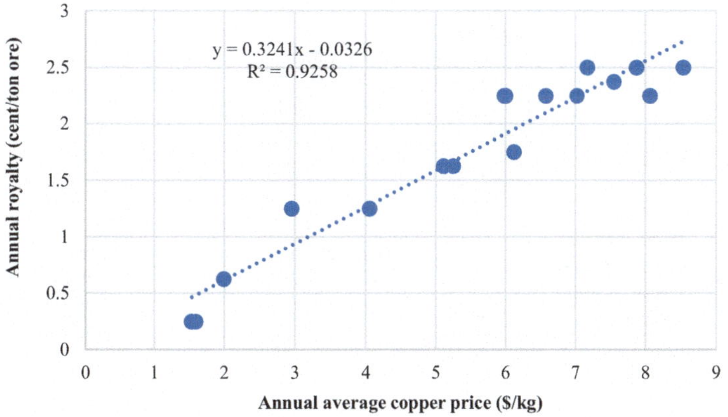

Fig. 3. The correlation between the price and the royalty data

This correlation can be used to understand how the royalty changes with a change in the copper price. Remember that other above mentioned independent factors have been considered implicitly in the data. As it is expected, with increasing the price, the governments have a general tendency to set a higher royalty rate on mining operations to participate in utilizing more benefits of the high commodities price.

3.2 Simulation-Based Procedure

In order to assess the effects that the royalty has on open pit mine design elements, we suggest the simulation-based procedure shown in Fig. 4. As shown in the figure, the historical data of the price help us simulate a series of real happened price scenarios. Then, these price scenarios are used to generate the corresponding royalty scenarios based on the described correlation between price and royalty. For each royalty scenario, the corresponding economic block model and UPL are calculated. This procedure is repeated for a sufficient number of times (for example, 100 times). Finally, the changes in the design elements are assessed using the outputs created during the simulation.

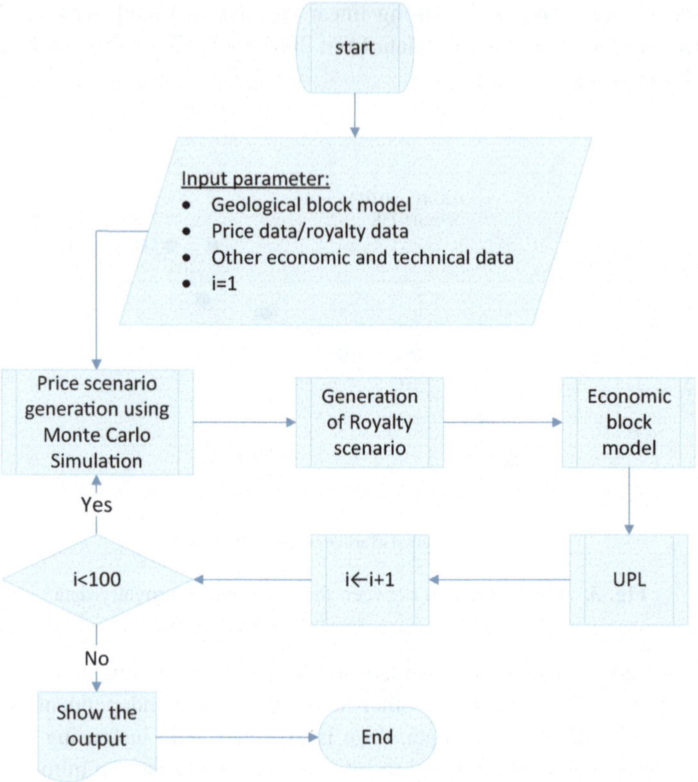

Fig. 4. The proposed simulation-based procedure

3.3 Block Economic Value Calculation

As shown in Fig. 4, the royalty is imposed on the extraction of ore material. Actually, there is an extra cost of royalty that a given ore block must pay. The cost basically changes the traditional way of ore-waste classification and thereafter the ultimate pit limit, mining sequence, and the net present value [11]. The traditional Block Economic Value (BEV) formula can be rewritten as in Eq. 1:

$$BEV(r) = \begin{cases} \left[\bar{g} \times M \times y \times (P - s) - (C_p \times M) - (C_r(r) \times M)\right] - (C_o \times M) & \text{if } \bar{g} \geq g_c(r) \\ \text{or} & \\ -(C_w \times W) & \text{if } \bar{g} < g_c(r) \end{cases} \quad (1)$$

Where BEV(r) is the block economic value in scenario r ($), *M* is tonnage of ore per block, $\bar{g}$ is estimated grade for block in %, *y* is recovery in %, P is the average commodity price in $/tonne metal, s is selling cost in $/tonne metal, *Cp* is processing cost in $/tonne ore, *Co is* mining cost in $/tonne ore, W is tonnage of waste per block, *Cw is* mining cost in $/tonne waste, and *Cr(r) is the* mining royalty in scenario r

(\$/tonne ore). The parameter $g_c(r)$ is also the break-even cut off grade in scenario r which can be calculated using Eq. 2:

$$g_c(r) = \frac{C_r(r) + C_p + C_o}{y \times (P - s)} \tag{2}$$

The royalty scenario (Cr(r)) is calculated based on the price scenario P(r) and the identified relationship between price and royalty described in Fig. 3.

4 Case Study: Sungun Copper Mine (SCM)

The proposed approach was applied to the Sungun Copper Mine (SCM) of Iran. The SCM is the second largest open pit mine of Iran in the northwest part of the country. The feasibility studies showed that there exist 806 million tonnes of geological resources with an average grade of 0.67% Cu. SCM is in operation now with an annual ore production of 14 million tonnes. The same technical and economic data used by Paricheh and Osanloo (2018) were used to define the economic block models and then to optimize the UPLs [12]. As well, the 9550 daily copper price data from Jan 1980 to the end of Oct 2017 were used [12]. The Monte Carlo simulation method was applied to generate 100 real price and the corresponding royalty scenarios. It is worth mentioning that in all scenarios, the copper price is considered constant (i.e., 3.57 \$/kg Cu) and the price scenarios are used just to simulate the royalty scenarios.

5 Results

The royalty in any form is a cost and thus will influence mine design elements that are set to optimize mine profitability. This section will illustrate the impact of royalties as an additional cost on mine life, mineable reserve, stripping ratio, break-even cutoff grade, revenue, and average grade using the proposed simulation-based procedure. The results have been shown in Fig. 5. It is worth noting that the royalty scenarios are sorted in ascending order in Fig. 5. As shown in the figure, the imposition of royalties less than 1.5 cent/tonne ore does not change the design elements significantly. Totally, the results show that mining royalty often does not influence mine design elements with a 72% probability of concurrence. Whereas, compared to the base case where the royalty is not considered, there is about 28% chance that the royalty influences the mine design elements. As shown in Figs. 5a, b and d, the mine life, mineable reserves and total revenue might be decreased by 6 months (1.06%), 6 million tonnes (1.07%) and \$ 60 million (0.73%), respectively. Inversely, as shown in Figs. 5c, e and f, stripping ratio, average grade and cutoff grade might be relatively increased by 0.23%, 0.34% and 0.42%, respectively.

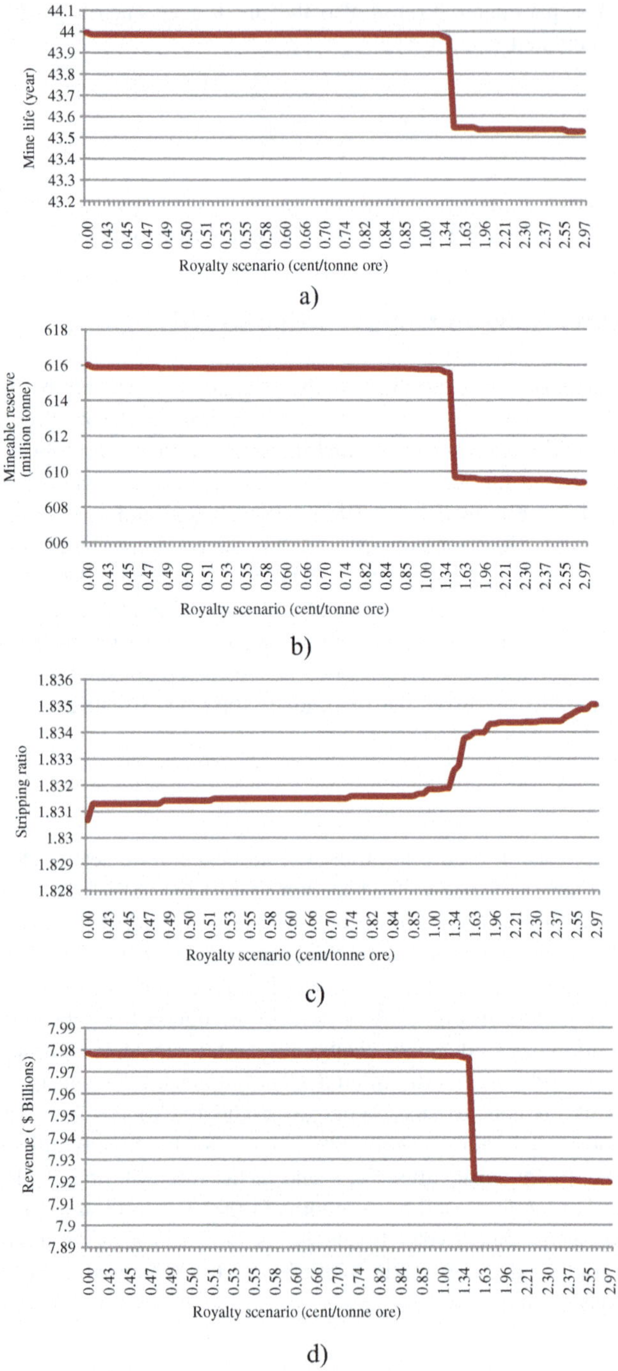

Fig. 5. The effects of royalty on mine design elements

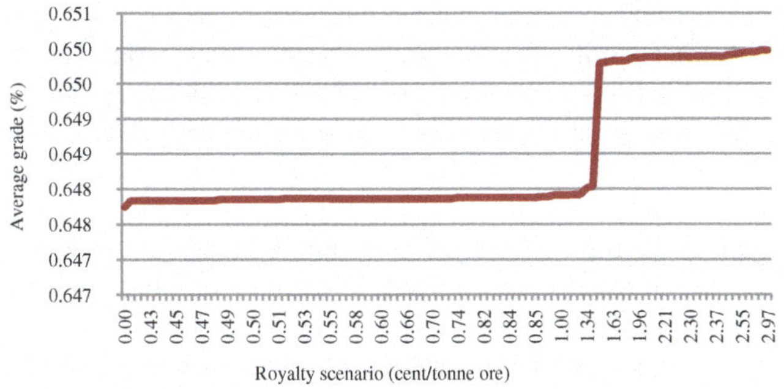

e)

f)

Fig. 5. *(continued)*

6 Discussion

From the government perspective, determining the optimal level of mining royalties which maximizes the net present value of the social benefits flowing from mining sector is a core subject. Unfortunately, in practice, it seems impossible to determine the optimal level of mining royalties for a specific mine, which would require not only knowledge of economic behavior of a mining company, but also knowledge of mining's contribution to the sustainable development of the country or the region in which it occurs. Even worse, it is more difficult when a government wants to impose a uniform royalty tax on all minerals or mines. The above reveals that mineral royalties may have a negative effect on economic growth and development both where mining royalties is too high and too low. Therefore an approach that considers these side effects is necessary because where mining royalties are too high, the investment will decrease as investors shift their budgets to other alternatives, and if taxation is too low the nation will lose revenue helpful for improving the public welfare. First and foremost, to reach an optimum level of the royalty rate, we need to consider the likely consequences of mining royalties on mine design elements. The generic approach presented here can cover this part of the problem completely. To somehow consider the other party or the contribution of royalty on sustainable development of the local community, it seems reasonable to make a comparison between the given mine and fiscal systems of other similar mining sectors around the world. The historical data show that for all mineral types around the world, the royalties vary from 1% to 15% of the mineral value. The word 'value' refers to mine-mouth value, concentrate value, gross sales or Net Smelter Return (NSR). Exclusively, for copper deposits, the royalty varies between 3–5% of the mineral value [4]. While in SCM the royalty rate is at most about 3 cents per tonne ore or 0.3% of concentrate value (to convert an unit-based royalty to a value-based royalty, it just needs to divide a unit-based royalty rate by the value of 1 tonne of ore, here about $10 per tonne ore). Table 2 illustrates the comparison of the mine design elements for three alternative royalty rates. The first when no royalty rate is imposed (scenario 0). The second for when the most possible royalty based on the current government policy is imposed (scenario 100) and the last for when a royalty rate of 3% of the value is imposed (scenario 101). As shown in the table, in scenario 101, the mine life and mineable reserve would decrease by 4.4% than the base case where no royalty is imposed. Similarly, compared to the base case, the total revenue would decrease vy0.73% and 2.7% for the 100th and 101st scenarios, respectively. By this way, the mining company would lose $60 million (7980–7920) and $222 million (7980–7758) in scenario 100 and 101, respectively. On the other hand, the government would gain just $18.3 million (609 * 0.03) and $176.4 million (588 * 0.3), respectively.

Table 2. Sensitivity analysis for the mine design elements on different royalty rates

Design element	Scenario		
	0	100	101
Royalty rate (¢/ton)	0	2.97	30
Royalty rate (% of value)	–	0.3	3
Mine life (year)	44	43.53	42.06
Percentage of change	–	−1.06	−4.4
Mineable reserve (million tonne)	616	609	588
Percentage of change	–	−1.07	−4.41
Stripping ratio	1.83	1.835	1.84
Percentage of change	–	0.23	2.5
Average grade (%)	0.647	0.649	0.658
Percentage of change	–	0.34	1.7
Cutoff grade (%)	0.304	0.305	0.318
Percentage of change	–	0.45	4.4
Revenue ($ billions)	7.98	7.92	7.758
Percentage of change	–	−0.73	−2.7

7 Conclusion

The imposition of a special tax like royalty is sure to have far-reaching consequences and imposing inappropriate tax brings many disadvantages to both mining investors and local society. Especially, rising tax rates in order to make more money to the state undermine the company's incentive to invest in mining industries and more importantly have detrimental effects on the mine design elements. So far, there has not been a generalized world acceptable approach to examine the effects of royalty on mine design elements such as cut-off grade, strip ratio, and revenue as well as mineable reserve and its influence on Ultimate Pit Limit (UPL) determination. Therefore, the main purpose of this study was to provide a simulation-based procedure to firstly predict/simulate a series of real royalty scenarios and then to examine the effects that these royalties might have on mine design elements.

The possible effects of royalties on the design elements of the SCM are considered. The results showed that mining royalty often does not influence the mine design elements with a probability of 72%; whereas, there is about 28% chance that the royalty influences the mine design elements compared to the base case where the royalty is not considered. In the end, since royalty rate was about 10 times less than the average royalties imposed in other copper mines in the world, a higher level of royalty rate of 30 cent/tonne ore was suggested. This new royalty regime will increase the government's revenue up to $176 million, while reducing investor's revenue by $220 million. The government's royalty policymakers could apply the proposed approach to investigate what rate of royalty to levy considering goals of investors, governments and civil society by analysis of the consequences of these kinds of decisions.

References

1. Otto, J.M.: Position of the Peruvian taxation system as compared to mining taxation systems in other nations. Documento preparado para el Ministerio de Economía y Finanzas del Perú, Lima (2002)
2. Cawood, F.T.: The mineral and petroleum royalty bill-report to national treasury. J. South Afr. Inst. Min. Metall. **103**(4), 213–232 (2003)
3. Cawood, F.T.: An investigation of the potential impact of the new South African Mineral and Petroleum Resources Royalty Act. J. South Afr. Inst. Min. Metall. **111**(7), 443–453 (2011)
4. Otto, J.M., Andrews, C.B., Cawood, F., Doggett, M., Guj, P., Stermole, F., Stermole, J., Tilton, J.: Mining royalties: a global study of their impact on investors, government, and civil society. The World Bank (2006)
5. Grobler, J.: Mineral and Petroleum Resources Royalty Act: the impact on the fiscal and mining industry in South Africa. Doctoral dissertation, University of Pretoria (2014)
6. Birch, C.: Impact of the South African mineral resource royalty on cut-off grades for narrow, tabular Witwatersrand gold deposits. J. South Afr. Inst. Min. Metall. **116**(3), 237–246 (2016)
7. Tilton, J.E.: Determining the optimal tax on mining. In: Natural Resources Forum, vol. 28 (2), pp. 144–149. Blackwell Publishing Ltd., Oxford (2004)

8. Davis, G.A.: Mining Royalties: A Global Study of Their Impact on Investors, Government, and Civil Society, James Otto, Craig Andrews, Fred Cawood, Michael Doggett, Pietro Guj, Frank Stermole, John Stermole, and John Tilton. The World Bank, Washington, DC (2006). 296 pp. + xx (2007). Index ISBN-13: 978-0-8213-6502-1 ISBN-10: 0-8213-6502-9. Resources Policy, vol 32(3), pp 146-147
9. MacKie-Mason, J.K.: Some nonlinear tax effects on asset values and investment decisions under uncertainty. J. Public Econ. **42**, 301–327 (1990)
10. Samis, M., Davis, G.A., Laughton, D.: Using stochastic DCF and real option Monte Carlo simulation to analyze the impacts of contingent taxes on mining projects. In: Proceedings, Project Evaluation 2007, AusIMM, Melbourne, Australia, 19–20 June 2007, pp 127–138 (2007)
11. Whittle, J.: A decade of open pit mine planning and optimization-the craft of turning algorithms into packages. In: Proceedings APCOM 1999 Computer Applications in the Minerals Industries 28th International Symposium, pp. 15–24 (1999)
12. Paricheh, M., Osanloo, M.: A simulation-based risk management approach to locating facilities in open-pit mines under price and grade uncertainties. Simul. Model. Pract. Theory **89**, 119–134 (2018)

Mine Health, Safety and Environment

Risk Estimation Approach Considering Implementation of Automated Ventilation Systems into Kazakhstan Metal Mines

Sergei Sabanov(✉), Yerbol Tussupbekov, Bekbol Aldoamzharov, Abu-Saadi Karzhau, and Zarina Mukhamedyarova

School of Mining and Geosciences, Nazarbayev University, Nur-Sultan 010000, Kazakhstan
sergei.sabanov@nu.edu.kz

Abstract. With shifting from traditional to automated ventilation system it is necessary to estimate technical and financial risks which might influence on mining currently use in Kazakhstan. This requires development of a new approach to estimate technical applicability and financial viability of implementation of automated ventilation systems into operational metal mines.

Methodology use estimation of propagation of uncertainty sources through an operational risk framework that considers impacts on the existing mine ventilation system and automated ventilation benefits based on estimation of Net Present Value (NPV). Developed approach is considering risks of investment into reconstruction of the mine ventilation network that can be suitable for implementation of automated ventilation systems. The approach is also estimate risks and opportunities related with financial optimisation of ventilation network designed on outdated local regulations and on modern regulations. The approach can contribute to decision making on acceptance of global practices for mine ventilation design in Kazakhstan considering potential economical profit based on optimisation by implementation of automated ventilation. Results of this study will help to estimate ventilation efficiency in agreement with underground mine safety regulations and might assist Kazakhstan change the outdated regulations for the better.

Keywords: Risk estimation · Mine ventilation · Optimisation · Modeling

1 Introduction

1.1 Mine Ventilation Systems

Ventilation systems are responsible for 25–60% from the total mine energy consumption [1, 2]. According to Kocsis [1], conventional ventilation delivers excess air volumes to the mine sites rather than its "true" ventilation needs. As a result, inefficient use of ventilation systems demands for higher capital and operating costs and the outcome is lower profitability for the mining companies. This demonstrates that optimisation of a ventilation system is of great interest to save energy and budget [2]. In a normal ventilation situation, safety and feasibility are usually reflected in the required

E. Topal (Ed.): MPES 2019, SSGG, pp. 361–368, 2020.
https://doi.org/10.1007/978-3-030-33954-8_43

airflow quantity, the lower and upper limit of airflow quantity and the control variable, and the controllability of branches [3]. Thus, the main objectives are minimising power consumption, and accordingly, minimising the overall cost of ventilation. During emergencies, the economic factor is minor, and it is not easy to find an ideal objective function for ventilation optimisation [4, 5].

With shifting from traditional to automated ventilation system it is necessary to estimate technical and financial risks which might influence on mining currently use in Kazakhstan. Aim of this study was to develop risk estimation approach considering technical applicability and financial viability of implementation of automated ventilation systems into operational metal mines. Methodology use estimation of propagation of uncertainty sources through an operational risk framework that considers impacts on the existing mine ventilation system and automated ventilation benefits based on estimation of Net Present Value (NPV). Developed approach is considering risks of investment into reconstruction of a mine ventilation network that can be suitable for implementation of automated ventilation systems. The approaches are also estimate risks and opportunities related with financial optimisation of ventilation network designed on outdated local regulations and on modern regulations.

A case study was produced for a metal mine operating in Kazakhstan. The mine produces 2.6 million tonnes of ore per year. The main products are zinc and lead, and additional products include gold and copper. The mine has a more than 200-year operational history. Depending on the ore body configurations, several mining methods are used at the mine: sublevel caving; cut-and-fill stoping; sublevel stoping with partial shrinkage applied (ore veins); shrinkage stoping (narrow ore veins); and upward horizontal slicing with backfilling. The mine uses a boundary force-exhaust ventilation system [6]. There are four exhaust fans and one forced fan in operation to supply fresh air to a depth of 460 m for 18 levels (Fig. 1).

1. Shaft #1 exhaust fan 'GVHV-40-2200' supplying about 185 m^3/s
2. Shaft #2 ('№3') exhaust fan 'VUPD-2.8' supplying about 85 m^3/s
3. Shaft #3 ('Bel') exhaust fan 'VO-30VK' supplying about 100 m^3/s
4. Shaft #4 ('Vent Shurf') exhaust fan 'VOKD-1,8' supplying about 40 m^3/s
5. Shaft #5 ('Andr') forced fan 'VOD-30 M' supplying about 110 m^3/s

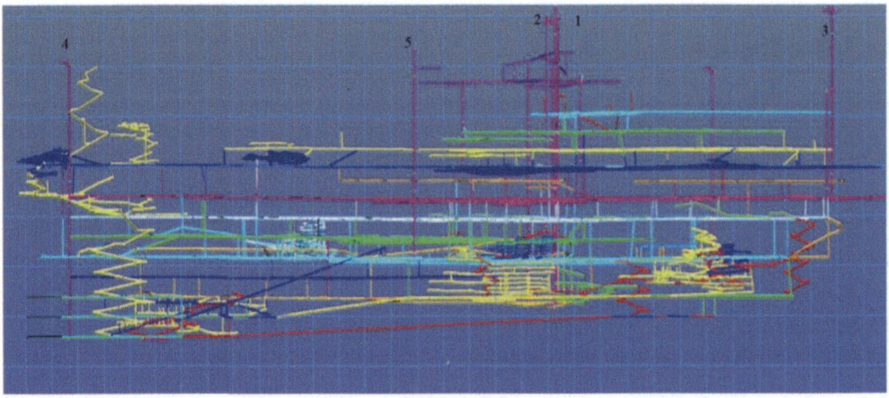

Fig. 1. Mine ventilation network layout.

There are five booster fans placed on levels 11, 14, and 16 to supplement the main fans. In order to reduce air leakage through the abounded open pit zone, inactive workings are being insulated, and positive pressure ventilation is produced by the forced fan 'VOD-30 M' installed on shaft #5 ('Andr'). At the moment, the mine ventilation network is a hardly controllable system, control of which is carried out with the help of continuously built and destructed airway stoppings (about 200 per year).

Accepted by local mining regulations, the method of determining ventilation requirements considers that the removal of diesel fumes is based on a diesel dilution rate of 0.112 m^3/s per kW [7]. This dilution rate is about two times higher than typically used global requirements and should be optimised in case of use modern diesel engines.

2 Materials and Methods

2.1 Ventilation System Optimisation

The optimum airway diameter can be calculated considering capital cost, operating cost, and total cost. The most relevant size of airway will stand for a minimum total cost. The capital cost function is:

$$P = F + f(S,L) \tag{1}$$

Where P is the capital cost, F is the fixed costs, and f(S, L) is the function of area S and length L. The calculation of the total operating cost for one fan running all days over the year is:

$$C = pA/1000\beta e * 24 * 365 \tag{2}$$

Where C is the operating cost (US$/year), p is the pressure, β is the efficiency, and A is the airflow. The fundamental indicators of optimum airway size are the amount of air passing through it and its installation cost [8].

Ventilation simulation software 'Ventsim' has financial simulation function which can help to optimise the size of vertical or horizontal airways by considering a variable cost of mining based on airways size, and then optimising airflow power costs through potential different sizes. This simulation method is based on variable airways size costs based on fixed and variable parameters and formulas. Increasing airway size is the easiest way to reduce frictional pressure losses and decrease ventilation costs in a mine. Increasing airway size however creates additional mining costs, and this is further exacerbated by the 'time value of money' which dictates that a dollar saved in mining costs now is worth more than a dollar saved in ventilation costs in the future. Another factor to consider is how long the airway is required to carry air, which affects how much ventilation cost can be saved in the future. Financial simulation also allows the simulation to adjust the flow through the airway based on the resistance of each size.

2.2 Risk Analysis Methodologies

Monte Carlo method is a statistical test method which be used to solve a mathematical physics problem in which it is difficult to determine the formula, through statistical sampling theory [9]. For a statistic with an uncertainty probability density function, the Monte Carlo method enables easier solving of the synthetic uncertainty which approaches the real solution [10]. A Monte Carlo simulation is a broad term for computational algorithms that generate random numbers (realisations) given a specific density function [11]. Monte Carlo simulation presents risk analysis by building models of possible outcomes.

3 Results and Discussion

3.1 Risk Estimation Approach

The main concept of the risk estimation approach is to consider ventilation modeling, event and operation risks modeling, and financial modeling (Fig. 2).

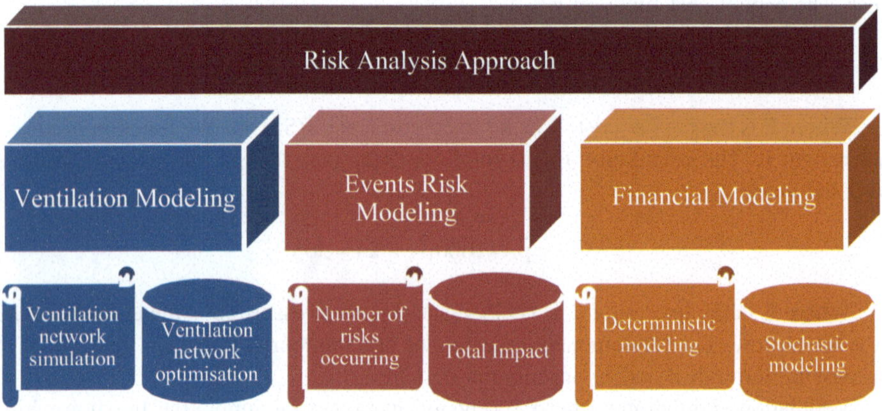

Fig. 2. Concept of the risk analysis approach.

Ventilation modeling is comprised of ventilation network simulation and optimisation steps. Financial simulation estimates an optimum ventilation infrastructure size, where high cost airflows can be optimised, taking into consideration mining development capital expenses and ventilation operational expenses.

Events risk modeling estimates the number of potential risks occurred and the total impact from these events. Events can be analysed based on information received from the mine's historical data. Financial modeling uses deterministic models for overall project cash flow estimation and stochastic cash flow models for investment analysis.

Financial risks for investments consider the reconstruction of the ventilation shafts, extension of the ventilation network, purchasing variable speed fans, and installation of automated ventilation control systems. Savings resulted from improving ventilation system efficiency by reducing airflow in areas without mining activity, and redistributing existing ventilation capacity.

Ventilation Network Modeling

The mine ventilation network was modeled based on 5 091 airways, with a total length of 74 969 m. This existing ventilation network was simplified by excluding closed airflows in abandoned mining areas.

To support mining equipment and leakages, the primary circuit airflow was calculated to be 520 m^3/s. However, there are periods during a shift when diesel equipment are not utilised (about 25% from full day time) and this is the potential time when the primary circuit airflow can be reduced to 275 m^3/s as it is not required to dilute diesel toxic gases from not operating diesel equipment. Financial simulation showed power consumption decrease on 110 kW and annual power savings is US$ 95 190. The most significant savings can be produced on the incline within a length of 320 m and on a part of the ramp within a length of 136 m (Fig. 3b). This saving derived from decreased friction costs by optimisation of airways size. To achieve this power savings, some capital investments for reconstruction of the incline and ramp are required.

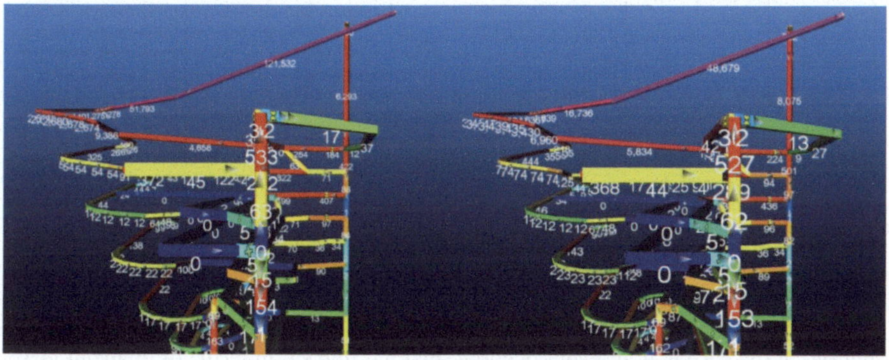

a. Original design (friction costs) b. Optimised design (friction costs)

Fig. 3. Ventilation optimisation

Events Risk Modeling

Events risk modeling comprises a fit comparison for the number of risks occurring and financial impact from these events with the use of the current ventilation system. Seven events with various probabilities of occurrence per year were considered:

1. High consumption of electricity;
2. Leakages;
3. Complicated ventilation network;
4. Regulators and doors;

5. Production scheduling;
6. Main fans; and
7. Mine ventilation management.

These events have historically happened in the studied mine, and therefore probabilities of their occurrence were estimated with help of 'Palisade@Risk' software on the received statistical data. Probability of the events occurrence used a Bernoulli distribution. The historical financial impact of these events was analysed by minimum, most likely and maximum expense values, using a Pert distribution. Figure 4 a demonstrates the probability of occurrence for these events within one year. As a result, from two to six events can occur with a probability of 90% (Fig. 4a). Moreover, there is only a 5% chance that more than six events per year will occur simultaneously. Thus, according to the output data, two events per year can occur with a probability of 11%, tree events with a probability of 27%, four events with a probability of 33%, and five events with a probability of 22% (Fig. 4a).

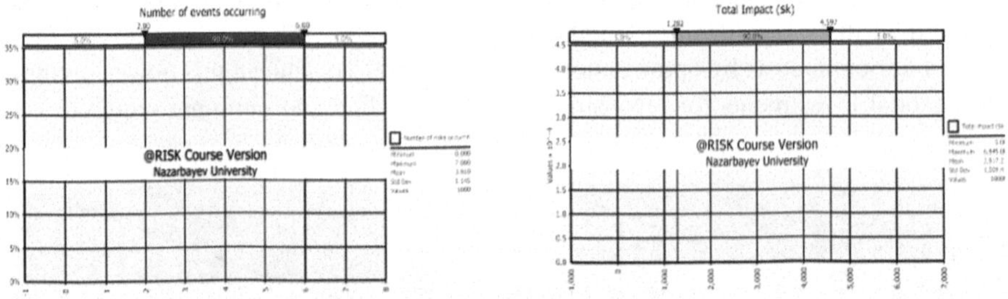

a. Number of events occurring   b. Fit comparison for total impact

Fig. 4. Events risk modeling results.

The next step analysed the impact from each event using a Pert probability distribution, in the case that an event occurs. The total impact, in terms of financial losses for the occurring events, was estimated with the help of a Monte Carlo simulation, and lies in the range of 1.3–4.6 million US$ in the corresponding defined value of 90% (Fig. 4b). For this degree of financial impact, risk avoidance and mitigation measures need to be considered.

Financial Modeling

The financial model comprises mining production for 10 years, and considered the overall optimisation of mine design and therefore production rescheduling. Based on savings derived from the optimised ventilation network, an NPV at a 10% discount rate can be increased on 1.9 million US$ according to deterministic modeling.

Stochastic modelling on this additional NPV for the optimised ventilation network was produced with the help of a Monte Carlo simulation (Fig. 5). For stochastic modeling, electricity pricing was used as a major variability, and it utilised a triangle distribution for each price each year of the mine's life.

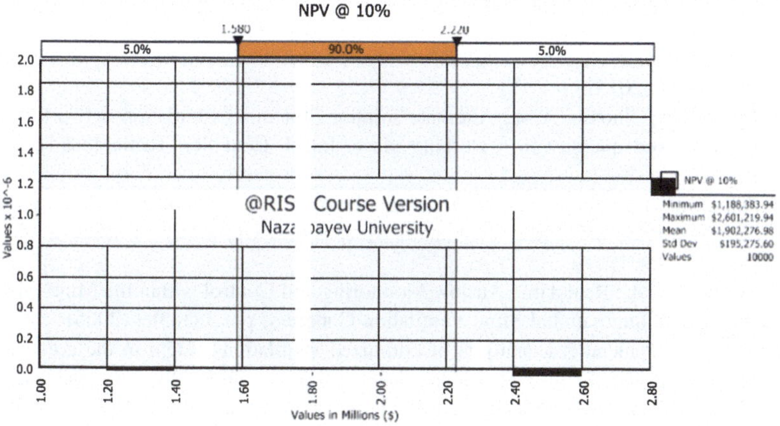

Fig. 5. NPV distribution for the optimised ventilation network.

After 10,000 simulated iterations, the Monte Carlo simulation produced a representation of possible outcomes for an additional NPV. Thus, the probability distribution for the optimised ventilation network demonstrates the additional NPV has a range of 1.6–2.2 million US$ in the corresponding defined value of 90% (Fig. 5). Moreover, there is only a 5% chance that the additional NPV will be less than 1.5 million US$.

4 Conclusion

Methodology use estimation of propagation of uncertainty sources through an operational risk framework that considers impacts on the existing mine ventilation system and automated ventilation benefits based on estimation of Net Present Value (NPV). Developed approach is considering risks of investment into reconstruction of the mine ventilation network that can be suitable for implementation of automated ventilation systems. The approach is also estimate risks and opportunities related with financial optimisation of ventilation network designed on outdated local regulations and on modern regulations. The approach can contribute to decision making on acceptance of global practices for mine ventilation design in Kazakhstan considering potential economical profit based on optimisation by implementation of auto-mated ventilation. Results of this study will help to estimate ventilation efficiency in agreement with underground mine safety regulations and might assist Kazakhstan change the outdated regulations for the better.

Acknowledgments. This study supported by Nazarbayev University Grant Program for Research Grant #090118FD5337.

References

1. Kocsis, C., Hardcastle, S.: Ventilation system operating cost comparison between a conventional and an automated underground metal mine. Technical Papers, 55, pp. 57–64 (2003)
2. De Souza, E.: Improving the energy efficiency of mine fan assemblages. Appl. Therm. Eng. **90**, 1092–1097 (2015)
3. El-Nagdy, K.A., Shoaib, A.M.: Alternate solutions for mine ventilation network to keep a preassigned fixed quantity in a working place. Int. J. Coal Sci. Technol. **2**(4), 269–278 (2015)
4. Wu, Z.L., Li, H.S.: Simulation of mine ventilation under the influence of mine fires. In: Proceedings of the US Mine Ventilation Symposium, University of Utah, 21–23 June, pp. 359–363 (1993)
5. Wu, H., Gillies, A.: Real-Time Airflow Monitoring and Control within the Mine Production System. Eighth International Mine Ventilation Congress, pp. 383–389 (2005)
6. Sabanov, S.: Financial risk analysis of optimized ventilation system in the gold mine. In: Proceedings of the 27th International Symposium on Mine Planning and Equipment Selection - MPES 2018. Springer Nature (2019). https://doi.org/10.1007/978-3-319-99220-4_3
7. Kazakhstan mining regulations requirements for underground mine ventilation (2004). http://egov.kz/cms/ru/law/list/P090002207_
8. McPherson, M.: Subsurface Ventilation and Environmental Engineering. Chapman & Hall, London (2009)
9. Wang, X., Xiong, J., Xie, J.: Evaluation of measurement uncertainty based on monte carlo method. In: MATEC Web of Conferences, ICCEMS 2018, vol. 206, p. 04004 (2018)
10. Ni, Y.: Practical Evaluation of Uncertainty in Measurement, 5th edn. pp. 35–54. China Zhijian Publishing House, Standards Press of China, Beijing (2015)
11. Pyrcz, M.J., Deutsch, C.V.: Geostatistical Reservoir Modeling. OUP, USA (2014)

Barite in Sediments from Underground Waters of Hard Coal Mines of the Upper Silesian Coal Basin (Poland)

Hubert Makuła(✉) and Zbigniew Bzowski

Department of Environmental Monitoring, Central Mining Institute,
Plac Gwarkow 1, 40-166 Katowice, Poland
{hmakula, zbzowski}@gig.eu

Abstract. The article presents the results of research on sediments from underground waters of hard coal mines of the Upper Silesian Coal Basin (USCB) carried out in the Department of Environmental Monitoring of the Central Mining Institute in Katowice (Poland). In the sediments, barite containing radium was found. This barite, as a mineral originating from underground waters, co-occurs with barite coming from gangues, present in hard coal deposits. The conducted mineralogical, physicochemical and radiometric studies allowed one to assess the properties of the underground water sediments. The assessment of the properties of the examined sediments takes into consideration the concentration of barite in order to separate its forms of occurrence, particularly in a relation to the concentration of radium.

Keywords: Sediments coal mines · Barite · USCB

1 Introduction

The co-occurrence of groundwater containing radium isotopes with hard coal deposits is a very frequent phenomenon in the Upper Silesian Coal Basin (USCB), which has been described in the literature [3, 4, 7, 9]. During exploitation of hard coal deposits occurring in the Upper Carboniferous deposits of the Upper Silesian Coal Basin, underground water is removed from post-mining excavations. Exploitation of hard coal requires the removal of groundwater to the surface and drainage to surface watercourses due to the water hazard.

The necessity to limit the flow of drained groundwater to surface water, forces the accumulation of these waters in settling tanks for their purification from the suspension. Such tanks may be underground tanks located in inactive excavations or in surface, adapted, artificial ponds, or anthropopressively transformed water reservoirs. Both in the conditions of underground and surface settling tanks, the underground water supply system enables the formation and sedimentation of material containing elevated concentrations of various elements, including barium sulphate. Depending on the origin of groundwater from the hydrogeological levels of the Upper Carboniferous period of the Upper Silesian Coal Basin, sediments formed therefrom differ in their mineral composition [3]. Generally, the basic mineral and chemical composition of sediments in

E. Topal (Ed.): MPES 2019, SSGG, pp. 369–375, 2020.
https://doi.org/10.1007/978-3-030-33954-8_44

underground water settling tanks drained from hard coal mining excavations is similar to the composition of Carboniferous gangue rocks, mainly sandstones, occurring in the Upper Carboniferous deposits. In addition, the mineral composition of the sediments supplements the phases associated with the crystallization process of barite forms and other sulphates formed in sediments.

In the case of sediments formed as a result of precipitation, the mechanism of formation of barite as a mineral precipitated out of aqueous solutions is recognized [10, 13]. Similarly, the process of crystallization under laboratory conditions of barite from sediments containing Pb, Sr and Ra [2, 14] is recognized. Ra^{2+} ions may also precipitate out together with calcium carbonates ($CaCO_3$), calcite or/and aragonite [3]. One may assume that the process of precipitation of barite out of groundwater removed from hard coal mines in the Upper Silesian Coal Basin can be very efficient and can lead to the concentration of both barium and radium isotopes much greater than those found in the Earth's crust.

There are many doubts about the actual way of radium isotopes accumulation in sediments containing barite. It is necessary to recognize the forms of barite in sediments, especially in relation to radium isotope content. On this basis, one should consider whether this barite as a radium carrier can be a raw material used both for obtaining radium and purifying barite.

2 Materials and Methods

For the research on the mineral composition and the content of metals, 16 samples of sediments from the hard coal mines of the Upper Silesian Coal Basin, drained to the surface from underground water that differ significantly in chemical composition, were collected.

Mineralogical studies were performed by means of DSH power method in the geometry of the Bragg-Brentano using diffractometer D8 Discover by Bruker, CuKα radiation, Ni filter and LYNXEYE_XE detector. The mineral composition was determined and calculated on the basis of the licensed standards ICDD (International Centre for Diffraction Data) in PDF-4+ 2018 and 2019 RDB database and databases: ICSD (Inorganic Crystal Structure Database) and NIST (National Institute of Standard and Technology). For registration and diagnostic purposes, the program DIFFRAC v.4.2. Bruker AXS was used. The quantitative calculations of individual crystalline phases were made on the basis of the Rietveld methodology [1, 8, 11, 12] using the Topas v.4.2. Bruker AXS programme. The content of amorphous substance in the tested samples was calculated using ZnO as an internal standard.

The basic chemical composition of the examined fluidized ashes was determined by fluorescence X-ray spectrometry with dispersive wavelength (XRF). XRF Primus 2 by Rigaku Corporation was used for the determinations in Department of Environmental Monitoring Central Mining Institute in Katowice.

The measurements of radium isotopes activity concentration were undertaken using a high-resolution HPGe gamma-spectrometer with a detection limit of less than 1 Bq/kg, the given error quotes to 1-sigma reliability according to the Laboratory of Radiometry in Central Mining Institute certified internal procedure.

3 Research Results

3.1 Mineral Composition of Sediments

The following crystalline mineral phases were found in the examined samples of sediments from mining underground water:

- quartz SiO_2,
- feldspar: albite $Na[AlSi_3O_8]$ – ortoclase or microcline $K[AlSi_3O_8]$, clay minerals: kaolinite $Al_4[Si_4O_{10}](OH)_8$, chlorite $(Mg_5Al)[AlSi_3O_{10}](OH)_8$, muscovite $KAl_2[AlSi_3O_{10}](OH)_2$ – biotite $K(Mg,Fe,Mn)_3[AlSi_3O_{10}](OH)_2$, illite (K,H_3O^+) $Al_2[AlSi_3O_{10}](OH)_2$, montmorillonite $Ca_{0,2}(Al, Mg)_2[(OH)_2Si_4O_{10}]\cdot 4H_2O$ and mixed-layers minerals,
- sulphates: barite $BaSO_4$, gypsum $CaSO_4 \cdot 2H_2O$,jarosite $KFe_3(SO_4)(OH)_6$,
- carbonates: siderite $FeCO_3$,calcite $CaCO_3$, dolomite $CaMg(CO_3)_2$ – ankerite $Ca(Fe, Mg)(CO_3)_2$,
- other: halite NaCl, pyrite FeS_2,hematite α-Fe_2O_3, goethite α-$FeO(OH)$.

The composition is supplemented with an amorphous substance in the form of hard coal.

Quartz, clay minerals, calcite and barite were found in all the studied sediments, while feldspars, dolomite and hematite usually accompany them. Barite in the examined samples is present in amounts from about 2 to over 40%. The amounts were mineralogical documented (Fig. 1) and confirmed by barium numbers in the examined sediment samples (Table 2). Examples of diffractograms with different barite content in sediments from underground water are shown in Fig. 1.

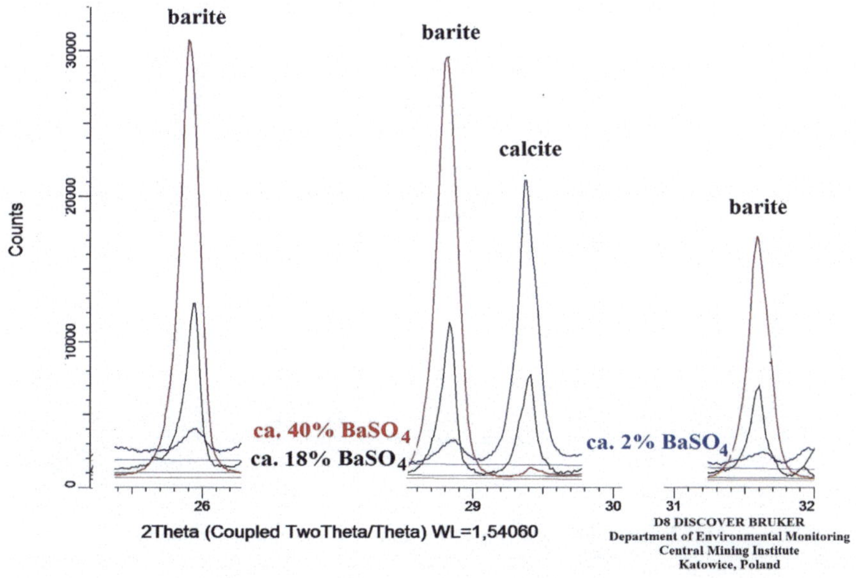

Fig. 1. The examples of diffractograms of sediments from underground water with different barite content.

Apart from barite, gypsum and jarosite were found in the examined sediments, which are probably products of secondary crystallization from sulphate waters. The genesis of halite in the studied sediments, as a result of evaporation of salt water in settling tanks or at the stage of preparation (drying) of test samples is similar. Sulphides, oxides and iron oxyhydroxide are also present in the examined sediments.

3.2 Forms of Barite

The barite in underground water sediments may occur as a mineral originating from the USCB gangue rocks, activated at the stage of mining works related to the exploitation of hard coal and this is the first form of barite (Fig. 2). The second form is the barite precipitated out of the water drained into the settling tank. Regarding phases, the barite may be a chemical compound of $BaSO_4$ or in phases containing such components as Sr, Pb and Ra. Table 1 presents diagnostic data from the PDF 4+ ICDD database for phases marked with the XRD method. Sediments showed the presence of barite, lead barite and strontium barite (Fig. 2).

After analysis of the obtained results of mineralogical studies, a general scheme of the occurrence of barite phases in the studied sediments can be summarized (Fig. 3). In the diagram, the documented XRD phases are marked in blue, while the phases discussed in the literature are green. In the absence of standards $(Ba,Pb,Ra)SO_4$, $(Ba,Sr,Ra)SO_4$ and $(Ba,Sr,Pb,Ra)SO_4$ in the PDF 4+ ICDD database (XRD), marked in brown on the scheme (Fig. 3), there was no occurrence in the examined sediments, which does not mean that these phases do not occur.

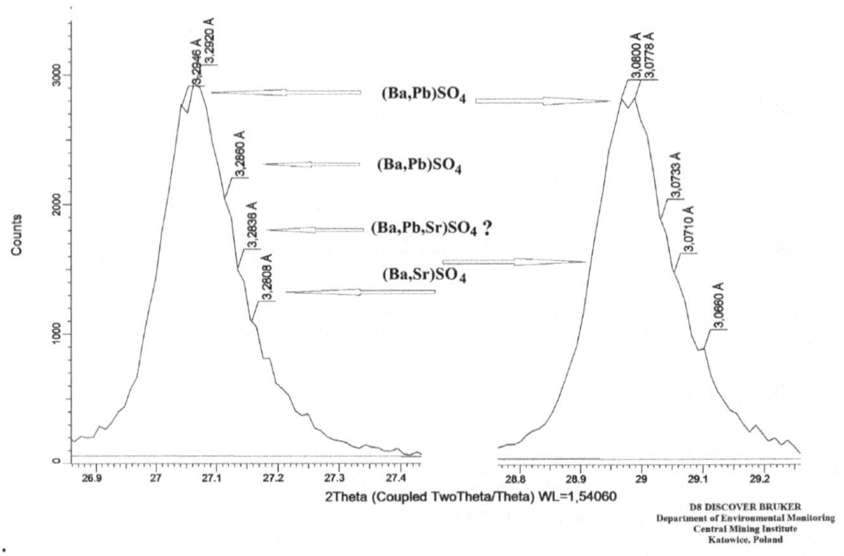

Fig. 2. Established forms of barite in the examined sediment.

Table 1. Diagnostic data from the PDF 4+ ICDD database for phases marked with the XRD method.

$BaSO_4$	$Ba_{0,8}Pb_{0,2}SO_4$	$Ba_{0,69}Pb_{0,31}SO_4$	$Ba_{0,75}Sr_{0,25}SO_4$	$Ba_{0,5}Sr_{0,5}SO_4$
d values (Å)				
3,435	3,417	3,413	3,406	3,371
3,310	3,296	3,287	3,281	3,245
3,095	3,080	3,075	3,070	3,038

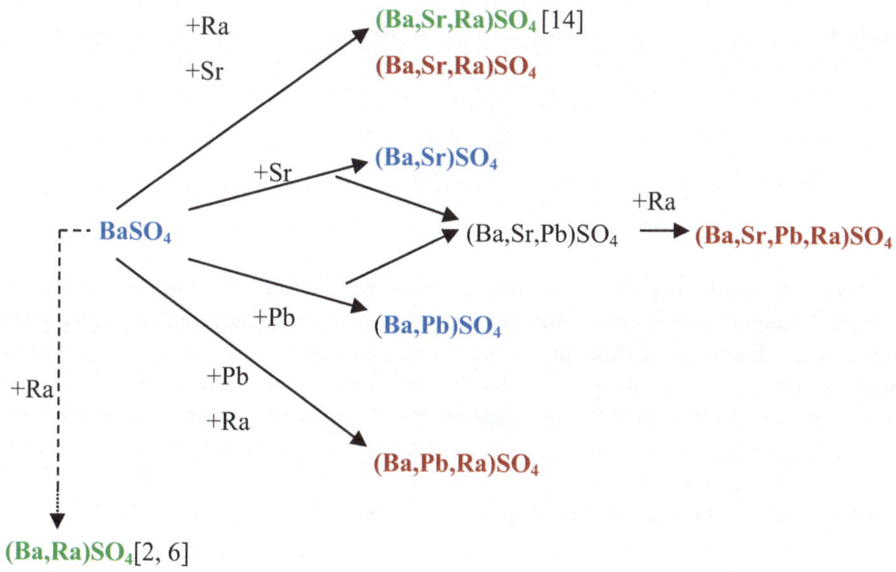

Fig. 3. Scheme of occurrence of barite phases in the studied sediments.

It is controversial that the crystalline phase $(Ba,Sr,Pb)SO_4$ occurs in the studied sediments, because there is only one standard $Ba_{0,977}Pb_{0,13}Sr_{0,01}SO_4$ in the PDF 4+ ICDD database, whose diagnostic data poorly identify peaks on the obtained diffraction patterns. Perhaps in the studied sediments there is a crystalline phase $(Ba,Sr,Pb)SO_4$, with a high proportion of strontium and smaller proportion of lead, which currently impedes to find evidence in the form of diagnostic data in XRD studies.

3.3 Metal Contamination

The results of indications of barium and strontium content as well as metal contamination of sediment samples from underground water are summarized in Table 2. The data in this table are supplemented with the results of measurements of natural radioactivity of the studied sediments. Table 3 presents the observed relationships between the content of Ba, Sr, Pb and ^{226}Ra in the studied sediments.

Table 2. Ba and Sr contents as well as metal impurities of examined sediment samples from underground water.

N = 16	Ba	Sr	Cu	Ni	Pb	Zn	^{226}Ra
	in %		in mg/kg (ppm)				in Bq/kg
Average	21,9	2,30	45	93	252	315	30 750
Min.	0,95	0,04	<5	<5	70	72	1 900
Max.	44,1	6,50	120	180	750	1080	68 000
Std. dev.	16,7	2,48	42	50	205	297	21 400

Table 3. Relations between the contents of Ba, Sr, Pb and ^{226}Ra at the level of significance $\alpha = 0.005$ in the examined sediments from underground water.

N = 16	Ba		
Sr	0,89	Sr	
Pb	0,26	0,18	Pb
^{226}Ra	0,97	0,79	0,40

From the results presented in Table 3, one may conclude that the correlations of barium, strontium and radium content are confirmed by strontium and radium diadochies in the barite. It is difficult to assess the position of lead in barite. Research indicates its occurrence along with other components in the studied sediments. One may assume that the reason for the decrease in the value of correlation coefficients Pb to Ba, Sr and Ra is the co-occurrence of lead along with other elements. In clay minerals and feldspars, lead can occur in the K – Pb – Ca system resulting from the similar values of the ionic radius of these elements.

4 Summary

The conducted mineralogical, physicochemical and radiometric research allowed to present the physicochemical properties of sediments, taking into account the barite content and the forms of its occurrence.

Established forms of crystalline phases of barite present in sediments from underground water settling tanks of the Upper Silesian Coal Basin indicate that the associated radium is present in sulphates sediment from these waters. The correlations of ^{226}Ra concentration with barium and strontium content indicate the connection of radium with strontium barite and strontium barite with lead. It cannot be ruled out that radium can be bound in other sulphates (gypsum, jarosite), which crystallize during the drying of the sediments. In addition, Ra^{2+} ions may also precipitate out together with calcite or/and aragonite [3].

In barite being present in the sediments, one cannot directly establish barium, strontium, lead and radium diadochy with X-ray diffraction method (XRD). However, the presented results of research on barite forms in sediments from underground water and the correlations found in barium, strontium and radium indicate that such diadochy exists in barite of the studied sediments. Therefore one should develop conditions and evaluate the possibility of removal of barite from sediment by flotation [5]. This will allow removal of barite and radioactive radium from the sediments, and obtaining the clay-quartz-carbonate material for commercial use.

References

1. Bish, D.L., Post, J.E.: Quantitative mineralogical analysis using the rietveld full-pattern fitting method. Am. Mineral. **78**(9–10), 932–940 (1993)
2. Brandt, F., Curti, E., Klinkenberg, M., Rozov, K., Bosbach, D.: Replacement of barite by a (Ba, Ra) SO4 solid solution at close-to-equilibrium conditions. Geochim. Cosmochim. Acta **155**, 1–15 (2015)
3. Bzowski, Z., Michalik, B.: Mineral composition and heavy metal contamination of sediments originating from radium rich formation water. Chemosphere **122**, 79–87 (2015)
4. Chalupnik, S., Wysocka, M.: Radium removal from mine waters in underground treatment installations. J. Environ. Radioact. **99**, 1548–1552 (2008)
5. Gurpinar, G., Sonmez, E., Bozkurt, V.: Effect of ultrasonic treatment on flotation of calcite, barite and quartz. Miner. Process. Extr. Metall. **113**, 91–95 (2004)
6. Heberling, F., Metz, V., Böttle, M., Curti, E., Geckeis, H.: Barite recrystallization in the presence of ^{226}Ra and ^{133}Ba. Geochim. Cosmochim. Acta **232**, 124–139 (2018)
7. Lebecka J.: Radioactive contamination in Upper Silesia caused by mine waters and sediments precipitated. Wiadomości Górnicze, no. 6, Katowice (1991)
8. Mahieux, P.Y., Aubert, J.E., Cyr, M., Coutand, M., Husson, B.: Quantitative mineralogical composition of complex mineral wastes – Contribution of the rietveld method. Waste Manag **30**, 378–388 (2010)
9. Michalik, B.: Radioactive contamination of environment caused by activity of underground mines. Prace Naukowe GIG, no. 883, Katowice (2011)
10. Pina, C.M., Becker, U., Risthaus, P., Bosbach, D., Putnis, A.: Molecular-scale mechanism of crystal growth in barite. Nature **395**, 483–486 (1998)
11. Rietveld, H.M.: A profile refinement method for nuclear and magnetic structures. J. Appl. Crystallogr. **2**, 65–71 (1969)
12. Rodriguez-Carvajal, J.: Structural analysis from power diffraction data the rietveld method. In: Proceedings of Ecole Cristallographie et Neutrons, Paris (1997)
13. Ruiz-Agudo, C., Putnis, C.V., Ruiz-Agudo, E., Putnis, A.: The influence of pH on barite nucleation and growth. Chem. Geol. **391**, 7–18 (2015)
14. Vinograd, V.L., Kulik, D.A., Brandt, F., Klinkenberg, M., Weber, J., Winkler, B., Bosbach, D.: Thermodynamics of the solid solution – aqueous solution system (Ba, Sr, Ra) SO4 + H2O: I. The effect of strontium content on radium uptake by barite. Appl. Geochem. **89**, 59–74 (2018)

Field Scale Assessment of Artificial Topsoil: A Victorian Coal Mine Experience

Anna Birjak[1(✉)], Alena Walmsley[1], Nicole Anderson[2], Jon Missen[2], and Mohan Yellishetty[1]

[1] Department of Civil Engineering, Monash University, Melbourne, VIC, Australia
annabirjak@gmail.com
[2] AGL Loy Yang, Loy Yang, VIC, Australia

Abstract. The Latrobe Valley is host to several open-cut coal mines which pose an environmental risk if poorly managed. To reduce risks associated with acid mine drainage and fire, progressive rehabilitation of batters is recommended. A lack of topsoil in the Latrobe Valley has led an industrial symbiosis to generate artificial topsoil's (ATS). Based on previous experiments, three ATS were created by combining overburden (OB), brown coal (BC), fly ash (FA) and one of the two waste products from a local paper mill – Effluent Sewage Recovery (ESR) or Enviroshield (ES). The ATS were compared to a local topsoil (TS). The study analyzed the physical and chemical properties of the topsoil's, seed germination and grass establishment. It was found that all ATS had similar properties and were alkaline compared to an acidic TS. The ATS were enriched in salts and found to be sodic to highly sodic and highly saline while TS was non-saline but slightly sodic. The ATS were also low in nitrogen and, due to the high pH of the soil, nutrients such as phosphorus, iron and manganese were not bioavailable. For the duration of the study, germination and grass establishment on ATS was not significant, however, grass was beginning to establish on the TS. As such, it is recommended further investigation be undertaken to determine an appropriate seed mix for the ATS which is tolerant to alkaline and saline soils. Further research is being conducted to determine the source of the alkalinity and salinity in the ATS.

Keywords: Artificial soil · Coal mining · Rehabilitation

1 Introduction and Background

Located in Central Gippsland, the Latrobe Valley is host to a large brown coal reserve which has been progressively mined since 1921 to serve the State's energy needs [13]. As these mines approach closure and the energy sector transitions to a less carbon intensive network, mine rehabilitation is becoming increasingly important.

Prior to rehabilitation, overburden (OB) dumps containing sulfide minerals, such as pyrite, were exposed to the atmosphere. When sulfide minerals are exposed to oxygen and water, they can react to form sulfuric acid, thereby lowering the pH of the soil leading to the mobilization of metals – a process commonly referred to as acid mine drainage (AMD) [6]. Additionally, the presence of coal in the batters heightens the risk

E. Topal (Ed.): MPES 2019, SSGG, pp. 376–389, 2020.
https://doi.org/10.1007/978-3-030-33954-8_45

of fire which could be potentially devastating as evidenced by the 2014 Hazelwood Fires. The fire that broke out at Hazelwood lasted for 45 days generating significant pollution which had adverse effects on the environment and nearby residents [22]. Interestingly, it was observed that rehabilitated sections of the mine did not catch fire. As such, progressive rehabilitation of batters which utilized OB in the process is recommended to mitigate the risks associated with AMD and the potential for a fire to ignite.

The progressive rehabilitation of batters involves:

- The infill of batters with OB to achieve a desired profile
- Application of topsoil
- Application of a seed mix to promote vegetation.

Currently, there is an insufficient supply of topsoil available in the Latrobe Valley to effectively rehabilitate mining affected land. To address this, an industrial symbiosis combining paper mill waste products and mining waste products has been proposed to create three artificial topsoil's (ATS) for the purpose of slope rehabilitation.

The ATS was created using OB, fly ash (FA), brown coal (BC) and one of the two waste products from a local paper mill - Effluent Sewage Recovery (ESR) or Enviroshield (ES). Laboratory based studies have been conducted to determine the optimum ratio of components (Table 1). The purpose of this study is to analyze the physical and chemical properties of the ATS compared to a local topsoil (TS) on the field scale. The success of the ATS will be defined by the successful germination of seeds and establishment of grass.

Table 1. The ratios of overburden (OB), fly ash (FA), effluent sewage recovery (ESR), Enviroshield (ES) and brown coal (C) in different ATS mixes.

	OB	FA	ESR	ES	C
Mix A	1	0.6	**0.2**	–	0.1
Mix B	1	0.6	**0.3**	–	0.1
Mix C	1	0.6	–	**0.2**	0.1

2 Materials and Methods

2.1 Site Description

The study was conducted between February 2018 and April 2018 at Loy Yang A, an open-cut coal mine located 12 km east of Morwell in the Latrobe Valley, Victoria (−38.23, 146.55; 56 m above sea level).

The area receives 745.2 mm of rainfall annually with spring being the wettest season [4]. During winter and, to a lesser extent spring, frost is common in the area which could have negative implications for seed germination [3].

2.2 Experimental Design

A batter located on the west side of the mine was cut down, filled with OB and compacted to create a 20° slope dipping to the east. The experiment was carried out on a section of the slope that was 280 × 70 m. The slope was split into 12 subplots, each 23 m wide. The 3 ATS and TS were applied to the slope in a randomized design (Fig. 1).

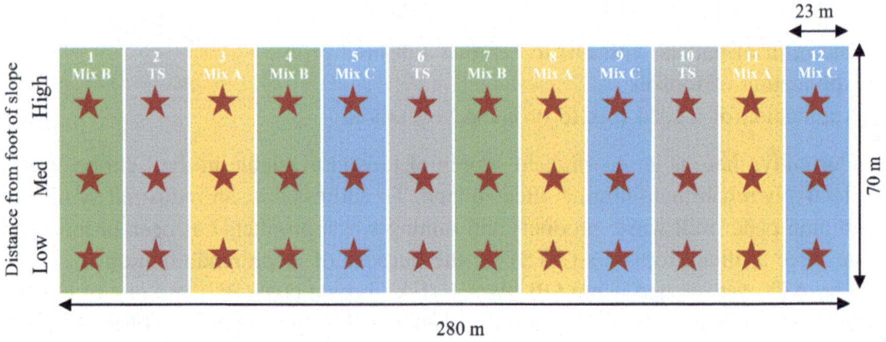

Fig. 1. Configuration that the topsoils were applied to the slope where red stars represent the location at which samples were taken

The ATS were mixed together on site using a row turner. The ATS and TS were then stockpiled at the foot of the slope. Dozers were used to push the soil from the foot of the slope up. This resulted in the soils being thicker at the foot of the slope compared to the top of the slope.

All subplots were sown with an identical pasture mix consisting of six different grass and legume species as specified in Table 2. At the time of sowing, fertilizer was applied at a rate of 270 kg/ha with a Nitrogen:Phosphorus:Potassium ratio of 11:15:14.

Table 2. Application rate, pH and salinity tolerance of the pasture mixed utilized in the field trial [1].

Plant species and variety	Application rate (kg/ha)	pH tolerance	Salinity tolerance, EC_e (ds/m)
Kingsgate perennial ryegrass	30	5.1–8.4	2.25–4.5
Riverina sub clover	6	6–7.3	0–1.10
Collamon sub clover	6	6–7.3	0–1.10
Uplands cocksfoot	4	5.5–7.5	2.25–4.5
Yarck 11 cocksfoot	4	5.5–7.5	2.25–4.5
Nomad white clover	2	6–7.5	0–1.10

2.3 Soil Monitoring

Monitoring occurred on a bimonthly basis with a total of two monitoring events undertaken. The first monitoring event occurred on the 28th of February, 2018 and the second on the 10th of April, 2018. Three samples were taken per subplot. Samples were taken in the middle of the subplot with distance from the foot of the slope varying from low (15 m) to medium (35 m) to high (55 m) (Fig. 1). Monitoring involved the sampling of soils followed by the analysis of their physical and chemical properties.

To measure the bulk density of the soil, samples were taken using a solid steel ring with a known volume of 100 cm^3. The weight of the samples was recorded, this is known as the wet weight. The soil was then dried in a conventional oven for 48 h at 105 °C. Following drying, the weight of the soil was recorded, this is referred to as the dry weight. Bulk density was then determined by dividing the dry weight of the soil by the volume of the steel ring used to collect the sample.

Soil samples were taken using a hand auger and targeted the top 100 mm of topsoil. The chemical properties of the soil were analyzed by Environmental Analysis Laboratory at Southern Cross University, using standard soil analytical methods. Major soil macro- and micronutrients were analyzed, as was pH and conductivity.

3 Results and Discussion

The properties of each topsoil were compared based upon the distance from the foot of the slope at which the sample was taken e.g. Mix B at a low distance compared to a Mix B at a medium distance from the foot of the slope (Fig. 1). This analysis was undertaken to determine if there were any significant lateral variation across the slope as the result of leaching. A similar approach was taken to determine if there was variation between subplots of the same mix e.g. Mix B on subplot 1 as opposed to on Mix B on subplot 4 (Fig. 1). Overall, the lateral variation was deemed insignificant and the properties of soils were analyzed based purely upon the mixture. It is assumed that this will change with time as leaching depletes the top of the profile of cations while it enriches the bottom layers of soil.

3.1 Texture and Bulk Density

Soil texture indicates the relative proportions of silt, clay and sand in the soil. Texture is indicative of various chemical properties such as cation exchange capacity (CEC). The results of the study show that the ATS formed clay loams while the TS formed a loam. These two textures are very similar with loams containing more silt and clay loams containing more clay. The implications of a higher clay content are an increased buffering capacity and increased CEC.

Bulk density (BD) affects the infiltration rate of water, root penetration and the ability for a soil to store water or air. The critical values at which BD restricts root penetration in loams and clay loams is 1.6 g/cm^3 [11]. Despite compaction applied by the dozers during spreading, the BD of all topsoil's was less than 1.6 g/cm^3. Thus, indicating there should be no negative impacts of BD in relation to root penetration.

The variation in BD over time relates to the stability and homogeneity of the soils. The BD of the TS was consistent over time while that of the ATS varied but appeared to converge at around 1.07 g/cm^3. This BD would provide optimal conditions for plant growth in a clayey soil [10].

3.2 PH

pH is one of the most important chemical properties of soil. pH influences nutrient availability, the solubility and reactivity of heavy metals and plays an important role in microbial activity and plant growth [5, 20].

Australian soils vary widely in pH dependent upon their regional setting and factors such as rainfall, leaching and subsurface lithology [5]. However, pH can be characterized on a more local scale with the Latrobe Valley being considered to host moderately to strongly acidic soils [25]. This local characterization of soil is likely to remain true on site as OB used to fill in the slope is characterized as being acidic [23].

The results from this study show the pH_w of the ATS varied from 8.4–8.8 while the pH_w of TS varied from 5.2–5.4 (Fig. 2). Based upon column leachate studies, the pH of the ATS was not anticipated to be so alkaline. The most alkaline component of the ATS was the FA with a pH_w of around 9 which should have been counteracted by the acidity of the OB and BC which had pH_w of around 5.5 and 4.0, respectively.

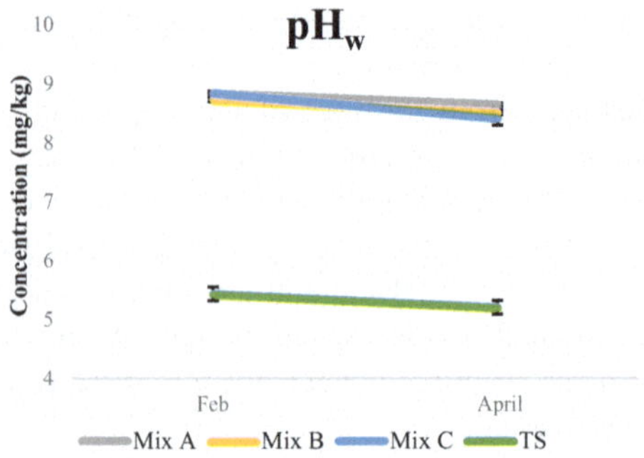

Fig. 2. Average concentration of pH for each topsoil as determined using a 1:5 water method.

Interestingly, the pH was observed to decrease for all soils between the two sampling events in February and April. This was unexpected as the soils were assumed to have a high buffering capacity due to the high CEC and high concentrations of organic matter (OM) within the soils. The rate of change was greatest in Mix C followed by Mix A, Mix B and, lastly, the TS. The decrease in pH is coupled with a decrease in OM. The reduction in OM is likely related to micro-organisms which breakdown OM and, in the process, create acid thereby lowering the pH of soils [27].

Although the ATS are still alkaline, the reduction in pH will be beneficial as it now approaches the upper tolerance limit of species within the pasture mix. However, the bioavailability of macro- and micronutrients such as phosphorus, iron and manganese which are required for healthy growth will be reduced given the alkalinity of the ATS. Finally, pH is intrinsically linked to sodium concentration with a pH greater than 8.5 indicating significant quantities of exchangeable sodium present which generates sodic soils and negatively affects soil structure [17]. This is further discussed below.

3.3 Salinity and Sodicity

Salinity and sodicity relate to the measurements of soluble salts present in the soil where salinity focuses on the whole salt while sodicity relates specifically to the sodium cation. Salinity and sodicity are important to predict the behavior of the soil in relation to flocculation and dispersion [17]. Soils with high concentrations of sodium will promote the generation of sodic soils leading to dispersion. Sodic soils are defined by an Exchangeable Sodium Percentage (ESP) greater that 6% where ESP is the concentration of sodium cations divided by the concentration of all cations pre-sent. Dispersion of soils due to high ESP can be counteracted if the EC of the soil is high and the soil will flocculate (Figs. 3 and 4).

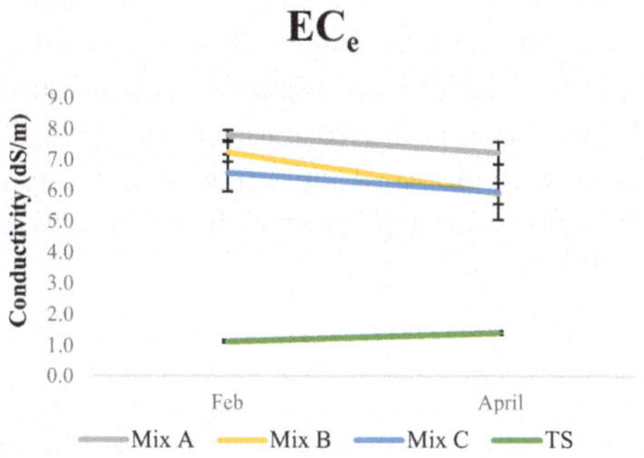

Fig. 3. Average EC_e for each topsoil determined using the measured EC_w and conversion rates for clay loam and loams of 8.6 and 9.5.

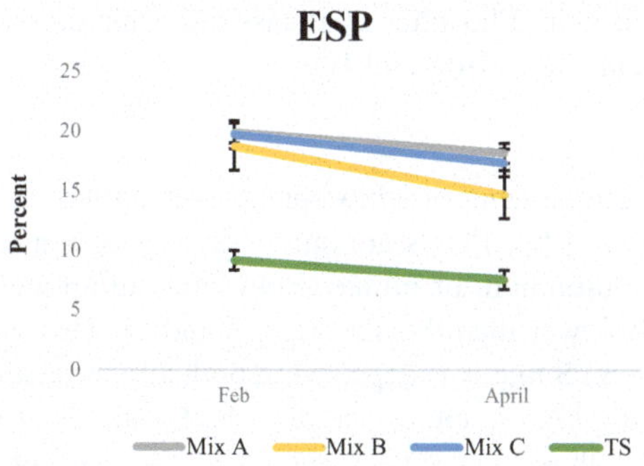

Fig. 4. Percent of exchangeable odium cations relative to other cations in the topsoil.

As expected, salinity and sodicity vary greatly across Australia. Northcote & Skene (1972) mapped soil morphology across the continent and defined 25 different salt-affected soil classes. Northcote & Skene (1972) estimated that 193 to 257M ha was affected by sodicity while 39M ha was affected by salinity. This study formed the backbone of Australia's evolving study of salt affected soils with many follow up studies focusing on the impacts of salt affected soils on agricultural land [19, 21]. In addition, a more recent study has estimated the quantity of sodic and saline land in Australia to be 340M and 66M ha, respectively [20].

The Latrobe Valley occurs within a saline province which is defined as an area where salinity has been identified but not necessarily mapped [26]. Additionally, the area has been characterized as having sodosols which are soils with a strong change in sodicity between topsoil and subsoil with subsoils being sodic to highly sodic [24]. On site, subsoils will consist of OB which has been characterized as been low in salinity but having a variable ESP. As such, it is expected that soils will display sodosol properties [23].

The results from this study characterize the ATS as moderately to highly saline and sodic to highly sodic while the TS was non saline but sodic (Figs. 6 and 7).

Although issues with salinity were anticipated, column studies predicted ESP levels less than 5% [15]. It is assumed that the high ESP levels of this study are a due to the heterogeneity of FA and OB. The EC was relatively consistent between February and April (Fig. 6) while ESP was seen to decrease for all soils (Fig. 7). This could indicate leaching of sodium ions prior to the April sampling event. Currently, the high levels of salt within the soil will likely restrict plant growth with the exception of some saline tolerant species such as Perennial Ryegrass.

3.4 Nutrients

Nutrients are essential for plant growth and metabolism. Dependent on the quantity required for healthy plant growth, nutrients can be split in to macro- and micronutrients. Although nutrients are necessary, some nutrients, such as iron, become toxic as when in excess.

This study focusses on the following nutrients:

- Macronutrients: Nitrogen, Phosphorus, Potassium, Calcium and Magnesium
- Micronutrients: Manganese, Iron and Boron.

Nitrogen

Nitrogen is a major component of chlorophyll and amino acids and is therefore essential for plant growth [29]. Plants are only able to access nitrogen from soil in its mineralized forms of ammonium or nitrate. Both ammonium and nitrate levels of the ATS were significantly lower than the TS (Figs. 5 and 6). This was unexpected as the only component of the ATS that was deprived in both ammonia and nitrate was the OB with all other parts being high in either one of or both sources of nitrogen. Despite the application of fertilizer, there was little variation in the concentration of mineralized nitrogen in the ATS between February and April. It is assumed that ammonium applied to the ATS reacted with the alkaline soils and formed ammonia gas which vaporized.

Given the acidity of the TS, this process did not occur and the levels of mineralized nitrogen increased. It is assumed this trend will continue in the TS as clovers began to establish. Clovers are part of the legume species which are known to facilitate symbiotic bacteria such as Rhizobium which fix nitrogen from the atmosphere into its mineral forms leading to increased levels of bioavailable nitrogen [16].

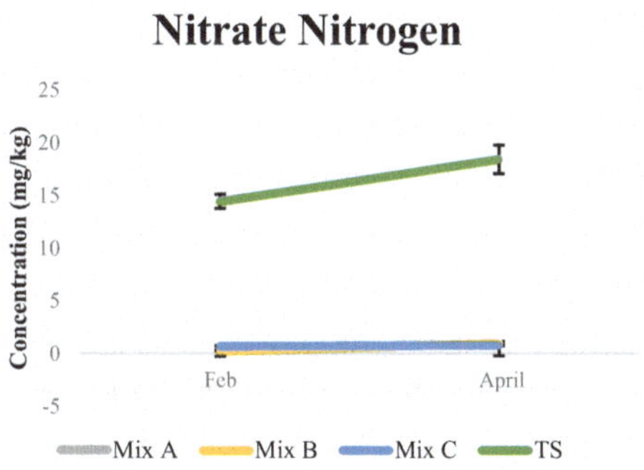

Fig. 5. Average concentration of Nitrate Nitrogen for each topsoil.

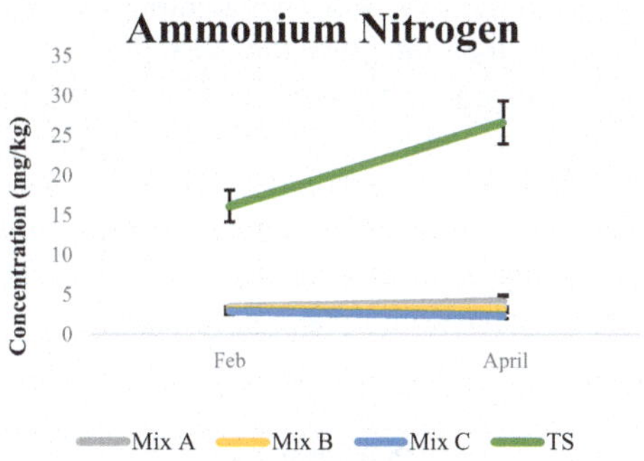

Fig. 6. Average concentration of Ammonium Nitrogen for each topsoil.

Phosphorus

Phosphorus is important to plants as it enables cell division, tissue development and energy conversion [9]. Similar to nitrogen, it was observed that the concentration of phosphorus in the ATS was significantly less than TS. Based upon column studies, it was expected that the ATS would have phosphorus at levels at least 10 times higher than what was detected in the field [15]. As pH controls phosphorus bioavailability, it is assumed the extremely low levels of phosphorus are related to the alkaline pH of the ATS.

The levels of phosphorus in the ATS were seen to double and triple between February and April while the levels in the TS decreased slightly (Fig. 7). The increase in phosphorus levels in the ATS is related to a decrease in pH coupled with the application of fertilizer which promoted the availability of phosphorus in the soil.

While, the slight decrease in phosphorus levels in the TS has been attributed to the germination and grass establishment which would consume phosphorus.

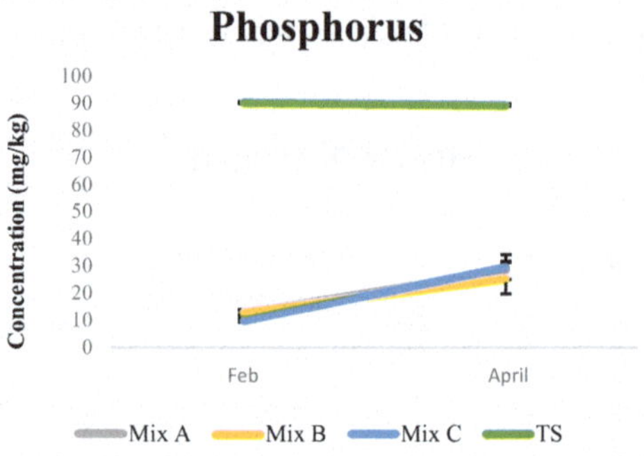

Fig. 7. Average concentration of Phosphorus for each topsoil.

Potassium

Potassium plays a vital role in plants by regulating CO_2 uptake and triggering the activation of enzymes [30]. Although the ATS were enriched in cations relative to the TS, the levels of potassium in the soils were comparable with the exception of Mix C which was significantly lower than the other topsoil's (Fig. 8). The variation between the ATS can be attributed to the nutrient source. Mix A and B utilized ESR which had a much higher potassium concentration relative to the ES used in Mix C. Despite this, all soils present levels of potassium which would enable plant growth. However, potassium levels must not be looked at alone as uptake of potassium will be prevented if other cations, specifically sodium, are in excess [8]. As such, the sodicity of the ATS must be addressed to ensure potassium uptake is sufficient.

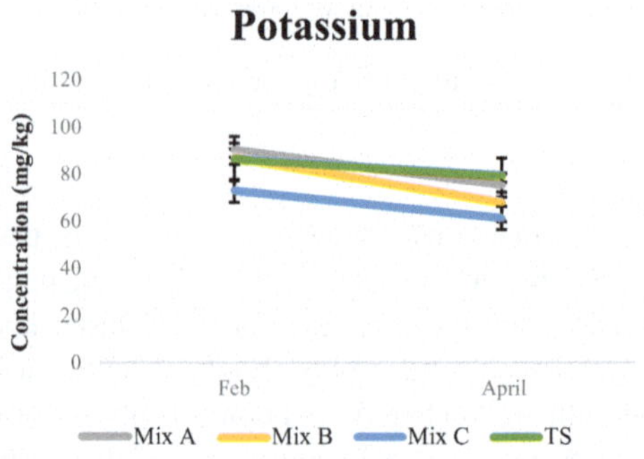

Fig. 8. Average concentration of Potassium for each topsoil.

Calcium

Calcium is utilized by plant to build and give structure to cell walls [28]. As expected, given cations were enriched in ATS relative to TS, levels of calcium within the ATS were significantly higher compared to the TS (Fig. 9). Levels of calcium were relatively consistent between February and April with the exception of Mix A. The concentration of Mix A was seen to decrease by 15%. However, as calcium isn't considered leachable, this has been attributed to the heterogeneity of calcium within the soil.

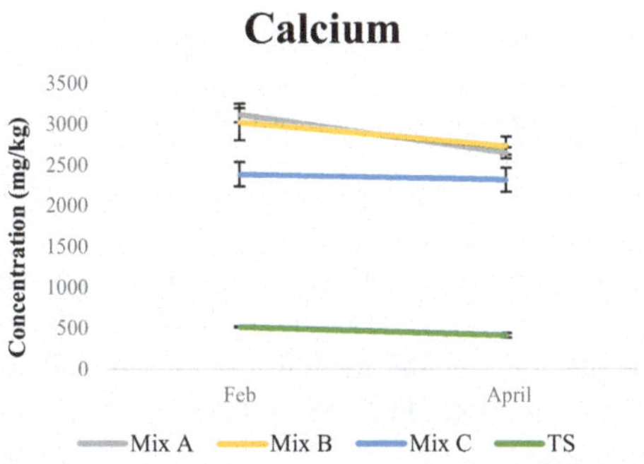

Fig. 9. Average concentration of Calcium for each topsoil.

Calcium is not considered toxic to plants so the high levels present in the ATS should not prohibit grass establishment [12]. The levels of calcium present in the ATS and TS should be sufficient to enable germination and grass establishment. However, it is important to note that the uptake of calcium maybe reduced by the high levels of sodium in the ATS [12].

Magnesium

Magnesium is essential to plant survival as it is utilized to produce chlorophyll [7]. As the ATS were enriched in cations relative to the TS it follows that the concentration of magnesium was also significantly higher in the ATS compared to TS (Fig. 10). Although the level of magnesium in the ATS is high, this won't necessarily correspond with efficient uptake. Cation competition, specifically with calcium and sodium, is likely to limit the ability for plants to uptake magnesium [7].

The concentration of magnesium in the ATS reduced between sampling events while the concentration within the TS remained constant. It is assumed the reduction in the ATS relates to the base components of each mix. Although all the concentration of magnesium decreased for all ATS the effects were much more noticeable for Mix A and B. As such, it assumed the reduction in magnesium is related to the use of ESR as a source of nutrients compared to ES.

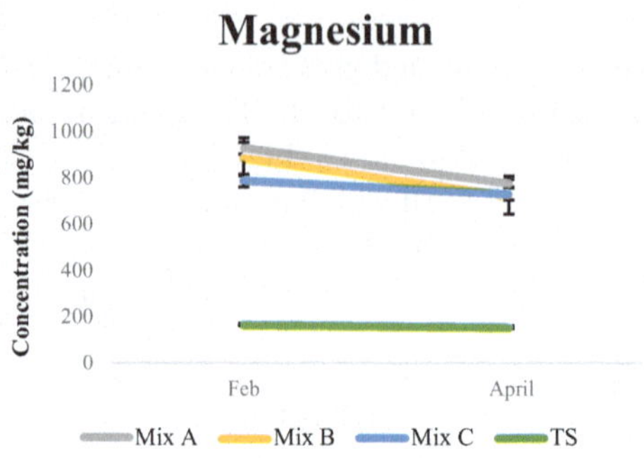

Fig. 10. Average concentration of Calcium for each topsoil.

Manganese

Manganese supports photosynthesis and is essential for growth however, when in excess, manganese can become toxic [14]. The bioavailability of manganese is controlled by pH. At high pH manganese availability is reduced while at low pH manganese availability can become toxic [14]. High OM and low nitrogen will also reduce the availability of manganese.

In the field trial, bioavailable manganese was seen to be significantly lower in the ATS compared to the TS (Fig. 11). With the exception of Mix A, the concentration of manganese was seen to increase between sampling events. As such, it is assumed this difference relates to pH with a decrease in pH increasing the bioavailability of manganese. However, this effect was negligible for Mix B and C based on the low initial concentration of manganese. As such, further reductions in pH coupled with increases in nitrogen would promote manganese concentration in ATS to sufficient levels.

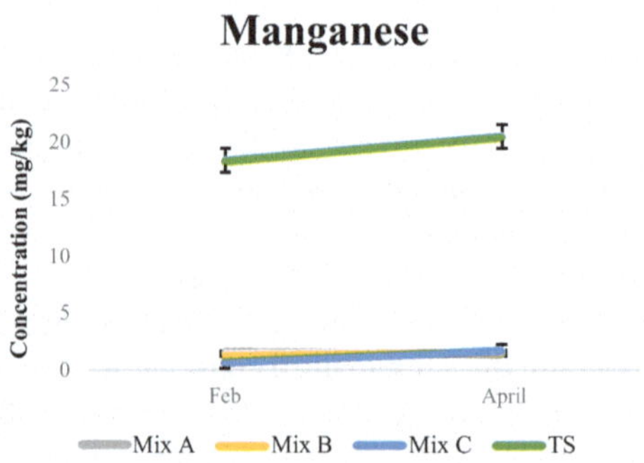

Fig. 11. Average concentration of Manganese for each topsoil.

Iron

Iron is used by plants to develop chlorophyll, enzymes and proteins however at high concentrations iron can become toxic [2]. Iron availability is heavily dependent on pH with high pH reducing availability and low pH increasing availability. Again, iron levels in ATS were seen to be very low compared to TS. This is likely due to the high pH of ATS making iron unavailable to plants (Fig. 12). Interestingly, although pH dropped between sampling events an increase in iron concentrations was only observed in Mix C. It's assumed these small changes in concentration are better explained by heterogeneity of the topsoil.

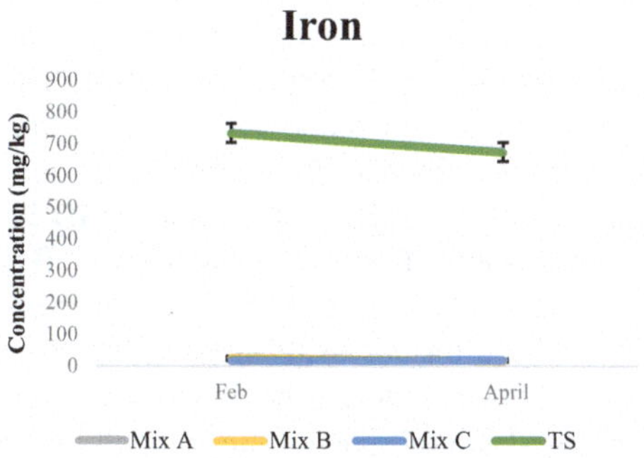

Fig. 12. Average concentration of Iron for each topsoil.

Boron

Boron plays a structural role in plant function and, although it is necessary for plant growth, in can be toxic when in excess [18]. Boron availability is again dependent on pH with boron being most available between pH of 5.5–7.5 or greater than 10.5. As such, boron toxicity can become an issue in alkaline soils. As expected, the concentration of boron in ATS was greater than in the TS (Fig. 13). The concentration was seen to decrease between February and April which corresponds with a decrease in pH. To ensure boron doesn't occur at toxic levels with the ATS, pH must be reduced.

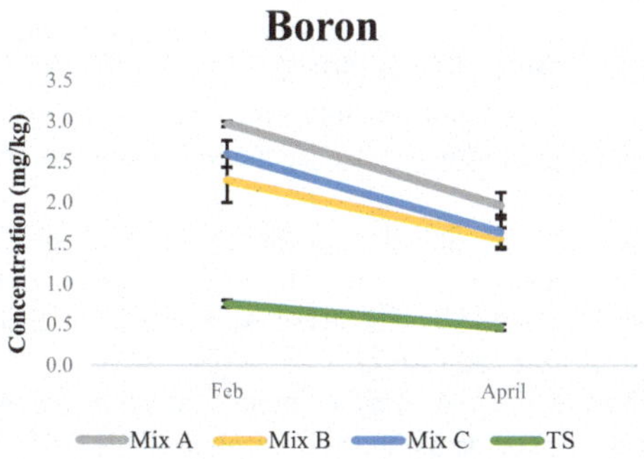

Fig. 13. Average concentration of Boron for each topsoil.

4 Conclusions

Industrial symbiosis was practiced to address the growing demand for topsoil within the Latrobe Valley. Three ATS were created and compared to a local topsoil. The ATS aimed to prevent AMD and reduce risks associated with fire whilst prompting seed germination and grass establishment. Overall, there was little significant variation between the three ATS. All ATS were alkaline and presented saline and sodic properties. The effect of alkalinity on nutrient availability was ultimately negative as it either made nutrients unavailable or in the case of boron, the availability was in excess leading to potential issue with toxicity. The properties of the TS were very different to the ATS as it was slightly acidic, non-saline and slightly sodic. Nutrients in the TS were sufficiently available and allowed for germination and grass establishment. Although the high alkalinity of the ATS will prevent AMD initially, its inability to establish grass will lead to erosion. If erosion is sufficient, OB will be exposed to the atmosphere which could potentially lead to AMD if left untreated.

Currently, the ATS has been unsuccessful in establishing grass cover. Going forward, it is recommended further monitoring be undertaken over the winter months to establish what effect increased rainfall has on the soils ability to support grass establishment. If the grass in unable to establish, it is recommended that the field trial be resown with an alkaline and saline tolerant seed mix. Additionally, further investigate is recommended to determine the root cause of the high pH and salinity of ATS.

Acknowledgements. We would like to thank Craig Skinner and Jason Muldoon for their help in the field. This research was funded by AGL Loy Yang and Industry Linkage SEED Grant 2428459 of Faculty of Engineering, Monash University.

References

1. Agriculture Victoria: Salinity and the Growth of Forage Species (2008). http://agriculture.vic.gov.au/agriculture/farm-management/soil-and-water/salinity/salinity-and-the-growth-of-forage-species. Accessed 4 Apr 2018
2. Briat, J., Dubos, C., Gaymard, F.: Iron nutrition, biomass production, and plant product quality. Trends Plant Sci. **20**(1), 33–40 (2015)
3. Bureau of Meterology: Annual and monthly potential frost days (2016). http://www.bom.gov.au/jsp/ncc/climate_averages/frost/index.jsp?period=an&thold=lt2deg#maps. Accessed 5 Oct 2017
4. Bureau of Meterology: Climate statistics for Australian locations (2017). http://www.bom.gov.au/climate/averages/tables/cw_085280_All.shtml. Accessed 5 Oct 2017
5. de Caritat, P., Cooper, M., Wilfor, J.: The pH of Australian soils: field results from a national survey. Soil Res. **49**(2), 173–182 (2011)
6. Evangelou, V.P., Zhang, Y.L.: A review: pyrite oxidation mechanisms and acid mine drainage prevention. Crit. Rev. Environ. Sci. Technol. **25**(2), 141–199 (1995)
7. Farhat, N., Elkhouni, A., Zorrig, W., Smaoui, A., Abdelly, C., Rabhi, M.: Effects of magnesium deficiency on photosynthesis and carbohydrate partitioning. Acta Physiol. Plant. **38**(6), 1–10 (2016)

8. Gourley, C.: Potassium. In: Peverill, K., Sparrow, L., Reuter, D. (eds.) Soil Analysis: An Interpretation Manual. CSIRO, Melbourne (1999)
9. Hopkins, B., Ellsworth, J.: Phosphorus availability with alkaline/calcareous soil. In: Western Nutrient Management Conference, vol. 6, pp. 88–93 (2005)
10. Hunt, N., Gilkes, R.: Farm Monitoring Handbook - A practical down-to-earth manual for farmers and other land users. University of Western Australia: Nedlands WA and Land Management Society: Como WA (1992)
11. Jones, C.: Effect of soil texture on critical bulk densities for root growth. Soil Sci. Soc. Am. J. **47**, 1028–1211 (1983)
12. Kirkby, E., Pilbeam, D.: Calcium as a plant nutrient. Plant, Cell Environ. **7**(6), 397–405 (1984)
13. Minerals Council of Australia (n.d) Brown Coal - Lignite. http://www.minerals.org.au/file_upload/files/resources/victoria/minerals_fact_sheets/Minerals_-_Fact_Sheets_-_Brown_Coal_-_Lignite.pdf. Accessed 15 Oct 2017
14. Mulder, E., Gerretsen, F.: Soil manganese in relation to plant growth. Ind. Eng. Chem. **4**, 221–277 (1952)
15. Mundodi, L.: Innovative Way to Mine Rehabilitation through Waste Utilization: A Case Study from Latrobe Valley. Monash University, Melbourne (2015)
16. Nesheim, L., Boller, N.: Nitrogen fixation by white clover when competing with grasses at moderately low temperatures. Plant Soil **133**(1), 47–56 (1991)
17. Northcote, K., Skene, J.: Australia Soils with Saline and Sodic Properties. CSIRO, Canberra (1972)
18. Reid, R.: Understanding the Boron transport network in plants. Plant Soil **385**(2), 1–13 (2014)
19. Rengasamy, P.: Transient salinity and subsoil constraints to dryland farming in Australian sodic soils: an overview. Aust. J. Exp. Agric. **42**(3), 351–361 (2002)
20. Rengasamy, P.: World salinization with emphasis on Australia. J. Exp. Bot. **57**(5), 1017–1023 (2006)
21. Rengasamy, P., Olsson, K.: Irrigation and sodicity. Aust. J. Soil Res. **31**(9), 821–837 (1993)
22. State Government of Victoria (2014) Hazelwood Mine Fire Inquiry, State Government of Victoria, Victoria
23. Taylor, M., Yellishetty, M., Panther, B.C.: Geotechnical and hydrogeological evaluation of artificial soils to remediate acid mine drainage and improve mine rehabilitation - an australian case study. In: Drebenstedt, C., Singhal, R. (eds.) Mine Planning and Equipment Selection. Springer, Cham (2014)
24. Victorian Government: Map of Victorian Sodic Soils (2014). http://vro.agriculture.vic.gov.au/dpi/vro/vrosite.nsf/pages/grains-soil-map-victoria. Accessed 25 Apr 2018
25. Victorian Government: Surface Soil pH (2015). http://vro.agriculture.vic.gov.au/dpi/vro/vrosite.nsf/pages/vic-soil_surface-soil-ph_map. Accessed 23 Mar 2018
26. Victorian Government: Victoria's Salinity Provinces (2017). http://vro.agriculture.vic.gov.au/dpi/vro/vrosite.nsf/pages/lwm_salinity-provinces. Accessed 25 Apr 2018
27. Wershaw, R.: Model for humus in soils and sediments. Environ. Sci. Technol. **27**(5), 814–816 (1993)
28. White, J., Broadley, R.: Calcium in plants. Ann. Bot. **92**(4), 487–511 (2003)
29. Zhao, D., Reddy, K., Kakani, K., Reddy, V.: Nitrogen deficiency effects on plant growth, leaf photosynthesis, and hyperspectral reflectance properties of sorghum. Eur. J. Agron. **22** (4), 391–403 (2005)
30. Zörb, C., Senbayram, M., Peiter, E.: Potassium in agriculture – status and perspectives. J. Plant Physiol. **171**(9), 656–669 (2014)

Application of Sodium Silicate Chemical Grouting to Tropical Regions

Hideki Shimada[1(✉)], Sugeng Wahyudi[1], Takashi Sasaoka[1], Akihiro Hamanaka[1], Yasuharu Toshida[2], and Tomohiko Abe[3]

[1] Kyushu University, Fukuoka 819-0395, Japan
shimada@mine.kyushu-u.ac.jp
[2] Toso Sangyo Co. Ltd., Chiyoda, Tokyo 102-0067, Japan
[3] Nittoc Construction Co. Ltd., Chuo, Tokyo 104-0061, Japan

Abstract. Recently, infrastructure at ASEAN countries like Indonesia is rapidly growing due to satisfy the high rates of economic growth in this region. The increasing of infrastructure is followed by increasing demand on chemical grouting. Unfortunately, there is no any established chemical grouting considering to the typical tropical soil of this region, which is acid or alkaline. Four decades ago, a chemical grouting has been developed in Indonesia. However, this chemical grouting has completely been prohibited by the government after the chemical is identified generates polymeric chemicals pollution. Based on this experience, this paper investigates the applicability of the chemical grouting of sodium silicate chemicals in tropical region with Indonesia as a case study. In the experiment, the chemical grouts were injected into Indonesian sands. After solidification of the chemical grouts, coefficient of permeability was measured by falling head hydraulic conductivity test. As the conclusion, it is indicated that the sodium silicate chemical grouting is useful to improve tropical acid soil environmentally friendly.

Keywords: Chemical grouting · Sodium silicate · Alkaline

1 Introduction

Infrastructure development in Indonesia has already evolved to an advanced level by solving complex problems [1]. In the infrastructure development area, construction work is increasing along with government expenditure budget [2]. Infrastructure projects such as train railways, highways, ports, and tunnels are sometimes required to improve the ground conditions for the safety and smooth construction. Therefore, it is desired to introduce the chemical grouting which is widely used in Japan as ground improvement in underground construction (Fig. 1). The chemical grouting is one of the ground improvements that improves the strength and permeability of the ground. This method has several superiorities in terms of the application: the machine equipment is compact, it can be constructed even in a narrow space, and the cost is relatively small.

In 1975, the use of polymeric chemicals in the construction of dams in Indonesia resulted in health damage. It is prohibited to use all chemical grouting in the country since the accidents. In order to reduce environmental issues, sodium silicate chemicals

E. Topal (Ed.): MPES 2019, SSGG, pp. 390–398, 2020.
https://doi.org/10.1007/978-3-030-33954-8_46

as lowest toxicity material is commonly used in the chemical grouting. However, there is no application of sodium silicate chemicals in Indonesia.

From these backgrounds, this research discusses the applicability of the chemical grouting of sodium silicate chemicals in Indonesia from the aspects of its functions. In order to discuss its applicability, the chemical grouts injecting test were carried out by using the samples of sands which were collected from Indonesia. Solidification of the chemical grouts were also evaluated by means of various laboratory tests.

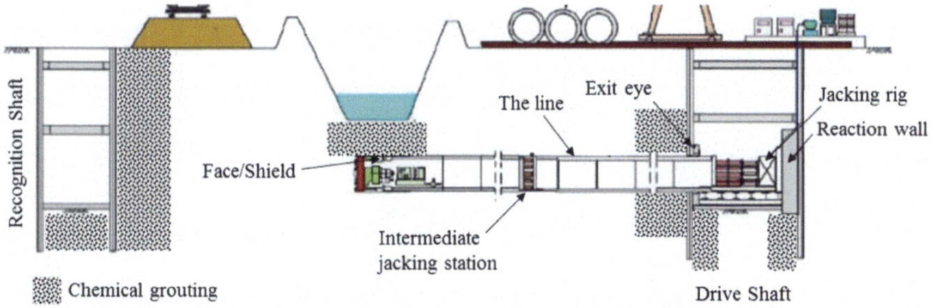

Fig. 1. Chemical grouting application on infrastructure development.

2 Material

2.1 Chemical Grout

The chemical grouts consist of two kinds of liquids (Liquid A and Liquid B). Liquid A is mainly composed of sodium silicate. Liquid B is composed with water and reactant. A certain amount of time is required for the chemical grouts to be solidified after mixing the liquids A and B with each other. This time is called gelling time. Solidification of chemical grouts occurs by reaction of sodium silicate of liquid A and acid contained in reactant of liquid B to form a polymer. The reaction formula is shown as (1).

$$Na_2O \cdot nSiO_2 + 2NaHCO_3 \rightarrow 2Na_2CO_3 + nSiO_2 + H_2O \tag{1}$$

As shown in Eq. (1), gelling time depends on the amount of sodium silicate contained in liquid A and the amount of acid contained in liquid B. In this study, chemical grouts with different gelling time are prepared by arranging the ratio of liquids A and B: chemical grout with gelling time of several seconds (Type 1) and several minutes (Type 2) were used. Table 1 shows the components of reactants contained in liquid B. Tables 2 and 3 show the composition of chemical grout and the results of gelling time measurement respectively. Based on the results of gelling time, it is expected that Type 1 can be used for improving the ground with high permeability due to rapid gelling time. On the other hands, Type 2 can be used to improve the ground with low permeability because it has a longer gelling time.

From the above, this study focuses on the construction as shown in Fig. 2. Chemical grout with rapid gelling time is injected as primary injection for sealing around the injection pipe and reducing the influence of ground water by roughly grouting. After that, chemical grout with a longer gelling time is injected as secondary injection for fully ground improvement in a wide area where the primary injection is performed.

Table 1. Reactants composition of liquid B.

	$NaHCO_3$	$KHCO_3$
Reactant I	60–80%	10–30%
Reactant II	40–60%	40–70%

Table 2. Chemical grout composition (Type 1).

Liquid A (50 mL)		Liquid B (50 mL)		
Water	Sodium silicate	Water	Sodium silicate	
			I	II
51.8%	48.2%	51.8%	48.2%	51.8%

Table 3. Chemical grout composition (Type 2).

Liquid A (50 mL)		Liquid B (50 mL)		
Water	Sodium silicate	Water	Sodium silicate	
			I	II
51.8%	48.2%	93.8%	–	6.2.%

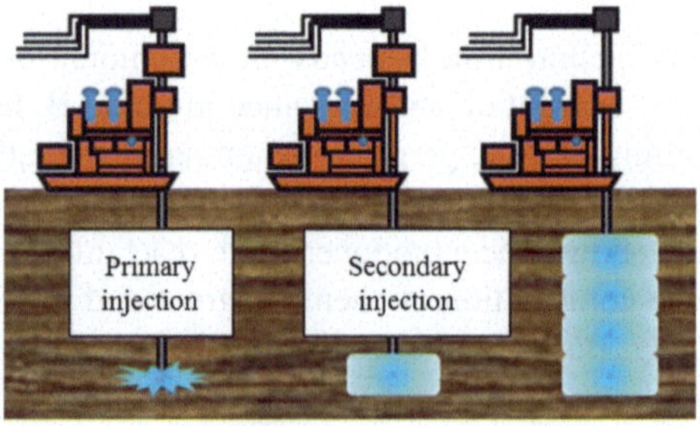

Fig. 2. Chemical grouts are injected into the grounds.

2.2 Grouted Soil

In addition to Volcanic Sand collected from the lahar deposit at the foot of the Gunter volcano in Java Island, Clayey Sand simulating the clay-containing soil in Indonesia, and Acid Sand which is adjusted pH by using sulfuric acid were used. Furthermore, Toyoura quartz sand was used as the soil which serves as a standard for improving effects when applying chemicals. The soil properties and particle size distribution are shown in Tables 4 and 5. These sands were mixed with each of two types of chemical grouts and solidified. They are called as Sand gel. A sand gel was prepared by mixing a soil sample and each chemical grout in a mold having an inner diameter of 5 cm and a height of 10 cm. In each sand gel, 300 g of soil sample and 100 mL of chemical grout were used.

Table 4. Permeability of pH of each sand.

	pH	Permeability (cm/sec)
Toyoura Quartz Sand	7.45	6.45×10^{-3}
Volcanic Sand	6.94	3.42×10^{-3}
Clayey Sand	6.27	5.43×10^{-3}
Acid Sand	4.07	8.87×10^{-3}
	2.3	

Table 5. Grain size distribution of each sand.

	Coarse Sand	Medium Grain Sand	Fine Grain Sand	Clay
The Particle Size	4 mm – 2 mm	2 mm – 425 μm	425 μm – 75 μm	Less than 75 μm
Toyoura Quartz Sand	0%	5%	95%	0%
Volcanic Sand	5%	50%	45%	0%
Clayey Sand	0%	40%	40%	20%
Acid Sand	5%	50%	45%	0%

3 Laboratory Tests

3.1 Hydraulic Conductivity Improvement Analysis

The falling head hydraulic conductivity test to measure the coefficients of permeability was conducted in order to discuss the function of the chemical grouts from the aspect of improving permeability [3]. Sand gels were cured in water for some curing times in order to understand changes of the coefficients of permeability with time elapsed. Curing times were 1, 7, 14, 28 days. From the construction guidelines of the chemical grouting, if the hydraulic conductivity is higher than 1×10^{-4} cm/s, it was judged that there was a water-blocking effect [4].

4 Result and Discussion

4.1 Application to Toyoura Quartz Sand

Figure 3 shows the results of permeability test in Toyoura quartz sand. From the results, it was confirmed that Type 1 and Type 2 maintain the water-blocking effect for a long period of time. In addition, it was confirmed that coefficient of permeability in Type 1 and Type 2 has a value of about 2.5×10^{-5} cm/s on the 28 days after curing in water.

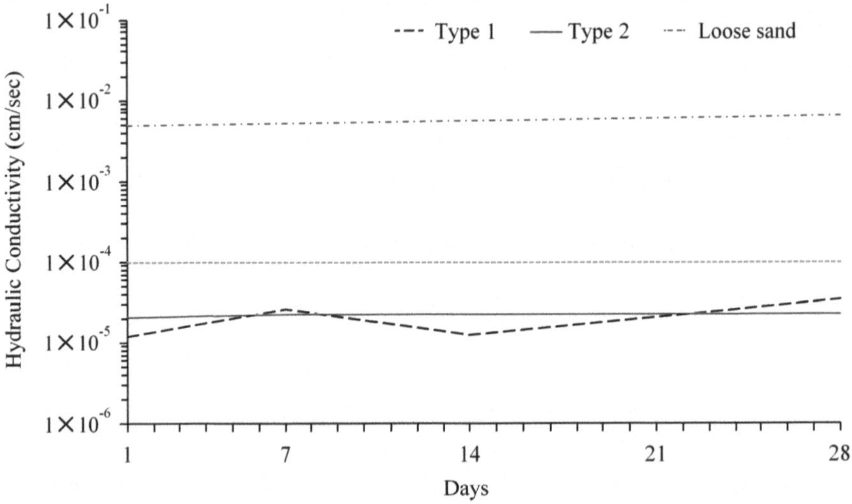

Fig. 3. Hydraulic conductivity performance of sand gels in Toyoura quartz sand.

4.2 Application to Volcanic Sand

Figure 4 shows the results of permeability test in Volcanic sand. From the results, it was confirmed that both of Type 1 and Type 2 in Volcanic sand satisfy the reference value of 1×10^{-4} cm/s which is the construction guideline of the chemical grouting. This fact can be explained that the permeability of the volcanic sand is high because the ratio of a large particle size such as coarse sand and medium grain sand is higher than other soils. Soil having higher permeability allows to penetrate chemical grouts largely. As a result, it seems that it was able to demonstrate the original improvement effect of the chemical grouts. That is, when sodium silicate-based chemical grouting is applied to Volcanic sand, sufficient improvement effect can be expected.

4.3 Application to Clayey Sand

Figure 5a shows the results of permeability test in clayey sand. In case of Type 1, the water-blocking effect was similar to that of Toyoura quartz sand. On the other hands, the coefficient of permeability increased with time elapsed and on 28 days after curing in water for Type 2, meaning that the sand gel was not satisfied 1.0×10^{-4} cm/s defined as the construction standard of the chemical grouting. As described above in Fig. 5b, the permeability of sand gel of both of Type 1 and Type 2 have to be satisfied 1.0×10^{-4} cm/s to ensure the water-blocking effect. Therefore, when sodium silicate-based the chemical grouting is applied to soil containing clay content, sufficient consideration is required if the construction period is prolonged.

It can be considered that the silica elution, which is one of deteriorating factors of sand gel, easily occurred due to the water absorbing action of the clay as a cause of the increase coefficient of the permeability of sand gel. Silica elution is a phenomenon in which the network of gels formed by solidified of the chemical grouts of Eq. (1) is destroyed by elution of unreacted silica due to penetration of water. Therefore, as a result of measuring the silica elution amount from each sand gel, it was confirmed that the silica leaching amount was larger in clayey sand than Toyoura quartz sand as shown in Fig. 5b. It is considered that the silica elution tends to occur easily in clayey sand because the clay easily absorbs water, meaning that the water contained clay affects silica elution. Furthermore, it can be confirmed that the silica elution amount is larger in Type 2 than Type 1. It is considered that clay absorbs more water before solidification in Sand gel of Type 2 because the gelling time of Type 2 is longer than that of Type 1. From the above, it was found that the water-blocking effect of the chemical grout is lowered due to the water absorbing action of the clay. For this reason, it is considered necessary to select a chemical grout taking account of the influence of clay when sodium silicate-based chemical grouting is adopted to clayey soil.

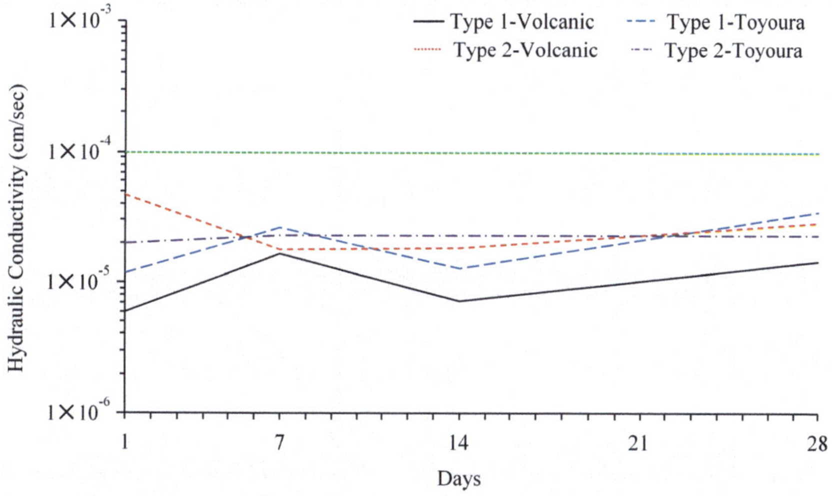

Fig. 4. Hydraulic conductivity performance of sand gels in volcanic sand.

4.4 Application to Acid Sand

Figure 6 shows the results of the permeability test in acid sand. It was confirmed that the water-blocking effect was confirmed for a long period of time in soil pH equals 2.30 and 4.07 when Type 2 was applied as in the case of applying to Toyoura quartz sand. On the other hand, it was confirmed that the standard value was not satisfied on sand gel to which Type 1 was applied on 7 days after curing in water. This is believed to be due to solidification of the chemical grouts before it penetrates the entire sand gel because the gelling time of the chemical grout is shortened due to the low pH of soil. Table 6 shows the gelling time of Type 1 and Type 2 when it is applied to acidic sands. Normally, a chemical grout forms a polymer by a reaction between sodium silicate and an acidic salt and solidifies, but the reaction between sodium silicate and sulfuric acid is occurred when it is applied to soil whose pH is adjusted with sulfuric acid, as shown in the Eq. (2). Therefore, it is considered that the solidification proceeded faster than usual, and the gelling time became shorter.

$$Na_2O \cdot nSiO_2 + H_2SO_4 \rightarrow nSiO_2 + Na_2SO_4 + H_2O \quad (2)$$

In Type 1, the gelling time could not be accurately measured because it was around a few seconds, but in Type 2, the gelling time is about 50% of normal in soil pH = 4.07 and about 20% of normal in soil pH = 2.30. That is, when the instantaneous type is applied in the case where the pH of the soil is acidic, the original effects of chemical grouting cannot be ensured due to its rapid gelling under acidic conditions. Therefore, it is considered necessary to select a chemical grout in consideration of shortening of gelling time when sodium silicate-based chemical grouting is applied to soil with low pH.

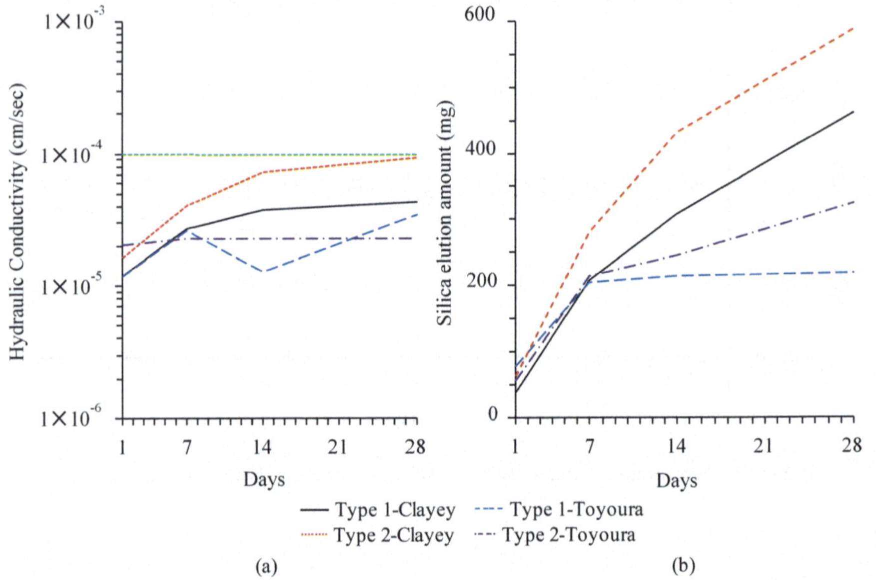

Fig. 5. (a) Hydraulic conductivity performances of sand gels in clayey sand; (b) Silica elution amount.

Table 6. Gelling time in soil with pH = 2.3 and 4.07.

pH	7.45	4.07	2.30
Type 1	±11 s	Unmeasurable	Unmeasurable
Type 2	±160 s	±86 s	±32 s

※Gelling Time at pH = 7.45 is the case of Toyoura

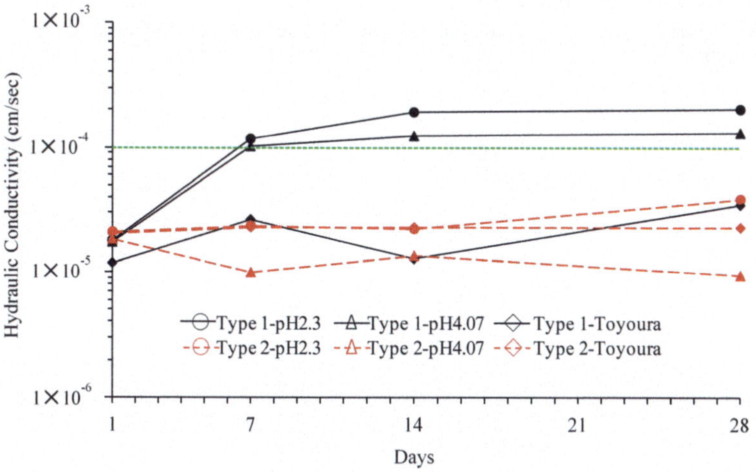

Fig. 6. Hydraulic conductivity performances of sand gels in acid sand.

5 Conclusion

In this study, we investigate the applicability of injection method for Indonesian soil conditions by conducting the falling head hydraulic conductivity test using sand gel prepared by soil samples simulating soil of Indonesia. The major findings are following.

- In the volcanic sand, the sodium silicate-based chemical grouting fully demonstrated the water-blocking effect.
- When applying the sodium silicate-based chemical grouting to clayey soil, it is necessary to select a chemical grout taking account of the influence of clay content.
- When applying the sodium silicate-based chemical grouting to soil with a low soil pH, it is necessary to select a chemical grout taking into consideration the effect of acid.

References

1. Arc Center of Excellence in Population Ageing Research: Asia in the Aging Century, Part 1 – Population trends (2013)
2. Pricewaterhouse Coopers LLP: Summary of South East Infrastructure Spending, Outlook to 2025 (2014)
3. American Society for Testing and Materials: Standard Test Method for Expansion and Bleeding of Freshly Mixed Grouts for Preplaced-Aggregate Concrete in the Laboratory, ASTM Standard C 940-10a (2010)
4. Japan Chemical Grouting Association: Design and Policy for Chemical Grouting Construction (1989)

Health and Safety in Brazilian Mines: A Statistical Analysis

Pedro Henrique Alves Campos(✉), Renan Collantes Candia, Luciano Fernandes Magalhães, Pedro Benedito Casagrande, Gilberto Rodrigues da Silva, and Viviane da Silva Borges Barbosa

Universidade Federal de Minas Gerais,
Antônio Carlos Av. 6627, Belo Horizonte, MG, Brazil
pedrocampos@demin.ufmg.br

Abstract. The extractive industry has always been considered one of the most dangerous for workers. However, there is a deficiency in finding updated materials that quantitatively describe work accidents in this segment of industry in Brazil. Through data provided by the Social Security, this work compares absolute numbers and other indicators of accidents and diseases in the mineral industry with data from other economic sectors. In addition, a detailed study on mining is done in order to check out specific sectors that are most responsible for these findings. The results show that the extractive industry is among the four economic sectors with respect to the highest occurrence of accidents, and is the first activity regarding mortality. The sectors which most contribute negatively to these results are the mining sectors of tin, coal, manganese and stone, sand and clay. In the study on the evolution between the years 2009 to 2016, it is observed a gradual decrease in the number of accidents and in the mortality rate of this industry in general, which indicates a greater concern of the companies on the safety and health subject.

Keywords: Work accident statistics · Occupational safety · Mine safety · Brazilian mines

1 Introduction

In the context of globalization, the issue of occupational safety and health poses a challenge for both governments and organizations, considering the social cost of occupational accidents.

According to Candia [1], despite the fact that in recent years significant reductions in the rate of injuries and accidents in mining have been observed, both the number and the degree of severity are still high. In this sense, the governments of developed countries consider strategic the improvement of the health and safety of the human resources involved in the productive processes.

Historically, the mineral extractive industry has been considered one of the economic activities that most exposes its workers to risks and damages to health [2]. Ramazzini, in his book *De Morbis Artificum Diatriba*, dated 1700, attributed this to two main causes: to the harmful substances that were manipulated, exhaled and

E. Topal (Ed.): MPES 2019, SSGG, pp. 399–407, 2020.
https://doi.org/10.1007/978-3-030-33954-8_47

absorbed during the work; and to the effort and inadequate postures to which the workers were subject during the extraction process [3]. Although Georgius Agrícola, in his book De Re Metallica, of 1556, already described the most common accidents and illnesses among miners [4], there were no laws or regulations to worker protection. In Brazil, the importance of prevention regarding occupational accidents went through milestones, such as the creation of the Ministry of Labor, Industry and Commerce (1930), the Regional Labor Stations (1940), the Consolidation of Labor Laws (1943), and the Regulatory Norms (1978).

According to the Law 8213 [5], "Work accidents are those caused by the exercise of work at the service of a company or a domestic employer or by the exercise of work of the special insured … causing bodily injury or functional disturbance that causes death or the temporary or permanent loss or reduction of the capacity to work."

Therefore, it is considered a work accident not only typical accidents (those occurring within the company or in the exercise of work), but also road accidents and diseases triggered by professional practice. These data are obtained from the Work Accident Communication (WAC).

With regard to the consequences, accidents can be classified as simple medical care, incapacity with leave of less than 15 days, incapacity with leave longer than 15 days, permanent disability and death. These data are obtained from the Single System of Benefits (SSB) [6].

Absolute numbers of causes and consequences of accidents, although important, are not good references for the evaluation of the risk exposure and damage to workers for each economic activity. Indicators of accidents supply this demand, being indispensable for the determination of prevention programs. Among them, the incidence rate of accidents, mortality rate and lethality rate stand out.

This study aims to promote discussion as well as bring some general statistics about work accidents in Brazilian mines. Other studies goes deeper into specific mining industries, refer to [7–9].

2 Methodology

The present study uses descriptive statistics to evaluate the current state of occupational accidents related to the mineral extraction industry in Brazil. The data were collected from the Statistical Yearbook of Accidents at Work (SYAW), made available on the Brazilian Social Security website. As this data began to be published since 2009, the analysis of this work comprises the period between 2009 and 2016, the last year with the information already disclosed. The SYAW uses the National Classification of Economic Activities (NCEA) for the stratification of economic activities. Table 1 shows the sections of the NCEA code. The focus of this work is the mining activities (extraction and beneficiation), which belong to section B - Extractive Industries. This section also includes oil and gas data that, although not considered mineral industries, were accounted for the analysis for the richness of the study. The detailing of the code with its subdivisions can be found on the Social Security website.

Table 1. NCEA code with sections and description

Section NCEA	Description
A	Agriculture, livestock, forest production, fishing and aquaculture
B	Extractive industries
C	Transformation industries
D	Electricity and gas
E	Water, sewage, waste management and decontamination activities
F	Construction
G	Trade, repair of motor vehicles and motorcycles
H	Transportation, storage and mail
I	Accommodation and food
J	Information and communication
K	Financial activities, insurance and related services
L	Real estate activities
M	Professional, scientific and technical activities
N	Administrative activities and complementary services
O	Public administration, defense and social security
P	Education
Q	Human health and social services
R	Arts, culture, sports and recreation
S	Other service activities
T	Domestic services
U	International organizations and other extraterritorial institutions

3 Results

The first evaluations of occupational accidents in the extractive industry aim at making a comparison with the indicators presented by other economic activities. Taking the average number of occurrences between 2009 and 2016, section B accounts for almost 1% of total accidents in Brazil. It is also noted a large proportion of occurrences, 8%, classified as 'Ignored/Zero', which had no linkage to an economic activity (Fig. 1). The existence of this class shows that the Social Security data entry system has deficiencies. Because it was an inconclusive class, it was decided to exclude it in the following analyzes.

Absolute numbers can mask the reality of the risks and damages of an economic activity, since it is natural that a section with a high number of employment is more susceptible to the occurrence of accidents. This is the situation in Sections C, G, and F.

Although it is not among the first five sections of the NCEA with the highest absolute numbers of accidents at work, when comparing the incidence rate, the extractive industries section appears in 4th place, behind sectors E, Q and C, as seen in Fig. 2. The incidence rate is defined by Eq. 1:

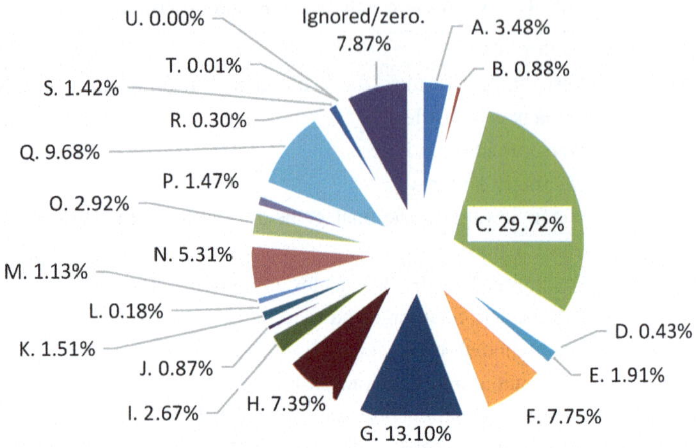

Fig. 1. Proportion of accidents and work diseases (2009–2016) by NCEA section.

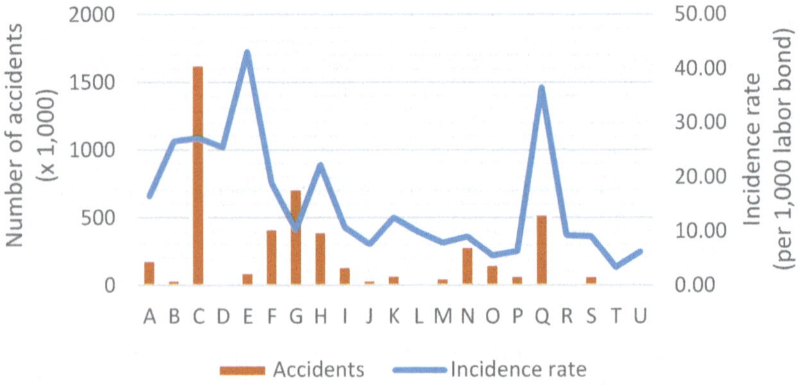

Fig. 2. Absolute number of accidents and incidence rate by NCEA section (2009–2016).

$$\text{Incidence rate (per 1,000 workers)} = \frac{\text{Number of annual accidents in work}}{\text{Average annual number of labor bond}} \times 1,000 \tag{1}$$

The incidence rate is an indicator of the intensity of accidents at work. By expressing the relationship between the working conditions and the average number of workers exposed to those conditions, it becomes a more adequate and realistic tool of analysis than the absolute numbers of accidents.

However, the perception of the importance of Occupational Health and Safety in the extractive industry is better evidenced by the number of mortality (Eq. 2) and lethality rate (Eq. 3), indexes in which this economic activity is in 1st and 2nd place, respectively. It is noticed again that the exclusive analysis of the absolute number of deaths, which is low in this industry in relation to the others, could lead to a skewed conclusion (Fig. 3).

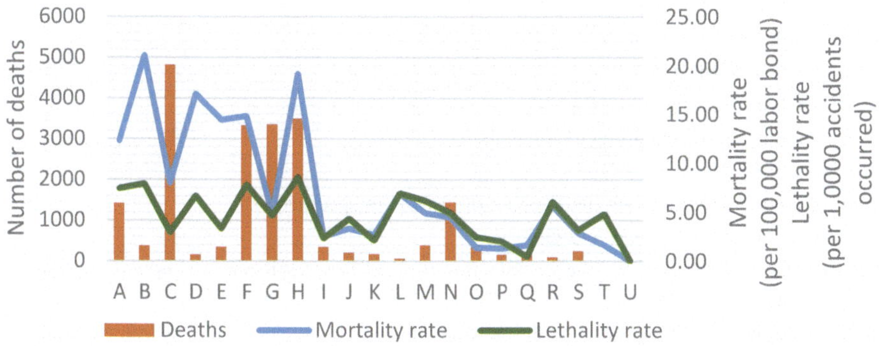

Fig. 3. Absolute number of deaths, mortality rate and lethality rate by NCEA section (2009–2016).

$$\text{Mortality rate}^{(\text{per } 100{,}000 \text{ workers})} = \frac{\text{Annual number of deaths}}{\text{Average annual number of labor bond}} \times 100{,}000 \qquad (2)$$

$$\text{Lethality rate}^{(\text{per } 1{,}000 \text{ accidents})} = \frac{\text{Annual number of deaths}}{\text{Number of recorded and unregistered accidents}} \times 1{,}000 \qquad (3)$$

In order to make a more detailed analysis of the mineral sectors with the greatest influence on the numbers already presented, the statistical analysis becomes stratified by the NCEA economic class. Table 2 shows the classes that compose section B - Extractive Industry. These classes comprises mining and beneficiation processes. In addition, oil and gas data were maintained for the richness of the study comparison. The details of each class can be found on the website of the National Classification Commission (NCC).

Figure 4 shows the number of accidents and the incidence rate of the Extractive Industry. It is observed that, despite the large number of accidents in the extraction of stone, sand and clay, oil and natural gas and their support activities, and iron ore; their incidence rates are not as high as that of other sectors. That is, because the sectors that generate the most jobs, they end up presenting high values of occupational accidents; however, the intensity with which accidents occur is not as high as the incidence rate shows. However, the mineral coal, radioactive minerals and tin sectors may be overlooked as to the absolute value of accidents, but the high incidence rate reflects the greater risk of accidents occurring in these environments. The manganese ore extraction class stands out with the lowest value.

Table 2. Classes that compose the Extractive Industry, according to the code NCEA.

Section B – Extractive Industry	
1	Extraction of coal
2	Extraction of oil and natural gas
3	Extraction of iron ore
4	Extraction of aluminum ore
5	Extraction of tin ore
6	Extraction of manganese ore
7	Extraction of ore from precious metals
8	Extraction of radioactive ore
9	Extraction of non-ferrous metal ores not elsewhere specified
10	Extraction of stone, sand and clay
11	Extraction of minerals for the manufacture of fertilizers, fertilizers and other chemical products
12	Extraction and refining of sea salt and rock salt
13	Extraction of gemstones (precious and semi-precious stones)
14	Extraction of nonmetallic minerals not specified above
15	Support activities for the extraction of oil and natural gas
16	Activities to support the extraction of minerals, other than oil and natural gas

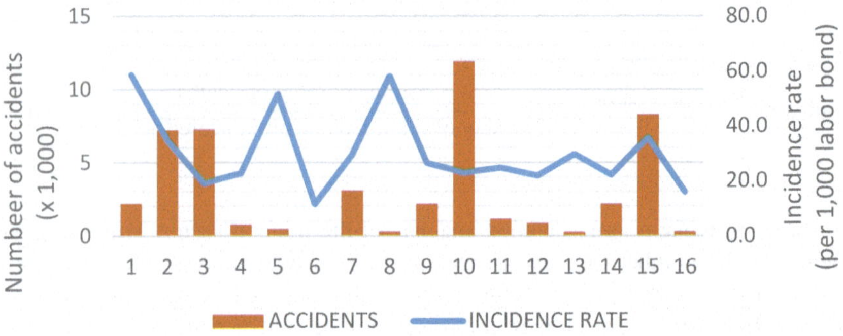

Fig. 4. Absolute number of accidents and incidence rate by sector of the Extractive Industry (2009–2016).

Figure 5 shows that mining of tin, coal, manganese and stone, sand and clay have the highest mortality rates, while manganese and stone, sand and clay are the most lethal. In these cases, oil and natural gas stand out with low numbers.

Finally, the evolution over the years 2009 to 2016 of accident numbers, their stratification by cause and consequence, as well as incidence rates, mortality and lethality can be visualized in Figs. 6 and 7. It is noticed that, influenced mainly due to the decrease in typical work accidents, the total number of accidents decreased over the years, as well as accidents which resulted in medical care, absence for more than 15 days and permanent incapacity. There is also a tendency to reduce the mortality rate, which is not accompanied by the lethality rate.

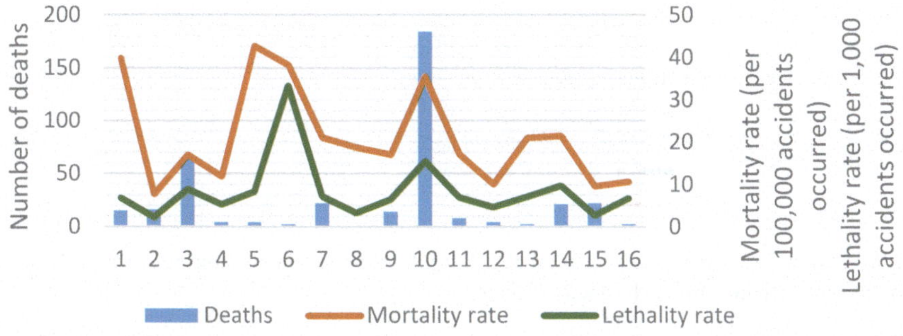

Fig. 5. Absolute number of deaths, mortality rate and lethality by sector of the Extractive Industry (2009–2016)

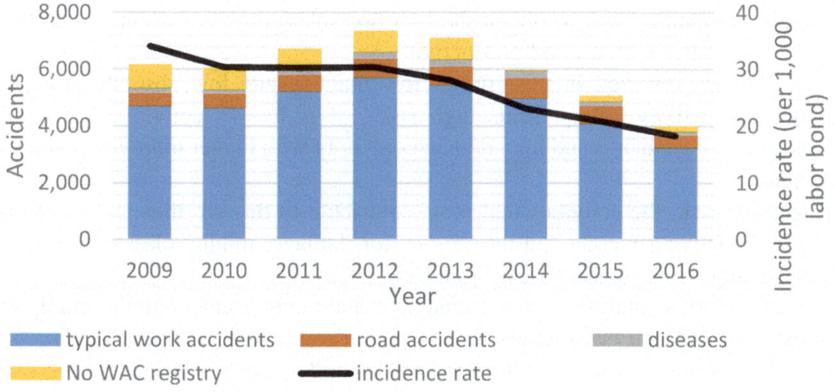

Fig. 6. Evolution of the number of accidents stratified by cause and incidence rate in the Extractive Industry.

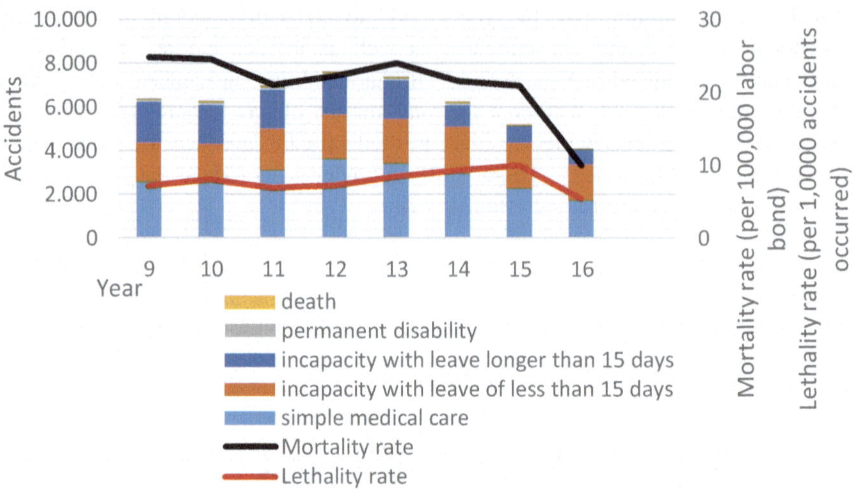

Fig. 7. Evolution of the absolute number of accidents stratified by consequence and rate of mortality and lethality in the Extractive Industry.

4 Conclusion

The Brazilian extractive industry has its detractors as a result of the legacy that was left behind by the ambitious and inhumane way that this activity was developed in the past. Nowadays, companies seek to obtain the best possible indexes of occupational health and safety, together with a good productivity.

The statistics revealed in this study show that, despite low absolute number of accidents, the activity still presents high risk - it is the 4th activity regarding the incidence rate of accident - and high damage - activity with higher mortality and the 2nd most lethal.

Regarding risk, the activities that most contribute to this are those related to coal mining, radioactive minerals and tin ore. As for damage, mining classes of tin, coal, manganese and stone, sand and clay stand out negatively.

Recent historical analysis shows a gradual reduction in accident and mortality rates in the extractive industry as a whole. Large mining companies have already absorbed a culture of prevention, working with good health and safety practices, making use of technology, skilled labor and making investments in the area.

Although the statistics shown here are revealing, it is important to mention that the numbers presented may be underestimated, as underreporting of accidents by companies is not unusual.

References

1. Candia, R.C., et al.: Análisis de la accidentalidad por caída de rocas en la minería de los Estados Unidos de Norte América. In: Segundo Congresso Iberoamericano en Minería Subterránea y a Cielo Abierto Umining. Santiago de Chile (2018)
2. Saleh, J.H., et al.: Safety in the mining industry and the unfinished legacy of mining accidents: safety levers and defense-in-depth for addressing mining hazards. Saf. Sci. **49**(6), 764–777 (2011)
3. Ramazzini, B.: As doenças dos trabalhadores – tradução de Raimundo Estrela, 4th edn, 321p. Fundacentro, São Paulo (2016)
4. Fundacentro: Introdução à Higiene Ocupacional. Fundacentro, São Paulo (2004)
5. Brasil: Lei n° 8.213, de 24 de julho de 1991. Dispõe sobre os planos de benefícios da previdência social e dá outras providências. Diário Oficial da União, Brasília, 24 de julho de (1991)
6. Ministério da Fazenda: Anuário Estatístico de Acidentes do Trabalho – AEAT 2017, vol. 1, Brasília (2017)
7. Vieira, A.V.: Avaliação do programa de proteção respiratória em uma mina subterrânea de ouro. (Dissertação), São Paulo (2004)
8. Zenaro, A.M.: Prevenção dos riscos de acidentes do trabalho na produção de pedra britada na mineração a céu aberto. (Dissertação), São Paulo (2017)
9. Gruenzner, G.: Avaliação da poeira de sílica: um estudo de caso em uma pedreira na região metropolitana de São Paulo. (Dissertação), São Paulo (2003)

Sustainable Development of Mining Resources

Socio-Economic Impact of Mine Closure and Development of Exit Strategy for Rural Mining Areas in Zambia: A Case Study of Kalumbila District

Bunda Besa(✉), Jimmie Kabwe, Jewette Masinja, and Webby Banda

School of Mines, The University of Zambia, P.O. Box 32379, Lusaka, Zambia
bbesa@unza.zm, bundabesa@yahoo.com

Abstract. Lumwana and Sentinel mines both located in Kalumbila District of Zambia are operational and are planned to continue operating until 2038 and 2033, respectively. It is anticipated that the closure of these mines would have a significant impact on the well-being of the local communities similar to the experiences of host communities in other areas where mines have been closed in Zambia. The local communities in Kalumbila District currently rely heavily on these mines for their socio-economic development in that the mines provide employment, education, social facilities and health services. In order to avoid negative experiences of these communities, an appropriate exit strategy is required for Lumwana and Sentinel mines, as part of their mine closure plan. This paper proposes a Community Exit Strategy (CES) that would mitigate the socio-economic impact of mine closure on the communities of Kalumbila District. The proposed exit strategy will alleviate the socio-economic impact of mine closure on communities. This is in line with Sustainable Development Goal 11.

The first step in developing the proposed Community Exit Strategy is re-establishing communication among the three key stakeholders i.e., government, local communities and investors. This should be done to provide a basis for the key stakeholders to have a common and clear understanding of their objectives and expectations. It is then important to carry out an assessment of the livelihood of the local community through a demographic study of the four main villages in the area namely; Mumena, Matebo, Mukumbi and Musele. This is aimed at determining the socio-economic characteristics of the population in these areas. Thereafter, an investigation of the potential socio-economic activities that could be undertaken in the area after mine closure will be carried out. The findings of the investigation will form the basis upon which the proposed community exit strategy can be implemented.

Keywords: Socio-economic · Mine closure · Kalumbila · Community · Natural resources · Zambia

E. Topal (Ed.): MPES 2019, SSGG, pp. 411–421, 2020.
https://doi.org/10.1007/978-3-030-33954-8_48

1 Introduction

Zambia's economy is significantly driven by mining. Mining alone accounts for 12% of its Gross Domestic Product (GDP), 70% of total export value and 62% of foreign direct investment [1]. Traditionally, Zambia's economy has benefited greatly from copper mining and still remains one of the key industries for economic growth [2]. The country's copper production increased by about 4% to 4.5% in 2018 and is projected to increase by the same margin in 2019 [3]. However, the country's reliance on copper mining puts it at risk to fluctuating world copper prices. Therefore, the Zambian government has been developing an economic diversification program to minimise the country's dependency on the copper industry. The program comprises of initiatives that encourage the expansion of alternative industries like agriculture, hydropower and tourism [3].

Zambia is among the highest copper producing countries in the world even though historically it has experienced ups and downs in production output. For instance, it produced 700,000 tonnes of copper in the 1970s and dropped to 255,000 tonnes in the late 1990s (Fig. 1). This was attributed to the lack of recapitalisation in mining assets.

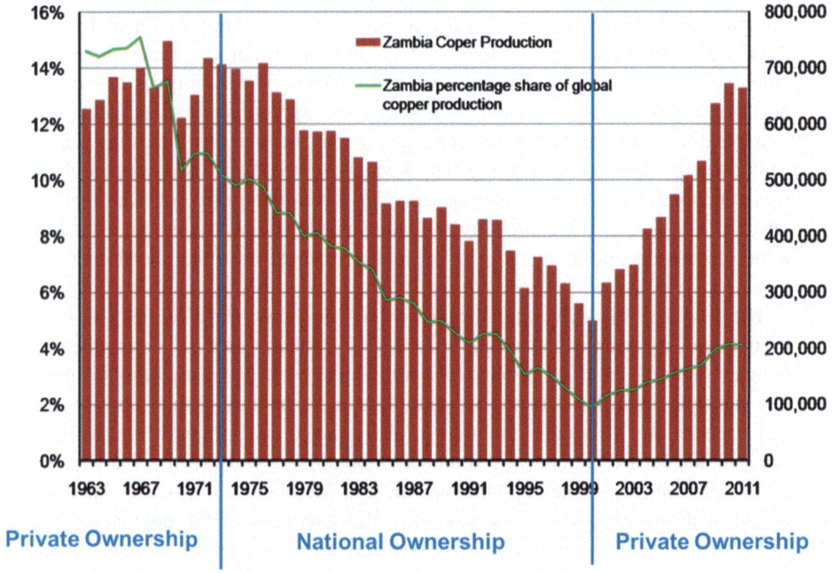

Fig. 1. Zambia's share of global production of copper between 1963–2011, reflecting ownership change [4].

Privatisation of the mines in the 2000s led to a revival of the industry due to an increase in investment by foreign investors [2]. It was reported that 715,000 tonnes of copper was produced equivalent to 4.4% of global output in 2011 making Zambia the sixth world largest copper producer [5]. Due to a number of expected expansion plans, Zambia was seen as a major growth area for copper production and was anticipated to rank under the top five copper producers globally going forward [5]. It has not quite made the top five yet, but has steadily increased its copper output, producing 799,329 tonnes and 861,946 tonnes in 2017 and 2018, respectively [6].

Copper mining in Zambia is currently focused in the rural areas of North Western province. In 2018, the province produced 575,077 tonnes out of a total production of 861,946 tonnes representing 67% of the country's annual production [6]. Three of the largest open pit copper mines in Zambia namely, Sentinel, Lumwana and Kansanshi operate in rural areas of the North-Western Province. Sentinel and Lumwana are both located in Kalumbila District (Fig. 2) while Kansanshi mine is located in Solwezi District (Fig. 3). These mines are major providers of income, employment and services such as education and health [7]. The closure of any of these mines would therefore, have a significant impact on the well-being of the community and the nation at large. It has to be noted that mines usually close for two main reasons i.e., either the reserves are depleted or they become unprofitable or legal non-compliance.

Fig. 2. Kalumbila Mine located in Kalumbila District in rural area of North-Western province

Fig. 3. Kansanshi Mine located in Solwezi District in rural area of North-Western Province

When mine closure occurs, various stakeholders raise post-mining environmental concerns in relation to the rehabilitation of the affected areas. Usually, minimal attention is paid to the socio-economic impacts of mine closure on the local communities [8]. Mine closure can have massive impact on a country's economy, depending on the operational size of the mine, however, for those people in the local communities the impact is traumatic [9]. The social services that the mines provide to the communities usually become unsustainable after mine closure. Local municipalities and the government do not usually have the ability to create alternative economic initiatives and hence cannot continue providing the services that were initially provided by the mines [10]. These issues lead to the economic, emotional and spiritual collapse of the communities which are usually then characterised by poverty, substance abuse, crime and violence [9]. This paper proposes a Community Exit Strategy as a partial solution to the socio-economic impacts of mine closure on the local community whose principle objectives would be:

(i) To re-establish communication among the three key stakeholders (government, investor and local community) in order to provide a basis for the key stakeholders to have a common and clear understanding of their objectives and expectations [11];

(ii) To carry out an assessment of the livelihood of the local community;
(iii) To investigate the potential socio-economic activities that can be undertaken in the area after mine closure in line with pre-mining land uses; and
(iv) To establish the roles and responsibilities of all the stakeholders in the implementation process.

2 Mine Closure

Mine closure is the process of terminating mining operations either temporarily or permanently. Bainton and Holcombe [12] describe it as a process that marks an important moment at the end of the mine life that involves more than closing down the metallurgical and mine site. It is more than a managerial-technical-aspect within the mine life cycle but also a social incident in the lives, of individuals, households, families, communities and local government [13]. Mine closure plans often focus only on environmental and physical aspects, such as land rehabilitation and asset removal, rather than social, cultural and economic aspects [8]. The intensified attention on sustainable development and its use in mining in the modern era has assisted in putting into context the necessity of addressing the similarly significant social aspects of mine closure [14]. Closure Plans must be developed in such a way that they comply with rules that support sustainable development. Scheduling for closing a mine must therefore, also concentrate on the various socioeconomic aspects of closure and not just on rehabilitating the environment [15].

There is insufficient literature regarding mine closure in Zambia. This is attributed to the fact that most of the mines that have been closed have been reopened such a Maamba Collieries Limited. Other mining companies have shut down certain mines but opened new ones within the same locality. In Luanshya town for example, Baluba mine was shut down due to plummeting copper prices and energy shortages while Muliashi mine remained operational despite being in the same area and owned by the same company [16]. Another reason, is the absence of strong provisions of this requirement in the Zambian body of law. Most of the available literature addresses the health issues related to the closure of Kabwe and impact of privatisation of Zambia Consolidated Copper Mines (ZCCM) [17]. It was noted that there was insufficient information regarding the social impact of mine closure in both developed and developing countries. Bainton and Holcombe [12] equally explained that compared with the substantial body of literature detailing the socio-economic, cultural, and political impacts of mining, there are relatively few publications that specifically address the social aspects of mine closure and associated issues of planning and managing the social domain towards the end of the project life-cycle.

3 Social Impacts of Mine Closure

Social impacts of mine closure are often neglected despite being as serious as environmental and economic ones [13]. The community challenges resulting from poorly carried out mine closure may cause serious problems for mining corporations [18]. For communities, on the other hand, it may cause a lot of stresses due to the risk of economic and social meltdown [18]. Constructing and commissioning a mining project pulls local people, businesses, and government agencies into an economic process. Each of these parties will, to different degrees, become dependent on the mine [19]. Some of these dependencies are mutual, whereas others are at the convenience of one or the other party [19]. Some mining companies add value to nearby communities through building or funding schools, clinics, hospitals, social centres, self-help schemes, roads, houses or sports facilities [20]. The people living in an area where mining is undertaken are given the chance to work with the mine to guarantee sustainable benefits from its operations. However, the employees and the communities at large are usually not ready for the impacts of mine closure [14]. This is due to the anticipated benefit losses that may result from the outcome, and sometimes, because of lack of awareness raising by the mine operators who are planning the closure.

Mine closures can have severe impacts on local economies, add to impoverishment, they prompt the loss of crucial services such as health and education, and cause migration which may consequently affect employment, and housing prices ([8, 12]) as well as closure of businesses that were established to service the mine operations. Ineffective mine closure activities, in reality, leads to interruption of community cohesiveness and it can also cause displacement of communities [21]. In developing nations, schools and clinics run by mining companies end up in the hands of the municipalities and that is usually a problem [14] since these may often not have capacity to sustain these services, as experience in Zambia has shown. Haney and Shkaratan [10] states that accommodation and public services are also negatively impacted by mine closure. For example, mine workers lose their right to housing, hence, the houses are abandoned and then inhabited by illegal occupants. In addition the structures and facilities previously owned by closed mines are usually vandalized hence compromising the safety of the mining communities [10]. Sometimes, abandoned mining tailings disposal site become illegal mining sites by the local impoverished community, who work these facilities at great risk to other persons.

Mine closure leads to social transformations in local communities associated with job loss such as escalation in poverty [22]. For instance, the closure of the Kabwe Lead Mines in Zambia in 1994 resulted in nearly 1,200 direct job losses and an additional 5,000 jobs by contractors [23]. The mine closure of Kabwe Lead Mine led to rapid decline of the local economy with minimal diversification being developed [24]. Luanshya Copper Mine in Zambia dismissed 1,600 employees when it closed its Baluba mine due to plummeting copper prices and energy shortages [16]. The closure of Zambia's Bwana Mkubwa mine in 2008 led to 400 employees losing their jobs [25]. Job losses raise the need to search alternate forms of livelihood and employment putting former mine workers to a difficult task of relocation and they may be unable to find employment with their skills as they may not be useful in the new environment

([12, 26, 13]). It is even more challenging for retrenched miners to get other jobs in rural areas [25]. Khanna [21] discovered that mining communities in Romania, Russia and Ukraine have a challenge with the number, quality, stability and emolument structure of jobs. Khanna [21] further explained that the basic situation on local labour markets led to a worsening of living standards for many due to qualitative transformation in the type of jobs. It also caused the development of individuals that are typically vulnerable in these extremely competitive employment markets. In most cases, mining communities are not prepared beforehand for the loss of employment and the ensuing poverty due to mine closure and hence suffer from shock. Post mine closure, mining communities appear to be fragile or vulnerable and their capacity to respond adequately to social instability, alienation and apathy on a community platform weakens [21].

Mine closure also affects local and regional governments which in turn affects local communities. Kemp et al. [8] note that payment of taxes and royalties usually cease or are significantly reduced upon mine closure. This further reduces local expenditure of governments and other beneficiaries. Governments that benefit from taxes, royalties and infrastructure provided by mines would equally have to adjust [14]. Municipal budgets experience a double impact as they lose tax revenues and at the same time cost obligations increase [21]. The difference between the mine closure impact on central governments and local authorities, is that the central governments, may often have alternative income sources that can be used to counter the loss in income, whereas the local authorities, may not always have that option.

4 Proposed Community Exit Strategy (CES) for Kalumbila Community

The closure of a mine in any jurisdiction ideally requires a CES that addresses the anticipated impact of closure on the local community. There are three key stakeholders in such a strategy each with very specific expectations. To initiate a successful closure, the impact on all three stakeholders must be fully taken into account. Masinja and Simukanga [11] described these stakeholders as:

- The government who lose tax revenues and at the same time cost obligations increase;
- The investor who incurs the mine closure costs; and
- The local community who lose employment, crucial services such as health and education and local economic meltdown risk.

In support, Limpitlaw [18] states that in order to properly close a mine, a trilateral discussion aimed at finding solutions is needed among mines, governments and communities. Despite closure concerns being different for each mine, it is feasible to create a number of principles to help the mine, municipality and community taking part in the closure process to guarantee full value for everyone involved [14]. Therefore, the proposed community exit strategy was generated using a series of steps that aim at meeting the expectations of all the three stakeholders. It was developed based on the landscape approach proposed by Sayer et al. [27]. In that proposal, ten principles were explicated to support implementation of the landscape approach. The ten principles placed emphasis on adaptive management, stakeholder involvement and multiple objectives. In the proposed community exit strategy, one or two of the ten principles were utilized.

The first step of the proposed community exit strategy is to re-establish communication among the three key stakeholders i.e. the Government of the Republic of Zambia (GRZ), mine owners and community in Kalumbila district. The purpose is to revisit the contractual agreement that was made among them prior to granting mining rights to the Kalumbila mining company if indeed this did happen. This is intended to bring out the obligations and expectations of all the stakeholders in relation to mine closure. Manjunath et al. [13] states that one of the problems faced during the mine closure process is miscommunication among the various stakeholders over concerns such as what the local community expects and if the mining company can meet those expectations. Manjunath et al. [13] claims that mitigating the socio-economic impacts of mine closure is not more about preparing for the inevitable end of the life of mine but more about the government making sure that the mining company does not avoid meeting its obligations. Owen and Kemp [19] suggest that mining companies should be ready to labour as the intermediary between the government and local communities as they are well situated in the direct impact zones. Manjunath et al. [13] and Owen and Kemp [19] both understand the necessity of communication during the mine closure process but contradict each other on who should be the mediator. However, emphasis should be placed on the purpose of the communication and not who meditates it. Therefore, Sayer et al. [27] provides a couple of principles that enables effective communication among the stakeholders. In principle number two, Sayer et al. [27] states that difficulties of implementation of a landscape approach can be solved through negotiations based on trust which in turn is achieved if objectives and values are shared. In other words, a common problem has to be identified as each stakeholder would only participate if it is in their best interest. Sayer et al. [27] further states that initially, reaching an agreement on the primary objectives may not be easy but concentrating on easy-to-reach intermediate goals may encourage stakeholders to work together. Principle number five emphasizes that a landscape approach needs to identify its stakeholders and understand their concerns and aspirations. All stakeholders should be acknowledged even if negotiations may not include all of them but the solutions should be fairly beneficial to all. The proposed CES in Kalumbila area recognizes its stakeholders from the onset and prioritizes initiating communication as the first step to achieving its purpose.

The second step is to carry out an assessment of the livelihood of the local community in Kalumbila area. This is intended to get an understanding of what people in Kalumbila area surrounding the mine do other than working in the mine. It will also provide a background on what socio-economic activities they were engaged in before the mining operations began in the area. This is important because any landscape approach that will be adopted post mine closure will have to be in line with what the local community are familiar with or are currently engaged in. For example, if pineapple farming is the norm in the area then it would be difficult for the local community to adopt sunflower farming even though it might be more lucrative. As part of the assessment, a demographic study of the Kalumbila area will be conducted to determine the socio-economic characteristics of the population. This will give an insight of the socioeconomic impact of mine closure on the local community and possible mitigation measures that can be put in place. Bainton and Holcombe [12] acknowledge that optimising mine closure activities on a social point of view would improve instead of reducing local capital to develop the platform for a more sustainable post-mining future. In other words, investment in improving social livelihood of the Kalumbila community after mine closure would not necessarily result in loss of capital for the mine but minimize the dependency on the mine by the community. This would be done through execution of activities that meet social requirements, build local capacity for self-development and foster resilience to change [8].

The third step involves conducting an investigation of the potential socio-economic activities that can be undertaken in Kalumbila area after mine closure in line with pre-mining land uses. Sayer et al. [27] in principle number four explains that landscapes and their constituents have several uses providing a variety goods and services that are of value to stakeholders differently. Hence, the investigation has to be carried out in line with the second step discussed above in terms of a landscape approach that the local community is familiar with. However, the third step is carried out to explore other socio-economic activities that are still within the comfort zone of the local community but have not been fully exploited. This is also intended to determine what alternative economic activities other than mining can be carried out in the area in the wider perspective of local economic development. Limpitlaw [18] in support states that linking local economic development and mine closures plans can guarantee that land uses after mine closure are in sync with surrounding development projects. Effective management and resourcing of those activities in an integrated way is likely to build the basis for long-term development [12]. The third step of the proposed Kalumbila community exit strategy is crucial as it would provide useful information that would prevent economic meltdown of the area in question.

Lastly, the outcomes of the three steps discussed above will form the basis upon which the proposed CES will be proposed. This will also require establishing the roles and responsibilities of all the stakeholders in the implementation process. In this case principle number seven by Sayer et al. [27] is particularly useful as it points out that resource access and land use determines the social and preservation outcomes. Therefore, the rights and responsibilities of different stakeholders have to be clear and acceptable by all. A legitimate system for conflict resolution is required when conflicting claims are made among the stakeholders. In principle number ten, Sayer et al. [27] states that the complicated and evolving nature of landscape approaches needs experienced and effective representation and institutions that are capable of solving all problems raised by the approach. In other words, stakeholders have to possess certain skills in order to effectively participate in various roles and responsibilities. The proposed community exit strategy will take into account the skills of the stakeholders in the various roles they will hold during the implementation process. Figure 4 summarises the steps that have to be taken to develop the proposed Community Exit Strategy for Kalumbila area.

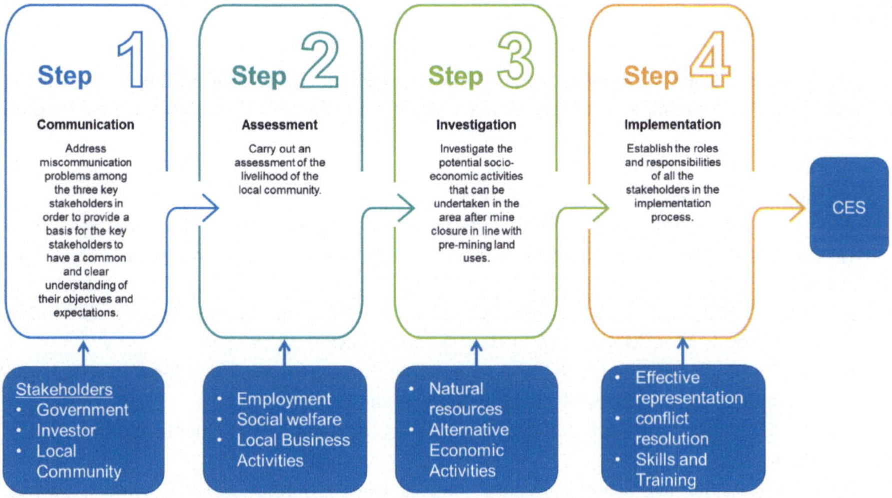

Fig. 4. Community Exit Strategy (CES) flow chart.

5 Key Considerations of the Proposed Community Exit Strategy

The following are the key considerations of the proposed community exit strategy in Kalumbila area;

(i) The strategy needs validation with good quality empirical data. A case study research will be carried out on Kalumbila District as it hosts two of Zambia's major copper mines that are both currently in operation and will be due for closure over the next 10 to 20 years.

(ii) The strategy will be used as a tool for mine closure planning across the mining industry with respect to the socio-economic aspect of the closure plan; and

(iii) The strategy will have quite wide applications, across different jurisdictions. This means that the collection of data from different jurisdictions will be in order.

6 Conclusion

This paper has proposed a Community Exit Strategy for Kalumbila mine that incorporates the expectations of the three stakeholders, government, investor and local community during the mine closure process. However, the research is still on going and will require good quality empirical data to validate it and establish its usefulness through a case study of Kalumbila district.

References

1. World Bank: How Can Zambia Benefit More from Mining? The World Bank Group (2016). http://www.worldbank.org/en/news/feature/2016/07/18/how-can-zambia-benefit-more-from-mining
2. World Bank: What would it take for Zambia's copper mining industry to achieve its potential? The World Bank Group (2011)
3. Africa Development Bank: Zambia Economic Outlook. Africa Development Bank Group (2019). https://www.afdb.org/en/countries/southern-africa/zambia/zambia-economic-outlook/
4. Sikamo, J., Mwanza, A., Mweemba, C.: Copper mining in Zambia, history and future. Southern Afr. J. Min. Metall. **116**, 491–496 (2016). ISSN 2225-6253
5. KPMG: Zambia mining guide. KPMG International Cooperative (2013). https://assets.kpmg/content/dam/kpmg/pdf/2013/08/zambian-country-guide.pdf
6. Zambia Chamber of Mines (2019). http://mines.org.zm/the-2018-copper-production-hits-861946-tonnes/
7. Barrick: Sustainability report summary. Barrick Gold Corporation (2017). https://barrick.q4cdn.com/788666289/files/sustainability/2017-Sustainability-Report-Summary.pdf
8. Kemp, D., Clark, P., Zhang, T.: Estimating socio-economic impacts of mine closure, vol. 8, pp. 1–19. Research paper, Centre for Social Responsibility in Mining, the University of Queensland, Australia (2008)
9. Marais, L.: The impact of mine downscaling on the free state goldfields. Urban Forum **24**(4), 503 (2013)
10. Haney, M., Shkaratan, M.: Mine closure and its impact on the community: five years after mine closure in Romania, Russia and Ukraine, vol. 42, pp. 1–77. Social Development Department, World Bank (2003)
11. Masinja, J., Simukanga, S.: An extractives industry contract negotiating model. In: Mineral Processing Conference, SAIMM 2014 (2014)
12. Bainton, N., Holcombe, S.: A critical review of the social aspects of mine closure. Resour. Policy **59**, 468–478 (2018)
13. Manjunath, A., Partha, S.A., Biswajit, P.: Environmental and socio-economic impacts due to mine closure. In: ResearchGate, Conference Paper (2014)
14. Laurence, D.: Optimisation of the mine closure process. J. Cleaner Prod. **14**, 285–298 (2005)
15. Ackerman, M., Van der Waldt, G., Botha, D.: Mitigating the socio-economic consequences of mine closure. SAIMM J. Southern Afr. Inst. Min. Metall. **118**(4), 439–447 (2018)
16. ZCCM-IH: China's CNMC says followed the law in closing Luanshya Copper Mines (2015). http://www.zccm-ih.com.zm/chinas-cnmc-says-followed-the-law-in-closing-luanshya-copper-mines/
17. Yabe, J., Nakayama, S.M.M., Ikenaka, Y., Yohannes, Y.B., Bortey-Sam, N., Oroszlany, B., Muzandu, K., Choongo, K., et al.: Lead poisoning in children from townships in the vicinity of a lead–zinc mine in Kabwe, Zambia (2014)
18. Limpitlaw, D.: Mine closure as a framework for sustainable development. School of Mining Engineering, University of the Witwatersrand, Johannesburg (2004). https://www.researchgate.net/publication/242116913
19. Owen, J., Kemp, D.: Mine closure and social performance: an industry discussion paper. Centre for Social Responsibility in Mining, Sustainable Minerals Institute, the University of Queensland, Brisbane (2018)

20. Cooke, J.A., Limpitlaw, D.: Post-mining site regeneration: Review of good practice in South Africa. Unpublished report for the Global Mining Consortium, Global Centre for Post-Mining Regeneration, Cornwall, UK (2003)
21. Khanna, T.: Mine closure and sustainable development: Results of the workshop organized by the World Bank Group Mining Department and Metal Mining Agency of Japan. Mining Journal Books, London (2000)
22. Stacey, J., Naudé, A., Hermanus, M., Frankel, P.: The economic aspects of mine closure and sustainable development. Literature overview of lessons for the socio-economic aspects of closure, vol. 1, pp. 1–32. The Coaltech Research Association, CSMI: WITS (2010)
23. BMR: The Kabwe Mine. BMR Group PLC (2016). http://www.bmrplc.com/kabwe_mine.php
24. MMMD: Environmental and Social Management Framework: Zambia Mining Environment Remediation and Improvement Project, pp. 12–15. Ministry of Mines and Mineral Development (2016)
25. Allafrica: Zambia: Bwana Mkubwa Mine Resumes Operations (2010). https://allafrica.com/stories/201001110091.html
26. Leteinturier, B., Laroche, J., Matera, J., Malaisse, F.: South Afr. J. Sci. **97**, 624–627 (2001)
27. Sayer, J., Sunderland, T., Ghazoul, J., Pfund, J., Sheil, D., Meijaard, E., Venter, M., Boedhihartono, A.K., et al.: Ten principles for a landscape approach to reconciling agriculture, conservation, and other competing land uses. PNAS **110**(21), 8349–8356 (2013)

Towards Low-Carbon Economy: A Business Model on the Integration of Renewable Energy into the Mining Industry

Kateryna Pollack(✉), Jan C. Bongaerts, and Carsten Drebenstedt

Technical University Bergakademie Freiberg, Gustav-Zeuner-Straße 12, 09599 Freiberg, Germany
eszharan@gmail.com

Abstract. Availability of technological solutions together with priorities decarbonizing the mining sector lead renewable energy (RE) to become an attractive energy source for the mining industry. Pilot projects were commissioned in several countries. However, there is a lack of knowledge on business models towards RE integration into the mining sector. The target of this paper is to present the results of a business model as a decision-making tool on comparison of different energy systems. Four scenarios are performed. These scenarios intend to compare fossil fuel systems and hybrid systems in the sense of no return. Part of that, a comparison of costs of ownership of each scenario has been performed as a basis for decision-making. Furthermore, the scenarios take into account a PV yield for different locations, degradation rate of PV modules, and carbon dioxide (CO_2) emissions impact. The key results of this business model give recommendations on the energy systems being integrated in mining operations. Additionally, the annual average savings of the cost of ownership of hybrid energy system and the CO_2 emissions impact including a CO_2 emission price fluctuation have been analyzed. This business model can be generalized to all mining settings, with specific practical implications for off-grid mines. This study aims to attract attention of policy makers, decision makers, and experts on fossil fuel and RE technologies in order to give alternatives in powering the mining industry and can be applied worldwide.

Keywords: Renewable energy · Mining industry · Hybrid energy system · CO_2 emissions · Levelized cost of electricity (LCOE)

1 Introduction

The mining industry is highly dependent on energy prices (Zharan and Bongaerts 2018). It appears, the share of energy costs accounts for 30–35% of all operating costs in mining operations (Slavin 2017), (Vyhmeister et al. 2017), (Cormack et al. 2017). In a case of metals mining, this rises to between 20% and 40% (Anon 2016). In the case of the Bisha Mine producing high-grade copper, gold, silver and zinc, the diesel price used for 2017 budget cost estimation is $0.55/L, delivered to site and the cost of electricity is $0.24/kWh. The share of energy costs for the Bisha Mine varies between 16% and 20% of all operating costs between 2017 and 2021 (Nevsun Resources Ltd.

E. Topal (Ed.): MPES 2019, SSGG, pp. 422–430, 2020.
https://doi.org/10.1007/978-3-030-33954-8_49

2016). Average cost savings by integrating a solar PV system powering a mine can be in the range of 25–30%. In very isolated locations with higher diesel prices, the savings can amount to more than 70% (Danvest 2015). In this matter, the rewards of replacing even a part of fossil energy by RE can bring substantial benefits.

The key targets towards energy transition have been set up (Gielen et al. 2019) as follows. Ensure reliable, universal, and affordable, access to innovative energy services; double the worldwide rate of improvement in energy efficiency; and increase substantially the share of RE in the worldwide energy mix. To reach these targets, CO_2 emission should be regulated under 1,100 gigatons; further, around 50%, 85%, and 34% of global natural gas, coal, and oil reserves should stay unburnable before 2050 (Shen et al. 2018).

PV modules have been the fastest developing RE technology converting solar energy into electricity (Tanesab et al. 2017). The diesel-PV hybrid system may produce about 1.8 GWh of electricity per year, or about 60% of the mine's annual daytime power needs (Zharan 2017). A hybrid PV-diesel system is more secure in electricity production than a stand-alone diesel system and a stand-alone PV system. Moreover, there is a great market potential for a HES, for instance, present fossil fuel system can be upgraded by new solar PV system. In 2020, the market potential for installed solar PV capacity is evaluated to be about 1,700 GWp up to 3,400 GWp (Breyer et al. 2011). Additionally, a HES not only imply greater savings on fuel than stand-along systems with lower CO_2 emissions and also savings on maintenance (Oviro and Jen 2018). Consequently, a HES can be an effective solution to increase RE penetration with respect to the mining industry.

Moreover, an investigation on a PV-diesel system with/without storage appears in many studies. Synchronized use of a PV and diesel generator (DG) with an energy storage could be a sufficient combination to electrify remote areas (Jamshidi and Askarzadeh 2019). This mitigates the intermittent nature of RE resources, ensuring an overall balance to energy supply and the improvement of system efficiency.

The research objectives of this paper are as follows: (i) to perform the business model within four scenarios and give recommendations based on the results of these scenarios, (ii) to analyze the outcomes of four scenarios towards saving the cost of ownership and the break-even point (the year of no return), (iii) to investigate the CO_2 emissions price fluctuation impact on break-even times of a diesel system and a hybrid diesel system.

2 Methodology

The mathematical model has been developed prior based on the LCOE approach (Pollack and Bongaerts 2019) and aims to find the year of no return of conventional and hybrid energy systems. In this paper, the outcomes of a business model on the integration of RE into the mining industry are performed and analyzed. Since an analytical determination of the break-even time is not possible, an attempt is made to perform a numerical calculation using the Gnuplot software of the basis of data on a hybrid system and a conventional system. All the data for this numerical calculation is contained in (Pollack and Bongaerts 2019).

The comparative analysis of four scenarios towards saving the cost of ownership was carried out applying the mathematical modeling method. The CO_2 emissions impact including the CO_2 emission price fluctuation was calculated applying the mathematical modeling method and the method of minimum and maximum prices of CO_2 emissions.

The methodological approach in this paper can be generalized to all mining settings, with specific practical implications for off-grid mines (Zharan and Bongaerts 2017). Obviously, it delivers a framework for decision-making on energy supply and energy procurement within the context of energy costs (represented by LCOE, CO_2 emissions) and break-even times.

3 Results

The description of each scenario is as follows. The total capacity of each system for each scenario is 10 MW. Further, the hybrid system consists of 9 MW of a fossil fuel system and 1 MW of a solar PV system. Scenario 1 covers a diesel genset system and a hybrid solar PV – fixed tilt + diesel genset system. Scenario 2 contains a diesel genset system (with low CAPEX of the diesel part of the hybrid system) and a hybrid solar PV – fixed tilt + diesel genset system (low CAPEX of the diesel part of the hybrid system). Scenario 3 includes a conventional diesel combined cycle (CC) system and a hybrid solar PV – fixed tilt + conventional diesel CC system. Finally, scenario 4 emphasizes a conventional natural gas combined cycle (CGCC) system and a hybrid solar PV – fixed tilt + CGCC system. Table 1 shows these scenarios including the capacities and technology of each fossil fuel and hybrid systems.

Table 1. Scenarios comparing fossil fuel system and a hybrid system

Scenario	Fossil fuel system	Capacity	Hybrid system	Capacity
Scenario 1	Diesel	10 MW	PV – diesel	1 + 9 MW
Scenario 2	Diesel (low CAPEX)	10 MW	PV – diesel (low CAPEX)	1 + 9 MW
Scenario 3	Conventional diesel CC	10 MW	PV – diesel	1 + 9 MW
Scenario 4	CGCC	10 MW	PV – natural gas	1 + 9 MW

3.1 Comparative Analysis of Four Scenarios in Terms of Savings of the Cost of Ownership

Moving to the analysis of the results of this business model, Fig. 1 shows the comparative analysis of four scenarios. The **green dotted curve** represents the break-even time of conventional and hybrid systems without degradation rate for high PV yield (2200 kWh/kWp). The **green solid curve** represents the break-even time with degradation rate for high PV yield. The **yellow solid curve** represents the break-even time with a degradation rate for average PV (1550 kWh/kWp). Finally, the **blue solid curve** represents the break-even time with a degradation rate for low PV yield (1250 kWh/kWp) for each scenario respectively. In this business model all scenarios were performed for off-grid mines.

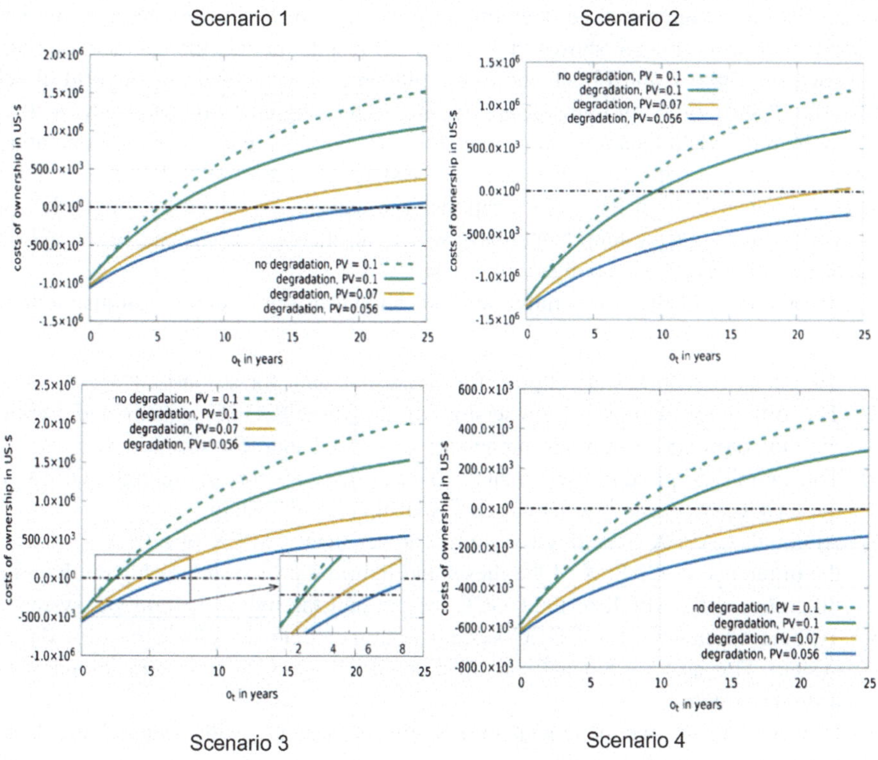

Fig. 1. Comparative analysis of four scenarios

Furthermore, Fig. 1 shows the approximate operating cost savings after replacing the fossil fuel system by the hybrid system. For each scenario these savings are different.

Table 2. Results on the comparative analysis of four scenarios

Scenario	PV yield	Approximate operating cost savings, $/year
Scenario 1	High	280.000,00
	Average	150.000,00
	Low	75.000,00
Scenario 2	High	234.000,00
	Average	75.000,00
	Low	n/a
Scenario 3	High	470.000,00
	Average	234.000,00
	Low	180.000,00
Scenario 4	High	94.000,00
	Average	37.500,00
	Low	n/a

The results on the approximate operating cost savings for high, average, and low PV yield for four scenarios are shown in Table 2. All results include the degradation rate.

Based on Table 2 and Fig. 1, the most sufficient energy system is a hybrid diesel CC system (Scenario 3) in terms of approximate operating cost savings in $/year. It is seen, in high PV yield locations it accounts for 470.000,00 $/year. The lowest operating cost savings are for the hybrid natural gas system (Scenario 4) and it accounts for 94.000,00 $/year in high PV yield locations. That can be explained by low OPEX of the hybrid natural gas system. Nevertheless, with an increase of RE part of the hybrid system the operating cost savings will expand.

The outcomes of the comparative analysis shown in Fig. 1 can be summarized as follows:

(1) The break-even times are shorter for scenario 1 than for scenario 4;
(2) The difference in cost of ownership of the diesel/CGCC system and hybrid diesel/natural gas system are higher for scenario 1 than for scenario 4.
(3) The natural gas price is lower than the diesel price and, correspondingly, OPEX of the CGCC system is lower than OPEX of the diesel system;
(4) Although CAPEX of a diesel system is lower than CAPEX of a CGCC system, the difference in CAPEX of the diesel and hybrid diesel system is higher than the differences of CAPEX of the CGCC system and the hybrid natural gas system;
(5) Turbine efficiency of a CGCC system is higher than the efficiency of a diesel system. This also has a beneficial effect on OPEX of a CGCC system compared to a diesel system;
(6) However, these two beneficial OPEX effects coincide with longer break-even times for a CGCC system;
(7) Therefore, in terms of CAPEX and break-even time, the hybrid diesel system is recommended for the mining industry;
(8) Nevertheless, in terms of OPEX, the hybrid natural gas system is recommended;
(9) In sum, operators can choose among high OPEX and shorter break-even times for a diesel system and for low OPEX and longer break-even times for a CGCC system.

3.2 CO_2 Emissions Impact Including a CO_2 Emission Price Fluctuation

In order to evaluate the influence of the CO_2 emission price fluctuation on the year of no return of fossil fuel and hybrid systems this section is performed. Figure 2 shows the CO_2 emission price per day. The data of Fig. 2 has been adapted from the CO_2 European allowance chat and transformed into $/t in a three-year period of time (t) (between March 2016 and March 2019). The solid line shows the CO_2 emission price over time (t). The dotted line shows the average CO_2 price over time (t). Further, this price varies from 4.2 $/t – the minimum CO_2 emission price, 25 $/t – the maximum CO_2 emission price, and finally 12.15 $/t – the average CO_2 emission price over time (t). The average CO_2 emission price has been calculated as the integral of all prices over time divided by the number of days.

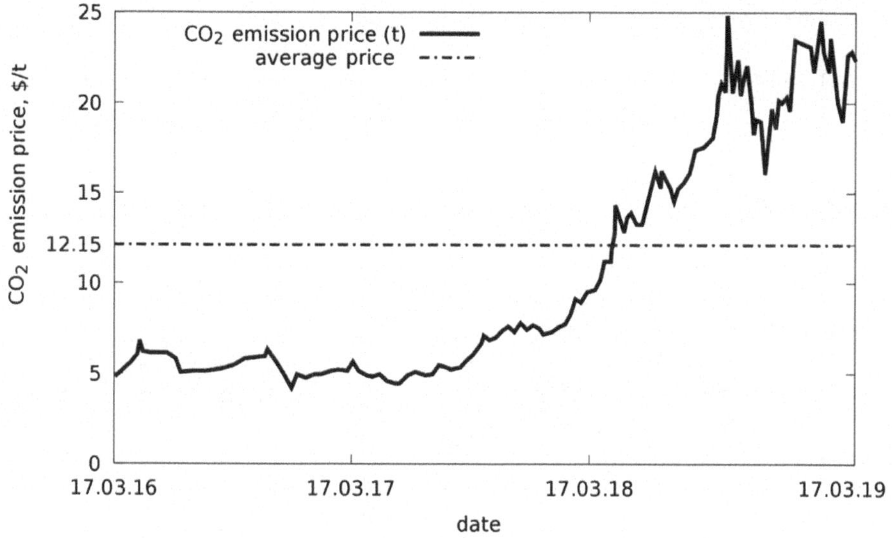

Fig. 2. CO_2 emission price per day (adapted from (Marketinsider 2018))

Table 3 shows the results on the CO_2 emission price fluctuation. As it is seen in Table 3, for high, average, and low PV yields the total CO_2 emission price fluctuation impact is in amount of 2 years, 4 years, and 6 years respectively.

Table 3. Results on the CO_2 emission price fluctuation

Shape of curve in Fig. 3	PV yield	CO_2 emission price	Break-even times in years
Green dotted	High	CO_2 max	4
Green solid	High	CO_2 avg	5
Green dashed	High	CO_2 min	6
Yellow dotted	Average	CO_2 max	8
Yellow solid	Average	CO_2 avg	10
Yellow dashed	Average	CO_2 min	12
Blue dotted	Low	CO_2 max	13
Blue solid	Low	CO_2 avg	16
Blue dashed	Low	CO_2 min	19

Figure 3 shows the CO_2 emissions price fluctuation impact on break-even times of a diesel system and a hybrid diesel system – scenario 1. It is seen in Fig. 3, the higher PV yield the lower the CO_2 emission price fluctuation impact, further, the lower PV yield the higher the CO_2 emission price fluctuation impact on the break-even times of two systems.

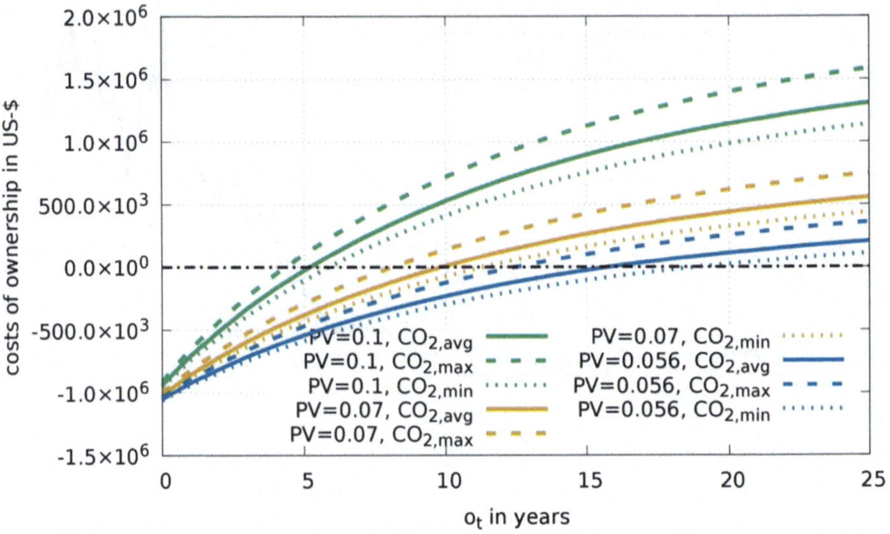

Fig. 3. CO_2 emissions price fluctuation impact on break-even times of a diesel system and a hybrid diesel system – scenario 1

Summing up the results on the CO_2 emissions price fluctuation impact, depending on the PV yield, the CO_2 emissions cost has a different influence on the break-even times of the all given hybrid systems. Taking into account a CO_2 emissions impact in this model the hybrid system becomes even more economically attractive than without CO_2 emissions impact. In this matter, it is suggested to encourage decision-makers and stakeholders to adopt hybrid systems into mining processing as an efficient solution towards reaching the emissions targets of the mining companies and contributing to low-carbon economy transition.

4 Conclusions

Summing up the outcomes of this paper, the following three research objectives have been reached. **First research objective:** in scenario 1, it is recommended to invest in a hybrid diesel system when the lifetime of a mine located in high PV yield is more than 5 years. In scenario 2, it is recommended to invest in a hybrid diesel system when the lifetime of a mine located in high PV yield is more than 10 years. In scenario 3, if the lifetime of a mine for any location is more than 7.5 years, it is recommended to invest in a hybrid diesel CC system. Finally, in scenario 4, it is recommended to invest in a hybrid natural gas system when the lifetime of a mine located in high PV yield is more than 10 years.

Second research objective: moreover, if a mine has a diversity of hybrid PV systems, a variety of break-even times will be confirmed. Taking into account the degradation of hybrid system, PV yield, and CO_2 emissions costs the results of four scenarios have been analysed in Sect. 3.1. Indeed, the main influence on the costs of ownership has PV yield. According to high, average, low irradiations the variation in break-even times of two systems vary. The costs of ownership are significantly lower with increasing PV yield resulting that location of a mine has a high impact on a break-even time of two systems. Scenario 3 shows, that the most sufficient energy system is a hybrid diesel CC system in terms of approximate operating cost savings, in high PV yield locations it accounts for 470.000,00 $/year. Scenario 4 shows the lowest operating cost savings for the hybrid natural gas system and it accounts for 94.000,00 $/year in high PV yield locations. That can be explained by low OPEX of the hybrid natural gas system. Nevertheless, with an increase of RE part of the hybrid system the operating cost savings will expand.

Third research objective: CO_2 emissions costs has a little impact on the break-even times of two systems. However, it is assumed to have an increase in CO_2 emissions costs in the future. Therefore, it is expected that CO_2 emissions will have more significant impact on break-even times in the future. The CO_2 emissions price fluctuation impact on break-even times of a diesel system and a hybrid diesel system has been analzyed. Depending on the PV yield, the CO_2 emissions cost has a different influence on the break-even times of the all given hybrid systems. Taking into account a CO_2 emissions impact in this model the hybrid system becomes even more economically attractive than without CO_2 emissions impact.

References

Anon: Energy and Mines Renewables in Mining Rankings and Awards (2016)

Breyer, C., Goerig, M., Schmid, J.: Fuel-parity: impact of photovoltaics on global fossil fuel fired power plant business, p. 12 (2011)

Cormack, D., Wood, M., Swart, A., Davidse, A.: Renewables in Mining: Rethink, Reconsider, Replay. Deloitte Touche Tohmatsu (2017)

Danvest: THEnergy study: Low-load Gensets for Solar–Diesel Hybrid Plants in the Mining Industry (2015)

Gielen, D., et al.: The role of renewable energy in the global energy transformation. Energy Strategy Rev. **24**, 38–50 (2019)

Jamshidi, M., Askarzadeh, A.: Techno-economic analysis and size optimization of an off-grid hybrid photovoltaic, fuel cell and diesel generator system. Sustain. Cities Soc. **44**, 310–320 (2019)

Marketinsider: CO_2 European Emission Allowances (2018)

Nevsun Resources Ltd.: Independent Technical Report 2016 Resources and Reserves Update Bisha Mine, Eritrea, SRK Consulting (Canada) Inc. (2016)

Oviro, P., Jen, T.: The energy cost analysis of hybrid systems and diesel generators in powering selected base transceiver station locations in Nigeria. Energies **11**(687), 2–20 (2018)

Pollack, K., Bongaerts, J.C.: Mathematical model on the integration of renewable energy in the mining industry. Int. J. Energy Sect. Manage. (2019)

Shen, W., Qiu, J., Dong, Z.: Electricity network planning targeting low-carbon energy transition. Global Energy Interconnection **1**(4), 487–499 (2018)

Slavin, A.: New renewable energy for mine project - LAMGOLD Essakane to benefit from largest hybrid plant in Africa (Case Study). Energy and Mines, pp. 547–559, 6 March 2017

Tanesab, J., Parlevliet, D., Whale, J., Urmee, T.: Seasonal effect of dust on the degradation of PV modules performance deployed in different climate areas. Renewable Energy **111**, 105–115 (2017)

Vyhmeister, E., et al.: A combined photovoltaic and novel renewable energy system: an optimized techno-economic analysis for mining industry applications. J. Clenear Prod. **149**, 999–1010 (2017)

Zharan, K.: Renewable energy for the mining industry: trends and developments. In: Matasci, L. (ed.) Boosting Resource Productivity by Adopting, p. 429. A World Resources Forum Production, St. Galen (2017)

Zharan, K., Bongaerts, J.C.: Decision-making on the integration of renewable energy in the mining industry: a case studies analysis, a cost analysis and a SWOT analysis. J. Sustain. Min. **16**(4), 162–170 (2017)

Zharan, K., Bongaerts, J.C.: Survey on integrating of renewable energy into the mining industry. J. Environ. Acc. Manag. **6**(2), 149–165 (2018)

A Valuation Approach to Investigate the Sustainability of Sorkhe-Dizaj Iron Ore Mine of Iran

Mahdi Pouresmaieli[1(✉)] and Morteza Osanloo[2]

[1] Department of Mining and Metallurgical Engineering, Amirkabir University of Technology, Tehran, Iran
mpouresmaieli@gmail.com
[2] Department of Mining and Metallurgical Engineering, Amirkabir University of Technology, PO Box: 15875-4413, 424, Hafez Ave., Tehran, Iran
Morteza.osanloo@gmail.com

Abstract. Mining activities affect environment, economic and society in its surrounding region. This study aims to assess the sustainability of Sorkhe-Dizaj open-pit iron ore mine which is located 300 km in north-west of Tehran-capital of Iran. The annual ore production of Sorkhe-Dizaj open-pit mine is 400,000 tonne, extracted an area of 16.67 Km^2. In this research, modified Folchi method has been used to assess the sustainability of Sorkhe-Dizaj mine. Folchi method is a 2D matrix to quantify the environmental impacts related to mining activities during the mine life-cycle. In its matrix, the rows represent 10 impacting factors of mining activities and columns represent 11 environmental components. The developed Folchi model covers economic and social aspects of mining operations in addition to environmental factors. The developed Folchi matrix has 28 impacting factors and 12 components. The results of developed model showed that, human health and safety, social relationship, water and air quality, flora, fauna and economy have high negative effects on sustainability of mine site among 12 components.

Keywords: Sustainable development · Semi-quantitative · Mining activities

1 Introduction

With the advancement of technology, minerals have become one of the fundamental human needs throughout their life. As a result, nowadays, the demand for minerals, especially metals, has increased. Moreover, due to the increase in mining activities in recent years, sufficient attention should be devoted to the issues related to sustainability in mining activities, including pollution control. Paying attention to the environment is one of the most significant concerns regarding the sustainability of a country [1, 2].

Iran is quite rich with respect to ore reserves, including lead, zinc, iron, and chromite reserves and a wide variety of building stones and industrial minerals, all of which can be found all over the country.

At present, Iranian mines are owned by private and public sectors. The privately-owned mines are smaller than the state-owned mines. So, considering their sizes, the

E. Topal (Ed.): MPES 2019, SSGG, pp. 431–446, 2020.
https://doi.org/10.1007/978-3-030-33954-8_50

former ones bring about fewer environmental hazards compared to the latter ones. Processing would do greater serious harm to soil and air (environment) than ore extraction [3]. Some scientific engineering studies, such as Folchi [4], Philips [5, 6], Osaloo and Rahmanpour [7], have drawn their attention to the exploration of the destructive effects of processing on the environment and made their best to propose some effective solutions. In 2007, Kerkeler et al. examined the environmental and economic impact of the phosphate ore processing [8]. Moyé et al. assessed groundwater resources and probed the environmental effect of abandoned mines in recent years [9]. Antoniadis et al. studied the areas around a highly-contaminated former mining region and focused on the bioavailability and risk assessment of potentially toxic elements of the edible vegetables and soils in the gardens of those areas [10]. Birich and Dimitrakopoulos investigated the effect of grade uncertainty, as a closely associated factor to sustainable mining, on mine life [11, 12]. Bouchard et al. undertook a study on the role of safety in mining activity [13]. Wright et al. [14] designed a study to find out the extent of the employees' knowledge about the mining project and the way it could affect this project. Ranjan et al. conducted a study in which they establish vegetation over the waste dump slope in a surface iron ore mine [15]. Bridge investigated the role of resource depletion in sustainable mining [16]. Silva et al. and Cuervo et al. devoted their attention to the investigation of dam failure and the causes of tailings dam breakage [17]. Concerning the leakage of tailings dams, Martín-Crespo et al. studied heavy metals leaking from dams [18]. Antunes et al. performed an environmental risk assessment by categorizing soils based on their toxicity profiles.

The abovementioned studies are very informative; however, in their approaches, they often confined themselves to assessing some limited aspects of the environmental impact. With the exception of the Folchi method, the current evaluation methods are limited in scope, measuring just one or two environmental aspects of mining and mineral dressing [19]. The Folchi approach comprises various environmental aspects of mining operations, such as fly rock, air noise, water and air pollution, etc. The process of data collection and data analysis using this technique is quite easy and simple [20]. The present study aimed to investigate a new approach to sustainability considering the important aspects of sustainability pillars.

2 The Environmental Impact of Mining Operations

Open-pit mining and ore processing have some unwanted impacts on surrounding mine land. As well, it causes water pollution, subsidence, dust and noise pollution, etc., can exert an influence on the environment [21]. When there is a permeable zone beneath a deposit, precipitation from that deposit can be propagated due to groundwater inflow. The excavation operation can also enhance the rate of this permeation. Moreover, propagation can emerge from decantation areas [22]. The abovementioned processes can change the topography, vegetation, and ecology of the areas involved as well. Controlling noise and vibration, drilling, and blasting operations as well as utilizing mine heavy tracks, crushers, and mills are of great significance and deserve numerous attention [23]. Dust is produced as a result of drilling, blasting, loading, transport, and haulage; nonetheless, almost all stages of ore processing operations can contribute to dust pollution [24].

Another impact of such operations on the environment is water pollution [25, 26]. Similarly, contaminated water caused by mining operations has vital detrimental impacts on rivers, agriculture, fresh portable water, and ecosystems since it contains plenty of heavy metals and suspended solid particles, and its pH level is decreased. Drainage in mines brings about water-level reductions, leading to unfavorable changes in the adjacent lakes and aquatics [27]. Acid Mine Drainage (AMD) is the main source of water and land contamination [24].

Soil erosion is another impact of mining operation, especially surface mining, resulting in the massive degradation of lands [28].

The uneconomic materials obtained from mining operations are known as mine wastes, including natural materials remaining after blast procedures and fragmentation processes in mines and quarries or primarily crushed, milled and ground materials gained from the processing stage in plants. Furthermore, tailings mostly consist of chemical additives. While overburden is considered as mining wastes, various properties of materials may require different environmental treatments and dumping [22].

Lead, copper, zinc, aluminum, arsenic, sulfur, cadmium, cobalt, mercury, marcasite, and pyrite (FeS2) are deemed as heavy metals and toxic minerals and are potential threats to the environment [18, 29]. Particularly, AMD stems from wastes containing pyrite.

Pyrite can be frequently found along with valuable minerals in different mines, especially, in copper mines. In a majority of cases, pyrite, considered as gangue, is separated from ore during processing operations and finally is sent to tailings dams for disposal [30]. As a matter of fact, the air-oxidation of pyrite in tailings ponds creates acid mine drainage, leading to a substantial adverse effect on the environment [9, 10].

Some other minerals, known as salts, are generated from pyrite or marcasite (FeS2) oxidation and have strong acids and weak bases. They are found in the mining of certain minerals, such as copper and coal. The evaporation of pooled acidic mine waters along with the spoils containing sulfides produces acidic minerals which include Fe3+ sulfates to a large extent and Al3+ sulfates to a minor extent. The interaction of AMD with natural carbonates, including salts, changes pH levels in natural waters [31]. Acid mine drainage (AMD) generated from mine wastes interacts with natural carbonates, including salts and induces variations in pH levels in natural waters [32]. Open pit mining has a greater negative impact on the environment, compared to underground mining [33]. So, it is a matter of paramount importance to study pollution caused by open-pit mining and its negative impact on the pillars of sustainable mining and, in particular, on the environment.

3 Materials and Methods

3.1 The Developed Folchi Method

The Folchi method is based on experts' opinions and MADM methods. MADM models are used to evaluate, prioritize and choose between different options based on specific criteria which are also commonly used for weighting. These criteria usually explain the features of the options. As a matter of fact, it decides how to choose the best option among the available options, so that the options that you choose can bring you the most benefit and success [34]. In this approach, all the factors of sustainable development in mining operations are taken into account. It has 7 stages:

1. Characterizing the sustainable terms of geology, hydrology, economy, health, safety, etc.
2. Identifying the influential factors in modifying sustainable conditions in terms of mine life.
3. Determining the possible range of each influential factor.
4. Modifying each individual sustainable condition based on mining.
5. Correlating each parameter and each sustainable component.
6. Estimating the specific magnitude of each influential factor using the previously determined range.
7. Calculating the weighted sum of the sustainable impact on each sustainable component.

In this method, initially, some effective parameters in mining operations, including health and safety, use of territory, air quality, water quality, flora and fauna, social relationship, surface and underground, landscape, noise, economy, and soil degradation, were defined for a certain area. Then, the degree each of the mining indexes (directly or indirectly) influenced each of the sustainable parameters was estimated via a rating system based on various possible scenarios. Finally, the sum of effective parameters was measured and deemed as the overall effect of mining on each of the sustainable indexes. Using this method, the influential factors were as follows: 1. Alteration of the Areas' Potential Resource 2. Visibility of the Pit 3. Interference with Surface Water 4. Interference with Underground Water 5. Vehicular Traffic 6. Atmospheric Release of Gas and Dust 7. Fly-rock 8. Noise 9. Ground Vibration 10. The Degree of Local Business (direct/indirect) 11. Geological Structure 12. Engineering Geological and Hydrological Conditions 13. The Degree of Geological Disaster Damage 14. Resource Depletion 15. Environmental Pollution 16. Mine Size 17. Mining Method 18. Vegetation Coverage 19. Importance of the Area 20. Population Density 21. Acid Mine Drainage (AMD) 22. Tailing Leakage 23. Tailing Dam Failure 24. Waste Dumps 25. Rehabilitation 26. Employees' Skills 27. Safety 28. Geological Uncertainty (grade and tonnage).

The possible scenarios for each influential factor were then identified and a magnitude range was measured for each of these scenarios. Table 1 presents various possible scenarios along with their corresponding magnitudes for each influential factor. The error reported in Table 1 was ± 0.008. The magnitudes of the influential factors are provided in Table 2.

Table 1. The correlation matrix of the weighted impact of each influential factor on each sustainable component

Impact Factors	Human health and safety	Social relationship	Water quality	Air quality	Use of territory	Flora and fauna	Surface	Underground	Landscape	Noise	Economy	Soil degradation
Alteration of the Area's Potential Resources	0.225	0.238	0.405	0.339	0.385	0.417	0.375	0.492	0.506	0.455	0.323	0.476
Visibility of the Pit	0.337	0.238	0.27	0.339	0.513	0.278	0.5	0.328	0.633	0.455	0.215	0.317
Interference with Surface Water	0.449	0.357	0.676	0.169	0.385	0.417	0.5	0.328	0.38	0.227	0.323	0.476
Interference with Underground Water	0.449	0.357	0.676	0.169	0.256	0.278	0.25	0.656	0.253	0.227	0.323	0.476
Vehicular Traffic	0.337	0.476	0.135	0.508	0.256	0.417	0.25	0.164	0.253	0.909	0.323	0.317
Atmospheric Release of Gas and Dust	0.449	0.357	0.405	0.847	0.256	0.417	0.25	0.164	0.38	0.227	0.323	0.317
Fly-rock	0449	0.357	0.135	0.169	0.256	0.139	0.25	0.164	0.127	0.455	0.108	0.159
Noise	0.449	0.476	0	0.169	0.256	0.139	0.25	0.164	0.253	0.909	0.215	0.159
Ground Vibration	0.337	0.357	0.27	0	0.256	0.139	0.375	0.492	0.127	0.455	0.323	0.159
Employment of Local Work Force	0.225	0.476	0.27	0.339	0.385	0.278	0.25	0.164	0.253	0.682	0.538	0.317
Geological Structure	0.225	0.119	0.27	0.169	0.385	0.278	0.375	0.492	0.38	0	0.323	0.317
Engineering Geological and Hydrological Conditions	0.225	0.119	0.405	0.339	0.385	0.278	0.375	0.492	0.253	0	0.323	0.317
The Degree of Geological Disaster Damage	0.337	0.357	0.405	0.339	0.256	0.278	0.375	0.328	0.253	0	0.43	0.317
Resource Depletion	0.225	0.357	0.27	0.339	0.385	0.278	0.25	0.328	0.38	0.227	0.43	0.317

(continued)

Table 1. *(continued)*

Impact Factors	Human health and safety	Social relationship	Water quality	Air quality	Use of territory	Flora and fauna	Surface	Underground	Landscape	Noise	Economy	Soil degradation
Environmental Pollution	0.562	0.476	0.541	0.678	0.385	0.556	0.375	0.328	0.38	0.227	0.43	0.476
Mine Size	0.112	0.357	0.27	0.339	0.385	0.278	0.5	0.328	0.506	0.455	0.43	0.476
Mining Method	0.337	0.238	0.27	0.339	0.256	0.417	0.375	0.492	0.38	0.682	0.323	0.317
Vegetation Coverage	0.337	0.357	0.405	0.508	0.256	0.556	0.5	0.164	0.506	0.455	0.323	0.476
Importance of the Area	0.225	0.357	0.27	0.339	0.513	0.417	0.375	0.492	0.38	0.455	0.323	0.476
Population Density	0.449	0.476	0.405	0.678	0.513	0.417	0.375	0.328	0.38	0.909	0.323	0.476
AMD (Acid Mine Drainage)	0.449	0.357	0.676	0.339	0.385	0.556	0.5	0.492	0.38	0	0.323	0.635
Tailing Leakage	0.449	0.357	0.541	0.339	0.385	0.556	0.5	0.328	0.38	0	0.323	0.476
Tailing Dam Failure	0.562	0.476	0.541	0.339	0.513	0.556	0.5	0.328	0.506	0	0.43	0.476
Waste Dump	0.337	0.238	0.405	0.339	0.385	0.417	0.375	0.328	0.506	0	0.215	0.476
Poor Rehabilitation	0.449	0.476	0.405	0.508	0.385	0.556	0.375	0.164	0.506	0.455	0.323	0.476
Employee Skills	0.449	0.357	0.27	0.339	0.256	0.139	0.125	0.328	0.253	0.455	0.43	0.159
Safety	0.562	0.595	0.27	0.339	0.256	0.278	0.25	0.492	0.253	0.682	0.43	0.159
Geological Uncertainty (tonnage and grade)	0	0.238	0.135	0.339	0.513	0.278	0.25	0.656	0.253	0	0.860	0
TOTAL	10	10	10	10	10	10	10	10	10	10	10	10

Table 2. The range of magnitudes for the influential factors

Influential factors	Scenario		Magnitude
Alteration of the Area's Potential Resources	Parks, protected areas		8–10
	Urban areas		6–8
	Agricultural areas, wood		3–6
	Industrial areas		1–3
Visibility of the Pit	Visible from inhabited areas		6–10
	Visible from main roads		2–6
	Not visible		1–2
Interference with Surface Water	Interference with lakes and rivers		6–10
	Interference with irrelevant water systems		3–6
	No interference		1–3
Interference with Underground Water	Interference with the water table of superficial and permeable grounds		5–10
	Interference with the water table of deep and permeable grounds		2–5
	Interference with the water table of deep and un-permeable grounds		1–2
Vehicular Traffic	Increasing by 200%		6–10
	Increasing by 100%		3–6
	No interference		1–3
Atmospheric Release of Gas and Dust	Free emissions in the atmosphere		7–10
	Emission around the given reference values		2–7
	Emission below the given reference values		1–2
Fly-rock	No blast designs and no clearance procedures		9–10
	Blast design and no clearance procedures		4–9
	Blast design and clearance procedures		1–4
Noise	A peak air overpressure at a 1 km distance	<141 db	8–10
		<131 db	4–8
		<121 db	1–4
Ground Vibration	Cosmetic damage, above the tolerability threshold		7–10
	The tolerability threshold		3–7
	Values under the tolerability threshold		1–3
Employment of Local Work Force	Job opportunities	Low	7–10
		Medium	3–6
		High	1–2
Geological Structure	Active faults are developed, soft–hard clastic rocks		6–10
	Faults are comparatively developed, soft–hard clastic rocks		2–6
	No fault is developed, hard rocks		0–2
Engineering Geological and Hydrological Conditions	Complex conditions		6–10
	Moderate conditions		2–6
	Simple conditions		0–2

(*continued*)

Table 2. (*continued*)

Influential factors	Scenario	Magnitude
The Degree of Geological Disaster Damage	A major disaster event occurred, killing more than 3 people and bringing about economic losses	6–10
	A major disaster event occurred, killing no more than 3 people and bringing about economic losses	2–6
	Major disaster event (land collapses, landslides, mudflows, and ground fissures) did not occur, just a small event happened with no casualties, but with negligible economic losses	0–2
Resource Depletion	The resource is a strategic resource and has a major effect on unemployment in the area	8–10
	Either the resource is a strategic resource and doesn't have a major effect on unemployment in the area or the resource is not a strategic resource and has a major effect on unemployment in the area	3–7
	The resource is not a strategic resource and doesn't have a major effect on unemployment in the area	1–2
Environmental Pollution	Tailings dams and refuse dumps are not built; waste gas and dust pollution, discharge of wastewater, and solid waste pollution in the mining area	6–10
	Tailings dams and refuse dumps are built but interception facilities are not completed; waste gas and dust pollution, discharge of wastewater, and solid waste pollution in the mining area	2–6
	Tailings dams and refuse dumps are built and stacked orderly, interception facilities are completed; waste gas and dust pollution, discharge of wastewater, and solid waste pollution in the mining area	0–2
Mine Size	A large-scale mine (such as open pit iron mines/200mt.a)	6–10
	A moderate-scale mine (such as open pit iron mines/60–200mt.a)	2–6
	A small-scale mine (such as open pit iron mines/60mt.a)	1–2
Mining Method	Underground mining: The caving method	8–10
	Open pit mining on a large scale	2–7
	Open pit mining on a small scale, and underground mining: The supported and self-supported/naturally methods	1–2

(*continued*)

Table 2. (*continued*)

Influential factors	Scenario	Magnitude
Vegetation Coverage	The area is occupied and destroyed and the vegetarian coverage of the lands before mining activities was great; the area is harder to be reclaimed; Spring spots and exploited wells nearly dried up in the mining area; the landforms in the mining area are destroyed to a large extent; the land had a cultivated land use before mining	7–10
	The area is occupied and destroyed and the vegetarian coverage of the lands before mining activities was moderate; the area is hard to be reclaimed; the groundwater table is declined; the flux of spring spots and the exploited wells have significantly decreased in the mining area; the landforms in the mining area have changed to a moderate degree; the land had a forest land use before mining	3–6
	The area is occupied and destroyed and the vegetarian coverage of the lands before mining activities was small; the area is easy to be reclaimed; the groundwater table is declined; spring spots and exploited wells have no significant decrease in mining area; the landforms in the mining area have changed to a small degree; the land had an industrial land use before mining	1–3
Importance of the Area	The mining site is located in the center of the city which has a strong influence on major traffic roads and tourism scenic areas or major projects	6–10
	The mining site is located in scenic areas or in a city or a town close to major traffic roads and tourism scenic areas	2–6
	The mining site is located in the countryside or mountain areas far from major traffic roads and tourism scenic areas	0–2
Population Density	Density is high (population is more than 500 in per mining unit)	6–10
	Density is moderate (population is between 200 and 500 in per mining unit)	2–6
	Density is low (population is less than 200 in per mining unit)	0–2
AMD (Acid Mine Drainage)	Producing acidic contaminants less than 35 km far from the residential areas	7–10
	Producing acidic contaminants more than 35 km far from the residential areas	3–6
	Producing no acidic contaminants	1–3

(*continued*)

Table 2. (*continued*)

Influential factors	Scenario	Magnitude
Tailings Dam Failure	The safety factor is less than 1	9–10
	The safety factor is between 1 and 1.5	3–8
	The safety factor is more than 1.5	1–3
Waste Dumps	The waste dump slope is more than 36°	8–10
	The waste dump slope is between 32 and 36°	3–7
	The waste dump slope is less than 32°	1–2
Tailings Leakage	More than 66% of the elements of the soil around the mine are heavy metals	7–10
	33% to 66% of the elements of the soil around the mine are heavy metals	3–7
	Less than 33% of the elements of the soil around the mine are heavy metals	1–3
Poor Rehabilitation	The development of vegetation is slow; the coverage is less than 20%	7–10
	The development of vegetation is moderate; the coverage is 20%–60%	2–6
	The development of vegetation is fast; the coverage is more than 60%	0–2
Employees	The employees don't have enough knowledge and experience	8–10
	The employees are experienced	4–7
	The employees have high knowledge and a great deal of experience	1–4
Safety	The road slope is higher than 12°	7–10
	Semi-automatic machines with standard road widths are used and the road slope is between 8 and 12°	3–6
	Fully-automatic machines with standard road widths are used and the road slope is less than 8°	1–3
Geological Uncertainty (tonnage and grade)	Dilution is high and there have been a lot of unusual changes in the trend of rising prices in the last 30 years	7–10
	Dilution is low and there have been a lot of unusual changes in the trend of rising prices in the last 30 years	5–7
	Dilution is high and there have been a few unusual changes in the trend of rising prices in the last 30 years	3–5
	Dilution is low and there have been a few unusual changes in the trend of rising prices in the last 30 years	1–2

4 Implementation of the Framework on Sorkhe-Dizaj Iron Ore Mine

The earlier studies have confirmed 347 mining segments pertaining to the sustainability indicators [35]. These indicators must be technically capable of assessing the determined alternatives in terms of the sustainability criteria [36]. Regarding sustainability influential factors, 28 factors were utilized in this model. Considering the magnitude range of each factor, as indicated in Table 2, the influential factors in Sorkhe-Dizaj Iron Ore were identified, as shown in Table 3. The matrix of the influential factors in decreasing sustainability was then prepared based on the magnitude selected among the values presented in Table 1. Table 3 indicates the resultant matrix of this model.

5 Results and Discussion

As seen in Figs. 1 and 2, the results of the developed Folchi method revealed that six components of sustainable development, namely, human health and safety, social relationship, air quality, water quality, flora and fauna, and economy, had the most significant effect. Nonetheless, concerning the fact that a total impact rate of 100 received by any of the sustainable components is deemed as the main impact, it was found that these six sustainable components received low actual impacts. Moreover, considering other sustainable components, namely, use of territory, surface, underground, landscape, noise and soil degradation, no significant impact was observed. Based on these findings, it can be declared that there would be no change in the sustainability of the surrounding area due to future mining activity.

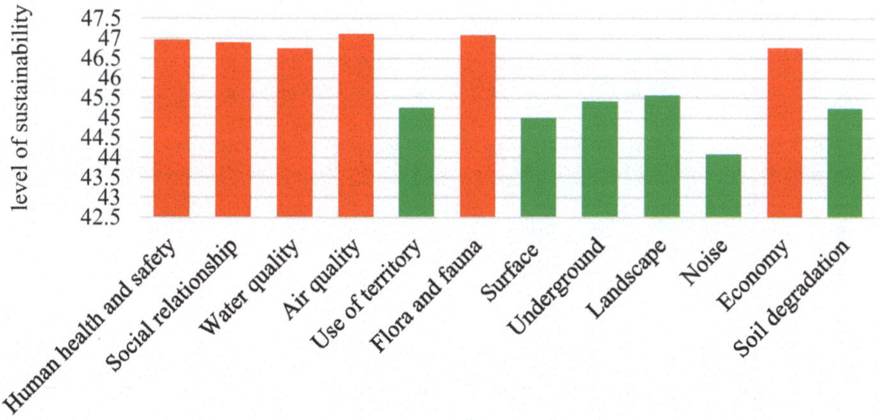

Fig. 1. The results of the quantitative analysis of sustainable impact assessment

Table 3. The magnitude of the influential factors and the impact of each influential factor on the sustainable component

Impact factors	Magnitude of impacting factor for Sorkhe-Dizaj iron ore mine	Human health and safety	Social relationship	Water quality	Air quality	Use of territory	Flora and fauna	Surface	Underground	Landscape	Noise	Economy	Soil degradation
Alteration of the Area's Potential Resources	**5**	1.124	1.191	2.027	1.695	1.923	2.084	1.875	2.459	2.532	2.273	1.613	2.381
Visibility of the Pit	**4**	1.348	0.952	1.081	1.356	2.051	1.111	2	1.312	2.532	1.818	0.86	1.27
Interference with Surface Water	**5**	2.247	1.786	3.379	0.848	1.923	2.084	2.5	1.64	1.899	1.137	1.613	2.381
Interference with Underground Water	**4**	1.798	1.428	2.703	0.678	1.026	1.111	1	2.623	1.013	0.909	1.29	1.905
Vehicular Traffic	**4**	1.348	1.905	0.54	2.034	1.026	1.667	1	0.656	1.013	3.636	1.29	1.27
Atmospheric Release of Gas and Dust	**6**	2.696	2.143	2.432	5.085	1.538	2.5	1.5	0.983	2.278	1.364	1.936	1.905
Fly-rock	**5**	2.247	1.786	0.676	0.848	1.282	0.695	1.25	0.82	0.633	2.273	0.538	0.794
Noise	**4**	1.798	1.905	0	0.678	1.026	0.556	1	0.656	1.013	3.636	0.86	0.635
Ground Vibration	**4**	1.348	1.428	1.081	0	1.026	0.556	1.5	1.967	0.506	1.818	1.29	0.635
Employment of Local Work Force	**4**	0.899	1.905	1.081	1.356	1.538	1.111	1	0.656	1.013	2.727	2.15	1.27
Geological Structure	**1**	0.225	0.119	0.27	0.17	0.385	0.278	0.375	0.492	0.38	0	0.323	0.318
Engineering Geological and Hydrological Conditions	**4**	0.899	0.476	1.622	1.356	1.538	1.111	1.5	1.967	1.013	0	1.29	1.27
The Degree of Geological Disaster Damage	**3**	1.011	1.071	1.216	1.017	0.769	0.833	1.125	0.984	0.76	0	1.29	0.953
Resource Depletion	**4**	0.899	1.428	1.081	1.356	1.538	1.111	1	1.312	1.519	0.909	1.72	1.27

(continued)

Table 3. (*continued*)

Impact factors	Magnitude of impacting factor for Sorkhe-Dizaj iron ore mine	Human health and safety	Social relationship	Water quality	Air quality	Use of territory	Flora and fauna	Surface	Underground	Landscape	Noise	Economy	Soil degradation
Environmental Pollution	**5**	2.809	2.381	2.703	3.39	1.923	2.778	1.875	1.64	1.899	1.137	2.151	2.381
Mine Size	**3**	0.337	1.071	0.811	1.017	1.154	0.833	1.5	0.984	1.519	1.364	1.29	1.429
Mining Method	**3**	1.011	0.714	0.811	1.017	0.769	1.25	1.125	1.475	1.139	2.045	0.968	0.953
Vegetation Coverage	**5**	1.686	1.786	2.027	2.543	1.282	2.778	2.5	0.82	2.532	2.273	1.613	2.381
Importance of the Area	**4**	0.899	1.428	1.081	1.356	2.051	1.667	1.5	1.967	1.519	1.818	1.29	1.905
Population Density	**2**	0.899	0.952	0.811	1.356	1.026	0.833	0.75	0.656	0.759	1.818	0.645	0.952
AMD (Acid Mine Drainage)	**7**	3.146	2.5	4.73	2.373	2.692	3.889	3.5	3.443	2.658	0	2.258	4.444
Tailing Leakage	**6**	2.696	2.143	3.243	2.034	2.308	3.334	3	1.967	2.278	0	1.936	2.857
Tailing Dam Failure	**3**	1.685	1.429	1.622	1.017	1.538	1.667	1.5	0.984	1.519	0	1.29	1.429
Waste Dump	**5**	1.686	1.191	2.027	1.695	1.923	2.084	1.875	1.64	2.532	0	1.076	2.381
Poor Rehabilitation	**8**	3.595	3.81	3.243	4.068	3.077	4.445	3	1.311	4.05	3.636	2.581	3.81
Employee Skills	**6**	2.696	2.143	1.622	2.034	1.538	0.833	0.75	1.967	1.519	2.727	2.581	0.952
Safety	**7**	3.933	4.166	1.892	2.373	1.795	1.945	1.75	3.443	1.772	4.773	3.011	1.111
Geological Uncertainty (tonnage and grade)	**7**	0	1.666	0.946	2.373	3.590	1.945	1.75	4.591	1.772	0	6.021	0
TOTAL		**46.96**	**46.9**	**46.76**	**47.12**	**45.25**	**47.09**	**45**	**45.41**	**45.57**	**44.09**	**46.78**	**45.24**

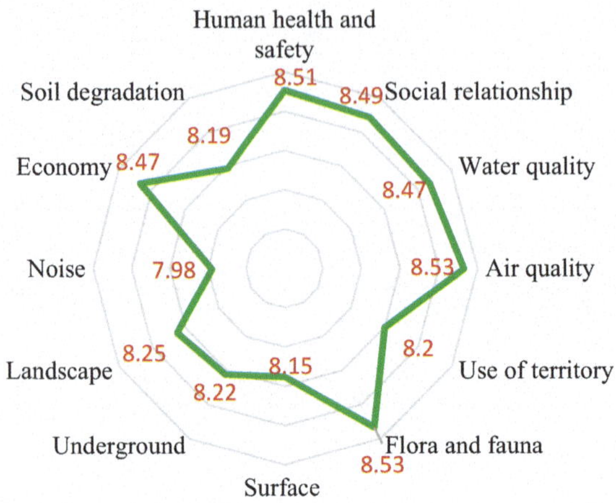

Fig. 2. The percentage of sustainable impact assessment for the components

6 Conclusion

The current study, investigating Sorkhe-Dizaj Iron Ore Complex, aimed at delving into the significance of each of the sustainable development factors. To this end developed Folchi method was applied. In the Folchi method, there were 28 sustainable influential factors impacting 12 sustainable components. The obtained result revealed that, based on the data analysis via the Folchi method, Sorkhe-Dizaj Iron Ore Complex contained six sustainable components, namely, human health and safety, social relationship, air quality, water quality, flora and fauna, and economy in mining had the most adverse impact on them.

Acknowledgment. We would like to acknowledge our special thanks to Prof. Sukumar Bandopadhyay, Prof. Javad Sattarvand, Prof. Erkan Topal, Prof. Konstantinos Fytas, Prof Bekir Genc Dr. Asiyeh Hekmat, Dr. Behzad Ghodrati, and Dr. Yousof Azimi, in addition to the engineers of Sorkhe-Dizaj mine and the engineers of Zanjan province organization of industry, mine and trade for responding to the project's questionary part.

References

1. Chakraborty, M.K., Ahmad, M., Singh, R.S., Pal, D., Bandopadhyay, C., Chaulya, S.K.: Determination of the emission rate from various opencast mining operations. Environ. Model Softw. **17**(5), 467–480 (2002)
2. Tadesse, S.: Environmental Policy in Mining: Corporate Strategy and Planning for Closure, pp. 415–422. (2000). ISBN 1-56670-365-4, A contribution to published book
3. Hansen, Y., Broadhurst, J.L., Petrie, J.G.: Modelling leachate generation and mobility from copper sulphide tailings–an integrated approach to impact assessment. Miner. Eng. **21**(4), 288–301 (2008)

4. Folchi, R.: Environmental impact statement for mining with explosives: a quantitative method. In: Proceedings of the Annual Conference on Explosives and Blasting Technique, vol. 2, pp. 285–296, February 2003. ISEE 1999
5. Phillips, J.: Using a mathematical model to assess the sustainability of proposed bauxite mining in Andhra Pradesh, India from a quantitative-based environmental impact assessment. Environ. Earth Sci. **67**(6), 1587–1603 (2012)
6. Phillips, J.: The application of a mathematical model of sustainability to the results of a semi-quantitative environmental impact assessment of two iron ore opencast mines in Iran. Appl. Math. Model. **37**(14–15), 7839–7854 (2013)
7. Rahmanpour, M., Osanloo, M.: A decision support system for determination of a sustainable pit limit. J. Cleaner Prod. **141**, 1249–1258 (2017)
8. Krekeler, M.P.S., Morton, J., Lepp, J., Tselepis, C.M., Samsonov, M., Kearns, L.E.: Mineralogical and geochemical investigation of clay rich mine tailings from a closed phosphate mine, Bartow Florida, USA. Environ. Geol. (2007). https://doi.org/10.1007/s00254-007-0971-8
9. Moyé, J., Picard-Lesteven, T., Zouhri, L., El Amari, K., Hibti, M., Benkaddour, A.: Groundwater assessment and environmental impact in the abandoned mine of Kettara (Morocco). Environ. Pollut. **231**, 899–907 (2017)
10. Antoniadis, V., Shaheen, S.M., Boersch, J., Frohne, T., Du Laing, G., Rinklebe, J.: Bioavailability and risk assessment of potentially toxic elements in garden edible vegetables and soils around a highly contaminated former mining area in Germany. J. Environ. Manag. **186**, 192–200 (2017)
11. Birch, C.: Optimization of cut-off grades considering grade uncertainty in narrow, tabular gold deposits. J. South Afr. Inst. Min. Metall. **117**(2), 149–156 (2017)
12. Goodfellow, R.C., Dimitrakopoulos, R.: Global optimization of open pit mining complexes with uncertainty. Appl. Soft Comput. **40**, 292–304 (2016)
13. Bouchard, J., Sbarbaro, D., Desbiens, A.: Plant automation for energy-efficient mineral processing. In: Energy Efficiency in the Minerals Industry, pp. 233–250 (2018)
14. Wright, M., Tartari, V., Huang, K.G., Di Lorenzo, F., Bercovitz, J.: Knowledge worker mobility in context: pushing the boundaries of theory and methods. J. Manage. Stud. **55**(1), 1–26 (2018)
15. Ranjan, V., Sen, P., Kumar, D., Saraswat, A.: Enhancement of mechanical stability of waste dump slope through establishing vegetation in a surface iron ore mine. Environ. Earth Sci. **76**(1), 35 (2017)
16. Bridge, G.: Mining and mineral resources. In: The International Encyclopedia of Geography (2017)
17. Cuervo, V., Burge, L., Beaugrand, H., Hendershot, M., Evans, S.G.: Downstream geomorphic response of the 2014 Mount Polley tailings dam failure, British Columbia. In: Workshop on World Landslide Forum, pp. 281–289 (2017)
18. Martín-Crespo, T., Gómez-Ortiz, D., Martín-Velázquez, S., Martínez-Pagán, P., De Ignacio, C., Lillo, J., Faz, Á.: Geoenvironmental characterization of unstable abandoned mine tailings combining geophysical and geochemical methods (Cartagena-La Union district, Spain). Eng. Geol. **232**, 135–146 (2018)
19. Antunes, S.C., Castro, B.B., Pereira, R., Gonçalves, F.: Contribution for tier 1 of the ecological risk assessment of Cunha Baixa uranium mine (Central Portugal): II soil ecotoxicological screening. Sci. Total Environ. **390**, 387–395 (2008)
20. Jimeno, L.C., Jimeno, L.E., Carcedo, F.J.A.: Drilling and blasting of rocks. Geo-mining Technological Institute of Spain, pp. 345–351 (1995)
21. Dudka, S., Adriano, D.C.: Environmental impacts of metal ore mining and processing: a review. J. Environ. Qual. **26**, 590–602 (1997)

22. Charbonnier, P.: Management of mining, quarrying and oreprocessing waste in the European Union. BRGM service EPI, France (2001)
23. Alamdarie, A.M.: Regional Tectonic and Structural Significance of Late Cretaceous-cenozoic Extension in Iran. New Mexico Institute of Mining and Technology (2017)
24. Rawat, N.S.: A study of physicochemical characteristics of respirable dust in an Indian coal mine. Sci. Total Environ. **23**, 47–54 (2003)
25. Shikazono, N., Zakir, H.M., Sudo, Y.: Zinc contamination in river water and sediments at Taisyu Zn–Pb mine area, Tsushima Island, Japan. J. Geochem. Explor. **98**, 80–88 (2008)
26. Chalupnik, S., Wysocka, M.: Radium removal from mine waters in underground treatment installations. J. Environ. Radioact. **99**, 1548–1552 (2008)
27. Ritcy, G.M.: Tailings management: problems and solutions in the mining industry. Environ. Int. **26**, 389–394 (1989)
28. Sengupta, M.: Environmental Impacts of Mining: Monitoring, Restoration, and Control, pp. 3–20. CRC Press, Boca Raton (1993). ISBN 0873714415
29. Daskalakis, D.K., Helz, G.R.: Solubility of CdS (Greenockite) in sulphidic waters at 25 °C. Environ. Sci. Technol. **26**, 2462–2468 (1999)
30. Wills, B.A.: Mineral Processing Technology: An Introduction to the Practical Aspects of Ore Treatment and Mineral Recovery, 7th edn. Elsevier, Butterworth Heinemann, London (2006)
31. Berner, E.K., Berner, R.A.: The Global Water Cycle, Geochemistry and Environment. Prentice Hall Inc., Englewood Cliffs (1987)
32. Shahriar, K., Samimi, Namin F.: A new approach to waste dump site selection according to the fuzzy decision-making process. Can. Inst. Min. Metall. Petrol. CIM **100**, 1–6 (2007)
33. Zhong, Z.: Overview of national mineral policy in China: opportunities and challenges for the mineral industries. Resour. Policy **23**, 79–90 (1998)
34. Chou, T.Y., Lin, W.T., Lin, C.Y., Chou, W.C., Huang, P.H.: Application of the PROMETHEE technique to determine depression outlet location and flow direction in DEM. J. Hydrol. **287**, 49–61 (2004)
35. Yaylacı, E.D.: Indicator-based sustainability assessment for the mining sector plans: case of Afsin-Elbistan coal basin. Int. J. Coal Geol. **165**, 190–200 (2016)
36. Yaylacı, E., Düzgün, S.H.: Evaluating the mine plan alternatives with respect to bottom-up and top-down sustainability criteria. J. Cleaner Prod. **167**, 837–849 (2017)

Drilling, Blasting and Excavation Engineering

Effect of Delay Time and Firing Patterns on the Size of Fragmented Rocks by Bench Blasting

Takashi Sasaoka[1(✉)], Yoshiaki Takahashi[2], Akihiro Hamanaka[1], Sugeng Wahyudi[1], and Hideki Shimada[1]

[1] Kyushu University, Fukuoka 819-0395, Japan
sasaoka@mine.kyushu-u.ac.jp
[2] The National Institute of Advanced Industrial Science and Technology, Tsukuba 305-8568, Japan

Abstract. This paper discusses the size control method of fragmented rocks induced by bench blasting in an open pit mine, especially the effect of delay time and firing pattern. Based on the results of a series of field tests, it can be said that the delay time and firing pattern have an impact on the size of fragmented rocks. The application of two directions of firing pattern that the order of ignition of blast holes is from the center to both ends of the row of blast holes can produce a more uniform size of fragmented rocks compared with that of a one-directional firing pattern which is the order of the ignition of blast holes from one side to the other. Due to the rock mass conditions and products specification, the size of fragmented rocks and its distribution can be controlled by applying an appropriate delay time and firing pattern.

Keywords: Delay time · Firing pattern · Rock fragmentation

1 Introduction

Rock blasting is the rock excavation and fragmentation technique most widely adopted in the various fields of the mining, civil and construction industries because of its efficient, and economical aspects [1]. On the other hand, in case that the blasting operation may have an obvious impact on the surrounding environment, such as ground vibration, fly rock, noise, etc. blasting standards or conditions are not appropriate [2]. Moreover, the size of fragmented rocks have an obvious impact on the open pit mining operations such as loading, hauling and crushing, and it may have a large influence of the total operation cost. Hence, one of the keys for success of safe and economical mining operations is to design an appropriate blasting standard based on the rock mass conditions.

From these points of view, this paper discusses the size control method of fragmented rocks induced by bench blasting in an open pit mine especially considering the effects of delay time and firing pattern without changing other blasting standards such as burden, spacing and/or powder factor.

E. Topal (Ed.): MPES 2019, SSGG, pp. 449–456, 2020.
https://doi.org/10.1007/978-3-030-33954-8_51

2 Overview of Field Experiments

The case study Mine A used in this research is located in the southern part of Kyushu Island in Japan. This mine is an open-pit metal mine and extracts a silicic-acid ore which contains gold.

A series of blasting tests were conducted at three faces (east, west, and north) in B pit. Table 1 represents the basic blasting design used in this pit. Ammonium Nitrate Fuel Oil (ANFO) was used as the explosive and was initiated hole by hole using electric detonators. Before blasting, a photograph of the bench face was taken for evaluation of the fracture/joint state of rock mass. After blasting, a photograph of the debris was taken for the rock fragmentation analysis, and then the rock samples were collected in order to the measure mechanical properties of rock at each face.

In this test, the delay time and the firing pattern (the order of the ignition) were changed and their effects on the size of fragmented rocks were discussed.

Table 1. Basic blasting standard at Mine A.

Burden (m)	2.5
Spacing (m)	1.5
Borehole diameter (mm)	76
Bench height (m)	10
Bench angle (°)	80
Drilling angle (°)	80
Drilling length (m)	12
The number of blast hole (hole)	10
Powder factor (g/t)	170
Charge quantity (kg/hole)	23.7

2.1 Evaluation of Fracture/Joint Conditions in Rock Mass

Before blasting, a photo of each face was taken by a digital camera in order to investigate and record the fracture condition of the rock mass. As shown in Fig. 1, courses of traverse were set at the face wall (every 2 m), and then the sizes and abundance ratios of each rock block separated by discontinuities and courses of traverse were calculated. This space of fracture/joint assumes the particle size of each rock block. The particle size at 50% of the gain size accumulation curve of rock blocks before blasting (Xb_{50}) was used as a representative parameter for evaluating the joint/fracture condition of rock mass at each face. Moreover, the directions of the major joint systems were measured by using a clinometer.

2.2 Fragmentation Analysis

Fragmentation assessment was achieved by the analysis of a scaled photograph taken from the fragmented rocks. Paley recommended a procedure for taking photographs of fragmented rocks as to minimize errors due to distortion [3]. Two balls with a diameter of 24 cm were used as a scale in the photograph. The balls were placed in the same vertical line down the fragmented rocks, preferably with one ball near the top of the fragmented rocks and the other near the bottom. The balls should not be placed randomly in the fragmented rocks nor in a horizontal line across them. The camera was held such that the long axis of the photograph was vertical. The photograph was then taken with the camera as perpendicular to the surface of fragmented rocks as possible. By having two balls on the surface of fragmented rocks, allowance was made for a variable scale within the photograph when the camera could not be positioned perpendicular to the surface of the fragmented rocks.

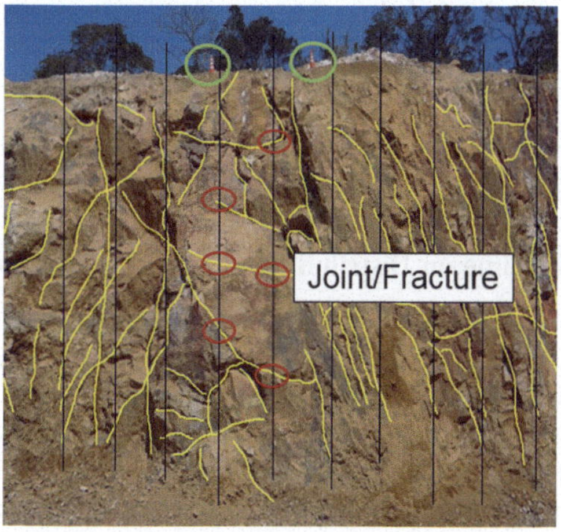

Fig. 1. Measurements of joints/fractures condition of rock mass at test face.

The scaled fragmentation photographs were manually digitized from the original photograph on the computer screen by software known as Split-Desktop, developed by Split Engineering, as illustrated in Fig. 2 [4]. The outlines of visible rocks above a certain minimum resolution, 3 mm in diameter on the photograph, were traced by a mouse. After the digital image was analyzed, the particle size distribution of fragmented rock was derived, as shown in Fig. 2. The representative particle size at 50% of the gain size accumulation curve, Xp_{50}, was used in this research.

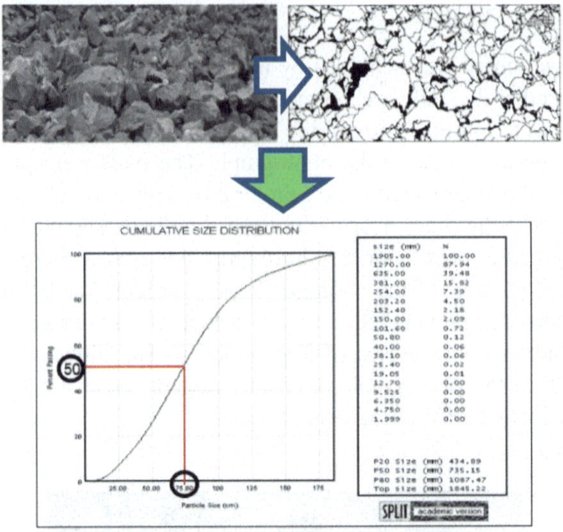

Fig. 2. Procedures of fragmentation analysis.

3 Results and Discussions

3.1 Effect of Delay Time on Distribution of Rock Fragmentation

Delay blasting is generally conducted in order to control blast-induced ground vibration. In addition, it influence on the fragmented effect since it creates a new free face. Figure 3 illustrates the blasting pattern discussed in this section. By using MS electrical detonators, two types of delay time: 25 ms and 50 ms were set in the field experiment. In addition, firing direction was also discussed. One was firing from edge to edge of the blasting hole and the other one was firing from the center to the edges of the holes.

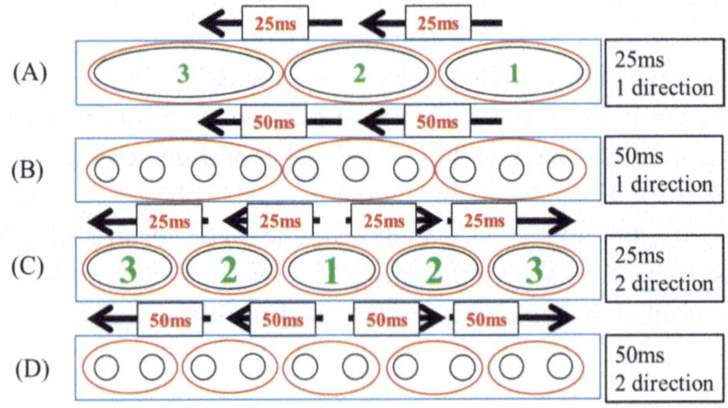

Fig. 3. The illustration of blasting pattern.

Example photographs of muck pile in each blasting pattern are shown in Fig. 4. As shown in these photographs, size distribution of rock fragmentation is different from each blasting pattern. In the case of blasting patterns (A) and (B) in Fig. 3, firing from edge to edge of the row, the size of rock fragmentation is obviously different depending bench face. In other words, the size tends to be big around the area of start of firing point and the one is likely to be small around the area of the end of firing point.

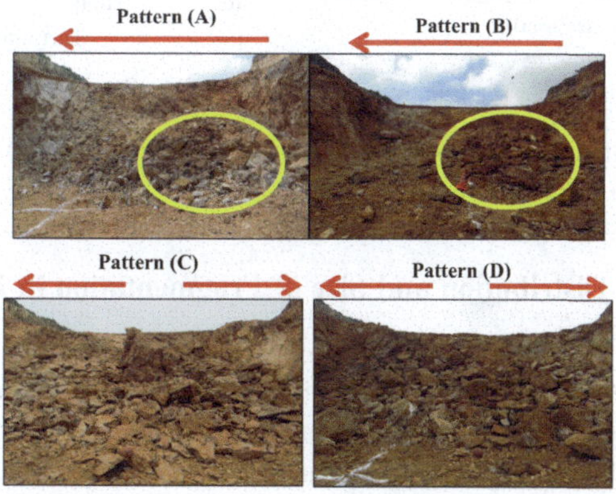

Fig. 4. The muck pile after blasting in each blasting pattern.

On the other hand, in the case of blasting patterns (C) and (D) in Fig. 3, firing from the center to the edges, overall of the size tends to be homogeneous. Hence, as a next step, the photograph of the muck pile is divided in to 3 parts as shown in Fig. 5; the photographs are analyzed by Sprit-Desktop software again. The percent passing of the size of fragmentation of patterns (A) and (C) is shown in Fig. 6(a) and (b), respectively. The results described above are successfully seen in these figures. In case of one direction firing pattern, the size tends to be big around the area of start of firing point since stress wave interference is hard to occur around the area and at the end of firing point, stress wave interference help to reduce the size of rock fragmentation. On the other hand, because stress wave interference equally occurred overall blasting area, resulting in homogenous size distribution.

Fig. 5. The divided photographs for discussing the effect of blasting pattern (a) dividing pattern for Pattern (A) and (B) and (b) dividing pattern for Pattern (C) and (D).

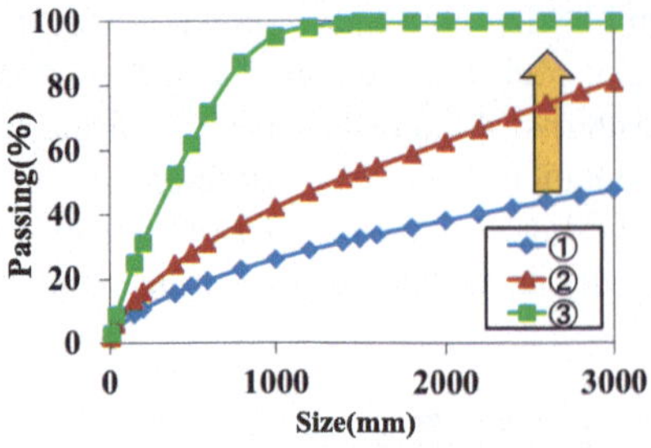

(a)

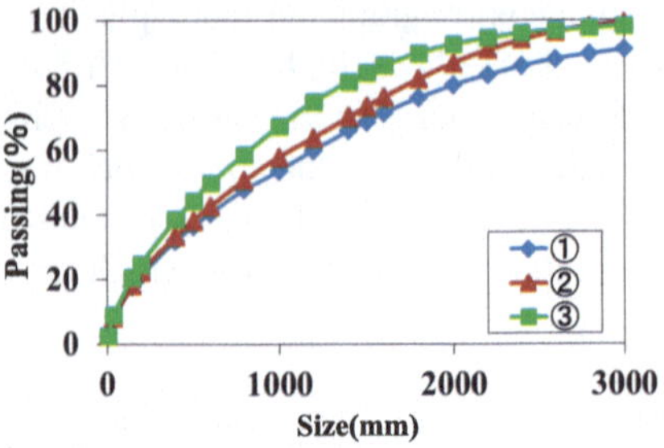

(b)

Fig. 6. The accumulation curves obtained from divided photograph (a) the result of Pattern (A) and (b) the result of Pattern (C).

3.2 Prediction of Distribution and Size of Fragmentation in Delay Blasting

Based on the discussion described above, the prediction of fragmentation size in delay blasting is established in this section. In order to access the distribution, the homogeneity of the distribution have to be quantitatively evaluated. Therefore, the uniformity coefficient is defined on a basis of uniformity coefficient which is generally used to classify the soil [5] as follow:

$$n = \frac{X_{p60}}{X_{p10}} \tag{1}$$

Where, n is uniformity coefficient, X_{P60} and X_{P10} are the particle size at 60% and 10% of the gain size accumulation curve, respectively. In the field of soil classification, the range of the size distribution is classified as wide when $n \geqq 10$. On the other hand, the soil is judged as uniform when n is less than 10. Representative distributions and their uniformity coefficients are shown in Fig. 7(a), (b) and (c). By visual observation, the distributions are divided into three ranks in this study. The divided rank is listed in Table 2. In addition, the result of uniformity coefficient in each firing pattern is shown

(a) (b) (c)

Fig. 7. The representative distribution of fragmented rock in uniformity coefficient (a) 2.1 (b) 15.1 (c) 20.3.

Table 2. The rank of distribution in each uniformity coefficient.

Rank	Uniformity coefficient
Good	0–10
Normal	10–20
Bad	20 ~

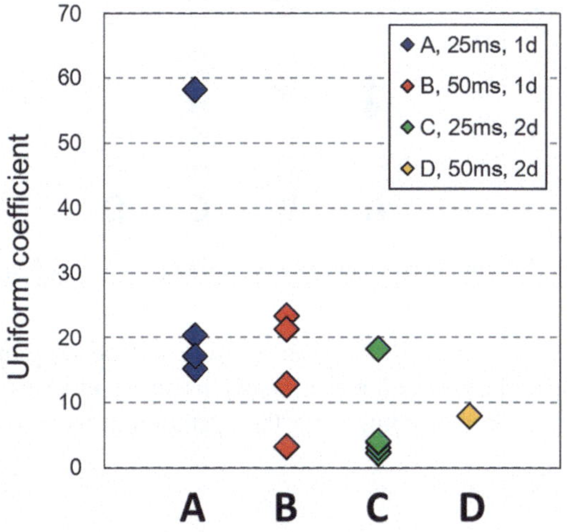

Fig. 8. Uniformity coefficient in each firing pattern.

in Fig. 8. Moreover, the averages of uniform coefficient of patterns A, B, C and D are 27.7, 15.1, 5.65 and 7.98. Based on the results, it can be said that two directions of firing pattern can make the distribution more uniform. This might be because the formation of stress wave interference and free face occur symmetrically in the case of two directions of firing pattern. On the other hand, the behavior of superposition of stress waves is different depending upon the place in the case of one direction of firing pattern, which result in un-uniform size distribution.

Furthermore, the effect of delay time and firing pattern is discussed. The relationship between firing pattern and X_{P50} is illustrated in Fig. 9. As can be seen in this figure, X_{P50} in the case of 50 ms looks like small. Hence, the average of X_{P50} of each firing pattern is calculate. The average X_{P50} of pattern A, B, C and D are 961.3 mm, 742.1 mm, 537.6 mm and 500.7 mm. This result suggested that the size of fragmented rock can be reduced by applying 50 ms of delay time in this mine. There might be certain delay time which can reduce the size of fragmented rock. On the contrary, the average X_{P50} of one direction (A and B) and two directions (C and D) of firing pattern are 851.2 mm and 528.3 mm, respectively. Moreover, the average X_{P50} of 25 ms (A and C) and 50 ms (B and D) of delay time are 707.0 mm and 661.7 mm, respectively. It can be seen that although delay time influence on the mean size of fragmented rock, the influence of firing direction is more significant than that of delay time. Two directions of delay time have good dependent advantage of both distribution and mean

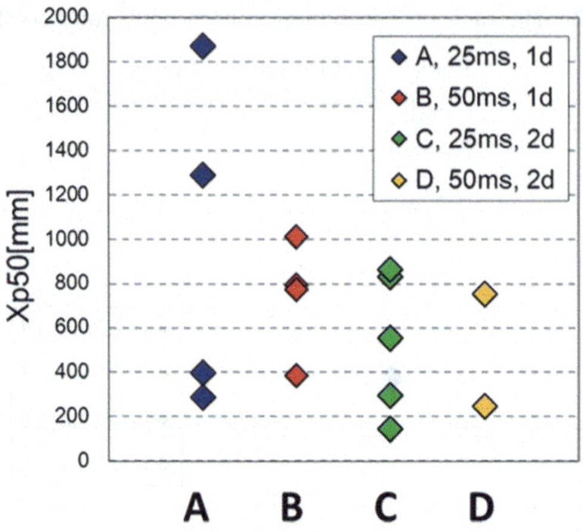

Fig. 9. The relationship between firing pattern and X_{P50}.

size of fragmented rock. Although the required size is depending upon the operation, two directions of firing pattern is better to apply basically in terms of uniformity and delay time should selected depending upon the operation in order to control the size of fragmented rock.

4 Conclusions

Delay time has an obvious impact on the size and distribution of blast-induced fragmented rocks. Firing direction strongly influence on both the distribution and size of blast-induced fragmented rocks. Homogeneous size of blast-induced fragmented rocks can be obtained by conducting two directions firing pattern and one direction firing pattern make the size distribution heterogeneous. On the other hand, delay time influence on the size of fragmented rocks, but the influence of the firing direction on the size is larger than that of delay time.

Acknowledgements. This project is sponsored by the limestone association of Japan. The authors would like to express their gratitude to staff of Mine A for providing their field assistance and kindly cooperation for this research.

References

1. Bhandari, S.: Engineering Rock Blasting Operations, p. 375. A.A. Balkema, Rotterdam (1997)
2. Reza, N.: Soil Dyn. Earthq. Eng. **43**, 133–138 (2012)
3. Paley, N.L.: Proceedings of the International Journal for Blasting and Fragmentation, vol. 2, no. 4, pp. 415–431 (1998)
4. Sprit Engineering LLC: Manual of Split-Desktop Software (2003)
5. Japanese Geotechnical Society: Standard of Soil Testing 2nd Edn., pp. 35–36 (2010)

Illumination of Contributing Parameters of Uneven Break in Narrow Vein Mine

Hyongdoo Jang[1(✉)], Sina Taheri[1], Erkan Topal[1], and Youhei Kawamura[2]

[1] Western Australian School of Mines, Curtin University, Kalgoorlie, WA, Australia
hyongdoo.jang@curtin.edu.au
[2] Graduate School of International Resource Sciences, Department of Earth Resource Engineering and Environmental Science, Akita University, Akita, Japan

Abstract. One of the principal challenge facing the stope production in underground mining is the overbreak and underbreak (UB: uneven break). Although the UB features a critical economic fallout to the entire mining process, it is much inevitable and usually left as an unpredictable phenomenon in underground mines. The complex mechanism of UB must be examined to minimize the UB phenomenon. In this study, the contribution of ten primary UB causative parameters is scrutinized investigating a published UB prediction ANN model. The inputs (UB causative factors) contributions to the output (percentage of UB) of the ANN model were analyzed using Profile methodology (PM). The results PM revealed the essential importance of geological parameters to UB phenomenon as the calculated contributions of adjusted Q-rate (GAQ) and average horizontal to vertical stress ratio (GSK) are 20.48% and 18.12% respectively. Also, the trends of the other eight UB causative factors were investigated. The findings of this study can be used as a reference in stope design and reconciliation processes to maximize the productivity of the underground mine.

Keywords: Overbreak · Underbreak · Narrow vein · ANN

1 Introduction

In the modern mining industry, many underground metalliferous mines are operating by various type of stoping methods due to the flexible applicability over various ground conditions and the ability to mechanization into large scale. Pakalins noted that in 1996, 50% underground metalliferous mines were operating by open stoping method and in 2011, over 60% of underground production in North America was produced by sublevel stoping method (Pakalnis and Hughes 2011). Recently, Jang et al. (2015a) noted that more than 85% of underground metalliferous mines rely on various stoping methods.

Despite the advantages of stoping method, mines are often suffering from unplanned dilution and ore-loss that inevitably occur at the stope production blasting. The unplanned dilution can be defined as the influx of over broken rocks from beyond the planned stope margin while the unplanned ore-loss is the remaining ore even after the production. In other words, the unplanned dilution and ore-loss are overbreak and

E. Topal (Ed.): MPES 2019, SSGG, pp. 457–466, 2020.
https://doi.org/10.1007/978-3-030-33954-8_52

underbreak from the blasting engineers' perspective and Jang et al. (2015b) noted them as uneven break (UB). Figure 1 schematically demonstrates the unplanned dilution and ore-loss.

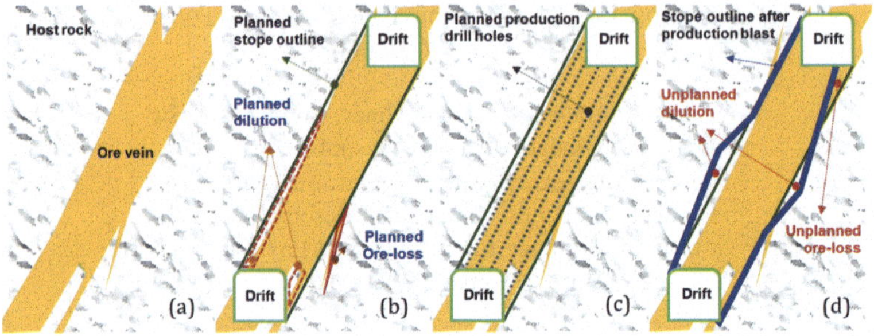

Fig. 1. Schematic view of dilution and ore-loss throughout stope production stages (a) an example of ore vein shape (b) stope production plan (c) production drilling plan (d) stope after production (Jang et al. 2016)

The overbreak and underbreak (UB: uneven break) directly influence the productivity of a mine. In fact, one of the primary tasks of mining engineers is to minimize UB phenomenon under a manageable level. Thus, understanding of the causative factors and complex UB causing mechanism is essential.

Table 1. The list of ten UB causative factors and corresponding percentage of UB

No.	Input parameters										Output
	Blasting					Geology		Stope design			
	BHL	BPF	BAN	BHD	BSB	GAQ	GSK	SPT	SAR	SOB	UBP
	m	Kg/t	°	mm	(S/B)	–	(H/V)	tons	(W/H)	–	%
1	15.20	0.56	2.00	89	1.25	37	1.85	18,480	1.65	1	−2.30%
2	13.50	0.42	55.00	89	1.50	30	1.99	4,200	2.67	1	4.70%
2	7.30	0.45	0.00	76	1.11	40	2.12	896	0.5	0	3.10%
⋮	⋮	⋮	⋮	⋮	⋮	⋮	⋮	⋮		⋮	⋮
1065	8.00	0.68	17.00	76	1.25	26	2.22	2,980	0.38	0	−18.60%
1066	16.60	0.91	0.00	89	1.15	14	3.15	16,607	0.83	1	10.10%
1067	20.20	0.90	0.00	89	1.20	19	3.25	6,804	0.37	0	12.70%
Min.	0.70	0.15	0.00	76	0.57	6.30	1.74	130	0.07	0	−65.40
Max.	25.80	3.00	170.20	89	1.50	93.30	14.38	51,450	4.17	1	92.00

***BHL**: Average length of blasthole, **BPF**: Powder factor, **BAN**: Angle difference between hole and wall, **BHD**: Diameter of blasthole, **BSB**: Space (S) and burden (B) ratio, **GAQ**: Adjusted Q rate, **GSK**: Average horizontal to vertical stress ratio, **SPT**: Tons of stope planned, **SAR**: Aspect ratio of stope, **SOB**: Opened or blind of the stope **UBP**: Percentage of uneven break (over and under break)*

The study aims to illuminate the complex causing mechanism of UB adopting an UB prediction model published by Jang et al. (2015a). The UB prediction model was developed by investigating 1067 stope production results from underground stoping mines in Western Australia. Each dataset consists of 10 input parameters (UB causative factors) and the corresponding output (measured percentage of overbreak). Five blasting related factors, i.e., average length of blasthole (BHL), powder factor (BPF), angle difference between hole and wall (BAN), diameter of blasthole (BHD), and Space (S) and burden (B) ratio (BSB), two geology parameters, i.e., adjusted Q rate (GAQ), average horizontal to vertical stress ratio (GSK), and two stope design factors, i.e., tons of stope planned (SPT), opened or blind of the stope (SOB) were used as input factors while the corresponding percentage of overbreak and underbreak (UBP) was set as the output in the model. The study relied on the historical data thus human errors, i.e., the drilling errors and the hole deviation were not considered. The influence of human errors in UB phenomenon was indirectly included in the study with the BHL as the drilling accuracy is generally expressed as a percentage of the hole depth (Stiehr and Dean 2011). The list of data sets is shown in Table 1.

2 Current over and Underbreak Management in Underground Stoping Mines

Minimising uneven break (UB) in stope production is one of the essential tasks because it is directly influencing not only to the productivity of stope production but also the profit throughout of mining operations. However, UB is challenging to manage due to its highly complex causing mechanism. Many engineers and scholars have been endeavoring to manage the unwanted phenomenon, but the industry often relies on historical stope reconciliation results and intuitions of experienced engineers.

Few empirical systems have introduced and the stability graph method (Mathews et al. 1981; Nickson 1992; Potvin 1988) is widely used to manage the stope overbreak. This method is plotting a stability number (N) against a hydraulic radius (HR: area/perimeter of the stope wall) to judge the stability of stope wall. The stability number is an integrated geotechnical factor that has been modified by Potvin (1988) and is defined as:

$$N' = Q' \times A \times B \times C \tag{1}$$

Where N′ presents the modified stability number, Q′ is the modified Q value, and A, B, and C are factors that stand for the stress, the joint orientation, and the gravity respectively.

The stability of a designed stope wall can be easily assessed with the stability graph method but cannot guarantee a reliable judgment. Considering the complexity of the overbreak and underbreak phenomenon during rock blasting, the stability of a stope wall cannot be adequately assessed with few geotechnical factors in the stability graph method (Jang 2014). For example, Fig. 2 demonstrates 134 cases of stability analysis results with measured overbreak in each data point. Most of the stopes were designed to be in the stable zone in the stability graph but overbreak were randomly occurred.

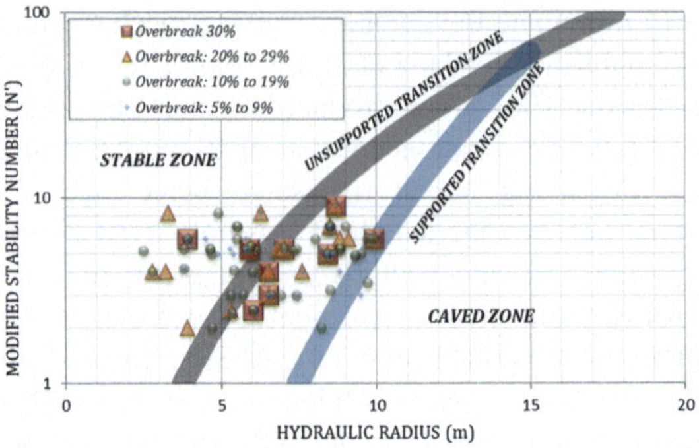

Fig. 2. Stability graph method analysis results and corresponding overbreak results (Jang, 2014)

Germain and Hadjigeorgiou (1997) conducted a study on stope overbreak investigating the powder factor of stope production blast and Q-value (Barton et al. 1974). The study used a linear regression analysis to find the relation between overbreak and the two variables but the result was statistically insignificant as the correlation coefficient of less than 0.3. The stress effects on stope overbreak were investigated by Stewart et al. (2005). The study found the critical influence of stress on the overbreak phenomenon as more than 50% of overbreak occurred where the induced stress was exceeded the damage criterion. Pakalnis (1986) developed an overbreak prediction model investigating 133 open stope production results in Canada. The model can predict the percentage of overbreak on stope wall investigating the hydraulic radius and rock mass rating (RMR) (Bieniawski 1973, 1974) in three different stope conditions, i.e., isolated, adjacent rib, and echelon. Lang (1994) introduced the critical span cave method to analyze the stability of the cut and fill stoping mines and updated including 292 case studies in the University of British Colombia. In the study, the critical span was defined as the diameter of the largest circle of unsupported back in the stope and the stability of a stope back was determined by plotting designed critical span of the stope with the RMR value. The method has been used in some Canadian mines but it requires careful consideration as it founded with regional data. Despite the endeavors of many researchers, the causing mechanism and the exact contributions of causing parameters of over and underbreak are not studied.

3 UB Prediction ANN Model

In this study, artificial neural network (ANN) is adopted to examine the contribution of UB causative parameters (independent variables) to the UB (a dependent variable). Artificial neural network (ANN) is a parallel computation inference model that is comprised of input, hidden, and output layers with a number of mathematical elements called the artificial neuron. The neurons in each layer are fully interconnected and the strength of the connection is expressed with a weight value. In the ANN model, dependent variables, i.e., inputs, are set in the input layer while the independent variable, i.e., output, is placed at the output layer. The hidden layer is located between the input and the output layer. The artificial neurons in the input and output layer are interconnected through neurons in the hidden layer.

The UB prediction ANN model consisted of ten inputs, forty hiddens, and one output neuron. The hyperbolic tangent function was used as the transfer function in the forward pass while Levenberg-Marquardt (LM) algorithm (Levenberg 1944; Marquardt 1963) was employed as the learning function in the backward pass of the model training stage. 70% of the collected datasets were randomly allocated to the training stage while 15% of each remaining datasets were used for the validation and the test stage. The model trained with 100 iterations and the root-mean-square error (RMSE) in training stage was 1.90E-2 and the test resulted with the correlation coefficient of 0.719.

4 Parameter Contribution Analysis of UB Prediction ANN Model

The UB prediction ANN model had achieved a statically significant result which facilitates to investigate the inputs and output sensitivity of the model. Often ANN is treated as 'a black box' as it is difficult to demonstrate how the model is optimized. Thus, some researches and engineers hesitate to adopt the ANN even it has been successfully utilized in many studies through various disciplines. To illuminate the model optimization process of ANN, various methods were developed e.g., Garson's algorithm (Garson 1991), connection weights algorithm (CWA) (Olden and Jackson 2002), Partial derivatives (PaD) method (Dimopoulos et al. 1995), Relative Strength of Effect (RSE) (Yang and Zhang 1998), and profile method (PM) (Lek et al. 1995). The Garson's algorithm and CWA can be categorized as the connection weights based algorithms while PaD, RSE, and PM can be called as the sensitively based algorithms.

In this study, the profile method is used as it investigates the entire range of inputs and output combinations. The percentage of a predicted uneven break when all inputs are set their 80th except GAQ (GAQ, 80th) was computed at the entire range of GAQ. The profiling process for GAQ was conducted in other ranges, i.e., 60th, 40th, and 20th, of remaining input variables. The profile method results of all ten input parameters of the UB prediction ANN model is shown in Appendix 1. Appendix 1-a, b, c, d, e, f, g, h, i, and j are profiling results of 10 inputs, and the overall sensitivity is demonstrated in Appendix 1-k.

5 Discussion

Through the PM application, the sensitivity of 10 UB causative factors to UB phenomenon has been analyzed. Two geological parameters, i.e., GAQ and GSK, revealed to have the relatively high effect to UB than other parameters by achieving a sensitivity of 20.48% and 18.12% respectively. Appendix 1-a shows the PM results of GAQ and the percentage of overbreak (UBP) decreased as the quality of rock becomes better. Appendix 1-b demonstrates the variance of UBP over the entire range of GSK which implies that UBP tends to decrease as the stope is getting deeper. The SAR (stope aspect ratio) shows 12.40% of contribution to UBP and the results are graphed in Appendix 1-c. The SAR shows a proportional relationship to UBP which indicates that the overbreak likely occurs in broader stope. The proportional trend of SAR over UBP in 60th and 80th are much steeper than other ranges which means the SAR (aspect ratio of stope) influences more in good rock condition. In succession, BSB revealed to have 10.41% of contribution to UBP which is shown in Appendix 1-d.

Table 2. Key findings of ten UB causative parameters investigated by the profile method (PM)

Factor	Con. %	Key findings
GAQ	20.48%	The better the rock quality, the less the percentage of overbreak
GSK	18.12%	The deeper the stope, the less the percentage of overbreak
SAR	12.40%	The wider the stope, the more the percentage of overbreak
BSB	10.41%	The longer the ring burden, the more the percentage of overbreak
SPT	8.20%	The bigger the stope size, the more the percentage of overbreak
SPF	7.26%	The overbreak increases when the powder factor is increased
BHD	6.99%	The bigger the blasthole dia., the more the percentage of overbreak
BHL	6.13%	The longer the length of blasthole, the less the percentage of overbreak
SOB	5.25%	The percentage of overbreak is increased when the stope is blinded
BAN	4.77%	The percentage of overbreak is increased with the parallel blasthole pattern than the fanned pattern

The UBP is generally increased with increasing BSB except for the range of 40th which can be interpreted as the more overbreak would occur when the ring burden is longer. Next contributor to UBP is SPT that shows the overall sensitivity of 8.20%. As observed in Appendix 1-e, except the remaining parameter range of 40th, UBP generally increased when stope size is bigger. Appendix 1-f is PM results of BPF (powder factor) with the overall contribution of 7.26%. The BPF shows relatively high sensitivity in the range of the remaining parameters in 60th and 80th. As the rock quality, i.e., GAQ is the highest contributor to UBP form this study, the result of BPF can be interpreted as the leverage of BPF to overbreak is higher in the poor rock than the excellent quality rock. The PM results of the blasthole diameter (BHD) is shown in Appendix 1-g which has 6.99% of overall contribution to UBP. The overbreak is likely increased in 89 mm (3.5 in.) blast hole than 76 mm (3 in.) expect the remaining parameter range of 20th. Appendix 1-h demonstrates the relations between the blasthole length (BHL) and UBP. The overall sensitivity of BHL to UBP is calculated as 6.13%, and it shows that the percentage of overbreak (UBP) is generally decreased in longer blasthole. The next influential contributor to UBP is SOB (opened or blind of the stope) with the overall contribution of 5.25% as shown in Appendix 1-i. The magnitude of overbreak is gradually increased in the blind stope except for the 20th of remaining parameter range. BAN (angle difference between hole and wall) has the overall contribution of 4.77% to the UBP, and the tendency is demonstrated in Appendix 1-j. Considering GAQ as the most influential contributor to UB phenomenon, the amount of overbreak is likely increased in the parallel drilling with the stope wall than the fanned pattern drilling. Finally, the overall contribution of the ten UB causative inputs to UBP is shown in Appendix 1-k. The key findings from the profile method (PM) application are summarised in Table 2.

6 Conclusion

The overbreak and underbreak (UB: uneven break) are unavoidable in underground stope production where the drill-and-blast is employed. A thorough examination of UB causative factors is essential to reveal the causing mechanism to minimize the unfavorable phenomenon. In this study, the contribution of ten UB causative factors to the percentage of the uneven break has been studied by investigating inputs and output sensitivities of a published UB prediction ANN model. The UB prediction ANN model is established using 1067 datasets and achieved the correlation coefficient of 0.71. The sensitivity of the ANN model inputs and output is investigated using profile method (PM) which can thoroughly study the entire ranges of inputs and output relations. The result of PM method shows that the uncontrollable geological parameters, i.e., GAQ and GSK, revealed to have more influence than other blasting and stope design parameters as the overall contribution of 20.48% and 18.12% respectively. In succession, the contribution of SAR and BSB to UBP were 12.40% and 10.41% respectively. The other parameters, i.e., SPT, SPF, BHD, BHL, SOB, and BAN demonstrated less than 10% of contribution to the percentage of uneven break (UBP).

The mechanism of the overbreak and underbreak due to the dynamic blasting loads is exceptionally complex, and the exact causing parameters and their contributions have not been clearly discovered. The study attempted to reveal the contributions of ten key causing factors to overbreak and underbreak phenomenon in the underground stope blasting. The findings through the study can be used as a reference for design, production, and reconciliation of the underground stope production.

Appendix 1

The result of sensitivity analysis of the UB prediction ANN model. (BHL: Average length of blasthole, BPF: Powder factor, BAN: Angle difference between hole and wall

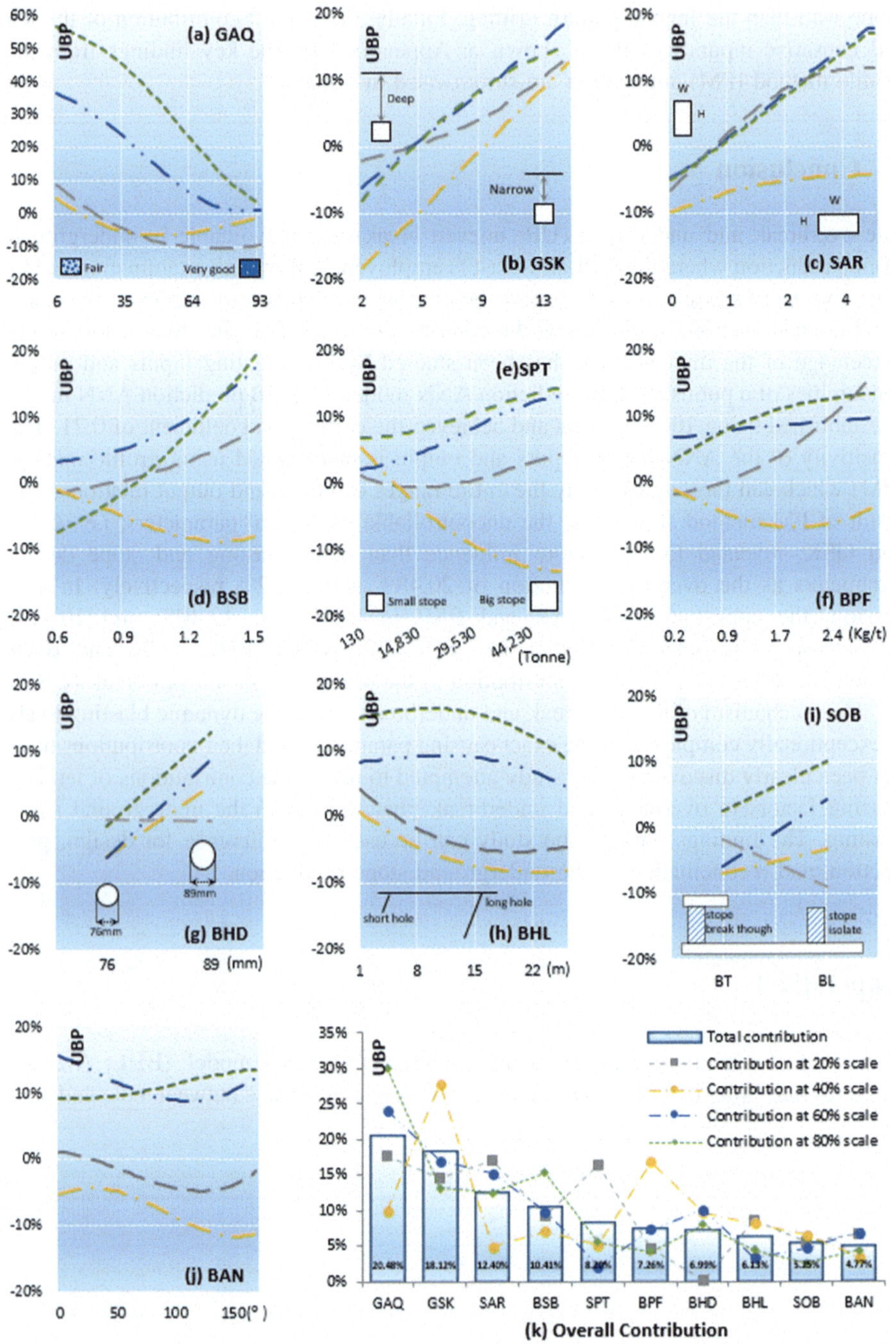
(a) GAQ
(b) GSK
(c) SAR
(d) BSB
(e)SPT
(f) BPF
(g) BHD
(h) BHL
(i) SOB
(j) BAN
(k) Overall Contribution
Total contribution
Contribution at 20% scale
Contribution at 40% scale
Contribution at 60% scale
Contribution at 80% scale
GAQ
GSK
SAR
BSB
SPT
BPF
BHD
BHL
SOB
BAN

References

Barton, N.R., Lien, R., Lunde, J.: Engineering classification of rock masses for the design of tunnel support. Rock Mech. **6**, 189–236 (1974). Southern Nevada: Geological Society of America Bulletin 88, 943–959

Bieniawski, Z.T.: Engineering classification of jointed rock masses. Civ. Eng. South Afr. **15**(12), 335–343 (1973)

Bieniawski, Z.T.: Geomechanics classification of rock masses and its application in tunnelling. Paper presented at the Advances in Rock Mechanics, Proceedings of the 3rd International Society of Rock Mechanics Congress, Denver, Colorado (1974)

Dimopoulos, Y., Bourret, P., Lek, S.: Use of some sensitivity criteria for choosing networks with good generalization ability. Neural Process. Lett. **2**(6), 1–4 (1995)

Garson, G.D.: Interpreting neural-network connection weights. AI Expert **6**(4), 46–51 (1991)

Germain, P., Hadjigeorgiou, J.: Influence of stope geometry and blasting patterns on recorded overbreak. Int. J. Rock Mech. Min. Sci. **34**(3), 115.e111–115.e112 (1997). https://doi.org/10.1016/S1365-1609(97)00219-0

Jang, H.: Unplanned dilution and ore-loss optimisation in underground mines via cooperative neuro-fuzzy network. Ph.D., Curtin University, Kalgoorlie, WA, Australia (2014)

Jang, H., Topal, E., Kawamura, Y.: Decision support system of unplanned dilution and ore-loss in underground stoping operations using a neuro-fuzzy system. Appl. Soft Comput. **32**, 1–12 (2015a)

Jang, H., Topal, E., Kawamura, Y.: Unplanned dilution and ore loss prediction in longhole stoping mines via multiple regression and artificial neural network analyses. J. South Afr. Inst. Min. Metall. **115**, 449–456 (2015b)

Jang, H., Topal, E., Kawamura, Y.: Illumination of parameter contributions on uneven break phenomenon in underground stoping mines. Int. J. Min. Sci. Technol. **26**(6), 1095–1100 (2016)

Jang, H.D.: Unplanned dilution and ore-loss optimisation in underground mines via cooperative neuro-fuzzy network. Ph.D. Dissertation, Western Australian School of Mines, Curtin University, Kalgoorlie, Western Australia, Australia (2014)

Lang, B.D.A.: Span design for entry-type excavations. Master, University of British Columbia, Vancouver, Canada (1994)

Lek, S., Belaud, A., Dimopoulos, I., Lauga, J., Moreau, J.: Improved estimation, using neural networks, of the food consumption of fish populations. Mar. Freshw. Res. **46**(8), 1229–1236 (1995)

Levenberg, K.: A method for the solution of certain problems in least squares. Q. Appl. Math. **2**, 164–168 (1944). doi:citeulike-article-id:1951284

Marquardt, D.W.: An algorithm for least-squares estimation of nonlinear parameters. J. Soc. Ind. Appl. Math. **11**(2), 431–441 (1963)

Mathews, K.E., Hoek, E., Wyllie, D.C., Stewart, S.: Prediction of stable excavation spans for mining at depths below 1000 m in hard rock. In: Paper presented at the CANMET DSS Serial No: 0sQ80-00081, Ottawa (1981)

Nickson, S.D.: Cable Support Guidelines for Underground Hard Rock Mine Operations. University of British Columbia, Vancouver, Canada (1992)

Olden, J.D., Jackson, D.A.: Illuminating the "black box": a randomization approach for understanding variable contributions in artificial neural networks. Ecol. Model. **154**(1), 135–150 (2002)

Pakalnis, R.C.T.: Empirical stope design at Ruttan mine. In: Paper presented at the Department of Mining and Minerals Processing, Vancouver (1986)

Pakalnis, R.C.T., Hughes, P.B.: Sublevel Stoping. In: Darling, P. (ed.) SME Mining Engineering Handbook, 3rd edn. Society for Mining, Metallurgy, and Exploration, Inc., United States of America (2011)

Potvin, Y.: Empirical open stope design in Canada. Ph.D., University of British Columbia, Vancouver, Canada (1988)

Stewart, P., Slade, J., Trueman, R.: The effect of stress damage on dilution in narrow vein mines. In: Paper presented at the 9th AusIMM Underground Operators Conference 2005 (2005)

Stiehr, J.F., Dean, J.: ISEE blasters' handbook, pp. 442–452. International Society of Explosives, Cleveland (2011)

Yang, Y.-J., Zhang, Q.: The application of neural networks to rock engineering systems (RES). Int. J. Rock Mech. Min. Sci. **35**(6), 727–745 (1998)

A Study on Rock Cutting Forces and Wear Mechanisms of Coated Picks by Lab-Scale Linear Cutting Machine

Sathish Kumar Palaniappan(✉), Samir Kumar Pal, and M. P. Dikshit

Department of Mining Engineering, Indian Institute of Technology Kharagpur, West Bengal 721302, India
sathishiitkgp@gmail.com

Abstract. Conical cemented carbide tip has facilitated extensive application in the field of mining engineering because of their exceptional combination of strength, hardness and high wear resistance. This paper focuses on different cutting forces and wear mechanisms in a conical pick which has been used in a surface miner machine for rock cutting. In the present work, an attempt has been made by coating aluminium titanium nitride (AlTiN) on the tip to calculate the cutting forces and to understand the wear mechanisms. Modified shaper machine was used to cut the rock sample linearly and their corresponding forces were measured by using a 3D strain gauge octagonal dynamometer. Coated tool possess highest hardness of 22.47 GPa with lowest wear rate of 2.54×10^{-4} cm^3/cm and 4.42×10^{-4} cm^3/cm, whereas uncoated tool with lowest hardness of 19.10 GPa depicts wear rate of 2.95×10^{-4} cm^3/cm and 5.65×10^{-4} cm^3/cm for 2 mm and 4 mm depths of cut. Parameters such as cutting forces, amount of coal removed, cutting efficiency and specific energy generated during cutting were also analysed for different depths of cut and the best compromise was found among them. The worn out surface has been critically examined using scanning electron microscopy (SEM) and energy dispersive X-ray (EDX) analysis. This study may be helpful for the selection of suitable surface miner machine and their cutting pick for hard rock cutting operation in mines.

Keywords: Coating · Cemented carbide · Wear analysis

1 Introduction

In recent years, mechanical cutting of coal/rock has emerged and evolved over time which provides increasing amount of excavation from soft to medium rocks, thus often replaces the traditional drilling and blasting method in both civil and mining industries. The important parameters like cutting rate, cutting forces, specific energy (SE), pick consumption and cutting vibrations restrict the applications of mechanical excavators. Harder rocks are always present in the mines. The presence of undesirable rock materials (i.e., hard rock) in-between coal is the main reason behind cutting bit damage. During cutting, the tool is hardly hit by coal/rock and the sudden high impact of harder

E. Topal (Ed.): MPES 2019, SSGG, pp. 467–475, 2020.
https://doi.org/10.1007/978-3-030-33954-8_53

rocks leads to cracking and crushing of the WC-Co grains, which renders the tool absolutely useless [1].

Especially in India, numerous trails of WC-Co tipped picks on coal and soft to medium rocks are very effective and proven satisfactory, but unsuccessful in hard rock cutting. Increase in strength of coal/rock dramatically reduces the performance of mechanical excavators. The working conditions of WC-Co picks are severe, complex and dynamic. Picks work under high impact and stress conditions often results in failure. Many researchers have paid increasing attention to pick failure and have noted that the failure styles mainly include premature wear, WC-Co tip drop off, tipping, fracture and normal wear. Hard rock cutting process by using worn picks end up in hazardous problems like respirable dust and frictional heat generation. Wear analysis of the cutting picks has been an interesting subject of research for a long back. Wear assessment to find the causes of wearing gives an idea to modify the materials concentration of the picks, so that the service life can be enhanced for the safe working environment. It is also important to monitor the progress of tool wear and correlate with force variations and SE consumed during the tests.

Based on the extensive literature survey, it is evident that several works had been carried out by the researchers on enhancing the performance of cutting picks [2]. In order to extend the role of cutting bits to harder rocks, many innovative attempts are being made by the manufacturers and researchers. Most of these attempts are concentrated on developing more resistant and longer lasting bits, optimizing the bit geometry and lacing patterns, and the addition of high pressure water jet assistance. Although rock cutting using mechanical tools has been extensively studied, most of the previous research had focused on either a single or a limited number of principal parameters [3]. Few studies have been conducted on the analysis of thermal effects and cutting chips in rock cutting although they are crucial and are indicators of the frictional heat, cutting efficiency and dust generation in a rock cutting operation [4]. One of the aspects of developing efficient bits for hard rock cutting is to understand the forces acting on bits during mining operation under varying cutting conditions and correlate them with progress of tool wear [5]. Knowing the magnitude of forces acting on the individual bits is also an important aspect of machine design, since it allows the engineers to estimate the cutterhead torque and machine power requirements for a particular application. Also, specific energy is a direct measure of cutting efficiency. Therefore, there is a need in the industry to understand the wear mechanisms and to estimate the accurate forces acting on bits and specific energy while cutting hard rocks. These may be measured by full-scale or small-scale laboratory linear cutting tests. Coatings enhance surface hardness, thermal and wear resistance of cutting tool [6], decreases frictional force and prevent chemical interactions between tool and work material [7]. The bulk material possesses high toughness with high thermal conductivity to evade the deformation of tool and its geometry. The combined properties of coating and bulk material can be attained based on their working condition, work material and their end applications. Coatings augment the cutting characteristics and determine the performance of bit based on working conditions and the several cutting parameters.

Till date, very little work has been done on coating the WC-Co tip of cutting picks to reduce the wear and to extend their application in cutting harder rocks. Coating was performed on cutting bits by employing appropriate materials and methods. However, the studies on the deposition of coating materials for enhancing the performance of coal/rock cutting bits constitutes in lesser dimension. Hence, this research work lies in experimental analysis of coated WC-Co bits to understand its performance for cutting high strength rocks and get a better insight into the wear phenomenon.

2 Lab-Scale Linear Rock Cutting Setup

2.1 Lab-Scale Linear Rock Cutting Setup

The experimental investigation of this study involved newly modified shaping machine (Make: AAR PEE, Model: RPS-1000) i.e., lab-scale linear rock cutting setup where various bits can be used to cut the hard rocks. For this purpose, the test rig can be raised, lowered or transverse relative to the cutting bit, which is having a stroke of 1010 mm, voltage of 440 V and power of 5.6 kW. The machine can accommodate a block of rock having a length of 28 cm, a width of 18 cm and a height of 10 cm. Especially, the cross-head of the shaper has been modified to perform hard rock cutting experiments. The cutting bit is attached to a strain gauge based triaxial force dynamometer having precision in the order of 1 kN and covering a range from 0 to 100 kN [8] along with a suitable tool holder. The major components of lab-scale linear rock cutting setup are visualized in Fig. 1.

Fig. 1. Components of lab-scale linear rock cutting setup.

Strain gauge based triaxial force dynamometer [9] was used in this experiment to resolve the forces acting on the bit during hard rock cutting into three mutually perpendicular components. Data acquisition system modules will be used to collect the data at the specified sampling rate. The calibration factors will be developed to determine the cutting, normal and side forces of cutting bits during the experiment.

2.2 Sample Preparation

WC-Co bits having a conical tip angle of 40° and tip width of 19.3 mm (Grade: YG11C and Code: W2919), purchased from Jinan Xinyu Cemented Carbide Co. Ltd., China were used in this study for cutting high strength rocks at different depths of cut. These bits were mainly used for coal-cutting bits, tricone bits to cut medium-hard to hard formations, drilling hard rock formations, auger tips and well drilling tools and at the same time for drilling stone walls, cement in construction and installation work, and for machining stone materials. Aluminium titanium nitride (AlTiN), a second generation supernitride coating material with nanocomposite structure was used in this study for deposition onto the WC-Co bit. This material is suited for processing difficult-to-machine materials and for hard processing.

In this deposition process, both the target and substrate were positioned at a distance of 5 cm by the respective holders which were mounted inside the vacuum chamber. The vacuum was generated inside the chamber with the help of rotary pump. When the vacuum pressure reached the preset value, the sputter-ion pump was started and the argon gas was introduced into the chamber, which acted as a working medium to deposit the target particles over the substrate. Coating on the substrate was performed at certain time duration of 15 min with some constant process parameters during coating, which were as follows, argon pressure of 0.7 bar, vacuum chamber pressure 6 Pa and process power of 80 W.

The rock samples were shaped approximately into a cubic form with a dimension of $28 \times 18 \times 10\ \text{cm}^3$ for proper fitting on the shaper machine. The rock samples used during the cutting tests have been collected and carefully selected. There were no obvious fractures and discontinuity planes within the structure of the samples. Before performing the tests, surface of hard rock samples were carefully trimmed to obtain flat surface and for getting the good results. Some mechanical properties of the rock sample are presented in Table 1.

Table 1. Mechanical properties of the rock sample.

Rock properties	Values
Uniaxial compressive strength (MPa)	83.77
Indirect tensile strength (MPa)	9.71
Shear strength (MPa)	11.64
Youngs Modulus (GPa)	24.04
Poisson's ratio	0.11

3 Experimental Design/Procedure

New WC-Co bits were used in this study to estimate the performance of cutting bits under hard rock cutting operation. Utmost care was paid to use sharp picks for each single cut throughout the cutting trials. Presently, the linear rock cutting tests of coated and uncoated bits were performed by using modified shaping machine on the block of fine grain sandstone having higher strength. Similar rock samples will also be collected to perform the experiments for optimizing the results. The design allowed the bit to be mounted at an attack angle of 45° during the linear cutting, which is within the range of angles [10] when mounting picks on continuous miners, road headers and shearers. The attack angle was not further increased because it has been observed that at an inclination higher than 45°, the tool tip and some parts of the tool body begin to rub against the rock sample. The rake and clearance angles were kept constant during linear cutting of hard rock because it was difficult to vary the orientation of the tool post of the shaper machine. Currently, the tests were undertaken at two different depths of cut i.e., 2 mm and 4 mm with cutting bits. Higher depths of cut will also be experimented in future for all types of rock samples. At each depth of cut, which constituted a pass across the surface of the sample, the cut spacing varied accordingly because of limitations in shaping machine. Length of the cuts was around 20 cm or cutting along the longest dimension of the sample. The weight loss or volume loss of the cutting bits after all the experiments were calculated to determine the wear rate. The same set of testing was performed atleast more than 3 times to ensure reliability and repeatability of data. The removed rock (debris) was carefully collected to calculate the yield for determining the SE. Some photographs for the linear rock cutting experiments performed so far can be seen in Fig. 2.

Fig. 2. Photographs for the linear rock cutting experiments performed.

The cutting forces (F_N, F_C, F_S) will be recorded in each cut by using triaxial force dynamometer and corresponding yield (Q) will be calculated for determining the SE i.e., dividing F_C by Q with the convenient units. The relation between depths of cut, cutting forces, bit wear and SE should be drawn. The data obtained by the experiment will be plotted on the graph for further detailed analysis. The morphological characterization of the tested bits will also be analyzed to determine the deformation of coating thickness, bit-rock interactions and to understand the wear mechanisms. Finally, the obtained results helps to optimize the coating materials for a cutting bit to excavate high strength rocks under desired operating parameters.

4 Results and Discussion

4.1 Scanning Electron Microscopy Analysis

Coated samples require special preparation for SEM analysis, except for cleaning and mounting on a specimen stub. The coated bit was cut into two halves through wire cut electrical discharge machine and the cut section was examined. Samples tend to charge when scanned by the electron beam, and especially in secondary electron imaging mode, this causes scanning faults and other image artifacts. Before experimenting for SEM analysis, a thin conductive layer of gold (Au) was sputter coated on the surface of bits. The thickness of AlTiN deposited WC-Co bits were measured from cross-sectional SEM morphology (Fig. 3). From the image, deposition of coating material on the substrate is evident. The coating thickness was found to be around 3 μm to 4.5 μm through ImageJ software.

4.2 Energy Dispersive X-Ray Analysis

Figure 3 also shows the EDX plot of AlTiN coated cutting bit. It confirms the presence of versatile coated elements such as Al, Ti, N. From the Fig. 3, the presence of surface (Al, Ti, N) and subsurface (substrate - WC) elements were also evident. The EDX plot clearly validates the deposition of AlTiN coating. It also includes the presence of subsurface tungsten element (base material). Conductive Au layer was also substantiated from the results.

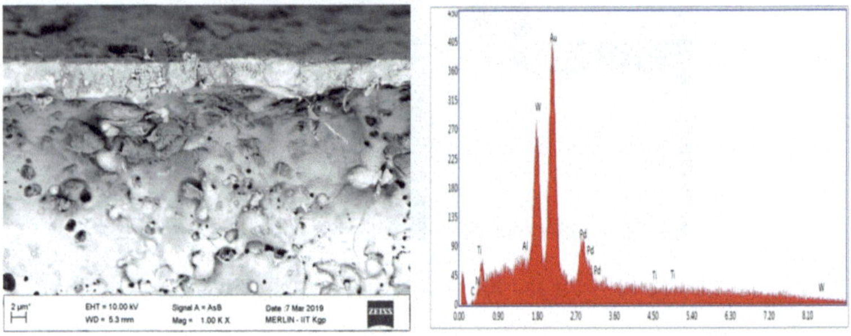

Fig. 3. Cross-sectional SEM image and EDX plot of AlTiN coated tool.

4.3 Hardness Testing

Hardness is an important parameter correlating with wear resistance of the material. In applications where the wear has to be restricted, high hardness materials were typically used. Micro-hardness tester is commonly used to measure micro-hardness of the samples. In this work, the hardness of coated and uncoated tools were measured by using Vickers hardness tester 400 series (Make: Wolpert Wilson) for a load of 1 kg with the time of indentation (dwell time) 10 s by right pyramid shaped diamond indenter. Hardness measured in micro Vickers HV was converted into GPa for further interpretation. Depth of indentation was chosen not to be more than one tenth of the individual layer thickness. The diagonal length of the indention marks on the substrate was measured at 500x magnification. The results were averaged for four measurements. Hardness value comparison of coated and uncoated bits is illustrated in Fig. 4. From the hardness measurement, it was observed that AlTiN coated cutting bits possess higher hardness value (22.47 GPa) compared to uncoated bits (19.10 GPa), which also reduces the wear rate at different depths of cut for the same rock samples. The reason for high hardness of AlTiN coated bit may be due to the formation of carbo-nitride hard compounds (CN), the presence of which is also confirmed through EDX analysis.

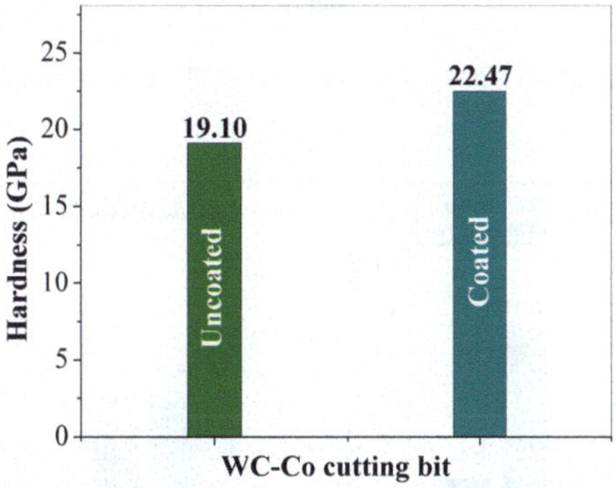

Fig. 4. Hardness value of AlTiN coated and uncoated WC-Co tool.

4.4 Wear Analysis of Cutting Bits

A brief wear assessment of the coated and uncoated bits after hard rock cutting has been done. The wear rate of uncoated and coated tool shows a similar trend with varying depths of cut. Coated bits shows decreased wear rate compared to uncoated one. The results were depicted for different depths of cut. Figure 5 shows the wear rate of uncoated and coated bits under two different depths of cut (i.e., 2 mm and 4 mm). For instance, cutting the harder rock at two different depths of cut depict decreases in wear rate with increase in hardness of the bit. Coated tool possess highest hardness of 22.47 GPa with lowest wear rate of 2.54×10^{-4} cm^3/cm and 4.42×10^{-4} cm^3/cm, whereas uncoated tool with lowest hardness of 19.10 GPa depicts wear rate of 2.95×10^{-4} cm^3/cm and 5.65×10^{-4} cm^3/cm for 2 mm and 4 mm depths of cut. The percentage decrease in wear rate of coated bit compared to uncoated bit is 14% for 2 mm depth of cut and 22% for 4 mm depth of cut. Figure 6 shows the relation between hardness and wear rate for coated and uncoated bit at two different depths of cut.

The force calculations and surface energy determination for all the tested uncoated and coated samples are not yet disclosed in this paper, because the results are found unique and are still in process, which will be published later. The correlation obtained between wear, forces and surface energies are possible and the optimization process will be carried out to come-up with better coating material for improving the performances of surface miner in hard rock cutting operation.

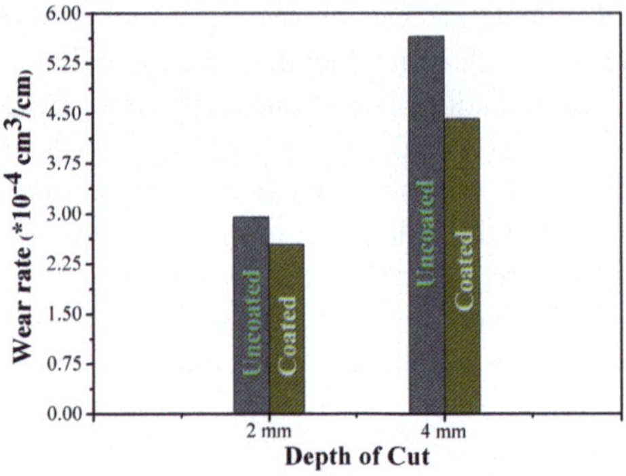

Fig. 5. Wear rate of AlTiN coated and uncoated bit at different depths of cut.

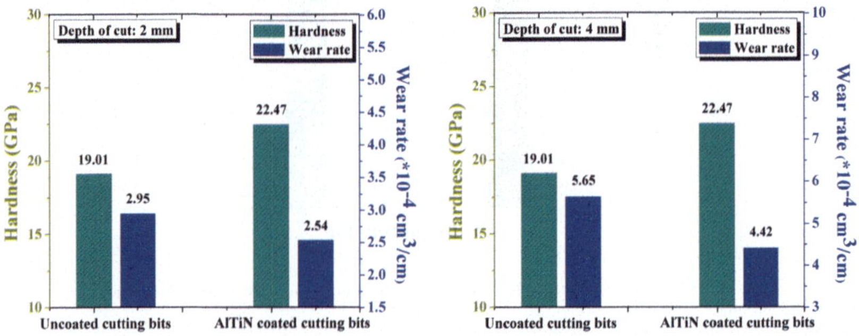

Fig. 6. Variation of wear rate with respect to hardness at 2 mm and 4 mm depth of cut.

References

1. Liu, Z.C., Roxborough, F.: Mining Science and Technology. Balkema Publications, Rotterdam (1996)
2. Jeong, H.Y., Jeon, S.W., Cho, J.W.: A study on punch penetration test for performance estimation of tunnel boring machine. Tunn. Undergr. Space Technol. **22**, 144–156 (2012)
3. He, X., Xu, C.: Specific energy as an index to identify the critical failure mode transition depth in rock cutting. Rock Mech. Rock Eng. **49**, 1461–1478 (2015)
4. Copur, H., Bilgin, N., Balci, C., Tumac, D., Avunduk, E.: Effects of different cutting patterns and experimental conditions on the performance of a conical drag tool. Rock Mech. Rock Eng. **50**, 1585–1609 (2017)
5. Dewangan, S., Chattopadhyaya, S.: Performance analysis of two different conical picks used in linear cutting operation of coal. Arab. J. Sci. Eng. **41**, 249–265 (2016)
6. Deketh, H.J.R.: Wear of Rock Cutting Tools. Balkema Press, Brookfields (1995)
7. Zhang, D., Shen, B., Sun, F.: Study on tribological behavior and cutting performance of CVD diamond and DLC films on Co-cemented tungsten carbide substrates. Appl. Surf. Sci. **256**, 2479–2485 (2010)

8. Balci, C., Bilgin, N.: Correlative study of linear small and full-scale rock cutting tests to select mechanized excavation machines. Int. J. Rock Mech. Min. Sci. **32**, 468–476 (2006)
9. Shaw, M.C.: Metal Cutting Principles. Clarendon Press, Oxford (1984)
10. Mostafavi, S.S., Yao, Q.Y., Zhang, L.C., Li, X.S., Lunn, J., Melmeth, C.: Effect of attack angle on the pick performance in linear rock cutting. In: 45th International Proceedings on US Rock Mechanics/Geomechanics Symposium, pp. 1–4. American Rock Mechanics Association, California (2011)

Rock Mechanics and Geotechnical Applications

Evaluation on the Instability of Stope Mining Influenced by the Risks of Slope Surface and Previous Mined-Out Activities

Naung Naung[1,2(✉)], Takashi Sasaoka[1], Hideki Shimada[1], Akihiro Hamanaka[1], Sugeng Wahyudi[1], and Pisith Mao[1]

[1] Department of Earth Resources Engineering, Kyushu University, Fukuoka, Japan
naung15r@mine.kyushu-u.ac.jp

[2] No. (2) Mining Enterprise, Ministry of Natural Resources and Environmental Conservation, Nay Pyi Taw, Myanmar

Abstract. Open stope mining is the most common mining method adopted in underground metal mines in Myanmar. However, the assessments on the stability of stope still remain quite limited for this method. Even though mining objective is to recover ore as much as possible from the vein, safety of workers and machinery in the advancing stopes must be ensured. The main reasons for the occurrence of underground mine's instability come from the stress redistribution of previous mined-out activities, the existence of discontinuities and void space, and the influence of on-going mining activities, etc. In addition to these problems, if the underground mining is conducted near the slope surface, the influence of slope surface should be taken into account to that mining activities. Therefore, considering the importance of rock stability, the investigations on the strength of rock mass near the slope surface are carried out and stability of advancing stope with previous mined-out activities are conducted at Modi Taung gold mine, one of the largest underground gold mines in Myanmar. From the preliminary outcome of this research, the rock mass near the slope surface can be more subjected by differential stress than inside of the rock mass. This result indicates that the condition of rock mass near slope surface is more unstable than that of the places far from slope surface. Moreover, the results highlight that the potential buckling failures from hanging wall and foot wall are more severe than the roof falls from the stope in this mine site.

Keywords: Instability · Previous mined-out · Slope surface · Stope mining

1 Introduction

Myanmar has more than 300 prospected gold deposits across the country [1]. Gold deposits are classified as either primary or alluvial types and the mining method used in these gold deposits are both surface and underground. Previously, most of the underground mines across the country are operating at the easily assessable shallow regions. However, in recent years, underground mining activities are going to move into deeper places to scrape the valuable mineral. As the depth of mines increase, the occurrence of higher induced stress is becoming a more common phenomenon

E. Topal (Ed.): MPES 2019, SSGG, pp. 479–487, 2020.
https://doi.org/10.1007/978-3-030-33954-8_54

causing some problems for instability and safety of mines. The stability of underground mines may be influenced by many factors such as the quality of rock, the existence of discontinuities and void space, in-situ stress conditions, the influence of previous mined-out and on-going mining activities, and mining depth, etc. [2]. In addition to these factors, if the underground mining is located near the slope surface, the influence of slope surface should be taken into account to that mining activities. Therefore, considering the importance of rock stability, the investigations on the strength of rock mass near the slope surface are carried out and stability of advancing stope with previous mined-out activities are conducted out at Modi Taung gold mine, one of the largest underground gold mines in Myanmar.

2 Brief Lithology and Mine Plan at Research Mine Site

Veins at Modi Taung gold mine are hosted by four main lithologies such as mudstone, sandstone-siltstone, limey sandstone or limestone, and igneous intrusions. Host rocks in Shwesin vein are predominantly mudstone but sandstone occupies short segments in all levels, and veins tend to occur along the inclined interface between sandstone and mudstone [3].

The accessible shallow area has been already mined out at natural slope of Shwesin vein system, hence, the deposits are left in deeper regions. Therefore, the mining plans to continue to deeper levels, in which separated into 6 blocks, i.e., block 1 to block 6. The overall mine plan is illustrated in Fig. 1. Overhand cut and fill mining method is adopted to extract the minerals at Shwesin vein and the dimension of stope is 2.5 m in height and 2 m in width. The waste rocks from excavation are only used to both fill the stope and provide permanent wall support for the lower mine out cavity. Vein dipping at Shwesin vein are up to 80° and vein thickness varies with elevation ranging from 40 cm to 140 cm [4].

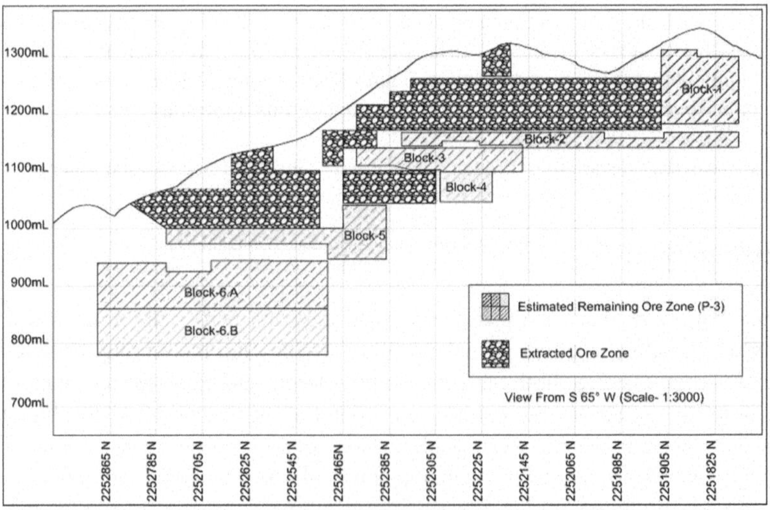

Fig. 1. Simplified mine plan at Shwesin vein (source: NPGPGL).

3 Numerical Model with FLAC^{3d} Software

In this study, the size of basic numerical model is 260 m × 260 m × 320 m with vein dip 70° as shown in Fig. 2. As the slaty mudstone is a dominant rock type in Modi Taung gold mine, and therefore, the hanging wall and footwall are assigned as homogenous model for simplification. The horizontal and vertical in-situ stresses are equally subjected to this model and all mechanical properties of rock used in this study are shown in Table 1.

In order to know the stability of mine opening, a factor of safety of 1.3 would generally be considered for a temporary mine opening while a value of 1.5 to 2.0 may be required for a permanent excavation [5]. In the simulations, the Mohr-Coulomb failure criterion is adopted, and the stope was considered in a stable condition when Mohr-Coulomb safety factor is greater than 1.3 and unstable condition will meet when safety factor is less than 1.3.

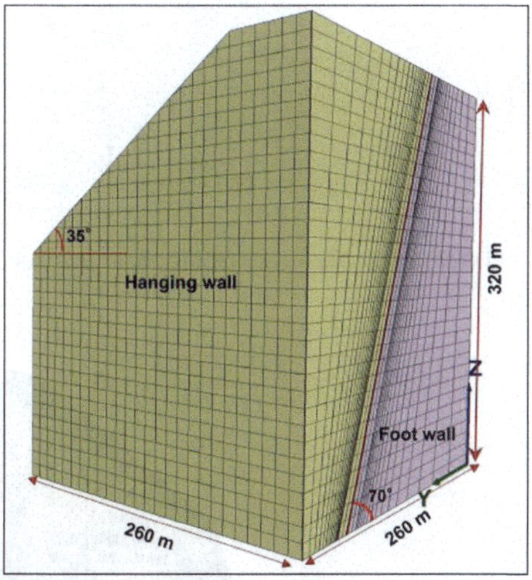

Fig. 2. Numerical model for research study

Table 1. Rock mass properties used in analyses.

	E [MPa]	ν [−]	σ_t [MPa]	φ [deg]	C [MPa]
Hanging wall	3786	0.23	0.035	44	0.761
Foot wall	3786	0.25	0.065	40	0.687
Vein	3786	0.22	0.034	45	0.77
Filling rock	153.1	0.321	–	20.5	0.200

In this research, two types of numerical investigations are carried out by using FLAC 3D software [6]. Firstly, the numerical simulations on the stability of advancing stope near the slope surface are carried out for understanding the condition of rock mass near the natural slope surface as shown in Fig. 3. Secondly, in order to evaluate the potential instability of stope mining which is surrounded by previous mined-out, the simulations are conducted in Block 3 which is separated with zone 1, 2 and 3 as shown in Fig. 4.

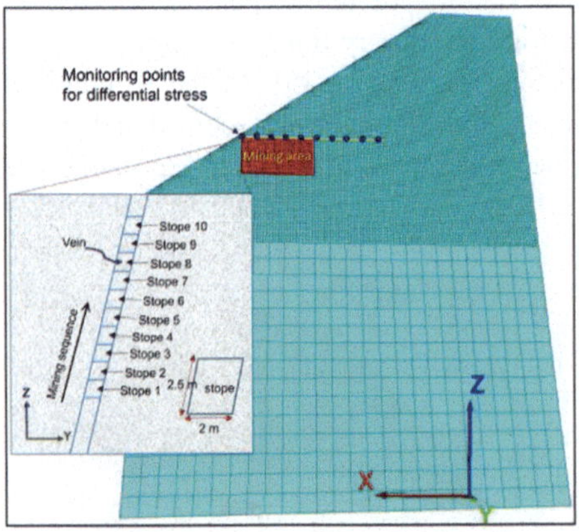

Fig. 3. Simulation model for rock stability near the slope surface.

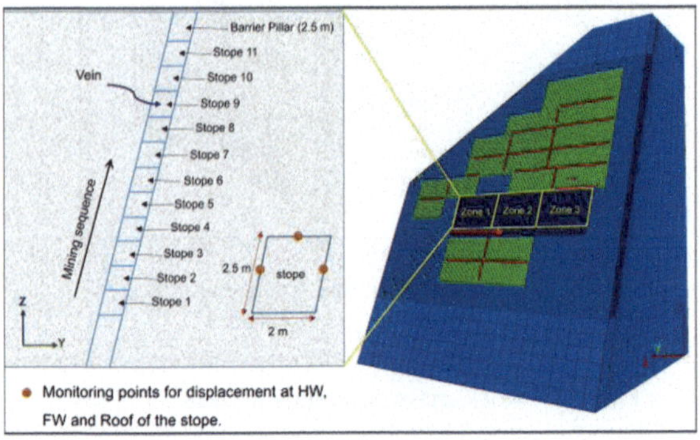

Fig. 4. Numerical simulations of zone 1, 2 and 3 in Block 3.

4 Results and Discussions

4.1 Instability of Rock Mass Affected by the Slope Surface

4.1.1 Condition of Rock Mass Near the Slope Surface Before Mining Activities

At first, the condition of rock mass near the slope surface is investigated by monitoring with the variation of differential stress before mining condition. As well known, differential stress is the difference between the major and minor principal stresses and rock mass can deform (break/flow) due to differential stress [7]. Even though the most serious part is occurred at the base of the slope surface as shown in Fig. 5, the instability of rock mass are gradually propagated into the rock mass. As can be seen in Fig. 6, the result shows that the differential stress near the slope surface is large and the stress gradually decrease started from 12.85 m distance from the slope surface. That trend indicates that the stability of rock mass near the slope surface is more severe than that of the places far from the slope surface.

4.1.2 Stability of Rock Mass Near the Slope Surface During Mining Activities

Next, in order to understand the stability of mining near the slope surface, the simulation is done during mining activities as shown in Fig. 3. In this simulation, the monitoring places are recorded at three cutting planes of stoping area; (1) edge of stope

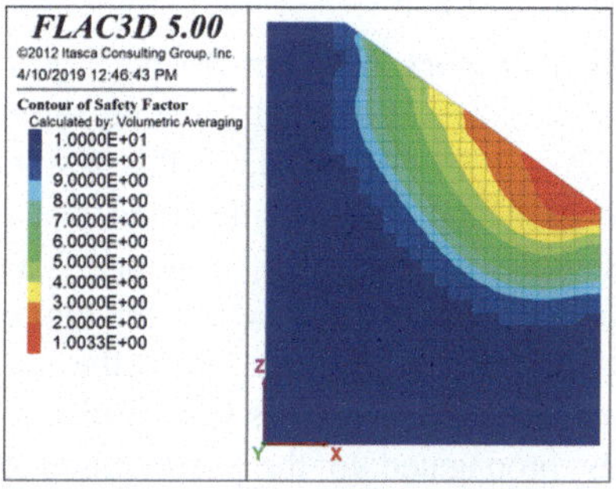

Fig. 5. Condition of rock mass near the slope surface at initial state.

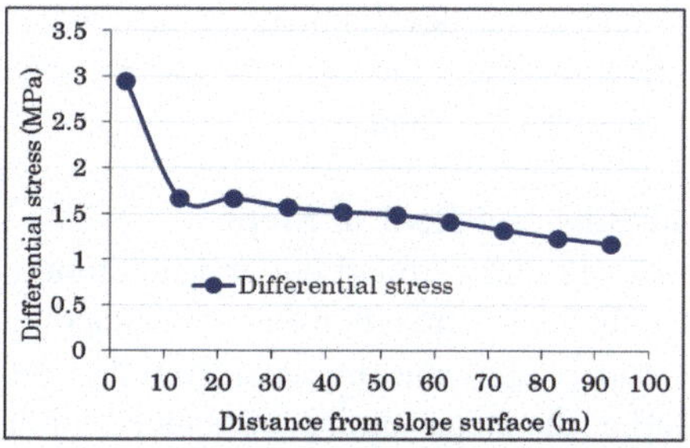

Fig. 6. Differential stress indicators near the slope surface.

near slope surface, (2) middle of stope, (3) edge of stope far from the slope surface. The simulation result is shown in Fig. 7. The results suggest that the stope instability are more likely to occur at the edge of stope near the slope surface. The yield elements are more propagating at the edge of stope near the slope surface than the place far from the slope surface. Therefore, according to this result, it can be said that the rock mass near the slope surface can be more affected by differential stress than that of inner part, and this result highlights the importance of rock stability when the underground mining activities operate near the slope surface.

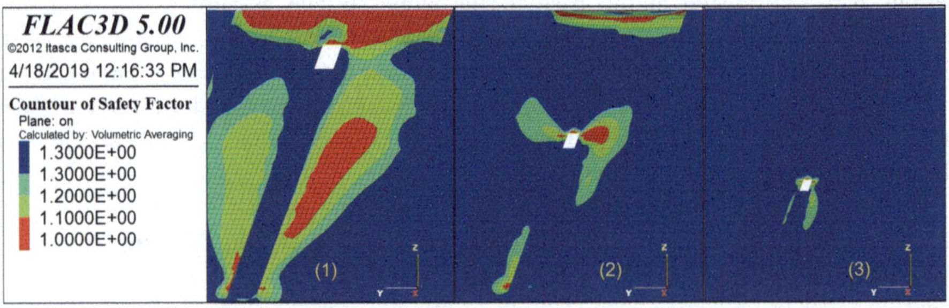

Fig. 7. Instability in stope mining near the slope surface; (1) edge of stope near slope surface, (2) middle of stope, (3) edge of stope far from the slope surface

4.2 Instability of Stope at Block 3

4.2.1 Potential Instability of Rock Mass in Block 3

Figure 8 shows the condition of differential stresses during the stope mining in planned-mining zones 1, 2 and 3 of Block 3. In this result, it can be seen that the differential stresses in zones 2 and 3 are higher than that of zone 1. These trends indicate the potential instability of rock mass in zones 2 and 3 reach more serious condition than that of zone 1. One of the reasons for this condition is the effect of mining activities repeatedly from zone 1 to 3. When the mining activities operate in zones 2 and 3, the influence of induced stresses from previous mined-out activities and mining zone 1 is already propagated to the surrounding of mining zone2 and 3. Afterwards, on-going mining activities are being carried out upwards repeatedly in the bound of the ore body in zones 2 and 3, therefore the surrounding rock masses are continuously disturbed, and the intense redistributed stresses are affected to the rock masses in later mining zones. The development of instability increasing around the stope can be clearly observed in Fig. 9 that describe the occurrence of failure zones and contour of safety factor in zones 1, 2 and 3.

4.2.2 Displacement of Ribs and Roof of Stope

Common mining instability in underground cut and fill stoping include roof falls and rib falls of the stope. To obtain a comprehensive understanding of the deformation at the roof and ribs of the stope, the monitoring data from the surface of roof and ribs of the stope is collected during the stope mining in planned-mining zones 1, 2 and 3 as shown in Fig. 4. All the results collected from hanging wall, footwall and roof of the stope are shown in Figs. 10, 11 and 12.

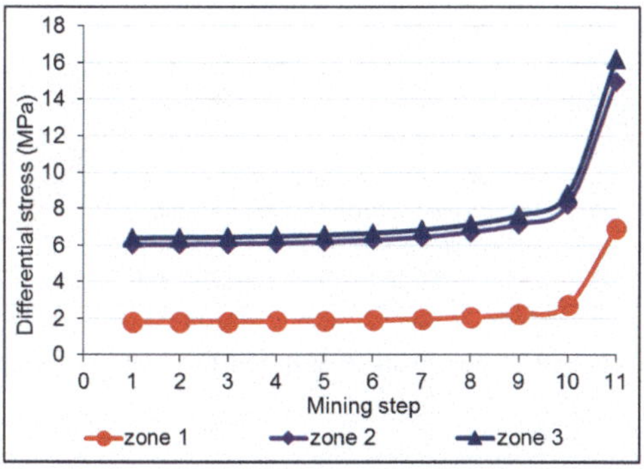

Fig. 8. Differential stress indicators during planned-mining's zones 1, 2 and 3.

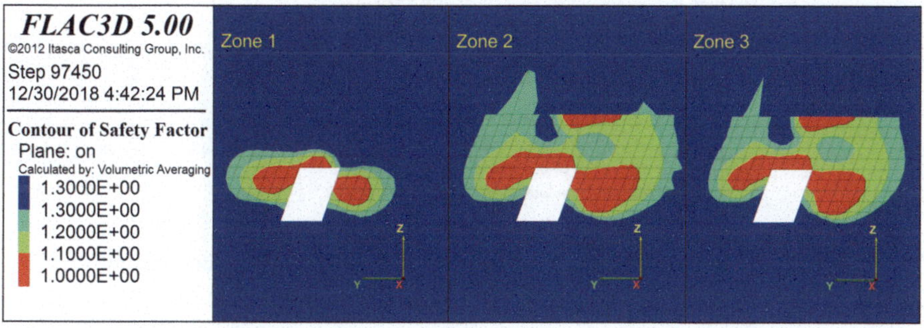

Fig. 9. Contour of safety factor in mining's zones 1, 2 and 3

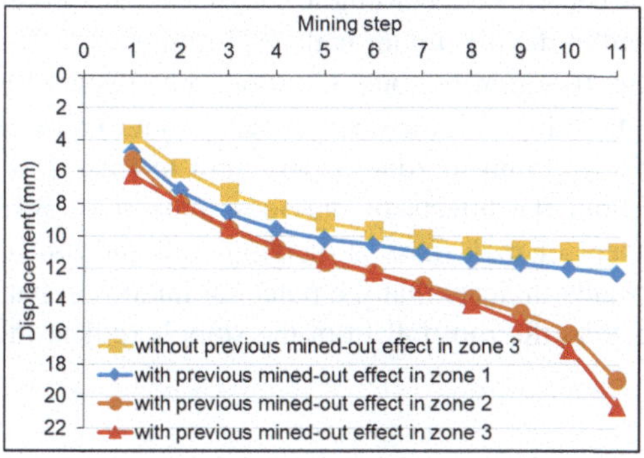

Fig. 10. Displacement at hanging wall of stope in the bound of stoping sequence

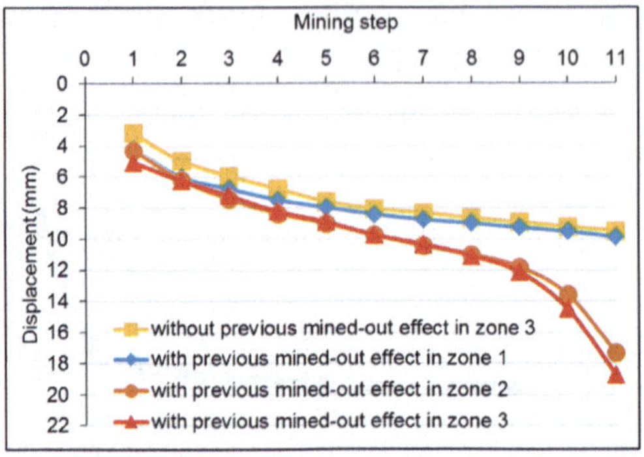

Fig. 11. Displacement at foot-wall of stope in the bound of stoping sequence

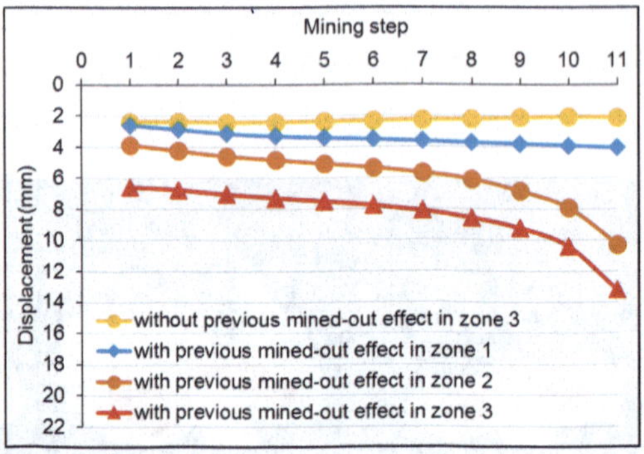

Fig. 12. Displacement at the roof of stope in the bound of stoping sequence

From the results of Fig. 10, it can be seen that the displacement at hanging wall of each stope is steadily decreased and it reached at the maximum with 20.6 mm in zone 3, 19 mm in zone 2, 12.3 mm in zone 1 and 10.9 mm in zone 3 without the effects of previous mined-out. Meanwhile, according to Fig. 11, the displacement at foot-wall of each stoping is reached at the maximum with 18.7 mm in zone 3, 17.3 mm in zone 2, 9.9 mm in zone 1 and 10.9 mm in zone 3 without the effects of previous mined-out. Furthermore, according to Fig. 12, the displacement from roof of each stope is reached at the maximum with 13.2 mm in zone 3, 10.2 mm in zone 2, 4 mm in zone 1 and 2 mm in zone 3 without the effects of previous mined-out. By seeing all of these graphs, the displacement of rock mass at hanging wall and foot-wall are higher than stope's roof. These results indicate that the potential failures at hanging wall and foot-wall are more severe than the roof fall from the stope's roof in this mine site.

5 Conclusion

From this paper, the result shows that the rock mass near the slope surface can be more affected by differential stress than that of inner part, and this result highlights the importance of rock stability whenever the underground mining activities operate near the slope surface. Moreover, if the rock masses are steadily disturbed by the induced stresses from previous mined-out activities, the intense redistributed stresses can be affected to the stability of rock masses in later mining zones. Thus, the potential instability will be more developed to later mining's zones. Furthermore, from the analysis on the displacement of stope, the results highlight that the potential failures from hanging wall and foot wall are more severe than the roof fall from the stope's roof in this mine site.

Acknowledgments. The authors would like to acknowledge Japan International Cooperation Agency (JICA) for supporting financial assistance of this research and field trip to mine site.

References

1. Swe, Y.M., Aye, C.C., Zaw, K.: Gold deposits of Myanmar. Geol. Soc. London, Mem **48**(1), 557–572 (2017)
2. Sunwoo, C., Jung, Y.-B., Karanam, U.M.R.: Stability assessment in wide underground mine openings by Mathews' stability graph method. Tunn. Undergr. Sp. Technol. **21**(3–4), 246 (2006)
3. IMHL, Production sharing proposal of IMHL, unpublished (2003)
4. Erskine, T.R.: Geology, Structure and Mineralisation Characteristics of the Modi Taung Gold Deposit, Myanmar, B.Sc Thesis, Univ. Tasmania, no., May 2014
5. Hoek, E., Kaiser, P.K., Bawden, W.F.: Support of Underground Excavations in Hard Rock. CRC Press, Boca Raton (2000)
6. Itasca Consulting Group, Inc.: FLAC — Fast Lagrangian Analysis of Continua, Ver. 5.0. Minneapolis: Itasca (2012)
7. Stress and Strain - Rock Deformation. http://www.columbia.edu/~vjd1/stress-strain_basic.htm. Accessed 08 Feb 2019

Application of Synthetic Nets as an Enabler of Optimised Pit Slopes at Skorpion Zinc Mine

Amory Mumba[1] and Bunda Besa[2(✉)]

[1] Skorpion Zinc Mine, Rosh Pinah, Namibia
[2] School of Mines, The University of Zambia, P.O. Box 32379, Lusaka, Zambia
bbesa@unza.zm

Abstract. Pit slope optimization is a fundamental design requirement for open pit mines. Several input parameters are incorporated into the slope design process including the slope angle, commodity price, recoveries and equipment size among others. The quality of the rock mass determines how steep the slopes should be designed for a safe and economic open pit mine. Despite rigorous efforts to optimize pit slopes, rock falls from extremely poor zones still occur. A complex and highly heterogeneous rock mass at Skorpion Zinc mine presents a challenge in steepening the hanging wall slopes for maximum resource recovery; competent limestone rock hosts intrusions and shear zones with extremely weathered sheared sericite schists. These ground conditions have resulted in significant rock falls from shear zones onto existing ramps and often-disrupting production. This paper presents how synthetic high wall safety nets have been applied at Skorpion Zinc mine in optimizing the pit slope angles, reducing clean-up cost and enhancing safety. Slope Stability analysis, using Rocscience software, was conducted on interim pit designs, incorporating angles of slopes with and without safety nets. Results show that the slopes were more stable when covered with synthetic nets than the ones left without support. Hence, the hanging wall inter-ramp slope angles were increased from the actual slope angle of 54° to 61° with an optimal angle being 61°. At this optimal slope angle, Probability of Failure (PoF) of <25% and Factor of Safety (FoS) of 1.2, there is a cost-saving of 8.42% of the total waste stripping i.e., a saving of US $9.6 m. The use of synthetic nets has therefore, enhanced the confidence in optimizing pit slope angles and justified the application in rock fall management at Skorpion Zinc Mine.

Keywords: Optimisation · Skorpion Zinc mine · Slope stability analysis · Rockfall · Synthetic net

1 Introduction

1.1 Location and General Open Pit Layout

Skorpion Zinc Mine is located in Southern Namibia, approximately 50 km north of the Orange River and 25 km north of Rosh Pinah. The mine is located in the Gariep Belt and is an oxide Zinc deposit (Fig. 1).

E. Topal (Ed.): MPES 2019, SSGG, pp. 488–498, 2020.
https://doi.org/10.1007/978-3-030-33954-8_55

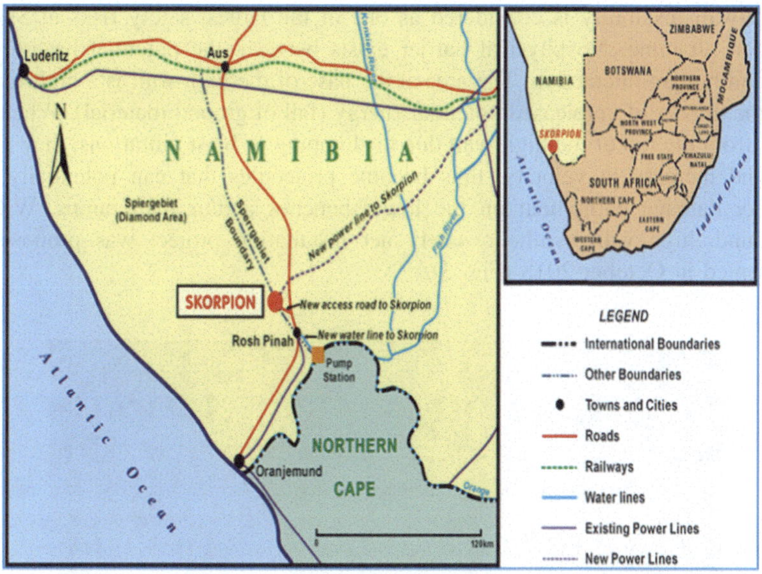

Fig. 1. Location of Skorpion Zinc Mine

With mining having commenced in 2002, the current open pit is deep and has progressed to the 26th bench from surface (i.e. 260 m depth, Fig. 2). The planned depth of the open pit is 310 m, which is based on a life of mine of approximately 2 years. Rock types hosting the ore deposit include volcaniclastic and clastic sediments (felsic tuff, arkose, sandstone, shale, schist, etc.), limestone and rhyolite. The volcaniclasic sediments have been strongly altered during metamorphism and hydrothermal alteration events, and all are deeply weathered.

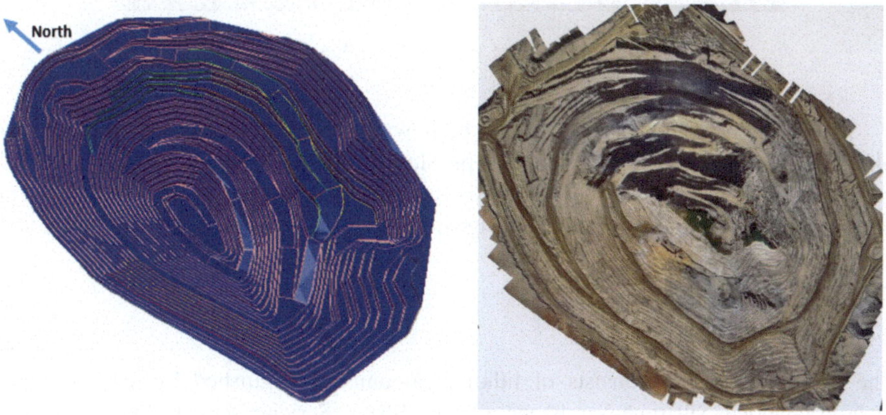

Fig. 2. Skorpion Zinc open pit mine

High-wall instability is considered as one of the largest safety risks at Skorpion Zinc open pit mine. No physical barrier exists between the potential hazard (loose rocks) and human interaction. The area at the base of the high-wall is a high-risk area due to the impact of unreleased potential energy (fall of ground material). When rocks unravel from the top of the high-wall due to slumping or blast vibrations, they pick up speed and increase in velocity, thus become projectiles that can potentially injure people or damage equipment on the lower benches and/or haul ramps. With this background, high wall synthetic safety net installation project was proposed and implemented in October 2018 (Fig. 3).

Fig. 3. Installation of high wall synthetic safety nets

This paper presents how synthetic high wall safety nets have been applied at Skorpion Zinc mine in optimizing the pit slope angles, reducing clean-up cost and enhancing safety. Slope stability analysis, using Rocscience software (SLIDE 2018) [1], was conducted on interim pit designs, incorporating angles of slopes with and without safety nets.

1.2 Geological Setting

The geological model consists of lithological units distinguished by the mine and include, Banded Siltstone-Sandstone-arkose (BSA), Sericite schists, Quartz sericite schist (QSS), Black argillaceous unit and Limestone (including a siliciclastic marker horizon). Many of the lithological units described in the mine are locally interbedded and contacts are gradational.

1.3 Current Geotechnical Issues

A complex and highly heterogeneous rock mass at Skorpion Zinc mine presents a challenge in steepening the hanging wall slopes for maximum resource recovery and reduced waste stripping; competent limestone rock hosts intrusions and shear zones with extremely weathered sheared sericite schists. These ground conditions (high heterogeneity) have resulted in significant rock falls from shear zones onto existing ramps and often-disrupting production and incurring clean-up costs up to 80% of unplanned/unscheduled works. Ongoing monitoring of wall conditions using the Movement and Surveying Radar (MSR) and survey prism monitoring is utilized in managing production and safety risk in affected zones. High wall synthetic safety nets as a failure management strategy have also been applied.

1.4 Pit Slope Optimisation

The term "optimisation" is derived from mathematics, particularly mathematical programming and basically refers to the study of problems in which one seeks to maximise or minimise some function. This approach may use a "heuristic" algorithm to arrive at a near optimal solution. Simply put, a heuristic is a problem solving technique in which the most appropriate solution of several, found by alternative methods, is selected at successive stages. The concept of optimisation in mining was first applied in the 1960s to pit design and addressed the problem of determining the optimal pit outline based on given technical and economic data. In the last decades of the 20th century, the growing processing power of computers has made possible to apply a number of methods for the determination of ultimate pits. These methods are: Lerchs-Grossmann method [2, 3], network or maximal flow techniques [4, 5], floating cone method [6, 7] the Korobov algorithm [8], the corrected form of the Korobov algorithm [9], parameterization techniques [4, 10] and dynamic programming method [11–15], which assumptions and algorithms determine the today's direction of the design in open pit mining. The algorithms of the above-counted methods are the core of the programs which are used in the computer designing methods. The main goal of these methods, almost all of which are based on block models, is the optimal open pit outline with the steepest dip of the final slopes under technological and physical constraints, and minimal costs of mining desirable blocks, in other words, the maximum net profit. The complexity of the geological conditions of a deposit and dynamism of the economic indicators define the choice of the most adequate method of design for the mining operation [16]. During the study of pit slope optimization at Skorpion Zinc mine i.e., with or without synthetic safety nets, the cross section through a rockfall prone zone as shown in Fig. 4 was used for the analysis.

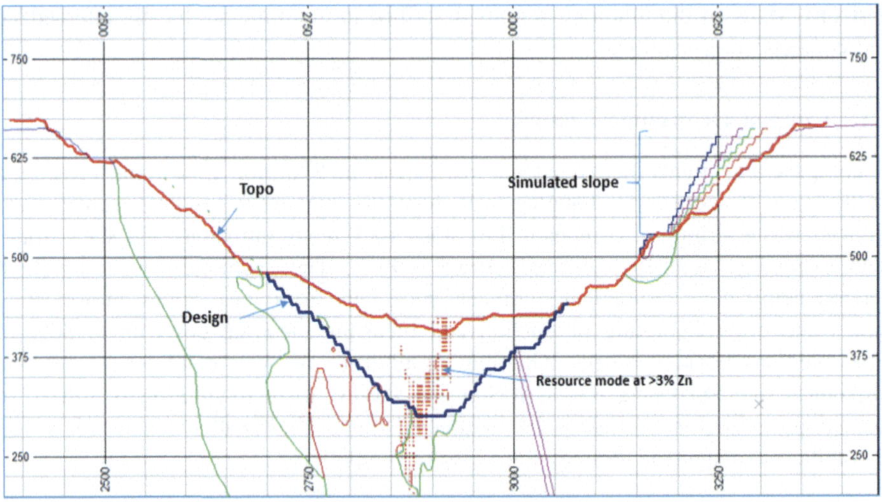

Fig. 4. Representative section across area of concern

1.5 Stability Analysis

Slope stability analysis was conducted to understand the effect of modifying slope angles in order to achieve safe and economic pit slopes. A sensitivity analysis was conducted using a Rocscience software (SLIDE 2018) [1] which uses Limit Equilibrium Algorithms to determine the factors of safety and probability of failure. The method used in the analysis is the Bishop's simplified method. In this method, circular slip surfaces are examined, and the one rendering the lowest safety factor is selected [17].

The shear strength parameters were obtained from the Skorpion Zinc Mine geotechnical database. Tables 1 and 2 present the input parameters used in the analysis while Figs. 5 and 6 show snapshots of the SLIDE 2018 [1] model used in the stability analysis. The Arkose (ARK) material analysed is a Calcerous greywacke while LMST is an altered carbonate with calcerous arenite.

Table 1. Input parameters used in the stability analysis with SLIDE 2018

Name	Property	Distribution	Mean	Std. Dev.	Rel. Min	Rel. Max
ARK	Cohesion (kPa)	Normal	43	28.2843	43	180
ARK	Phi (°)	Normal	30.914	2.8284	8	31
ARK	Unit weight (kN/m^3)	Normal	20	0.39156	19	30.3
LMST	Cohesion (kPa)	Normal	12235	2.8	12235	13200
LMST	Phi (°)	Normal	32	1	25	36
LMST	Unit weight (kN/m^3)	Normal	25	1.2	18	24

Table 2. Safety net properties

Support type	End anchored
Force application	Active
Force orientation	Parallel to reinforcement
Out of plane spacing	1 m
Anchor capacity	100 kN

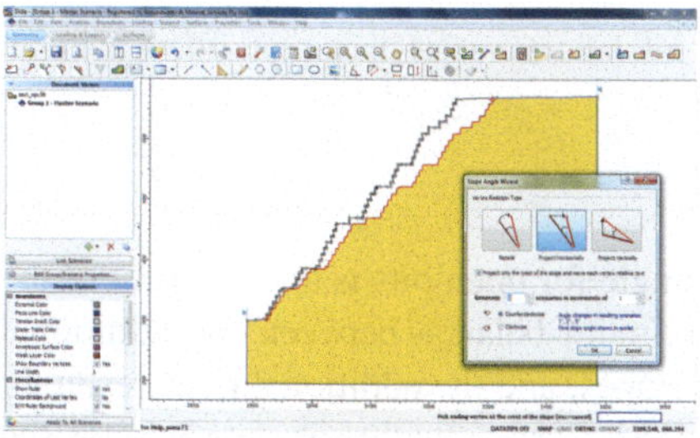

Fig. 5. Snapshot of the Slide 2018 model used in the stability analysis

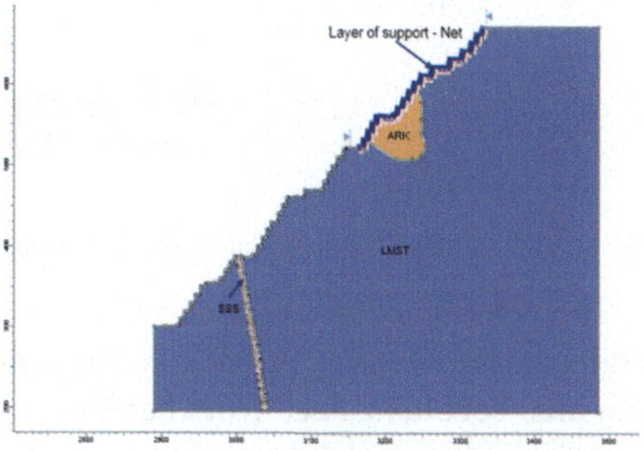

Fig. 6. The actual model used for stability analysis

The slope stability evaluation was carried out for one sector that is prone to rock falls and is supported by high wall synthetic safety nets. Tables 3 and 4 as well as Figs. 7 and 8 show results obtained from SLIDE 2018 [1] models.

Table 3. Stability analysis results from SLIDE 2018 – slope without net

Slope angle (°)	FoS	PoF
54	1.32	9.2
56	1.29	10.6
58	1.24	14.7
60	1.18	23.9
62	1.08	38.3
64	1.01	50.7
66	0.91	68.0

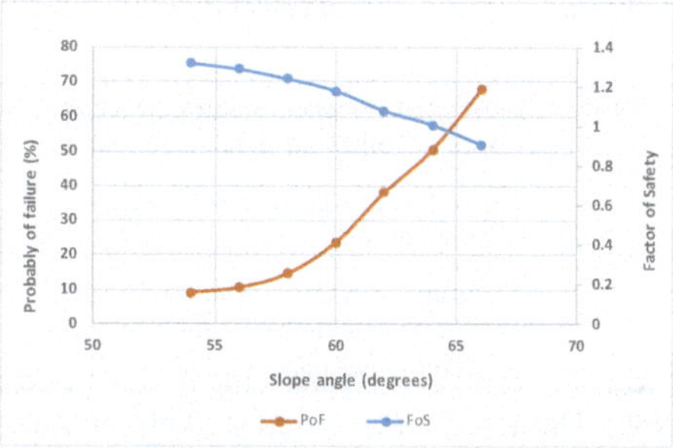

Fig. 7. Probability of failure versus factor of safety without synthetic safety nets

Table 4. Stability analysis results from SLIDE 2018 – slope with net

Slope angle (°)	FoS	PoF
54	1.48	3.9
56	1.319	8
58	1.295	10
60	1.273	11.7
62	1.121	27
64	1.034	36
66	0.966	52

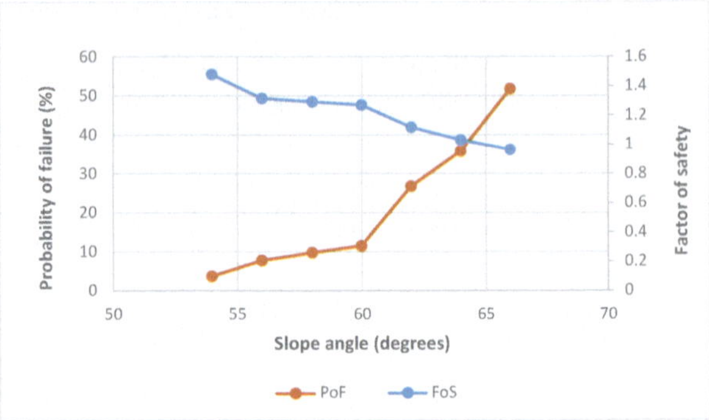

Fig. 8. Probability of failure versus factor of safety without synthetic safety nets

The determination of the acceptable slope angle for open pits is a key aspect of the mining business, as it implies seeking the optimum balance between the additional economic benefits gained from having steeper slopes, and the additional risks resulting from the reduced stability of the pit slopes [18]. The difficulty in determining the acceptable slope angle is related to geological uncertainties within the rock mass. These uncertainties are accounted for in the process of slope design and different methodologies have been used for this purpose. This paper reviews three acceptability criteria successfully applied at some mines and adopted at Skorpion Zinc. The three design acceptability criteria were adopted as a guide in optimizing the pit slopes. These are presented in Tables 5, 6 and 7.

Table 5. Slope design acceptability criteria – 1 [19]

Criteria	Type of slope	Consequence of failure	Acceptability Values	
			FoS	PoF (FoS<1.5)
1	Individual Benches: small (50m) temporary slopes not adjacent to haulage roads	Not Serious	1.3	20%
2	Any slope of a permanent or semi-permanent nature	Moderately serious	1.6	10%
3	Medium sized (50 -100m) and high slopes (<150m) carrying major haul roads or underlying permanent mine installations	Very serious	2	5%

Note; The actual criteria used at Skorpion Zinc mine is subject to a more thorough analysis of the consequences of failure.

Table 6. Slope design acceptability criteria – 2 [20]

Category	Description	Acceptable PoF
1	Critical slopes where failure may affect continuous operations and pit safety	<5%
2	Slopes where failures have a significant impact on costs and safety	<15%
3	Slopes where failure has no impact on costs and where minimal safety hazards exist	<30%

Table 7. Slope design acceptability criteria – 3 [21]

Slope Scale	Consequence of failure	FoS	PoF [Fos<1]
Bench	Low-High	1.1	25%-50%
	low	1.15-1.2	25%
Inter-ramp	Medium	1.2	20%
	High	1.2-1.3	10%
	Low	1.2-1.3	15%-20%
Overall	Medium	1.3	5%-10%
	High	1.3-1.5	<5%

1.6 Acceptability Criteria and Optimal Pit Slope

The shaded cells in Tables 5, 6 and 7 relate to Skorpion Zinc mine current assessment of the acceptability criteria that was applied to the Pit slopes. The probabilities of failure associated with acceptability criteria are in the range of 10%–25%. The corresponding factors of safety lie between 1.2 and 2.6. The slope performance with respect to acceptability criteria for Skorpion Zinc mine is presented in Tables 8 and 9. The unsupported slopes exceeded the minimum factor of safety and the PoF. However, the operation of the slope presented a risk that may have been unacceptable. Therefore, additional restraint to the slope and a comprehensive monitoring programme was implemented to reduce the risk level to within acceptable levels.

Table 8. Summary of acceptability criteria

Design	FoS (Supported)	PoF (%) (Supported)	FoS (Unsupported)	PoF (%) (Unsupported)	Compliance with acceptability criteria	
					FoS	PoF
54	1.48	3.9	1.32	9.2	Compliant	Compliant
56	1.319	8.0	1.29	10.6	Compliant	Compliant
58	1.295	10.0	1.24	14.7	Compliant	Compliant
60	1.273	11.7	1.18	23.9	Compliant	Compliant
62	1.121	27	1.08	38.3	noncompliant	noncompliant
64	1.034	36	1.011	50.7	noncompliant	noncompliant
66	0.966	52	0.91	68		

Table 9. Slope performance criteria

Criteria	Performance of slope with respect to acceptability criteria	Interpretation
1	Satisfies all three criteria	Stable slope
2	Exceeds minimum mean factor of safety but violets one or both probabilistic criteria	Operation of slope presents risk that may or may not be acceptable; level of risk can be reduced by comprehensive monitoring programme.
3	Falls below minimum mean FoS but satisfies both probabilistic criteria	Marginal slope; minor modifications of slope geometry required to raise mean FoS to satisfactory levels.
4	Fall below minimum mean FoS and violates both probabilistic criteria.	Unstable slope; major modifications of slope geometry required; rock improvement and slope monitoring may be necessary.

At FoS 1.2, Fig. 9 shows the optimal pit slope angle at Skorpion Zinc mine without application Synthetic safety net while Fig. 10 shows the optimal pit slope angle after application of Synthetic safety net.

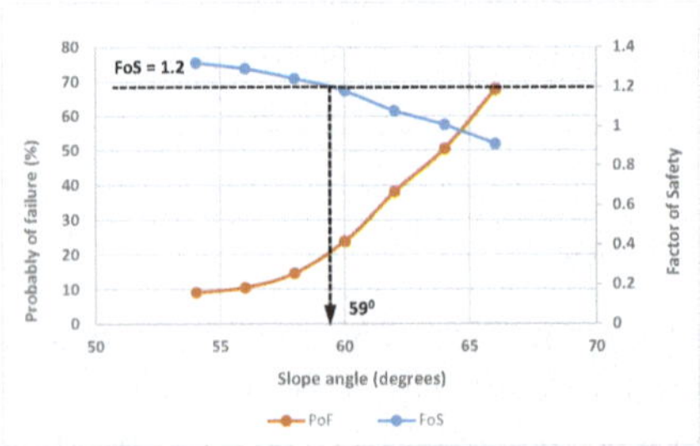

Fig. 9. Optimal pit slope angle without Synthetic safety net.

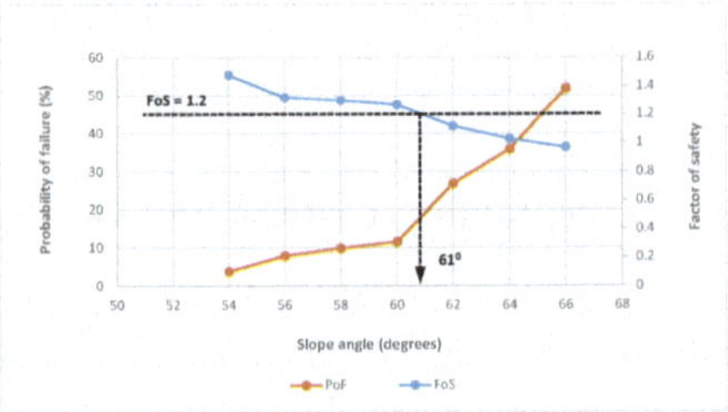

Fig. 10. Optimal pit slope angle after application of Synthetic safety net.

As can be seen from Figs. 9 and 10, at FoS of 1.2, the optimal pit slope for Skorpion Zinc mine improved from 54° to 61°. This is a huge improvement in terms of economic benefits of the pit.

1.7 Cost Analysis

Figure 11 shows the cost analysis for the use of synthetic safety nets versus non-use of the nets at Skorpion zinc mine. According to the figure, there is a US$9.6 m reduction in the total cost of clean-up operations when synthetic safety nets are used in pit design representing approximately 8.42% reduction.

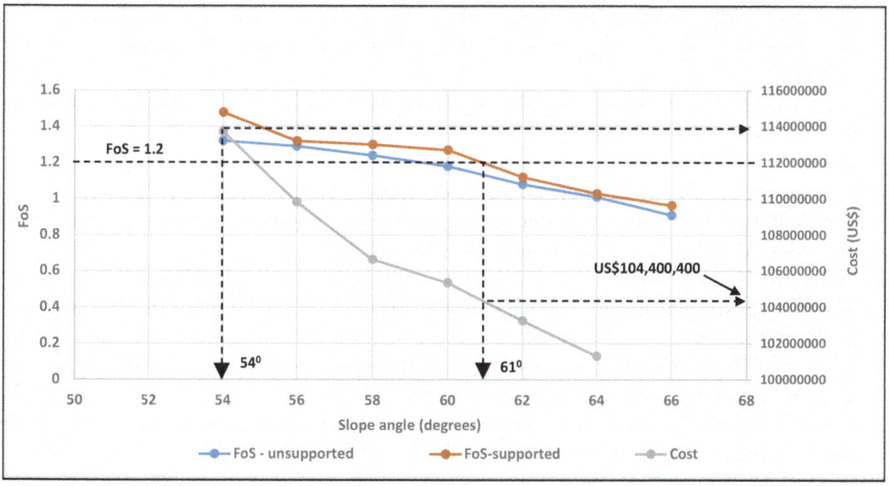

Fig. 11. Cost analysis for the use of synthetic safety nets versus non-use of the nets.

1.8 Conclusion and Recommendations

Results show that the slopes were more stable when covered with synthetic nets than the ones left without support. Hence, the hanging wall inter-ramp slope angles were increased from 54° to 61°. At optimal slope angle i.e. 61°, PoF of <25% and FoS of 1.2, there is a cost saving of 8.42% of the total waste stripping (i.e., US$9.6 m). The use of synthetic nets has, therefore, enhanced the confidence in optimizing pit slope angles and justified its application in slope management at Skorpion Zinc Mine. It was, recommended that all identified potentially unstable zones should be covered with high wall synthetic nets to maximise the slope angle and enhance safety.

References

1. Rocscience Inc.: Slide Version 8.0 - 2D Limit Equilibrium Slope Stability Analysis, Toronto, Ontario, Canada (2018). https://www.rocscience.com
2. Alford, G.G., Whittle, J.: Application of Lerchs-Grossman pit optimisation to the design of open pit mines. In: Open Pit Mining Conference, Calgary, Canada, pp. 201–207 (1986)
3. Lerchs, H., Grossmann, I.F.: Optimum design of open pit mines. CIM Bull. **58**, 47–54 (1965)
4. François-Bongarçon, D., Guibal, D.: Algorithms for parameterizing reserves under different geometrical constraints. In: Proceedings of 17th Symposium on the Application of Computers and Operations Research in the Mineral Industries (APCOM: AIME), New York, pp. 297–230 (1982)
5. Yegulalp, T.M., Arias, J.A.: A fast algorithm to solve the ultimate pit limit problem. In Proceedings of 23rd Symposium on the Application of Computers and Operations Research in the Mineral Industries (APCOM: AIME), Littleton, Colorado, pp. 391–397 (1992)

6. Berlanga, J.M., Cardona, R., Ibarra, M.A.: Recursive formulae for the floating cone algorithm. In: Mine Planning and Equipment Selection, Rotterdam, Netherlands, pp. 15–25 (1988)
7. Lemieux, M.: Moving cone optimizing algorithm. In: Weiss, A. (ed.) Computer Methods for the 80 s in the Mineral Industry, AIME, New York, pp. 329–345 (1979)
8. Korobov, S.: Method for determining optimal open pit limits. Ecole Polytechnique de l'Université de Montréal, Montreal, Technical report EP74-R-4, p. 24 (1974)
9. Dowd, P.A., Onur, A.H.: Open-pit optimization—part 1: optimal open-pit design. Trans. Inst. Min. Metall. (Sect. A: Min. Ind.) **102**, 95–104 (1993)
10. Matheron, G.: Paramétrage des contours optimaux (Centre de Géostatistique et de Morphologie mathématique), Internal report N-403; Note géostatistique 128, Fontainebleau, p. 54 (1975)
11. Johnson, T.B., Sharp R.W.: A three-dimensional dynamic programming method for optimal ultimate open pit design. Bur. Min. Invest. 7553 (1971)
12. Koenigsberg, E.: The optimum contours of an open pit mine: an application of dynamic programming. In: APCOM, New York, vol. 17, pp. 274–287 (1982)
13. Wilke, F.L., Wright, E.A.: Determining the optimal ultimate pit design for hard rock open pit mines using dynamic programming. Erzmetall **37**, 139–144 (1984)
14. Wright, A.: The use of dynamic programming for open pit mine design, some practical implications. Min. Sci. Technol. **4**, 97–104 (1987)
15. Yamatomi, J., et al.: Selective extraction dynamic cone algorithm for three-dimensional open pit designs. In: Proceedings of 25th Symposium on the Application of Computers and Operations Research in the Mineral Industries (APCOM) (Australasian Institute of Mining and Metallurgy), Brisbane, pp. 267–274 (1995)
16. Galić, I.: Mining design using specialized software, Master's thesis, Faculty of Mining, Geology and Petroleum Engineering, Zagreb, p. 95 (2002). (in Croatian)
17. Zhenhua, H., Seifollar, N.: Understanding and optimizing the geosynthetic-reinforced steep slopes **19**, 5802 (2014). Texas, USA (2015)
18. Terbrugge, P., et al.: A risk evaluation approach for pit slope design, Johannesburg, South Africa (2008)
19. Priest, S.D., Brown, E.T.: Probabilistic stability analysis of variable rock slopes. Trans. Inst. Min. Metall. Sect. A: Min. Ind. **92**, A1–12 (1983)
20. SRK Consulting: Acceptable probabilities of failure, mining rock slopes (Fitwaola report) (2011)
21. Read, J., Stacey, P. (eds.): Guidelines for Open Pit Slope Design, CSIRO Publishing (2009)

Numerical Investigation on Gate-Entry Stability of Trial Panel in Indonesia Longwall Coal Mine

Pisith Mao[1(✉)], Takashi Sasaoka[1], Hideki Shimada[1], Akihiro Hamanaka[1], Sugeng Wahyudi[1], Jiro Oya[2], and Naung Naung[1]

[1] Department of Earth Resources Engineering, Kyushu University, Fukuoka, Japan
mao.pisith.kh@gmail.com
[2] MM Nagata Coal Tech Co., Ltd., Tokyo, Japan

Abstract. Indonesia is known for its abundant coal reserve. However, weak geological condition is one of the most common challenges in Indonesian coal mine. The study area of this research is PT Gerbang Daya Mandiri (GDM) mine which is situated in East Kalimantan. Stratigraphy of this area comprises of claystone as dominant rock and multiple coal seams with the thickness ranging from 0.15 m to 9.8 m. In this mine, the longwall mining method attempts to be applied for coal seam extraction. Recently, GDM begun their development of trial panel for depth around 100 m to 150 m to understand the behavior of the rock for the real longwall operation. Based on rock mechanical analysis of core sample from drill holes, the competence of claystone is increasing according to the depth. This analysis also shows that for the depth up to 100 m, claystone is generally weaker than coal. Thus, by leaving part of coal seam on top and bottom of the excavated entry-gate would help to retain its stability. For 150 m depth on the other hand, the competence of claystone is increased to somewhat similar to those of coal. This paper uses numerical simulation by FLAC3D to investigate the optimum remain coal thickness (RCT) for stability of the entry-gate of trial panel for 100 m depth. This work also seeks to identify appropriate roof support for both 100 m and 150 m depths. The outcome results show that the optimum RCT is 1 m on roof and on the floor for 100 m depth. SS540 with 1 m spacing is appropriated for adopting for this trial panel except the case of high stress ratio, which required stronger support capacity by narrowing the support spacing or incorporating rock bolt.

Keywords: Gate-entry stability · Longwall mining · Weak rock · Remain coal thickness · FLAC3D

1 Introduction

Indonesia hosts an abundant portion of thick coal deposit, which is usually found in weak geological condition [1]. This weak geological condition has led to the limitation of excavation into a certain height [2]. As a result, some of coal thickness remains on top and bottom of the excavation. This can be beneficial for coal bed that is surrounded

E. Topal (Ed.): MPES 2019, SSGG, pp. 499–507, 2020.
https://doi.org/10.1007/978-3-030-33954-8_56

by weaker dominant rock as the harder remained coal can help improving the stability of any opening structure during the mining development and excavation.

Stability of gate-entry is widely considered to be one of the most important aspects in longwall mine design, especially during panel extraction. Gate-entry does not only serve as a way of material transportation but also an important component of ventilation system to get rid of dust and harmful gases and to provide fresh air to longwall face working area [3]. Many research works have been focused on developing mining technique for thick coal seam mine [2, 4, 5]. Some also included weak geological condition [1, 6]. However, few researchers have addressed the effect of remain coal thickness (RCT) on stability of gate-entry for longwall mining in thick coal seam.

This paper is a preliminary attempt to investigate the effect of RCT on gate-entry and incorporated it into longwall mine design in order to obtain a better optimization. To fully understand this effect, a study work is implemented on trial panel study developed in one coal mine in Indonesia using numerical analysis. This research work consists of modelling various case of RCT to estimate the optimum length of RCT on top and bottom of gate-entry for 100 m depth and investigating stability of gate-entry roof support for both 100 m and 150 m depth.

2 Characteristic of Study Area

PT Gerbang Daya Mandiri (GDM) is located in East Kalimantan Island, 15 km to the north from Samarinda city in the area called Kutai Kertanegara as shown in Fig. 1. General geological structure of that area is monocline, which contains many coal seams with the dip and thickness from 3° to 13° and from 0.15 m to 9.8 m respectively. The dominant rock in the area is claystone.

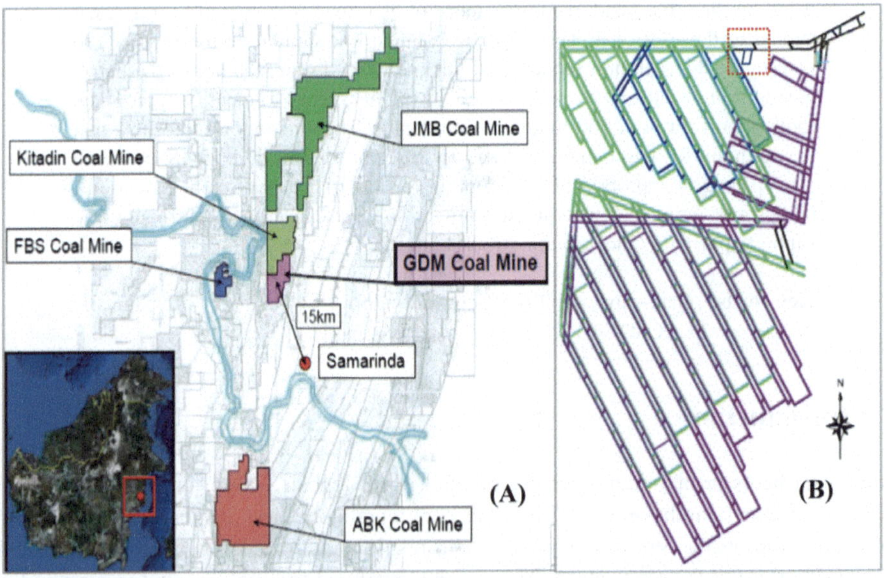

Fig. 1. Study area (A: Location map of GDM, B: Mine layout) [7]

This mine will adopt longwall mining method for coal extraction. GDM is developing mine in rock that falls into category of very weak or low strength rock [7]. The dominant rock is very weak, especially when it comes to the rock near to the surface. Recently, a trial panel is being developed in the depth around 100 m to 150 m in order to have a better understanding of rock behavior surrounding longwall panel. Due to the variation of coal thickness, thick coal can be found along this trial panel. As a result, the knowledge of RCT effect is crucial for developing and optimizing this trial panel.

3 Numerical Analysis

In order to investigate the behavior of the rock for this research, all numerical analysis was carried out using FLAC3D. The modeling is constructed based on the actual condition in the mine site and specific case study. Model validation is performed at the beginning of the numerical analysis to ensure the accuracy of the model by comparing the result form simulation to the actual result of field measurement. The excavation height is 3 m. Steel arch sets are used to support gate-entry roof.

3.1 Model Description

Simulation models were constructed based on specific case studies. General model dimension is illustrated in Fig. 2, which can be described as 60 m width and 300 m length with 100 m and 150 m of over-burden as well as 50 m of under-burden. For simplification, both over-burden and under-burden is considered to be dominant rock; claystone. Coal thickness is varied from 3 m to 8 m based on each individual case study. Mining height is 3 m.

Due to the fact that the competence of claystone increases with increasing the depth, two different mechanical properties of claystone are selected for 100 m and 150 m. Material type of steel SS540 is selected for steel arch support. Symmetrical model running is adopted for all simulations in order to reduce running time, which means only half of the panel is modelled in FLAC3D. All models are based on Mohr-Coulomb constitutive model. Mechanical properties of the rock and steel arch are listed down in Tables 1 and 2 respectively.

The first topic, which covers the optimum RCT in order to maximize the stability of gate-entry, is focused on 100 m depth only, due to the fact that the competence of claystone in 150 m depth is similar to those of coal. In the model, RCT on the roof and floor of the excavation area range from 0 m to 2.5 m as shown in Fig. 3. Stress ratios K, which is the ratio between horizontal and vertical stress, are 1, 1.5 and 2 and gate-entry is unsupported. Deplacement monitoring points are located in the roof and floor of the gate in the middle of the model.

The Second topic covers the stability investigation of gate-entry roof support. Stress ratio K are varied in 1 1.5 and 2. For this case, the results are obtained by monitoring maximum axial stress acting on steel arch support along the gate. The depth of the

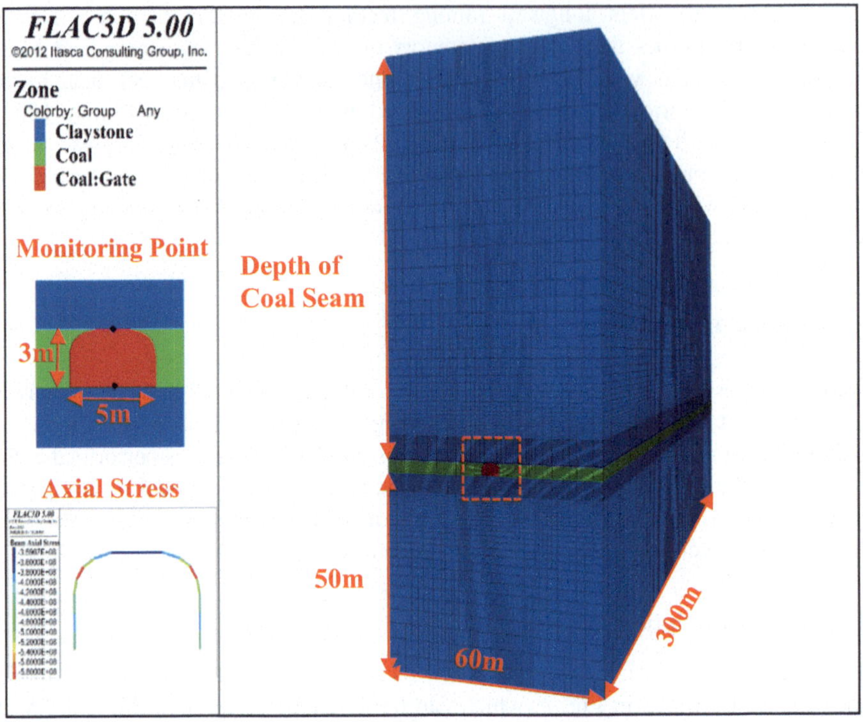

Fig. 2. General model dimensions

Table 1. Rock mechanical properties

Rock	UCS (MPa)	Density (kg/m^3)	E (MPa)	υ	C (MPa)	φ (°)	σ_T (MPa)
Claystone (100 m)	4.84	2,108.1	805.86	0.28	0.6	37.45	0.52
Claystone (150 m)	6.96	2,121.2	1,344.31	0.28	0.84	41.82	0.66
Coal	8.16	1,380	1,300	0.32	2.63	45.6	0.58
Goaf	–	1,700	15	0.25	0.001	25	–

Table 2. Steel arch support mechanical properties

Density kg/m^3	E (GPa)	υ	Cross Area (cm^2)	Yield Strength of SS540 (MPa)	I_y (10^{-8} m^4)	I_z (10^{-8} m^4)	J (10^{-8} m^4)
7,800	200	0.3	36.51	540	732	154	22

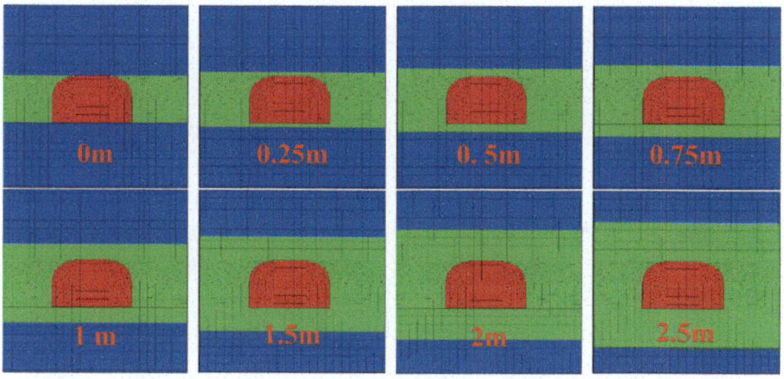

Fig. 3. Study case of remain coal thickness

overburden are 100 m and 150 m. The optimum thickness of RCT is only incorporated into the model for 100 m depth due to the fact that influence of RCT is only applicable to the depth 100 m. Steel arch SS540 roof support with 1 m spacing is adopted for 100 m depth. For 150 m depth coal seam thickness is considered to be 3 m. Two type of spacing 1 m and 0.5 m are introduced to the simulations for this depth.

3.2 Model Validation

Model validation is very essential for numerical study. The result from field monitoring displacement using extensometer and telltale on top roof of gate allows us to validate the simulated results. Extensometer N° 53 and telltale N° 39 are selected for model validation. Figure 4 shows location map of these devices. Both monitoring devices are installed on the main roadway roof, at the intersection with crosscut in the depth around 100 m. Simulation model was constructed according to the real rock condition where these devices are installed. According to 6 m geology borehole obtained from the installation process of these devices, the roof of the main gate surrounding this area comprises only claystone. Main roadway is supported by 1 m spacing steel arch. Due to the lack of information about in-situ stress, stress ratio K is assumed to be 1.

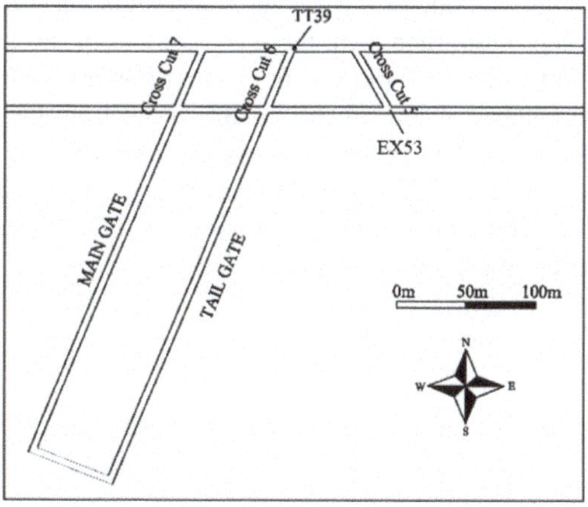

Fig. 4. Location map of field measurement

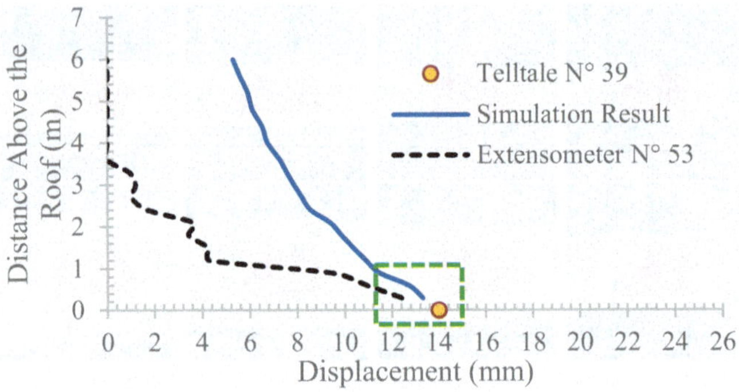

Fig. 5. Comparison between field measurement results and simulation result

The result of gate roof displacement is monitored until 6 m above the gate roof for both extensometer and simulation. Telltale on the other hand, only provides total displacement result of the gate roof. Figure 5 shows the comparison from model simulations result, extensometer N° 53 result and telltale N° 39. The conspicuous observation from these results comparison shows great agreement. The maximum displacement of the gate roof is approximately 13 to 14 mm. This evidence implies that this modelling configuration can be used effectively to predict the behavior of the rock during gate development.

4 Results and Discussions

4.1 Gate-Entry Stability Control by Leaving Remain Coal Thickness

The first set of model simulation is to highlight the effect of RCT on maximizing the stability of gate-entry. This correlation is shown in Figs. 6 and 7, which is the result of monitoring displacement of the roof and floor of the gate-entry. As expected, RCT can help to reduce displacement in gate-entry, which can maximize its stability.

Figure 6 is displacement monitored on the roof and Fig. 7 is monitored in the floor. Both figures show a big reduction of displacement when thickness of RCT is kept up to 1 m. When RCT is kept more than 1 m, the displacement of the gate is still reduced. However, the reduction value of displacement is rather minor compare to those of RCT less than 1 m. A pronounced statement can be drawn out from the observation of this result is that the displacement is significantly reduced for RCT up to 1 m. From this 1 m thickness, even a thicker RCT is left above or below gate excavation; only minor reduction can be seen. Thus, the optimum thickness of RCT to maximize stability of gate-entry is 1 m on the roof and the floor.

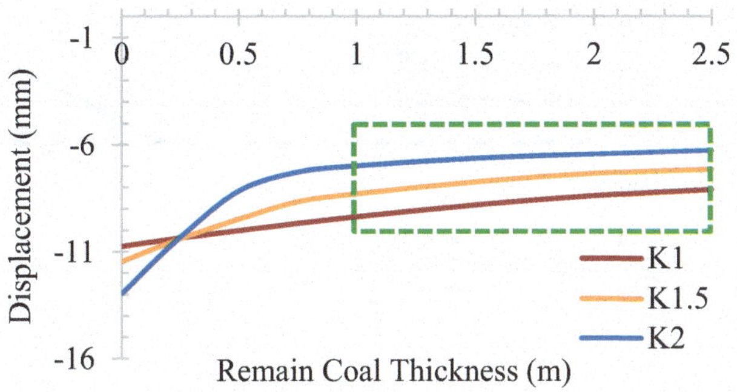

Fig. 6. Relationship between gate-entry displacement and RCT at the roof (K = σh/σv)

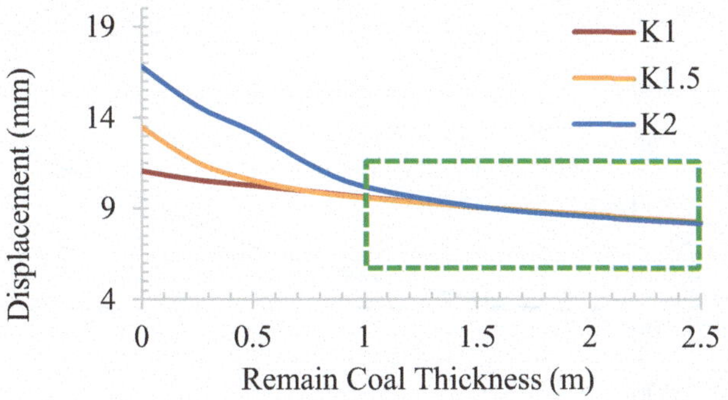

Fig. 7. Relationship between gate-entry displacement and RCT at the floor (K = σh/σv)

4.2 Investigating Stability of Roof Support During Gate-Entry Development

This part of the paper is dealing with the evaluation of steel arc capacity, which is suitable for adopting as support system in gate-entry. The assessment of gate entry in 100 m depth will also considers the effect of RCT.

In Fig. 8, which are results for 100 m depth, maximums axial stress along steel arch is compared with yield strength of steel arch support: ss540. The basic idea is that, if the maximums axial stress is higher than yield strength of support, steel arch will deform, and roof of gate-entry become unstable. These results highlighted that 1 m spacing ss540 is suitable for 100 m depth with stress ratio up to 1.6 for case where RCT is not available. In case where optimum RCT of 1 m is available, this support configuration can be use up to stress ratio of 1.9. In case of the higher stress ratio 2, it is recommended to narrow the spacing in order to increase the support capacity.

Figure 9 illustrates the simulated results for 150 m depth. From this results, it can be said that the stability of gate-entry can be maintained with 1 m spacing when the stress ratio is lower than 1.1. However, when the stress ratio is larger than 1.1, the spacing of steel arch should be changed from 1 m to 0.5 m. Moreover, it is recommended to incorporate additional support such as rock bolt or cable bolt with this steel arch when the stress ratio is larger than 1.6.

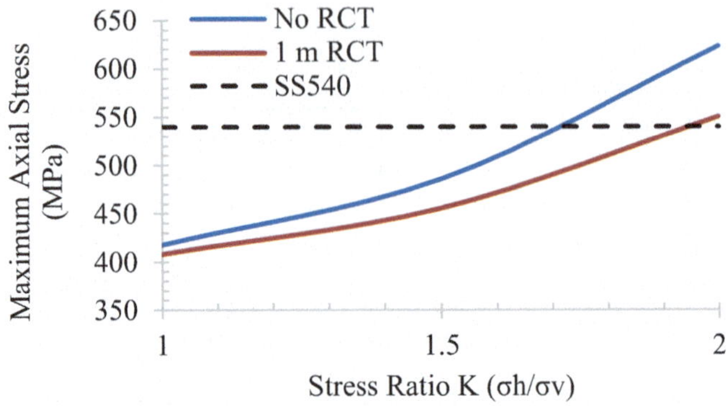

Fig. 8. Maximum axial stress on steel arch for 100 m depth (1 m spacing)

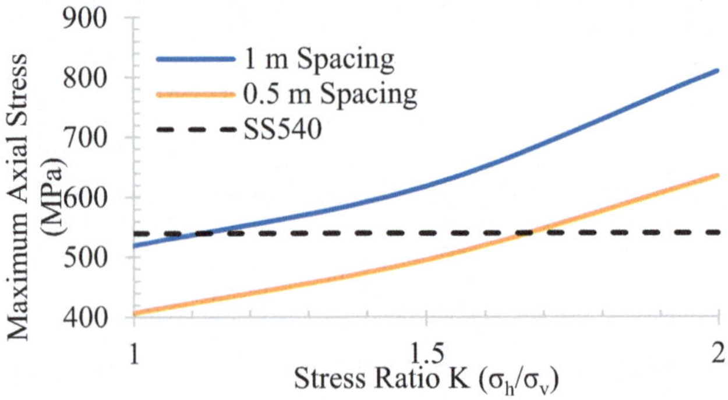

Fig. 9. Maximum axial stress on steel arch for 150 m depth

5 Conclusions

This research work leads us to the conclusion that:

Stability of gate-entry can be maximized by leaving RCT on top and bottom of the excavation area for 100 m depth. The optimum thickness of RCT should be kept around 1 m on the roof and on the floor of the gate-entry.

To maintain the stability of the gate-entry, 1 m spacing of SS540 can be effectively adopted to support gate-entry for this trail panel for 100 m depth up to stress ratio 1.6. The availability of optimum RCT could increase the capacity of this support to withstand up to stress ratio 1.9. A smaller spacing is recommended for this depth in case stress ratio is more than 2. For deeper depth of 150 m, 1 m spacing of steel arch is applicable for stress ratio less than 1.1. However, when stress ratio is larger than 1.1 spacing of steel arch should be changed from 1 m to 0.5 m spacing. Moreover, it is recommended to incorporate additional support such as rock bolt or cable bolt with this steel arch when stress ratio is larger than 1.6 for 150 m depth.

References

1. Sasaoka, T., Hamanaka, A., Shimada, H., Matsui, K., Lin, N.Z., Sulistianto, B.: Punch multi-slice longwall mining system for thick coal seam under weak geological conditions. J. Geol. Resour. Eng. **1**, 28–36 (2015)
2. Ozfirat, M., Simsir, F., Gonen, A.: A brief comparison of Longwall Methods used at mining of thick coal seams. In: The 19th International Mining Congress and Fair of Turkey, IMCET, Izmir, Turkey, pp. 9–12 (2005)
3. Brune, J., Sapko, M.: A modeling study on longwall tailgate ventilation. In: 14th United States/North American Mine Ventilation Symposium. Department of Mining Engineering, Salt Lake City, University of Utah (2012)
4. Hebblewhite, B.: Status and prospects of underground thick coal seam mining methods. In: The 19th International Mining Congress and Fair of Turkey, IMCET, Izmir, Turkey (2005)
5. Matsui, K., Sasaoka, T., Shimada, H., Furukawa, H., Takamoto, H., Ichinose, M.: Some considerations in underground mining systems for extra thick coal seams. Coal Int. **259**(2), 38–43 (2011)
6. Zarlin, N., Sasaoka, T., Shimada, H., Matsui, K.: Numerical study on an applicable underground mining method for soft extra-thick coal seams in Thailand. Engineering **4**(11), 739 (2012)
7. Pongpanya, P., Sasaoka, T., Shimada, H., Hamanaka, A., Wahyudi, S.: Numerical study on effect of longwall mining on stability of main roadway under weak ground conditions in Indonesia. J. Geol. Resour. Eng. **3**, 93–104 (2017)

Seismic Activity and Convergence in Deep Mining Field, Case Study from Copper Ore Mine, SW Poland

Anna Barbara Gogolewska[1(✉)] and Agnieszka Markowiak[2]

[1] Wrocław University of Science and Technology, 50-370 Wrocław, Poland
anna.gogolewska@pwr.edu.pl

[2] Graduate from Wroclaw University of Science and Technology, 50-370 Wrocław, Poland

Abstract. The copper ore deposit in SW Poland has been excavated for about fifty years by three deep mines owned by the KGHM Polish Copper JSC. Seismic activity and rock-bursts have been accompanying mining operations since the mines were launched. Therefore prevention measures have to be implemented to assess, mitigate and reduce the seismic and rock burst hazards. Seismic activity i.e. the energy and number of tremors is monitored permanently by means of the mine seismological network. The seismic hazard is also controlled with measurements of the convergence of mine workings. The results of the convergence and seismic activity measurements allow for continuous monitoring of rock mass behavior, as they may indicate increased energy accumulation in the rock mass. On this basis, procedures provoking the release of excess energy from the rock mass can be performed, which may limit the occurrence of uncontrolled, and therefore dangerous to people seismic phenomena. This threat results, among others, from the deposit geological structure and tectonics as well as the significant depth of exploitation. The necessity to carry out measurements for assessing the state of the rock mass is stated in legal regulations and internal mine ordinances. The purpose of the paper was to analyze and assess the relationship between the convergence and seismic activity. The investigation covered one mining panel in the Polkowice-Sieroszowice mine and the four-year period.

Keywords: Underground copper mining · Rock bursts · Seismicity assessment

1 Introduction

The KGHM Polish Copper JSC, which is situated in Lower Silesia province, SW Poland possess three deep underground mines extracting copper ores in difficult conditions of which the rock burst hazard is the most dangerous. Rock bursts occur due to, among others, the significant deposit depth and challenging geological-mining situation. Therefore, the mines have to develop and apply suitable preventive measures to assess the rock mass seismicity, mitigate, limit and monitor the hazard [3]. The appropriate assessment of this hazard may allow for its recognition and reduction. The convergence of mining workings is used to investigate the rock mass behavior since it

E. Topal (Ed.): MPES 2019, SSGG, pp. 508–515, 2020.
https://doi.org/10.1007/978-3-030-33954-8_57

may indicate the range of working squeezing. Linear (vertical and horizontal), surface and volume convergences are measured. Measurements performed in mines show that fluctuations in the convergence may indicate changes in the rock burst hazard [7]. Thus, the investigation of the relationship between the convergence and seismic activity appears to be justifiable. Assessing the state of the rock mass is necessary according to legal regulations and internal mine ordinances. Monitoring of the seismic hazard includes, among others, seismological observations of seismic activity and the convergence of mine workings. The convergence is defined as changes in dimensions of mined-out empty space as a result of its squeezing; an increase of the convergence may indicate the growth of the excavation compression. The linear one constitutes changes in vertical or horizontal dimensions of excavations. The vertical convergence is applied to evaluate the rock-burst hazard in KGHM's mines. It may reflect the relaxation of the rock mass strain, deformation of the surrounding rocks, deformation of the excavation walls and show the effects caused by various adjacent mining conditions [5, 6]. On recording changes in the convergence, one can assess the excavation deformation. Furthermore, it has been suggested that the growth of the seismic hazard may be indicated by the increase, disappearance or irregular progress of the convergence. The paper aimed to determine the relationship between changes in the vertical excavation convergence and seismic activity of a mining panel. The analysis helped assess the convergence dependence on the seismicity of the rock mass. The study was carried out in the D-IE mining panel of high seismic risk, in the Polkowice-Sieroszowice copper ore mine over the four-year period. The convergence was analyzed over the four-year period of high seismic activity. The changes in the 30-day (over the 30 days prior to the measurement day) and in the 24-hour vertical convergence were analyzed. The significant changes and rapid disappearance of the convergence were investigated. The relationship (linear correlation coefficients) between the vertical convergence and the seismic activity was determined.

2 Study Area Depiction

The KGHM Polish Copper JSC owns the Polkowice-Sieroszowice mine, Rudna mine and Lubin mine, which exploit copper ores within the Fore-Sudetic Monocline. The research area namely D-IE mining panel belongs to the G-54 mining division in the Polkowice-Sieroszowice mine. The Polkowice mine started in 1963 with the P-1 shaft construction and the first exploitation operations were conducted in 1968. The Sieroszowice mine was built in 1978 as a result of drilling the new excavations from the Rudna and Polkowice mines towards the Sieroszowice field. The Polkowice-Sieroszowice mine was established in 1996 by merging the Polkowice and Sieroszowice mines. Mining is conducted in the following mining areas: Sieroszowice (it exploits 85% of the deposit, 15% is exploited by Rudna mine), Polkowice; Radwanice-East; part of the Głogów Głęboki Przemysłowy (50% of this area is exploited by Rudna mine) [1, 4].

Currently, the mine is working on the mechanization and automation of the exploitation of the thin copper deposit. The excavation depth varies from 600 to 1200 m [1]. The mining conditions are good in most of the area occupied by the

Polkowice-Sieroszowice mine. However, strongly developed tectonics in the southern part of the deposit makes the exploitation difficult. The thickness of the deposit is relatively small and ranges from 0.5 to 4.5 m, hence the mine cannot use hydraulic backfill and has to implement selective mining techniques [1].

The D-IE mining panel is situated in the central part of the mine. The exploitation depth varies from 925 to 1050 m. The excavated deposit thickness changes from 2.0 to 3.2 m and the operational front length ranges from 230 to 700 m. The one-stage room-and-pillar method with roof self-deflection (J-UG-PS) is used for mining. Stone zones, which impede mining operations cover a significant part of the panel. Only slight faults were found in the area. Rock bursts, which constitute the most dangerous hazard in the panel are caused mainly by the significant deposit depth and the specific strength parameters of rocks. The compressive strength decreases from roof through deposit to floor i.e. from 153.3 MPa, through 101.6 MPa to 37.5 MPa respectively.

3 Convergence of Mine Excavations

As a result of mining exploitation, the original balance weight of the rock mass is violated. The induced stress arises in the rock mass around the mining site, which depends on the in-situ stress, strength properties of rocks, the excavation size and its location in relation to other ones. Such a transition from one to another equilibrium is manifested in the displacement of rock mass parts from the primary to the secondary position, usually to the inside of the excavation. This type of rock mass movement is called excavation squeezing, which can be measured by the convergence. Most publications on convergence relate to salt mines, in which the volume convergence is monitored [2, 5, 6]. In the copper mines the convergence measurements refer to the decrease in the distance between the roof and the floor of the excavation and are expressed in millimeters. It is claimed that the convergence changes are connected with the strain of the rock mass, the deformation of excavations and geological and mining conditions around the excavation [7]. In the Polkowice-Sieroszowice mine, the convergence is used to assess the rock burst hazard. Continuous measurements, enable one to observe changes in the squeezing rate of the mining excavations. The convergence is usually obtained by multiple measurements of the distance between the points stabilized on the excavation contour. Measurements are made at specific time intervals. The convergence is most often determined with a laser rangefinder. A benchmark is installed at a specific point (measuring mark) in the floor and marked with tape. The rangefinder is placed in the marked point so that the beam of the laser can be directed perpendicularly towards the roof, where there is a second marked point, e.g. an anchor rod to which a laser signal should be sent from the rangefinder set on the floor. This method of measurement allows for obtaining an electronic result, which is the vertical distance between two points [2].

The assessment of seismicity by means of convergence is based on specific guidelines. A drop in convergence below 20% of its 30-day average or the increase by more than 100% of its 24-hour average, persisting for three consecutive days may be connected with the increased seismic hazard. However, high convergence variability, its sudden increase or disappearance need not necessarily indicate an increase in seismic hazard, since certain operating conditions may change the convergence significantly. For example, stopping the operating front or reducing the opening width may trigger convergence changes in the working field. Therefore, this type of phenomenon should be considered when analyzing convergence measurement results. It is important to always carry out the analysis in relation to the experience gained during the operation of the tested field or division.

4 Convergence Related to Seismicity

Over the four-year period (2009–2012) of high seismic activity the convergence in the D-IE panel was measured daily at several measuring points established according to the progress of the front and the geological and mining situation in the field. The 24-hour average of convergence was calculated for given points and used to determine the 30-day average, which included last 30 days prior to the measurement. The vertical convergence i.e. decrease of the working height was monitored. It was mentioned above that increased energy accumulation in the rock mass can be determined on the basis of the convergence behavior. The seismic activity, i.e. the release of energy accumulated in the rock mass in the form of tremors is a direct cause of the rock burst hazard. High-energy tremors cause most rock bursts, which threaten people and equipment. Therefore, the analysis of the convergence in relations to the seismic activity was carried out. The average annual convergence, 24-hour convergence, 30-day convergence, number and energy of tremors and rock bursts were considered.

4.1 Annual and Monthly Convergence Increase

The average annual convergence along with the total energy and the number of tremors in years 2009-2012, in D-IE panel are shown in Table 1. There were 11 high-energy tremors and 8 rock bursts including 7 provoked ones.

Table 1. Convergence and seismic activity in D-IE mining panel over the 2009–2012 period.

Year	Convergence, mm	Energy of tremors, J	Number of tremors
2009	1449	$8.15 \cdot 10^7$	94
2010	1991	$2.55 \cdot 10^8$	147
2011	2337	$9.50 \cdot 10^7$	249
2012	1686	$1.02 \cdot 10^7$	151

The seismic activity was high related to both the energy and number of tremors. The increases of convergence remained at a similar level in subsequent years. In 2009 there were 94 tremors of $8.15 \cdot 10^7$ J total energy, while the convergence increase was 1447 mm. In 2010, the number of tremors increased to 147, and their energy amounted to $2.55 \cdot 10^8$ J, a large increase in the convergence, i.e. 1991 mm, was noted. In 2011, as many as 249 tremors were recorded, but their energy was lower than that of 2010 and amounted to $9.5 \cdot 10^7$ J, however there was a greater increase in the convergence, which reached 2337 mm. In 2012, the number of tremors was smaller than in 2011 and amounted to 151, and also their energy decreased to $1.0 \cdot 10^7$ J (due to the termination of mining); similarly, downward trend was observed in the convergence increase, which reached 1686 mm. A directly proportional relationship was observed between the annual number of tremors and the increase in the convergence; a linear correlation coefficient assumed 0.93124. However, a weaker directly proportional dependence was observed between the tremor energy and the convergence increase since the linear correlation coefficient reached 0.3433.

The average monthly convergence with the energy and number of tremors in a given month in years 2009–2012 were compared and the relationship of the convergence with each parameter was analyzed. The convergence connection with the monthly number of tremors is described by a linear correlation coefficient of 0.44496 which indicates a directly proportional relationship between these parameters. The linear correlation coefficient between the average monthly convergence and tremor energy was 0.18092, which indicates a weak relationship.

4.2 24-Hour and 30-Day Convergence Increase

In order to illustrate the 24-hour 30-day convergence changes related to the seismic activity in each year, four charts were made (Figs. 1, 2, 3 and 4). Each of them presents one year of analysis over the period 2009–2012 for the D-IE panel. As part of the analysis, the convergence growths preceding the occurrence of high-energy tremors were examined. Several examples (11) of this type of relationship were described below.

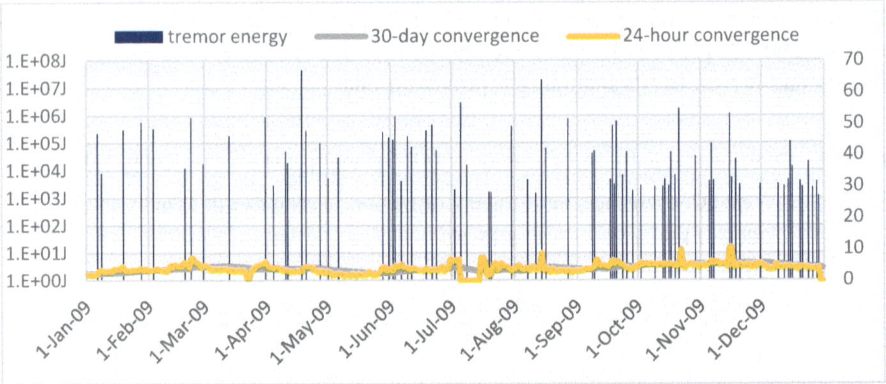

Fig. 1. Tremor energy vs. 30-day and 24-hour convergence in the D-IE mining panel in 2009.

The seismic activity was high in each year of the analysis but the average 24-hour convergence was different in particular years. Most of tremors and the highest energy occurred in 2011. In 2009, some of the rapid changes of the 24-hour convergence was connected with the increase of seismic activity. In 2009, the convergence fluctuated (increase and decreased) from 2 to 5 mm/24 h with a lot of rapid growths up to 10 mm.

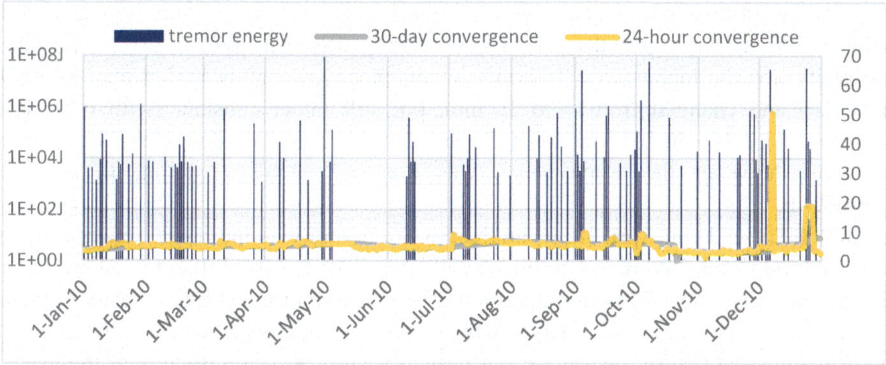

Fig. 2. Tremor energy vs. 30-day and 24-hour convergence in the DIE mining panel in 2010.

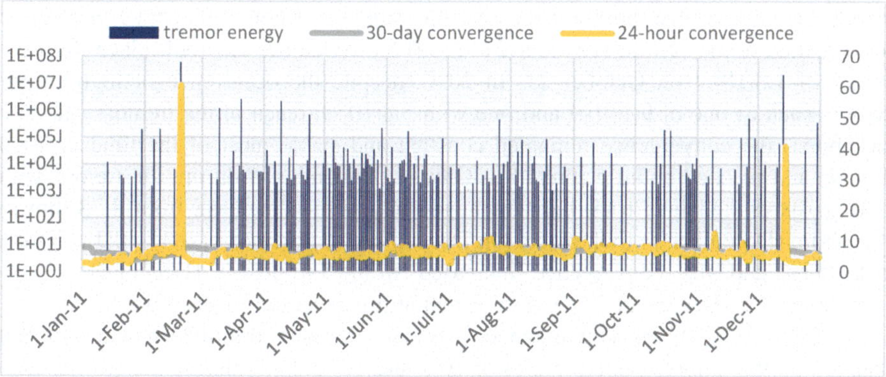

Fig. 3. Tremor energy vs. 30-day and 24-hour convergence in the DIE mining panel in 2011.

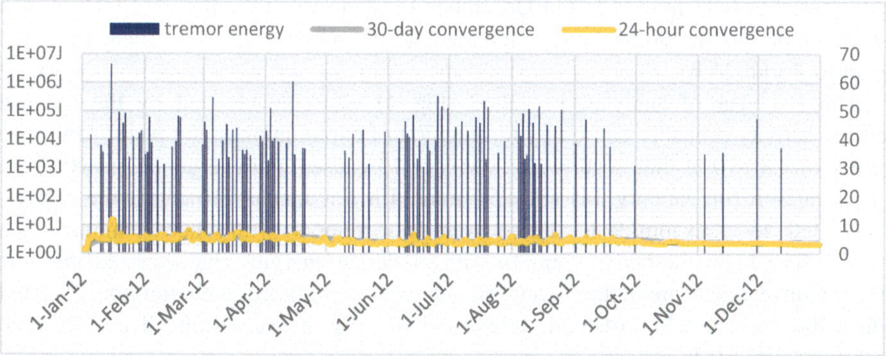

Fig. 4. Tremor energy vs. 30-day and 24-hour convergence in the D-IE mining panel in 2012.

In 2010 the 24-hour convergence was slightly higher (4–8 mm), rather constant and stable (with two high growths), but the seismicity was still high. In 2011 when the seismicity was the highest, the 24-hour convergence was slightly higher than in 2009 and 2010 and changed from 4 to 10 mm, but still rather constant (with two high growths) and stable. In 2012 the lowest seismic activity was recorded due to the mining cessation, however the convergence was not very stable changing from 3 to 6 mm/24 h.

In 2009 (Fig. 1) on April 18, when there was a tremor with the highest energy in the whole period (it resulted in a rock burst) with energy of $4.5 \cdot 10^7$ J, there were no major changes in the average 24-hour convergence before or after the event. On July 5, there was a tremor with energy of $3 \cdot 10^6$ J, which caused a rock burst, whereas the convergence maintained before the tremor at 4 mm/24 h (July 3–4), on the day of the event, increased to 7 mm and after (on July 7) dropped to 0. The subsequent high-energy tremor of $2.0 \cdot 10^7$ J, which occurred on August 14 was connected with the moderate change in the convergence from 3.3 mm (August 11) do 8.8 mm (August 14), after the tremor the convergence rapidly decreased 3.3 mm/24 h. Another moderate increase in convergence from 5.2 mm (October 21) to 10 mm (October 22) took place during a tremor of $1.8 \cdot 10^6$ J on October 21. In 2010, despite the occurrence of high-energy tremors such as one of $9.1 \cdot 10^7$J and one with $5.8 \cdot 10^7$ J, (each of the tremors caused a rock burst) the convergence remained constant and stable most of the time in 2010 (Fig. 2) and reached about 5 mm/24 h. Only on December 6 during the tremor with $3 \cdot 10^7$ J (it resulted in a rock burst) a huge convergence increase from 4.78 mm to 50.83 mm (December 7) took place. On December 24, a tremor of $3.41 \cdot 10^7$J caused a rock burst and the convergence increased from 5 mm/24 h (December 23) to 18.92 mm/24 h (December 24) and remained at this level until December 27.

In 2011 (Fig. 3), the convergence changed around 5–8 mm/24 h and was not changed by the tremors with energies of 10^5J and 10^6 J. Only during the two tremors with the highest energy did rapid convergence changes occur. On February 18, there was a tremor of $7.1 \cdot 10^7$ J, which caused a rock burst and the convergence increased to 60.83 mm/24 h (on the previous day the convergence was 5.83 mm/24 h); the next day it decreased to 6.33 mm/24 h. On December 13, there was a tremor of $2.1 \cdot 10^7$ J, and on December 14 the convergence increased from 4.57 to 41.29 mm, then decreased again to 4 mm/24 h.

The last year of analysis, i.e. 2012 (Fig. 4) was characterized by the reduced seismic activity probably caused by termination of exploitation in the panel. Only one tremor of $4.46 \cdot 10^6$ J that occurred on January 15 caused a convergence growth from 4.67 mm/24 h (on January 14) to 11.75 mm/24 h (on January 16), next the convergence fell to 4.33 mm/24 h. There were no tremors with energy higher than $1 \cdot 10^6$ J. The high number of tremors with moderate energies characterized the year 2012. Convergence anomalies occurred but were associated with tremors of 10^4 J which did not cause loss of work safety and stability of excavations. There were no rock bursts in 2012.

During four years, eleven high-energy tremors occurred of which only 3 were connected with the significant increases of the convergence.

The 30-day convergence was very stable since no differences in the convergence of particular parts of the front appeared. The mining method limited the stress and strain in the rock mass and hence the convergence was not negatively influenced. The preventive measures helped keep the convergence constant and stable. The 30-day average convergence did not show significant increase, decrease or vanishing, therefore no connection with the seismicity of rock mass was found in the D-IE mining panel.

5 Conclusions

Analysis of the correlation revealed the directly proportional relationship between the annual and monthly convergence and the number of tremors. However, a weaker linear correlation was obtained between the tremor energy and convergence. It was found that significant convergence increases was connected with 27% of high-energy tremors. Therefore, it may be concluded that the convergence changes could not help predict or assess the rock burst hazard. Moreover, the high seismic activity helped the energy accumulated in the rock mass release and hence reduce the strain and stress in rocks, which may increase the stability of the rock mass. It was noticed that, when small energy was accumulated in rocks, the excavation squeezing slightly fluctuated with seismic energy which could not significantly affect stress and strain in the rock mass. Other factors connected with the operations may also have had a significant impact on the convergence.

Acknowledgments. This manuscript publication has received funding from the European Institute of Innovation and Technology (EIT). This body of the European Union receives support from the European Union's Horizon 2020 research and innovation programme.

References

1. Bartlett, S., Burgess, H., Damjanović, B., Gowans, R., Lattanzi, C.: Technical report on the copper-silver production operations of KGHM Polish Copper JSC in the Legnica-Głogow Copper Belt area of southwestern Poland, pp. 13–35 (2013)
2. Bieniasz, J., Ciągło, W., Wojnar, W.: New method for measurements of salt room-and-pillar deformations with the use of laser. AGH University of Science and Technology Geodesy, Poland, 9(2/1) (2003)
3. Dubiński, J., Konopko, W.: Rock bursts – assessment – prediction – fighting. Main Mining Institute, Katowice, Poland (2000)
4. KGHM. http://www.kghm.com. Accessed 01 Aug 2019
5. Kortas, G.: Convergence as the measure of room workings squeezing. Mining Rev. **6**, 23–29 (2001)
6. Kortas, G.: On the convergence of workings and the purpose of working observations in salt mines. Geol. Geophys. Environ. **38**(1), 51–68 (2012)
7. Marcak, H.: Seismicity in mines due to roof bending. Arch. Min. Sci. **57**(1), 229–250 (2012)

Author Index

E. Topal (Ed.): MPES 2019, SSGG, pp. 517–519, 2020.
https://doi.org/10.1007/978-3-030-33954-8

The manufacturer's authorised representative in the EU is Springer Nature Customer Service Centre GmbH, Europaplatz 3, 69115 Heidelberg, Germany. If you have any concerns regarding our products, please contact ProductSafety@springernature.com

Printed and bound by CPI Group (UK) Ltd, Croydon, CR0 4YY
05/10/2026
02228145-0016